T0183187

# Remote Sensing Digital Image Analysis

John A. Richards

# Remote Sensing Digital Image Analysis

**Sixth Edition**

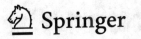 Springer

John A. Richards
ANU College of Engineering
and Computer Science
The Australian National University
Canberra, ACT, Australia

ISBN 978-3-030-82329-0          ISBN 978-3-030-82327-6   (eBook)
https://doi.org/10.1007/978-3-030-82327-6

1$^{st}$–5$^{th}$ editions: © Springer-Verlag Berlin Heidelberg 1986, 1993, 1999, 2006, 2013
6$^{th}$ edition: © The Editor(s) (if applicable) and The Author(s), under exclusive license to Springer
Nature Switzerland AG 2022
This work is subject to copyright. All rights are solely and exclusively licensed by the Publisher, whether
the whole or part of the material is concerned, specifically the rights of translation, reprinting, reuse
of illustrations, recitation, broadcasting, reproduction on microfilms or in any other physical way, and
transmission or information storage and retrieval, electronic adaptation, computer software, or by similar
or dissimilar methodology now known or hereafter developed.
The use of general descriptive names, registered names, trademarks, service marks, etc. in this publication
does not imply, even in the absence of a specific statement, that such names are exempt from the relevant
protective laws and regulations and therefore free for general use.
The publisher, the authors and the editors are safe to assume that the advice and information in this book
are believed to be true and accurate at the date of publication. Neither the publisher nor the authors or
the editors give a warranty, expressed or implied, with respect to the material contained herein or for any
errors or omissions that may have been made. The publisher remains neutral with regard to jurisdictional
claims in published maps and institutional affiliations.

This Springer imprint is published by the registered company Springer Nature Switzerland AG
The registered company address is: Gewerbestrasse 11, 6330 Cham, Switzerland

*This book is dedicated to the memory of the late David Landgrebe, Professor Emeritus of Purdue University.*

*As a teacher, mentor, friend and colleague to generations of engineers and scientists in remote sensing, Dave touched and influenced the lives and careers of many the world over.*

*The development of quantitative methods for the analysis of remote sensing image data owes much to Dave's leadership. He laid the foundations for the application of classification techniques to the earth sciences that we use today.*

*David Landgrebe was one of the true and inspirational pioneers of our field.*

# Preface

In a field like remote sensing image analysis, which changes so quickly, revising a long-standing textbook is a challenge. It is important to incorporate contemporary techniques, while not discarding procedures from the past which, while apparently superseded by newer methods, nevertheless retain value and are often simpler to use. Also, some processing operations from the past can become important again as data types and volumes change. Streaming methods for clustering are an example with the trend now to very large data sets.

As with the previous edition, judgements have had to be made about what to leave out, what to retain and what to add. Those judgements have been made against the intended purpose of the book. From the beginning, it has been designed as a teaching text for the senior undergraduate and postgraduate student, and as a fundamental treatment for those engaged in the application of digital image analysis in remote sensing projects or in remote sensing image processing research.

The presentation level is for the mathematical non-specialist. Because most operational users of remote sensing come from the earth sciences communities, the text is pitched at a level commensurate with their background. That is important because the recognised authorities in digital image analysis and machine learning tend to be from engineering, computer science and mathematics. Although familiarity with a certain level of mathematics and statistics cannot be avoided, the treatment here progresses through analyses carefully, with many hand-worked examples, so that any lack of depth in mathematical background should not take away from understanding the important aspects of image analysis and interpretation.

Although the principal focus of the treatment is on digital image interpretation and the analytical techniques that make that possible, the material is located within the domain of remote sensing applications. That means project objectives are as important as finding the best-performing algorithm. Algorithms need to be incorporated into methodologies that can generate optimal results from a careful combination of procedures, and in which the steps of choosing reference material to support the process and for assessing accuracy, may be just as important as algorithm performance. While algorithm performance is a key objective in the machine learning remote sensing research community, it is project outcomes that drive the remote sensing applications specialist. That is a key emphasis of this book.

Although the chapters can be used individually, the material is presented in a sequential manner. A person with little or no background in remote sensing image interpretation can start with the early chapters in order to appreciate key concepts in remote sensing and image formation, how errors arise in recorded imagery and how they can be corrected. The remaining chapters then work progressively through the major analytical methods fundamental to digital image analysis, finishing up with means by which methodologies can be devised to tackle remote sensing projects.

Over the years, many people have either directly or indirectly contributed to this book. The late David Landgrebe, to whom this edition is dedicated, was a friend and colleague who did much to shape my thinking about the application of quantitative methods in remote sensing. He pioneered many of the ideas that ended up in one way or another in parts of this book.

My colleague Associate Professor Xiuping Jia has been a great collaborator over the years, commencing when she undertook her Ph.D. Many of the methods presented here have been the result of a fruitful research partnership for which I express my sincere gratitude to her.

Dr. Terry Cocks, former Managing Director of HyVista Corporation Pty Ltd, Australia, kindly made available HyMap hyperspectral imagery of Perth, Western Australia, to allow many of the examples contained in this and the previous edition to be generated.

I am indebted to Jason Brown of Capella Space with whose encouragement this sixth edition was prepared; otherwise, it may not have happened. He also kindly provided the radar imagery used in Chap. 1.

Lastly, I acknowledge the dedication, support and encouragement of my wife Glenda. Her perseverance and understanding have been enormously important, and have made the job of writing this new edition fulfilling and satisfying, notwithstanding the demands it made on family time.

Canberra, Australia
June 2021

John A. Richards

# Contents

# Chapter 1
# Sources and Characteristics of Remote Sensing Image Data

**Abstract** The wavelength ranges commonly used for imaging the earth's surface are discussed, including reflected sunlight, thermal emission from the earth itself and the microwave radiation used in imaging radars. The idea of measuring energy coming from the earth's surface in a set of wavebands simultaneously is covered, leading to the concept of a multispectral image, or a hyperspectral image if the number of wavebands is very large. Remote sensing platforms and different sensor types are covered, along with the earth surface characteristics that can be detected with remote sensing instruments. Image scale is considered and the location of remote sensing within the fields of geographic information systems and digital earth models is introduced.

## 1.1 Energy Sources and Wavelength Ranges

In remote sensing energy coming up from the earth's surface is measured using a sensor mounted on a spacecraft or other elevated platform. That measurement is used to construct an image of the landscape beneath the platform, as depicted in Fig. 1.1.

In principle, any energy coming from the earth's surface can be used to form an image. Most often it is reflected sunlight so that the image recorded is, in many ways, similar to the view we would have of the earth's surface from an aircraft, even though the wavelengths used in remote sensing are often outside the range of human vision. The upwelling energy could also be from the earth itself acting as a radiator because of its own finite temperature. Alternatively, it could be energy that is scattered up to a sensor having been radiated onto the surface by an artificial source, such as a laser or a radar.

Provided an energy source is available, almost any wavelength could be used to image the characteristics of the earth's surface. There is, however, a fundamental limitation, particularly when imaging from spacecraft altitudes. The earth's atmosphere does not allow the passage of radiation at all wavelengths. Energy at some wavelengths is absorbed by the molecular constituents of the atmosphere.

© The Author(s), under exclusive license to Springer Nature Switzerland AG 2022
J. A. Richards, *Remote Sensing Digital Image Analysis*,
https://doi.org/10.1007/978-3-030-82327-6_1

**Fig. 1.1** Signal flow in a remote sensing system

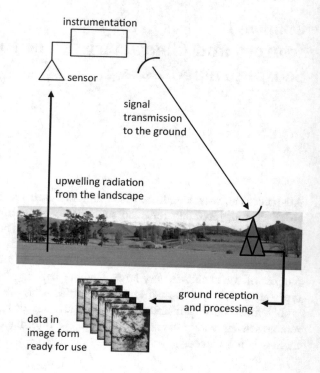

Wavelengths for which there is little or no atmospheric absorption form *atmospheric windows*. Figure 1.2 shows the transmittance of the earth's atmosphere on a path between space and the earth over a very broad range of the electromagnetic spectrum. The presence of a significant number of atmospheric windows in the visible and infrared regions of the spectrum is evident, as is the almost complete transparency of the atmosphere at radio wavelengths. The wavelengths used for imaging in remote sensing are clearly constrained to these atmospheric windows. They include the so-called *optical* wavelengths covering the visible and infrared, the thermal wavelengths and the radio wavelengths that are used in radar and passive microwave imaging of the earth's surface.

Whatever wavelength range is used to image the earth's surface, the overall system is a complex one involving the scattering or emission of energy from the surface, followed by transmission through the atmosphere to instruments mounted on the remote sensing platform. The data is then transmitted to the earth's surface, after which it is processed into image products ready for application by the user. That data chain is shown in Fig. 1.1. It is from the point of image acquisition onwards that this book is concerned. We want to understand how the data, once available in image format, can be interpreted.

We talk about the recorded imagery as *image data*, since it is the primary data source from which we extract usable information. One of the important characteristics of the image data acquired by sensors on aircraft or spacecraft platforms is

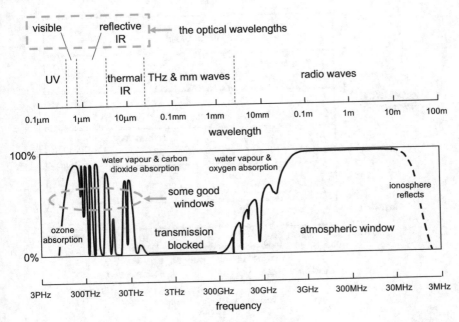

**Fig. 1.2** The electromagnetic spectrum and the transmittance of the earth's atmosphere, showing the positions of the atmospheric windows used in optical remote sensing

that it is readily available in digital format. Spatially it is composed of discrete picture elements, or *pixels*. Radiometrically—that is in brightness—it is quantised into discrete levels.

Possibly the most significant characteristic of the image data provided by a remote sensing system is the wavelength, or range of wavelengths, used in the image acquisition process. If reflected solar radiation is measured, images can, in principle, be acquired in the ultraviolet, visible and near-to-middle infrared ranges of wavelengths. Because of significant atmospheric absorption, as seen in Fig. 1.2, ultraviolet measurements are not made from spacecraft altitudes. Most common optical remote sensing systems record data from the visible through to the near and mid-infrared range: typically, that covers approximately 0.4–2.5 μm.

The energy emitted by the earth itself, in the thermal infrared range of wavelengths, can also be resolved into different wavelengths that help understand properties of the surface being imaged. Figure 1.3 shows why these ranges are important. The sun as a primary source of energy has a surface temperature of about 5950 K. The energy it emits as a function of wavelength is described theoretically by Planck's black body radiation law. As seen in Fig. 1.3 it has its maximal output at wavelengths just shorter than 1 μm and is a moderately strong emitter over the range 0.4–2.5 μm identified earlier.

The earth can also be considered as a black body radiator, with a temperature of 300 K. Its emission curve has a maximum in the vicinity of 10 μm as seen in Fig. 1.3. As a result, remote sensing instruments designed to measure surface temperature

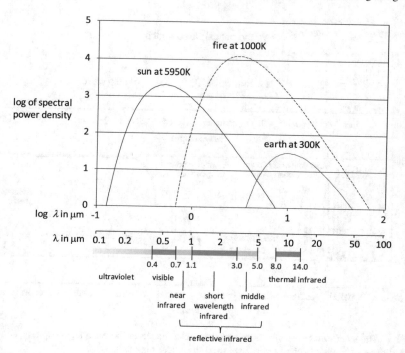

**Fig. 1.3** Relative levels of energy from black bodies when measured at the surface of the earth: the magnitude of the solar curve has been reduced as a result of the distance travelled by solar radiation from the sun to the earth; also shown are the boundaries between the different wavelength ranges used in optical remote sensing

typically operate somewhere in the range of 8–12 μm. Also shown in Fig. 1.3 is the blackbody radiation curve corresponding to a fire with a temperature of 1000 K. As observed, its maximum output is in the wavelength range 3–5 μm. Accordingly, sensors designed to map burning fires on the earth's surface typically operate in that range.

The visible, reflective infrared and thermal infrared ranges of wavelength represent only part of the story in remote sensing. We can also image the earth in the microwave or radio range, typical of the wavelengths used in mobile phones, satellite navigation systems, television, WiFi, Bluetooth and radar. While the earth does emit its own level of microwave radiation, it is often too small to be measured for most remote sensing purposes. Instead, energy is radiated from a platform onto the earth's surface. It is by measuring the energy scattered back to the platform that image data is recorded at microwave wavelengths.[1] Such a system is referred to as *active* since the energy source is provided by the platform itself, or by a companion platform. By comparison, remote sensing measurements that depend on an energy source such as the sun or the earth itself are called *passive*.

---

[1] For a treatment of remote sensing at microwave wavelengths see J. A. Richards, *Remote Sensing with Imaging Radar*, Springer, Berlin, 2009.

## 1.2   Primary Data Characteristics

The properties of digital image data of importance in image processing and analysis are the number and location of the spectral measurements (bands or channels), the spatial resolution described by the pixel size, and the *radiometric resolution*. These are shown in Fig. 1.4. Radiometric resolution describes the range and discernible number of discrete brightness values. It is sometimes referred to as *dynamic range* and is related to the signal-to-noise ratio of the detectors used. Frequently, radiometric resolution is expressed in terms of the number of binary digits, or bits, necessary to represent the range of available brightness values. Data with an 8 bit radiometric resolution has 256 levels of brightness, while data with 12 bit radiometric resolution has 4096 brightness levels.[2]

The size of the recorded image frame is also an important property. It is described by the number of pixels across the frame or *swath*, or in terms of the numbers of kilometres covered by the recorded scene. Together, the frame size of the image, the number of spectral bands, the radiometric resolution and the spatial resolution determine the data volume generated by a particular sensor. That sets the amount of data to be processed, at least in principle.

Image properties like pixel size and frame size are related directly to the technical characteristics of the sensor that was used to record the data. The *instantaneous field of view* (IFOV) of the sensor is its finest angular resolution, as shown in Fig. 1.5. When projected onto the surface of the earth at the operating altitude of the platform, it defines the smallest resolvable element in terms of equivalent ground metres, which is what we refer to as pixel size. Similarly, the *field of view* (FOV) of the sensor is the angular extent of the view it has across the earth's surface, again as

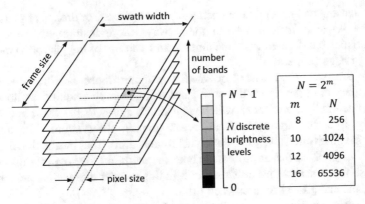

**Fig. 1.4**  Technical characteristics of digital image data

---

[2] See Appendix B.

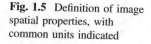

**Fig. 1.5** Definition of image spatial properties, with common units indicated

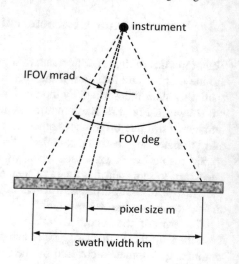

seen in Fig. 1.5. When that angle is projected onto the surface it defines the swath width in equivalent ground kilometres. Most imagery is recorded in a continuous strip as the remote sensing platform travels forward. Generally, particularly for spacecraft programs, the strip is cut up into segments, equal in length to the swath width, so that a square image frame is produced. For aircraft systems, the data is often left in strip format for the complete flight line flown in a given mission.

## 1.3  Remote Sensing Platforms

Remote sensing can be carried out using satellites, aircraft or drones as platforms to carry the imaging instruments. In many ways those instruments have similar characteristics but differences in the altitude and stability of the platform can lead to differing image properties.

There are two broad classes of satellite program: those satellites that orbit at geostationary altitudes above the earth's surface, generally associated with weather and climate studies, and those which orbit much closer to the earth and that are generally used for earth surface and oceanographic observations. The low earth orbiting satellites are usually in a sun-synchronous orbit. That means that the orbital plane is designed so that it precesses about the earth at the same rate that the sun appears to move across the earth's surface. In this manner the satellite acquires data at about the same local time on each orbit.

Low earth orbiting satellites can also be used for meteorological studies. Notwithstanding the differences in altitude, the wavebands used for geostationary and earth orbiting satellites, for weather and earth observation, are very comparable. The major distinction in the image data they provide generally lies in the spatial resolution available. Whereas data acquired for earth resources purposes has pixel sizes of the order of 1–10 m or so, that used for meteorological purposes (both at geostationary and lower altitudes) has a much larger pixel size, often of the order of 1 km.

The imaging technologies used in satellite remote sensing programs have ranged from traditional cameras to scanners that record images of the earth's surface by moving the instantaneous field of view of the instrument across the surface to record the upwelling energy. Typical of the latter technique is that used in the Landsat program in which a mechanical scanner records data at right angles to the direction of satellite motion to produce raster scans of data. The forward motion of the vehicle allows an image strip to be built up from the raster scans. That process is shown in Fig. 1.6. A dispersion device, such as a prism or diffraction grating, integrated with the sensor, separates the recorded signal into a number of wavebands by dispersing the radiation onto sets of detectors; there are as many separate images recorded of the region of the earth's surface as there are detectors and thus wavebands.

Some weather satellites scan the earth's surface using the spin of the satellite itself while the sensor's pointing direction is varied along the axis of the satellite. The image data is then recorded in a raster scan fashion.

With the availability of reliable detector arrays based on charge coupled device (CCD) technology, an alternative image acquisition mechanism utilises what is commonly called a "push-broom" technique. In this approach a linear CCD imaging array is carried on the satellite normal to the platform motion as shown in Fig. 1.7. As the satellite moves forward the array records a strip of image data, equivalent in width to the field of view seen by the array. Each individual detector records a strip in width equivalent to the size of a pixel. Because the time over which energy emanating from the earth's surface per pixel can be larger with push broom technology than with mechanical scanners, better spatial resolution is usually achieved.

Two dimensional CCD arrays are also available and find application in satellite imaging sensors. However, rather than record a two-dimensional snapshot image of the earth's surface, the array is employed in a push broom manner; the second dimension is used to record simultaneously a number of different wavebands for each pixel via the use of a mechanism that disperses the incoming radiation by wavelength. Such an arrangement is shown in Fig. 1.8. Often about 200 channels are recorded in this manner so that the reflection characteristics of the earth's surface are well represented in the data. Such devices are often referred to as *imaging spectrometers* and the data is described as *hyperspectral*, as against *multispectral* when of the order of 10 wavebands is recorded.

Aircraft scanners operate essentially on the same principles as those found with satellite sensors. Both mechanical scanners and CCD arrays are employed.

The logarithmic scale used in Fig. 1.3 hides the fact that each of the curves shown extends to infinity. If we ignore emissions associated with a burning fire, it is clear that the emission from the earth at longer wavelengths far exceeds reflected solar energy. Figure 1.9 re-plots the earth curve from Fig. 1.3 showing that there is continuous emission of energy right out to the wavelengths we normally associate with radio transmissions. In the microwave energy range, where the wavelengths

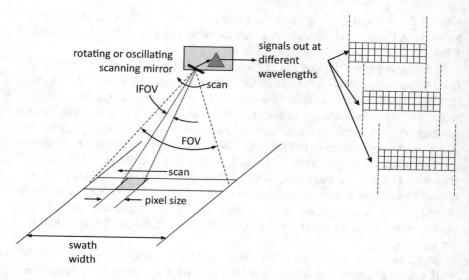

**Fig. 1.6** Image formation by mechanical line scanning, showing the received signal dispersed into several different wavelengths (or wavebands)

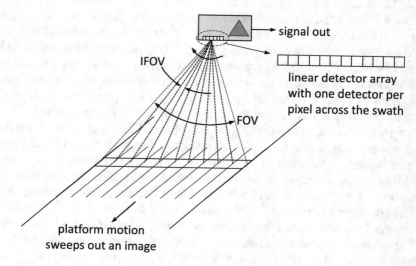

**Fig. 1.7** Image formation by push broom scanning

are between 1 cm and 1 m, there is, in principle, measurable energy coming from the earth's surface. As a result, it is possible to build remote sensing instruments that form microwave images the earth. If those instruments depend on measuring the naturally occurring levels shown in Fig. 1.9, then the pixels tend to be very large because of the extremely low levels of energy available. Large pixels are necessary to collect enough signal so that noise from the receiver electronics and the environment does not dominate the information of interest.

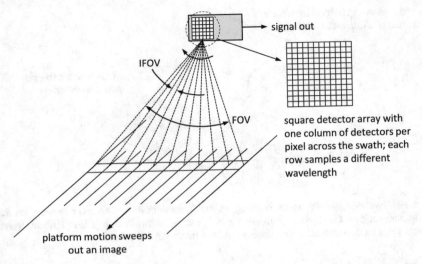

**Fig. 1.8** Image formation by push broom scanning with an array that allows the recording of several wavelengths simultaneously

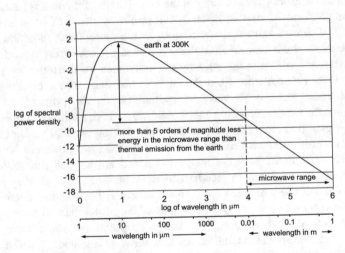

**Fig. 1.9** Illustration of the level of naturally emitted energy from the earth in the microwave range of wavelengths

More often, we take advantage of the fact that the very low naturally occurring levels of microwave emission from the surface permits us to assume that the earth is, for all intents and purposes, a zero emitter. That allows us to irradiate the earth's surface artificially with a source of microwave radiation at a wavelength of particular interest. In principle, we could use a technique not unlike that shown in Fig. 1.6 to build up an image of the earth at that wavelength. Technologically,

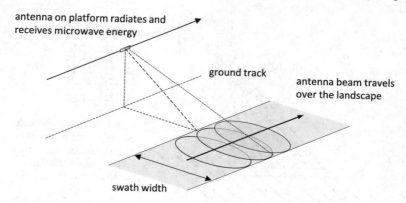

**Fig. 1.10** Synthetic aperture radar imaging; as the antenna beam travels over features on the ground many echoes are received from the pulses of energy transmitted from the platform, which are then processed to provide a very high resolution image of those features

however, the principle of synthetic aperture radar is used to create the image. We now describe that technique by reference to Fig. 1.10.

A pulse of electromagnetic energy at the wavelength of interest is radiated to the side of the platform. It uses an antenna that produces a beam that is broad in the across-track direction and relatively narrow in the along-track direction, as illustrated. The cross-track beamwidth defines the swath width of the recorded image. Features are resolved across the track by the time taken for the pulse to travel from the transmitter, via scattering from the surface, and back to the radar instrument. Along the track, features are resolved spatially using the principle of aperture synthesis, which entails recording many reflections from each spot on the ground and using signal processing techniques to synthesise high spatial resolution from a system that would otherwise record features at a detail too coarse to be of value. The technical details of how the image is formed are beyond the scope of this treatment but can be found in standard texts on radar remote sensing.[3] What is important here is the strength of the signal received back at the radar platform because that determines the brightness values of the pixels that constitute the radar image. As with optical imaging, the image properties of importance in radar imaging include the spatial resolution, but now different in the along and cross track directions, the swath width, and the wavebands at which the images are recorded.

Whereas there may be as many as 200 wavebands with optical instruments, there are rarely more than three or four with radar at this stage of our technology. However, there are other radar parameters. They include the angle with which the

---

[3] See Richards, loc. cit., I. H. Woodhouse, *Introduction to Microwave Remote Sensing*, Taylor and Francis, Boca Raton, Florida, 2006, F. M. Henderson and A. J. Lewis, eds, *Principles and Applications of Imaging Radar, Manual of Remote Sensing*, 3rd ed., Volume 2, John Wiley and Sons, N.Y., 1998, and F. T. Ulaby and D. G. Long, *Microwave Radar and Radiometric Remote Sensing*, The University of Michigan Press, Ann Arbor, 2014.

earth's surface is viewed out from the platform (the so-called *look angle*) and the polarisation of both the transmitted and received radiation. As a consequence, the parameters that describe a radar image can be more complex than those that describe an optical image. Nevertheless, once a radar image is available, the techniques of this book become relevant to the processing and analysis of radar image data. There are, however, some peculiarities of radar data that mean special techniques more suited to radar imagery are often employed.[4]

## 1.4  What Earth Surface Properties Are Measured?

In the visible and infrared wavelength ranges all earth surface materials absorb incident sunlight differentially with wavelength. Some materials detected by satellite sensors show little absorption, such as snow and clouds in the visible and near infrared. In general, though, most materials have quite complex absorption characteristics. Early remote sensing instrumentation, and many current instruments, do not have sufficient spectral resolution to be able to recognise the absorption spectra in detail, compared with how those features might appear in laboratory-recorded spectra. Instead, the wavebands available with some detectors allow only a crude representation of the spectrum, but nevertheless one that is more than sufficient for differentiating among most cover types. Even our eyes do a crude form of spectroscopy by allowing us to differentiate earth surface materials by the colours we see, even though the colours are composites of the red, green and blue signals that reach our eyes after incident sunlight has scattered from the natural and built environment.

More modern instruments record many, sufficiently fine spectral samples over the visible and infrared range that we can get very good representations of reflectance spectra, as we will see in the following.

### 1.4.1  Sensing in the Visible and Reflected Infrared Ranges

In the absence of burning fires, Fig. 1.3 shows that the upwelling energy from the earth's surface up to wavelengths of about 3 μm is predominantly reflected sunlight. It covers the range from the ultraviolet, through the visible, and into the infrared range. Since it is reflected sunlight the infrared is usually called reflected infrared, although it is then broken down into the near-infrared, short wavelength infrared and middle-infrared ranges. Together, the visible and reflected infrared ranges are called optical wavelengths as noted earlier. The definitions and the

---

[4] See Richards, loc. cit., for information on image analysis tools specifically designed for radar imagery.

ranges shown in Fig. 1.3 are not fixed; some variations will be seen over different user communities.

Most modern optical remote sensing instrumentation operates somewhere in the range of 0.4–2.5 μm. Figure 1.11 shows how the three broad surface cover types of vegetation, soil and water reflect incident sunlight over those wavelengths. By contrast, if we were to image a perfect reflector, the reflection characteristics would be a constant at 100% reflectance over the range. The fact that the reflectance curves of the three fundamental cover types differ from 100% is indicative of the selective absorption characteristics associated with their biophysical and biochemical compositions.[5]

It is seen in Fig. 1.11 that water reflects about 10% or less in the blue-green range of wavelengths, a smaller percentage in the red and almost no energy at all in the infrared range. If water contains suspended sediments, or if a clear body of water is shallow enough to allow reflection from the bottom, then an increase in apparent water reflection will occur, including a small but significant amount of energy in the near infrared regime. That is the result of reflection from the suspension or bottom material.

Soils have a reflectance that increases approximately monotonically with wavelength, however with dips centred at about 1.4, 1.9 and 2.7 μm owing to moisture content. Those water absorption bands are almost unnoticeable in very dry soils and sands. In addition, clay soils have hydroxyl absorption bands at 1.4 and 2.2 μm.

The vegetation curve is more complex than the other two. In the middle infrared range, it is dominated by the water absorption bands near 1.4, 1.9 and 2.7 μm. The plateau between about 0.7 and 1.3 μm is dominated by plant cell structure, while in the visible range of wavelengths plant pigmentation is the major determinant of shape. The curve shown in Fig. 1.11 is for healthy green vegetation. That has chlorophyll absorption bands in the blue and red regions leaving only green reflection of any significance in the visible. That is why we see chlorophyll pigmented plants as green. If the plant matter has different pigmentation, then the shape of the curve in the visible wavelength range will be different. If healthy green vegetation dies the action of chlorophyll ceases and the absorption dips in the blue and red fill up, particularly the red. As a result, the vegetation appears yellowish, bordering on white when completely devoid of pigmentation.

Inspection of Fig. 11.1 shows why the wavebands for different remote sensing missions have been located in the positions indicated. They are arranged so that they detect those features of the reflectance spectra of earth surface cover types that are most helpful in discriminating among the cover types and in understanding how they respond to changes related to water content, disease, stage of growth and so on. In the case of the Hyperion instrument the number of wavebands available

---

[5] For an easily read and comprehensive treatment see R. M. Hofer, Biological and Physical Considerations in Applying Computer-aided Analysis Techniques to Remote Sensor Data, Chap. 5 in P. H. Swain and S. M. Davis, eds, *Remote Sensing: The Quantitative Approach*, McGraw-Hill, N.Y., 1978.

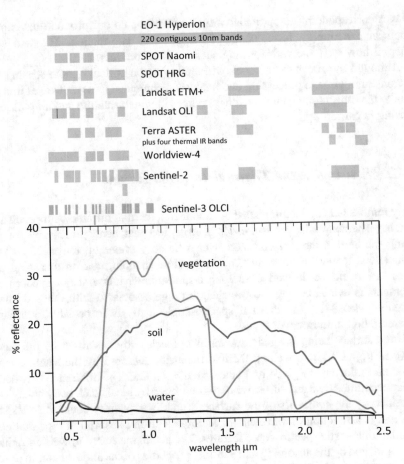

**Fig. 1.11** Spectral reflectance characteristics in the visible and reflective infrared range for three common cover types, recorded over Perth Australia using the HyVista HyMap scanner; shown also are the locations of the spectral bands of a number of common sensors, some of which also have panchromatic bands and bands further into the infrared that are not shown here

allows an almost full laboratory-like reconstruction of the reflectance spectrum of the earth surface material. We will see later that such a rendition allows scientific spectroscopic principles to be used in analysing what the spectrum tells us about a particular point on the ground.

It is important to recognise that the information summarised in Fig. 1.11 refers to the reflection characteristics of a single pixel on the earth's surface. With imaging spectrometers such as Hyperion we have the ability to generate full reflectance spectrum information for each pixel and, in addition, to produce a map showing the spatial distribution of reflectance information because of the lines and columns of pixels recorded by the instrument. With so many spectral bands available, we have the option of generating the equivalent number of images, or of combining the images corresponding to particular wavebands into a colour product that captures, albeit in a summary form, some of the spectral information. We will see in Chap. 3 how we cope with forming such a colour product.

Although our focus in this book will tend to be on optical remote sensing when demonstrating image processing and analysis techniques, it is of value at this point to note the other significant wavelength ranges in which satellite and aircraft remote sensing is carried out.

### 1.4.2 Sensing in the Thermal Infrared Range

Early remote sensing instruments that contained a thermal infrared band, such as the Landsat Thematic Mapper, were designed to use that band principally for measuring the earth's thermal emission over a broad wavelength range. Their major applications tended to be in surface temperature mapping and in assessing properties that could be derived from such a measurement. If a set of spectral measurements is available over the wavelength range associated with thermal infrared emission, viz. 8–12 μm, thermal spectroscopic analysis is possible, allowing a differentiation among cover types.

If the surface being imaged were an ideal black body described by the thermal curve in Fig. 1.3 the upwelling thermal radiance measured by the satellite is proportional to the energy given by Planck's radiation law. The difference between the radiation emitted by a real surface and that described by ideal black body behaviour is described by the *emissivity* of the surface, which is a quantity equal to or less than one, and is a function of wavelength, often with strong absorption dips that correspond to diagnostic spectroscopic features. The actual measured upwelling radiance is complicated by the absorbing and emitting properties of the atmosphere; in practice they are removed by correction algorithms, as is the wavelength dependence of the solar curve. That allows the surface properties to be described in terms of emissivity.

Figure 1.12 shows emissivity spectra in the thermal range for some common substances. Also shown in the figure are the locations of the wavebands for several remote sensing instruments that take sets of measurements in the thermal region. In Fig. 1.13 two examples are shown of identification in the thermal range, in one case using a thermal imaging spectrometer to detect fine detail.

### 1.4.3 Sensing in the Microwave Range

As noted earlier, microwave, or radar, remote sensing entails measuring the strength of the signal scattered back from each resolution element (pixel) on the earth's surface after irradiation by an energy source carried on the platform. Because of the wavelengths used, radar imaging can be carried out through cloud cover, and since

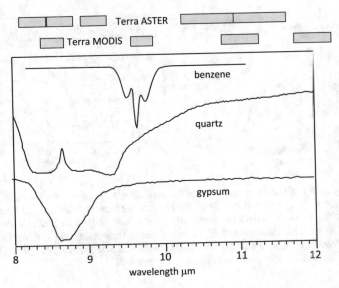

**Fig. 1.12** Some emissivity spectra in the thermal infrared range; not to scale vertically: the quartz spectrum was constructed using data kindly provided by the ASTER science team, NASA Jet Propulsion Laboratory, California Institute of Technology *pers comm* 2021; the gypsum curve was constructed from raw data kindly provided by Yoshiki Nimomiya of the Geological Survey of Japan, AIST *pers comm* 2021; the benzene spectrum was taken from D. Williams, Thermal multispectral detection of industrial chemicals, *pers comm* 2010 and 2021

it carries its own energy source, radar imaging can happen at any time of night or day. The degree of scattering from earth surface features is determined largely by two properties of the surface material: its geometric shape and its moisture content. Further, because of the much longer wavelengths used in microwave remote sensing compared with optical imaging, some of the incident energy can penetrate beyond the outer surface of the cover types being imaged. We will now examine some rudimentary scattering behaviours so that a basic understanding of radar remote sensing can be obtained.

Smooth surfaces act as so-called *specular* (mirror-like) reflectors in that the direction of scattering is predominantly away from the incident direction; as a result, they appear dark to black in radar image data. Rough surfaces act as *diffuse* reflectors in that they scatter the incident energy in all directions, including back towards the remote sensing platform. Consequently, they appear light in image data. Whether a surface is regarded as rough or not depends on the wavelength of the radiation used and the angle with which the surface is viewed (look angle). Table 1.1 shows the common frequencies and wavelengths used with radar imaging. At the longer wavelengths many surfaces appear smooth whereas the same surfaces can be diffuse at shorter wavelengths, as depicted in Fig. 1.14a. If the

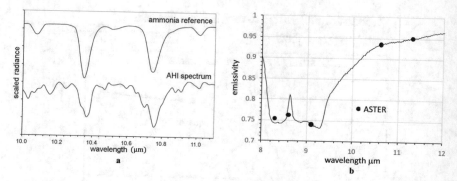

**Fig. 1.13 a** Ammonia spectrum recorded by the AHI thermal imaging spectrometer[6] compared with a laboratory reference spectrum (reproduced with permission from D. Williams, Thermal multispectral detection of industrial chemicals, *pers comm* 2010 and 2021) and **b** ASTER multispectral thermal measurements of Algodones dunes in California compared with a laboratory sample, constructed using data kindly provided by the ASTER science team, NASA Jet Propulsion Laboratory, California Institute of Technology *pers comm* 2021

**Table 1.1** Typical radio wavelengths and corresponding frequencies[7] used in radar remote sensing, based on actual missions; only the lower end of the K band is currently used

| Band | Typical wavelength (cm) | Frequency |
|------|-------------------------|-----------|
| P    | 66.7                    | 450 MHz   |
| L    | 23.5                    | 1.28 GHz  |
| S    | 12.6                    | 2.38 GHz  |
| C    | 5.7                     | 5.3 GHz   |
| X    | 3.1                     | 9.7 GHz   |
| Ku   | 2.16                    | 13.9 GHz  |

surface material is very dry then the incident microwave radiation can penetrate, particularly at long wavelengths, as indicated in Fig. 1.14b, making it possible to form images of objects underneath the earth's surface.

Another surface scattering mechanism is often encountered with manufactured features such as buildings. That is the *corner reflector effect* seen in Fig. 1.14c, which results from the right angle formed between a vertical structure such as a fence, building or ship and a horizontal plane such as the surface of the earth or sea. This gives a very bright response; the response is larger at shorter wavelengths.

Media such as vegetation canopies and sea ice exhibit *volume scattering* behaviour, in that the backscattered energy emerges from many, hard-to-define sites within the volume, as illustrated for trees in Fig. 1.14d. That leads to a light tonal appearance in radar imagery, with the effect being strongest at shorter wavelengths.

---

[6] Airborne Hyperspectral Imager, Hawaii Institute of Geophysics and Planetology (HGIP) at the University of Hawaii. This instrument has 256 bands in the range 8–12 μm.

[7] Wavelength in metres and frequency in megahertz are related by the expression $f(\text{MHz}) = 300/\lambda(\text{m})$

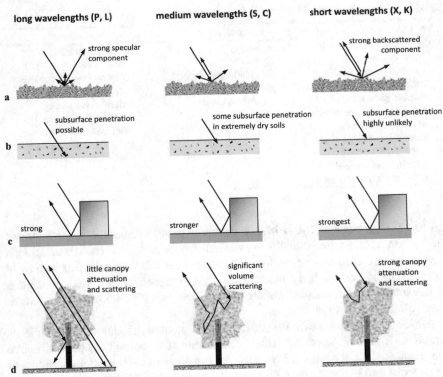

**Fig. 1.14** Common radar scattering mechanisms as a function of the wavelength of the irradiating energy **a** surface, **b** sub-surface, **c** corner reflector, and **d** volume scattering

At long wavelengths vegetation offers little attenuation to the incident radiation so that the backscatter is often dominated by the surface underneath the vegetation canopy. Significant forward scattering can also occur from trunks when the vegetation canopy is almost transparent to the radiation at those longer wavelengths. As a consequence, the tree trunk can form a corner reflector in the nature of that shown in Fig. 1.14c.

The radar response from each of the geometric mechanisms shown in Fig. 1.14 is modulated by the moisture contents of the materials involved in the scattering process. Moisture enters through an electrical property called *complex permittivity* which determines the strength of the scattering from a given object or surface. The angle with which the landscape is viewed also has an impact on the observed level of backscatter. Scattering from relatively smooth surfaces is a strong function of look angle, while scattering from vegetation canopies is weakly dependent on the look angle. Table 1.2 summarises the appearance of radar imagery in the different wavelength ranges.

It was mentioned earlier that the radiation used with radar has a property known as polarisation. It is beyond the level of treatment here to go into depth on the nature of polarisation, but it is sufficient for our purposes to note that the incident energy

**Table 1.2** Some characteristics of radar imagery

| Long wavelengths | Medium wavelengths | Short wavelengths |
|---|---|---|
| Little canopy response but good tree response because of corner reflector effect involving trunks; good contrast of buildings and tree trunks against background surfaces, and ships at sea; good surface discrimination provided wavelength not too long | Some canopy penetration; good canopy backscattering; fairly good discrimination of surface variations | Canopy response strong, poor surface discrimination because diffuse scattering dominates; strong building response, but sometimes not well discriminated against adjacent surfaces |

can be called horizontally polarised or vertically polarised. Similarly, the reflected energy can also be horizontally or vertically polarised. For each transmission wavelength and each look angle, four different images can be obtained as a result of polarisation differences. If the incident energy is horizontally polarised, depending on the surface properties, the scattered energy can be either horizontally or vertically polarised or both, and so on.

Another complication with the coherent radiation used in radar is that the images exhibit a degree of "speckle". That is the result of constructive and destructive interference of the reflections from surfaces that have random spatial variations of the order of one half a wavelength or so. Within a homogeneous region, such as a crop field, speckle shows up as a salt-and-pepper like noise that overlays the actual image data. It complicates significantly any analytical process we might devise for interpreting radar imagery that depends on the properties of single pixels.

A high quality space-acquired radar image is seen in Fig. 1.15. Recorded just north-west of Seville in Spain, it demonstrates a range of typical radar scattering mechanisms.

## 1.5   Spatial Data Sources in General and Geographic Information Systems

Other sources of spatial data exist alongside satellite or aircraft remote sensing imagery, as outlined in Fig. 1.16. They include simple maps that show topography, land ownership, roads and the like, and more specialised sources such as geological maps and maps of geophysical measurements such as gravity anomalies and magnetics. Spatial data sets like those are valuable complements to image data when seeking to understand land cover and land use. They contain information not available in remote sensing imagery and careful combinations of spatial data sources often allow inferences to be drawn about regions on the earth surface not possible when using a single source on its own.

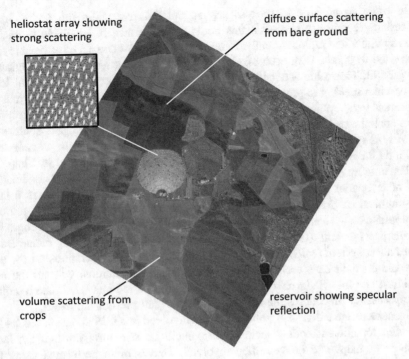

heliostat array showing
strong scattering

diffuse surface scattering
from bare ground

volume scattering from
crops

reservoir showing specular
reflection

**Fig. 1.15** X band HH SAR image recorded by Capella Space of the PS10 solar concentrator in Seville Spain. The image was acquired at 1942 GMT on 22 March 2021 with a look angle of 34.1°. Several different scattering types are identified on the image, which has a spatial resolution of better than 1 m. © *Capella Space Corp, All Rights Reserved*

**Fig. 1.16** A typical registered spatial data set such as might be found in a geographic information system; some data types are inherently numerical while others are often in the form of labels

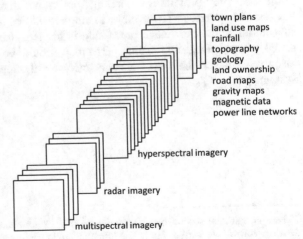

town plans
land use maps
rainfall
topography
geology
land ownership
road maps
gravity maps
magnetic data
power line networks

hyperspectral imagery

radar imagery

multispectral imagery

In order to be able to process any spatial data set using the digital image processing techniques treated in this book, the data must be available in discrete form spatially and radiometrically. In other words, it must consist of or be able to be converted to pixels, with each pixel describing the properties of a small region on the ground. The value ascribed to each pixel must be expressible in digital form. Also, when seeking to process several spatial data sets simultaneously they must be in correct geographic relation to each other. Desirably, the pixels in imagery and other spatial data should be referenced to the coordinates of a map grid, such as the UTM grid system. When available in this manner the data is said to be *geocoded*. Methods for registering and geocoding different data sets are treated in Chap. 2.

The amount and variety of data to be handled in a database that contains imagery and other spatial data sources can be enormous, particularly if it covers a large geographical region. Clearly, efficient means are required to store, retrieve, manipulate, analyse and display relevant data sets. That is the role of the geographic information system (GIS). Like its commercial counterpart, the management information system (MIS), the GIS is designed to carry out operations on the data stored in its data base according to a set of user specifications, without the user needing to be knowledgeable about how the data is stored and what data handling and processing procedures are utilised to retrieve and present the data.

Because of the nature and volume of data involved in a GIS attention has had to be given to efficient coding techniques to facilitate searching through the large numbers of maps and images often involved. That is often performed using the procedure known collectively as *data mining*.[8]

To understand the sorts of spatial data manipulation operations of importance in GIS one must take the view of the resource manager rather than the data analyst. While the latter is concerned with image reconstruction, filtering, transformation and interpretation, the manager is interested in operations such as those listed in Table 1.3. They provide information from which management strategies and the like can be inferred. To be able to implement many, if not most, of those, a substantial amount of image processing is needed. It is expected, though, that the actual image processing being performed would be largely transparent to the resource manager; the role of the data analyst will often be in the design of the GIS system.

---

[8] There is a special section on data mining in *IEEE Transactions on Geoscience and Remote Sensing*, vol. 45, no. 4, April 2007. The Introduction, in particular, gives a good description of the field. A more recent treatment is B. K. Wylie, N. J. Pastick, J. J. Picotte and C. Deering, Geospatial data mining for digital raster mapping, *GIScience and Remote Sensing*, vol. 56, no. 3, 2019, pp. 406–429.

**Table 1.3** Some typical GIS data operations

| |
|---|
| Intersection and overlay of spatial data sets (masking) |
| Intersection and overlay of polygons (grid cells, local government regions, etc.) on spatial data |
| Identification of shapes |
| Identification of points in polygons |
| Area determination |
| Distance determination |
| Thematic mapping from single or multiple spatial data sets |
| Proximity calculations and route determination |
| Searching by metadata |
| Searching by geographic location |
| Searching by user-defined attributes |
| Similarity searching |
| Data mining |

## 1.6   Scale in Digital Image Data

Because of IFOV differences the images provided by different remote sensing sensors are confined to application at different scales. As a guide, Table 1.4 relates scale to spatial resolution. That has been derived by considering an image pixel to be too coarse if it approaches 0.1 mm in size on a photographic product at a given scale. Landsat ETM+ data is seen to be suitable for scales smaller than about 1:250,000 whereas MODIS imagery is suitable for scales below about 1:10,000,000.

## 1.7   Digital Earth

For centuries we depended on the map sheet as the primary descriptor of the spatial properties of the earth. With the advent of satellite remote sensing in the late 1960s and early 1970s we then had available for the first time wide scale and panoramic earth views that supplemented maps as a spatial data source. Over the past four decades or so, with increasing geometric integrity and spatial resolution, satellite and aircraft imagery, along with other forms of spatial data, led directly to the construction of the geographic information system, now widely used as a decision support mechanism in many resource-related studies.

In twenty years the GIS notion has been generalised substantially through the introduction of the concept of the virtual globe.[9] This allows the user of spatial data to roam over the whole of the earth's surface and zoom in or out to capture a view at

---

[9] Perhaps the best-known examples are Google Earth and NASA's World Wind. See also Digital Earth Australia at http://www.ga.gov.au/dea.

**Table 1.4** Suggested maximum scales of hard copy products as a function of effective ground pixel size (based on 0.1 mm displayed pixel size): pan = panchromatic, XS = multispectral

| Scale | Nominal pixel (m) | Typical sensors with comparable pixel sizes |
|---|---|---|
| 1:5000 | 0.5 | WorldView2, 4 pan, Pléaides pan |
| 1:10,000 | 1 | Ikonos pan, SPOT Naomi pan |
| 1:50,000 | 5 | Ikonos XS, SPOT Naomi XS, SPOT HRG pan, TerraSAR-X |
| 1:100,000 | 10 | SPOT HRG XS, SPOT HRG pan, Alos PALSAR, AVNIR |
| 1:250,000 | 25 | Landsat ETM+, ASTER MIR, ALI, Landsat OLI |
| 1:500,000 | 50 | Landsat MSS, LISS |
| 1:5,000,000 | 500 | MODIS, OCTS |
| 1:10,000,000 | 1000 | MODIS, NOAA AVHRR, GMS visible, Sentinel OCLI |
| 1:50,000,000 | 5000 | GMS Thermal IR |

the scale of interest. Currently, there are significant technical limitations to the scientific use of the virtual globe as a primary mapping tool, largely to do with the radiometric and positional accuracy but, with further development, the current GIS model will be replaced by a virtual globe framework in which, not only positional and physical descriptor information will be available, but over which will be layers of other data providing information on social, cultural, heritage and human factors. Now known as *digital earth,* such a model puts spatial information and its manipulation in the hands of any user. It also allows the non-scientific lay user the opportunity to contribute to the information estate contained in the digital earth model. Citizen contribution of spatial data goes under the name of *crowdsourcing,* or sometimes *neogeography,* and will be one of the primary data acquisition methodologies of the future.

When combined with the enormous number of ground-based and spaceborne/ airborne sensors, the digital earth[10] concept promises to be an enormously powerful management tool for almost all of the information of value to us both for scientific and other purposes. The idea of the digital earth formed after a seminal speech given by former US Vice-President Al Gore in 1998.[11]

The digital earth paradigm is illustrated in Fig. 1.17. To make that work many of the image processing and analysis techniques presented in later chapters need to be employed.

---

[10] See the special issue on the sensor web of the *IEEE Journal of Selected Topics in Applied Earth Observation and Remote Sensing,* vol. 3, no. 4, December 2010.

[11] The content of Gore's speech is captured in A. Gore, The Digital Earth: Understanding our planet in the twenty-first Century, *Photogrammetric Engineering and Remote Sensing,* vol. 65, no. 5, 1999, p. 528.

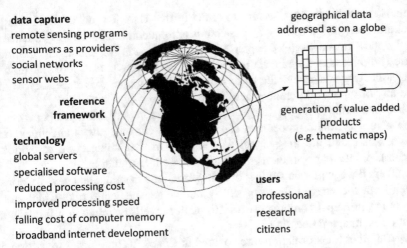

**data capture**
remote sensing programs
consumers as providers
social networks
sensor webs

**reference framework**

**technology**
global servers
specialised software
reduced processing cost
improved processing speed
falling cost of computer memory
broadband internet development

geographical data
addressed as on a globe

generation of value added
products
(e.g. thematic maps)

**users**
professional
research
citizens

**Fig. 1.17** Digital earth, showing the types of data gathering, the dependence on computer networks and social media, the globe as the reference framework, and the concept of inserting value-added products back into the information base, either as free goods or commercially available commodities

## 1.8    How This Book Is Arranged

The purpose of this chapter has been to introduce the image data sources that are used in remote sensing and which are the subject of the processing operations described in the remainder of the book. It has also introduced the essential characteristics by which digital image data is described. The remainder of the book is arranged in a sequence that starts with the recorded image data and progresses through to how it is utilised.

The first task that normally confronts the analyst, before any meaningful processing can be carried out, is to ensure as much as possible that the data is free of error, both in geometry and brightness. Chapter 2 is dedicated to that task along with the associated operation of registering images together, or to a map base. At the end of that chapter, we assume that the data has been corrected and is ready for analysis.

Chapter 3 then starts us on the pathway to data interpretation. It is an overview that considers the various ways that digital image data can be analysed, either manually or with the assistance of a computer. Such an overview is important because there is no single, correct method for undertaking image interpretation; it is therefore important to know the options available before moving into the rest of the book.

It is frequently important to produce an image from the recorded digital image data, either on a display screen or in hard copy format. That is essential when analysis is to be carried out using the visual skills of a human interpreter. Even when machine analysis is to be performed the analyst will still produce image products, most likely on a screen, to assist in that task. To make visual interpretation and recognition as easy as possible it is frequently necessary to enhance the visual appeal of an image. Chapter 4 looks at methods for enhancing the radiometric (brightness and contrast) properties of an image. It also looks at how we might join images side-by-side to form a mosaic in which it is necessary to minimise any brightness differences across the join.

The visual impact of an image can be also improved through operations on image geometry. Such procedures can be used to enhance edges and lines, or to smooth noise, and are the subject of Chap. 5. In that chapter we also look at geometric processing operations that assist image interpretation.

In Chap. 6 we explore a number of transformations that generate new versions of images from the image data recorded by remote sensing platforms. Chief among these is the principal components transformation, well-regarded as a fundamental operation in image processing.

Several other transformations are covered in Chap. 7. The Fourier transform and the wavelet transform are two major tools that are widely employed to process image data in a range of applications. They are used to implement more sophisticated filtering operations than are possible with the geometric procedures covered in Chap. 5, and to provide means by which imagery can be compressed into more efficient forms for storage and transmission.

At this stage the full suite of so-called image enhancement operations has been covered and the book moves its focus to automated means for image interpretation. Many of the techniques now to be covered come from the field of machine learning.

Chapter 8 is central to the book. It is a large chapter because it covers the range of machine learning algorithms commonly encountered in remote sensing image interpretation. Those techniques are used to produce maps of land cover, land type and land use from the data recorded by a remote sensing mission. At the end of this chapter the reader should understand how data, once corrected radiometrically and geometrically, can be processed into viable maps by making use of a small number of pixels for which the appropriate ground label is known. Those pixels are called training pixels because we use them to train the machine learning technique we have chosen to undertake the full mapping task. The techniques treated come under the name of supervised classification.

On occasions the user does not have available known samples of ground cover over which the satellite data is recorded—in other words there are no training pixels. Nevertheless, it is still possible to devise machine learning techniques to label satellite data into ground cover types. Chapter 9 is devoted to that task and covers what are called unsupervised classification and clustering.

Frequently we need to reduce the volume of data to be processed, generally by reducing the number of bands. That is necessary to keep processing costs in bounds, or to ensure some analysis algorithms operate effectively. Chapter 10 presents the techniques commonly used for that purpose. Two approaches are presented: one involves selecting optimal subsets of the existing bands, while the other entails transforming the data beforehand in an endeavour to make the task of discarding less useful (transformed) bands easier.

Chapter 11 brings together much of the material on classification and machine learning into a set of methodologies that are used to produce reliable classification and mapping results. Included here are the methods used to assess the accuracy of a classification exercise.

In Chap. 12 we look at techniques for performing a classification when several different types of image or spatial data are available. Those procedures can be numerical or statistical and can also be based on expert system methodologies.

A set of Appendices is given to provide supplementary material, including background mathematics and statistics.

## 1.9  Bibliography on Sources and Characteristics of Remote Sensing Image Data

This book is principally about the computer processing of remote sensing image data and is not a detailed treatment of remote sensing as a field. Should more background in remote sensing be needed then standard treatments include

F. Sabins, *Remote Sensing: Principles and Interpretation*, 3rd ed., Waveland, Long Grove IL, 1997

T. Lillesand, R.W. Kiefer and J. Chipman, *Remote Sensing and Image Interpretation*, 7th ed., J. Wiley and Sons, N.Y., 2015

Highlighted below are a number of image processing and analysis texts that will add further detail to the coverage in this book, often at a higher mathematical level. One of the first comprehensive texts on the computer processing of remotely sensed imagery is

P.H. Swain and S.M. Davis, eds., *Remote Sensing: the Quantitative Approach*, McGraw-Hill, N.Y., 1978

Even though much of the material has now been superseded, this standard book still has one of the best chapters on the spectral reflectance characteristics of earth surface cover types, information that is essential to understand when carrying out image interpretation.

A coverage of thermal remote sensing will be found in

C. Kuenzer and S. Dech, eds, *Thermal Infrared Remote Sensing*, Springer Science + Business Media, Dordrecht, 2013.

For examples of thermal spectral emission properties and spectra see

Y. Ninomiya and B. Fu, Regional lithological mapping using ASTER-TIR data: Case study for the Tibetan Plateau and the surrounding area, *Geosciences*, vol. 6, no. 3, 2016, pp. 39ff, https://doi.org/10.3390/geosciences6030039

Y. Ninomiya and B. Fu, Thermal infrared multispectral remote sensing of lithology and mineralogy based on spectral properties of materials, *Ore Geology Reviews*, vol. 108, 2019, pp. 54–72

D.J. Williams, A.N. Pilant, D.D. Worthy, B. Feldman, T. Williams and P. Lucey, Detection and identification of toxic air pollutants using airborne LWIR hyperspectral imaging, *SPIE 4th Int. Asia-Pacific Environmental Remote Sensing Symposium*, Honolulu, Hawaii, 8–11 November 2004, vol. 5655, pp. 1–8, 2005

G.C. Hulley and S.J. Hook, Generating consistent land surface temperature and emissivity products between ASTER and MODIS data for earth science research, *IEEE Transactions on Geoscience and Remote Sensing*, vol. 49, no. 4, April 2011, pp. 1304–1315

A standard treatment on the image enhancement procedures, both radiometric and geometric, is

R.C. Gonzalez and R.E. Woods, *Digital Image Processing*, 4th ed., Pearson Prentice-Hall, Upper Saddle River, N.J., 2018.

A simpler treatment of digital image processing techniques will be found in

K.R. Castleman, *Digital Image Processing*, 2nd ed., Prentice-Hall, Upper Saddle River, N.J., 1996.

for which a solutions manual is also available.

Like digital image processing there are many specialised and general texts on the pattern recognition or machine learning techniques that are fundamental to remote sensing image interpretation. Perhaps the most commonly used, with a broad coverage of techniques, is

R.O. Duda, P.E. Hart and D.G. Stork, *Pattern Classification*, 2nd ed., John Wiley & Sons, N.Y., 2001.

A more recent and mathematically detailed treatment is

C.M. Bishop, *Pattern Recognition and Machine Learning*, Springer Science+Business Media LLC, N.Y., 2006.

When the number of spectral bands recorded by a sensor exceeds about 100 there are special challenges for computer image interpretation. A book devoted to that problem is

D.A. Landgrebe, *Signal Theory Methods in Multispectral Remote Sensing*, John Wiley & Sons, Hoboken, N.J., 2003.

For a treatment of radar remote sensing that assumes little prior knowledge see

J.A. Richards, *Remote Sensing with Imaging Radar*, Springer, Berlin, 2009.

A good mathematical companion on matrix methods is

K.B. Petersen and M.S. Pedersen, *The Matrix Cookbook*, which can be found on a number of university web sites through web searching. It's original URL of matrixcookbook.com seems now to be incorrect but is recorded here for historical purposes.

## 1.10   Problems

1.1. Suppose a given set of image data consists of just two bands, one centred on 0.65 μm and the other centred on 1.0 μm wavelength. Suppose the corresponding region on the earth's surface consists of water, vegetation and soil. Construct a graph with two axes, one representing the brightness of a pixel in the 0.65 μm band and the other representing the brightness of the pixel in the 1.0 μm band. Show on this graph (which we might call the spectral space or spectral domain) where you would expect to find vegetation pixels, soil pixels and water pixels. Indicate how straight lines could, in principle, be drawn between the three groups of pixels so that, if a computer had the equations of those lines stored in its memory, it could use them to identify every pixel in the image.
Repeat the exercise for an image data set with bands centred on 0.95 and 1.05 μm.

1.2. Assume a frame of image data consists of a segment along the track of the satellite as long as the swath is wide. Compute the data volume of a single frame from each of the following sensors and produce a graph of average data volume per band versus pixel size.

NOAA AVHRR
Aqua MODIS
Landsat ETM+
SPOT HRG multispectral
GeoEye multispectral.

1.3. Determine a relationship between swath width and orbital repeat cycle for a polar orbiting satellite at an altitude of 800 km, assuming that adjacent swaths overlap by 10% at the equator.

1.4. A particular geosynchronous satellite is placed in orbit over the poles, rather than over the equator. How often does it appear over the same spot on the earth's surface, and where is that?

1.5. Geostationary satellites, with time, wander from the equatorial plane and have to be repositioned; this is called station keeping. If you could see the satellite overhead (for example, through a powerful telescope) what would the satellite path look like before correction?

1.6. Reconsider Problem 1.1 but instead of drawing lines between the classes in the spectral domain consider instead how you might differentiate them by computing the mean position of each class.

1.7. Imagine a scanner of the type shown in Fig. 1.6 is carried on an aircraft. In flight the aircraft can unintentionally but slowly change altitude and can be subject to cross winds. The pilot would normally compensate for the cross wind by steering into it. Describe the effect of these two mechanisms on the geometry of the recorded image data.

1.8. Using the results in Appendix A calculate the frame acquisition time for the following satellite sensors.

SPOT HRG
NOAA AVHRR
WorldView Pan.

1.9. Most remote sensing satellites are in orbits that pass over the regions being imaged at about mid-morning. Why is that important?

1.10. Derive a relationship between repeat cycle and swath width for a remote sensing satellite with an orbital period of 90 min, assuming a near-polar orbit. Choose swath widths between 50 and 150 km. For a swath width of 100 km how could a repeat cycle of 10 days be achieved? Would several satellites be a solution?

1.11. By examining Figs. 1.11 and 1.14 discuss how the combination of optical and radar imagery might improve the recognition of ground cover types.

1.12. Discuss the relative advantages of satellite and aircraft platforms for remote sensing image acquisition.

1.13. A particular satellite at an altitude of 800 km in near polar orbit carries a high resolution optical sensor with 1 m spatial resolution. If the orbit is arranged so that complete earth coverage is possible, how long will that take if there are 2048 pixels per swath width? See Appendix A for the relationship between altitude and orbital period.

1.14. A particular sensor records data in two wavebands in which the radiometric resolution is just 2 bits (see Appendix B). What is the theoretical maximum number of cover types that can be differentiated with the sensor?
If a sensor has $c$ channels and a radiometric resolution of $b$ bits show that the total number of sites in the corresponding spectral domain (see Problem 1.1) is $2^{bc}$. How many different sites are there for the following sensors?

SPOT HRV
Landsat ETM+
EO-1 Hyperion
Ikonos.

For an image of $512 \times 512$ pixels how many sites, on the average, will be occupied for each of these sensors?

1.15. Why is earth imaging from satellites not carried out at wavelengths of about 1 mm?

1.16. What imaging wavelengths would you use to map a fire burning on the earth's surface, superimposed on general landscape features?

1.17. Discuss the concept of map scale in the context of a virtual globe in the digital earth paradigm.
1.18. Many general purpose radar satellites operate at C band. Why?
1.19. Remote sensing satellites take about 90 min to orbit the earth and yet cover the whole earth in about one or two weeks. Is that possible because

   (a) the earth is rotating under the satellite,
   (b) the orbital plane is shifted after each completed orbit, or
   (c) both (a) and (b)?

1.20. Satellites are the preferred remote sensing platforms when (choose one and justify)

   (a) we want greatest control over the region being imaged,
   (b) we want cost-effective imaging over large areas, or
   (c) we want to minimise the effect of the atmosphere on the recorded imagery?

1.21. Suppose a particular sensor was capable of imaging in the following wavebands:

   1. 0.45–0.52 $\mu$m (blue)
   2. 0.52–0.60 $\mu$m (green)
   3. 0.63–0.69 $\mu$m (red)
   4. 0.76–0.90 $\mu$m (near IR)
   5. 1.40–1.45 $\mu$m (mid IR).

Which single band would be most effective for each of the following tasks?

   (a) Noting the loss of chlorophyll resulting from dying vegetation
   (b) Monitoring the loss of moisture from soil
   (c) Discriminating between vegetation and water
   (d) Monitoring the condition of vegetation
   (e) Discriminating between soil vegetation.

# Chapter 2
# Correcting and Registering Images

**Abstract** Sources of error and distortion in the recorded brightness values and in the geometry of remote sensing imagery are presented, along with detailed methods for their correction. Particular attention is given to the effect of the earth's atmosphere on recorded image data and those atmospheric constituents that have most influence. The use of control points and mapping functions to correct geometric errors is covered in detail, including as a means for registering images to a map base and to register sets of images to each other geographically. Mathematical models for common sources of geometric distortion are also treated. Examples of the main methods for correction and registration are given.

## 2.1 Introduction

When image data is recorded by sensors on remote platforms it can contain errors in geometry, and in the measured brightness values of the pixels. The latter are referred to as *radiometric errors* and can result from (i) the instrumentation used to record the data, (ii) the wavelength dependence of solar radiation and (iii) the effect of the atmosphere.

*Geometric errors* can also arise in several ways. The relative motions of the platform, its scanners and the earth can lead to errors of a skewing nature in an image product. Non-idealities in the sensors themselves, the curvature of the earth, and uncontrolled variations in the position, velocity and attitude of the remote sensing platform can all lead to geometric errors of varying degrees of severity.

It is usually important to correct errors in image brightness and geometry. That is certainly the case if the image is to be as representative as possible of the scene being recorded. It is also important if the image is to be interpreted manually. If an image is to be analysed by machine, using the algorithms to be described in Chaps. 8 and 9, it is not always necessary to correct the data beforehand; that depends on the analytical technique being used.

Some schools of thought recommend against correction when analysis is based on machine learning methods, because correction will not generally improve

© The Author(s), under exclusive license to Springer Nature Switzerland AG 2022
J. A. Richards, *Remote Sensing Digital Image Analysis*,
https://doi.org/10.1007/978-3-030-82327-6_2

performance; rather the (minor) discretisation errors that can be introduced into image data by correction procedures may lead to unnecessary interpretation errors and geometric correction could always be applied to the interpreted product after analysis is complete. With modern image data, this precaution is no longer as important, and most analysts would correct data before analysis. Also, commercial suppliers of image data regularly correct their products for errors in brightness and geometry before it is sold to the client.

Automated interpretation based on library searching or other similarity-based methods will always require radiometric correction. Generally, radiometric correction is also required before data fusion operations and when several images of the same region taken at different times are to be compared.

It is the purpose of this chapter to discuss the nature of the radiometric and geometric errors commonly encountered in remote sensing images and to develop computational procedures that can be used for their compensation. The methods to be presented also find more general application, such as in registering together sets of images of the same region but at different times, and in performing operations such as scale changing and zooming (magnification).

We commence with examining sources of radiometric errors, and methods for their correction, and then move on to problems in image geometry.

## 2.2  Sources of Radiometric Distortion

Mechanisms that affect the measured brightness values of the pixels in an image can lead to two broad types of radiometric distortion. First, the distribution of brightness over an image in a given band can be different from that in the ground scene. Secondly, the relative brightness of a single pixel from band to band can be distorted compared with the spectral reflectance character of the corresponding region on the ground. Both types can result from:

- instrumentation errors
- the spectral dependence of solar radiation
- the presence of the atmosphere as a transmission medium through which radiation must travel from its source to the sensors.

We now consider each of these and their correction mechanisms.

## 2.3  Instrumentation Errors

### 2.3.1  Sources of Distortion

Because the sensors used in remote sensing instruments often use sets of detectors within a band and, obviously, between bands, radiometric errors can arise from calibration differences among the detectors. An ideal radiation detector has a

**Fig. 2.1 a** Ideal linear
radiation detector transfer
characteristic, and
**b** hypothetical mismatches in
detector characteristics

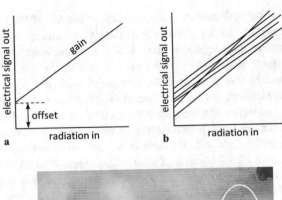

**Fig. 2.2** Reducing sensor
induced striping noise in a
Landsat MSS image:
**a** original image, and **b** after
destriping by matching sensor
statistics

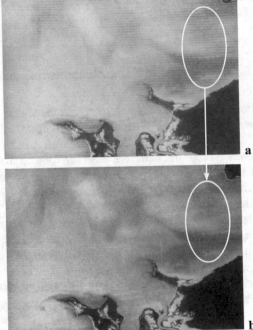

transfer characteristic such as that shown in Fig. 2.1a. It should be linear, so that
there is a proportionate increase or decrease of signal level with detected radiation.
Real detectors will have some degree of non-linearity. There will also be a small
signal out, even when there is no radiation in. Historically that is known as *dark
current* and is the result of residual electronic noise present in the detector at any
temperature other than absolute zero. In remote sensing it is usually called a detector
*offset*. The slope of the detector curve is called its *gain*, or sometimes *transfer gain*.

Most imaging devices used in remote sensing are constructed from sets of
detectors. In the case of the Landsat ETM+ there are 16 per band. Each will have
slightly different transfer characteristics, such as those depicted in Fig. 2.1b. Those
imbalances will lead to striping in the across swath direction similar to that shown
in Fig. 2.2a.

For push broom scanners, such as the SPOT HRG, there are as many as 12,000 detectors across the swath in the panchromatic mode of operation, so that longitudinal striping could occur if the detectors were not well matched. For monolithic sensor arrays, such as the charge coupled devices used in the SPOT instruments, that is rarely a problem, compared with the line striping that can occur with mechanical across track scanners that employ discrete detectors.

Another common instrumentation error is the loss of a complete line of data resulting from a momentary sensor or communication link failure, or the loss of signal on individual pixels in a given band owing to instantaneous drop out of a sensor or signal link. Those mechanisms lead to black lines across or along the image, depending on the sensor technology used to acquire the data, or to individual black pixels.

### 2.3.2 Correcting Instrumentation Errors

Errors in relative brightness, such as the within-band line striping referred to above and as shown in Fig. 2.2a for a portion of a Landsat Multispectral Scanner (MSS) image, can be rectified to a great extent in the following way. First, it is assumed that the detectors used for data acquisition in each band produce signals statistically similar to each other. In other words, if the means and standard deviations are computed for the signals recorded by each of the detectors over the full scene then they should almost be the same. This requires the assumption that statistical detail within a band doesn't change significantly over a distance equivalent to that of one scan covered by the set of the detectors (474 m for the six scan lines of Landsats 1, 2, 3 MSS for example). For most scenes this is usually a reasonable assumption in terms of the means and standard deviations of pixel brightness, so that differences in those statistics among the detectors can be attributed to the gain and offset mismatches illustrated in Fig. 2.1b.

Sensor mismatches of this type can be corrected by calculating pixel mean brightness and standard deviation within a band by using lines of image data known to come from a single detector. In the case of Landsat MSS that will require the data on every sixth line to be used. In a like manner five other measurements of mean brightness and standard deviation are computed for the other five MSS detectors. Correction of radiometric mismatches among the detectors can then be carried out by adopting one detector as a standard and adjusting the brightness values of all pixels recorded by each of the other detectors so that their mean brightnesses and standard deviations match those of the standard detector. That operation, which is commonly referred to as *destriping*, can be implemented by the operation

$$y = \frac{\sigma_d}{\sigma_i} x + m_d - \frac{\sigma_d}{\sigma_i} m_i \qquad (2.1)$$

where $x$ is the original brightness for a pixel and $y$ is its new (destriped) value in the band being corrected; $m_d$ and $\sigma_d$ are the reference values of mean brightness and standard deviation, usually those of a chosen band, and $m_i$ and $\sigma_i$ are the signal mean and standard deviation for the detector under consideration. Sometimes an independent reference mean and standard deviation is used. That allows a degree of contrast enhancement to be imposed during the destriping operation.

Figure 2.2 shows the result of applying (2.1) to the signals of the remaining five detectors of a Landsat Multispectral Scanner (MSS) image, after having chosen one as a reference. As seen, the result is good but not perfect, partly because the signals are being matched only on the basis of first and second order statistics. A better approach is to match the detector histograms using the methodology of Sect. 4.5.[1] It is also possible to correct errors in an observed image by using optimisation to match it to an assumed error-free image model,[2] and to use sub-space methods when the dimensionality is high.[3] More complex methods, however, are generally less suitable with large numbers of detectors.

Correcting lost lines of data or lost pixels can be carried out by averaging over the neighbouring pixels—using those on the lines on either side for line drop outs or the set of surrounding pixels for pixel drop outs. This is called *infilling* or sometimes *in-painting*.

## 2.4 Effect of the Solar Radiation Curve and the Atmosphere on Radiometry

We now examine the effect of environmental conditions on the radiometric character of recorded image data. To help focus on the important aspects, consider a hypothetical surface which will reflect all of the incident sunlight at all wavelengths. Assume, further, that there is no atmosphere above the surface, as depicted in Fig. 2.3a. A detector capable of taking many spectral samples will record the solar spectrum as shown.[4]

---

[1] This approach is demonstrated in M. P. Weinreb, R. Xie, I. H. Lienesch and D. S. Crosby, Destriping GOES images by matching empirical distribution functions, *Remote Sensing of Environment*, vol. 29, 1989, pp. 185–195, and M. Wegener, Destriping multiple sensor imagery by improved histogram matching, *Int. J. Remote Sensing*, vol. 11, no. 5, May 1990, pp. 859–875.

[2] See H. Shen and L. Zhang, A MAP-based algorithm for destriping and in painting of remotely sensed images, *IEEE Transactions on Geoscience and Remote Sensing*, vol. 47. no. 5, May 2009, pp. 1492–1502, and M. Bouali and S. Ladjal, Towards optimal destriping of MODIS data using a unidirectional variance model, *IEEE Transactions on Geoscience and Remote Sensing*, vol. 49, no. 8, August 2011, pp. 2924–2935.

[3] See N. Acito, M. Diani and G. Corsini, Subspace-based striping noise reduction in hyperspectral images, *IEEE Transactions on Geoscience and Remote Sensing*, vol. 49, no. 4, April 2011, pp. 1325–1342.

[4] If the spectral resolution of the detector were sufficiently fine then the recorded solar spectrum would include the Fraunhofer absorption lines associated with the gases in the solar atmosphere:

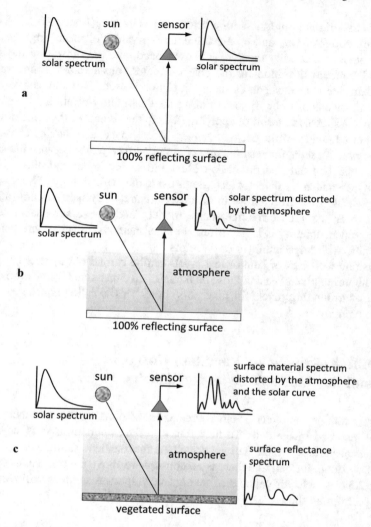

**Fig. 2.3** Distortion of the surface material reflectance spectrum by the spectral dependence of the solar curve and the effect of the atmosphere: **a** detection of the solar curve from a perfectly reflecting surface in the absence of an atmosphere, **b** effect of the atmosphere on detecting the solar curve, **c** detection of the real surface spectrum distorted by the atmosphere and the solar curve

Now suppose there is a normal terrestrial atmosphere in the path between the sun, the surface and the detector. The spectrum recorded will be modified by the extent to which the atmosphere selectively absorbs the radiation. There are well-known absorption features caused mainly by the presence of oxygen, carbon dioxide and water vapour in the atmosphere, and they appear in the recorded data as

See P. N. Slater, *Remote Sensing: Optics and Optical Systems*, Addison Wesley, Reading Mass., 1980.

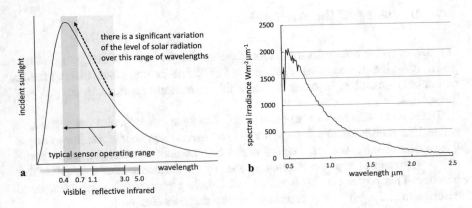

**Fig. 2.4 a** Showing the impact of the solar radiation curve at remote sensing wavelengths, and **b** the measured solar spectral irradiance of the sun above the earth's atmosphere,[5] over the wavelength range common in optical remote sensing

shown in Fig. 2.3b. The atmosphere also scatters the solar radiation, further complicating the signal received at the sensor. This reduces the solar energy that strikes the surface and travels to the sensor; energy also scatters from the atmosphere itself to the sensor superimposing onto the desired signal. We consider those additional complications in Sect. 2.6.

Figure 2.3c shows how the reflectance spectrum of a *real* surface might appear. The spectrum recorded is a combination of the actual spectrum of the real surface, modulated by the influence of the solar curve and distorted by the atmosphere. In order to be able to recover the true radiometric character of the image we need to correct for those effects.

## 2.5 Compensating for the Solar Radiation Curve

The wavelength dependence of the solar radiation falling on the earth's surface can be compensated by assuming that the sun is an ideal black body and able to be described by the behaviour of the Planck radiation law shown in Figs. 1.3 and 2.4a. For broad spectral resolution sensors that is an acceptable approach. For images recorded by instrumentation with fine spectral resolution it is important to account for departures from black body behaviour, effectively modelling the real emissivity of the sun, and using that to normalise the recorded image data. Most radiometric correction procedures compensate for the solar curve using the actual wavelength dependence measured above the atmosphere, such as that shown in Fig. 2.4b.

---

[5] Plotted, at lower spectral resolution, from the data in F. X. Kneizys, E. P. Shettle, L. W. Abreu, J. H. Chetwynd, G. P. Anderson, W. O. Gallery, J. E. A. Selby and S. A. Clough, *Users Guide to LOWTRAN 7*, AFGL-TR-0177, Environmental Research Paper No. 1010, 1988.

## 2.6    Influence of the Atmosphere

We now examine how solar irradiation produces the measured signal from a single pixel, using the mechanisms identified in Fig. 2.5. It is important, first, to define radiometric quantities in order to allow the correction equations to be properly formulated.

The sun is a source of energy that emits at a given rate of joules per second, or watts. That energy radiates through space in an inverse square law fashion so that at a particular distance the sun's emission can be measured as watts per square metre ($Wm^{-2}$), given as the power emitted divided by the surface area of a sphere at that distance. This power density is called *irradiance,* a property that can be used to describe the strength of any emitter of electromagnetic energy.

The power density scattered from the earth in a particular direction is defined by density per solid angle. This quantity is called *radiance* and has units of watts per square metre per steradian ($Wm^{-2}sr^{-1}$). If the surface is perfectly diffuse then the incident solar irradiance is scattered uniformly into the upper hemisphere, i.e., equal amounts are scattered into equal cones of solid angle.

The emission of energy by bodies such as the sun is wavelength dependent, as seen in Figs. 1.3 and 2.4, so that the term *spectral irradiance* can be used to describe how much power density is available in incremental wavebands across the wavelength range; that is actually the quantity plotted in Fig. 2.4b. Spectral irradiance is measured in $Wm^{-2}\mu m^{-1}$. Similarly, *spectral radiance* is measured in $Wm^{-2}\mu m^{-1}sr^{-1}$.

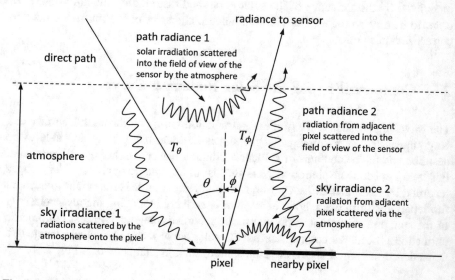

**Fig. 2.5** Effect of the atmosphere on solar radiation illuminating a pixel and reaching a sensor

Suppose in the absence of the atmosphere the solar spectral irradiance at the earth is $E_\lambda$. If the solar zenith angle (measured from the normal to the surface) is $\theta$ as shown in Fig. 2.5 then the spectral irradiance (spectral power density) at the earth's surface is $E_\lambda \cos \theta$. This gives an available irradiance between wavelengths $\lambda_1$ and $\lambda_2$ of

$$E_{os} = \int_{\lambda_1}^{\lambda_2} E_\lambda \cos \theta d\lambda \, \mathrm{Wm}^{-2}$$

For most instruments the wavebands used are sufficiently narrow that we can assume

$$E_{os} = E_{\Delta\lambda} \cos \theta \Delta\lambda = E(\lambda) \cos \theta \, \mathrm{Wm}^{-2} \tag{2.2}$$

in which $\Delta\lambda = \lambda_2 - \lambda_1$ and $E_{\Delta\lambda}$ is the average spectral irradiance over that bandwidth, centred on the wavelength $\lambda = (\lambda_2 + \lambda_1)/2$. $E(\lambda) = E_{\Delta\lambda}\Delta\lambda$ is the solar irradiance above the atmosphere at wavelength $\lambda$.

Suppose the surface has a reflectance $R(\lambda)$ in that narrow band of wavelengths, which describes the proportion of the incident irradiance that is scattered. If the surface is diffuse then the total radiance $L$ scattered into the upper hemisphere, and available for measurement, is

$$L = E(\lambda) \cos \theta R(\lambda)/\pi \, \mathrm{Wm}^{-2}\mathrm{sr}^{-1} \tag{2.3}$$

in which the divisor $\pi$ accounts for the upper hemisphere of solid angle. This equation relates to the ideal case of no atmosphere.

When an atmosphere is present there are two effects which must be taken into account that modify (2.3). They are the scattering and absorption by the particles in the atmosphere, for which compensation is needed when correcting imagery.

Absorption by atmospheric molecules is a selective process that converts incoming energy into heat; molecules of oxygen, carbon dioxide, ozone and water attenuate the radiation very strongly in certain wavebands.

There are two broad scattering mechanisms. The first is scattering by the air molecules themselves. That is called Rayleigh scattering, which has an inverse fourth power dependence on wavelength. The other is called aerosol or Mie scattering and is the result of the scattering of radiation from larger particles such as those associated with smoke, haze and fumes. Those particulates are of the order of one tenth to ten wavelengths. Mie scattering is also wavelength dependent, although not as strongly as Rayleigh scattering; it is approximately inversely proportional to wavelength. When the atmospheric particulates become much larger than a wavelength, such as those common in fogs, clouds and dust, the wavelength dependence disappears.

In a clear ideal atmosphere Rayleigh scattering is the only mechanism present. It accounts for the blueness of the sky. Because the shorter (blue) wavelengths are scattered more than the longer (red) wavelengths, we are more likely to see blue

**Fig. 2.6** Atmospheric and particulate scattering

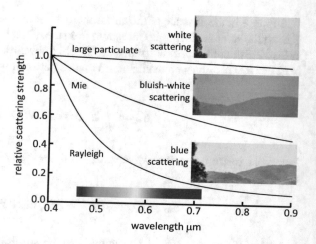

when looking in any direction in the sky. Likewise, the reddish appearance of sunset is also caused by Rayleigh scattering. That is the result of the long atmospheric path the radiation has to follow at sunset during which most short wavelength radiation is scattered away from direct line of sight, relative to the longer wavelengths.

Fogs and clouds appear white or bluish-white owing to the (almost) non-selective scattering caused by the larger particles. Figure 2.6 shows the typical scattering characteristics of different atmospheres.

We are now in the position to include the effect of the atmosphere on the radiation that ultimately reaches a sensor. We will do this by reference to the mechanisms shown in Fig. 2.5, commencing with the incoming solar radiation. They are identified by name in the following.

*Transmittance and Aerosol Optical Thickness (AOT).* In the absence of an atmosphere, transmission of the available solar irradiance to the surface at any wavelength is 100%. However, because of scattering and absorption, not all of the solar radiation reaches the ground. The amount that does, relative to that for no atmosphere, is called the transmittance. Let this be denoted $T_\theta$ in which the subscript indicates its dependence on the zenith angle of the source, which determines the path length through the atmosphere. In a similar way there is an atmospheric transmittance $T_\phi$, between the point of reflection and the sensor. Transmittance has a value between 0 and 1, although it is sometimes quoted as a percentage.

Transmittance can be written in terms of the Aerosol Optical Thickness (AOT), or Aerosol Optical Depth (AOD) $\tau$ of the atmosphere which is related to its scattering and absorbing properties at the time an image was recorded. Transmittance and optical thickness are related by

$$T = \exp(-\tau \sec \theta) \tag{2.4}$$

For an AOT near zero the transmittance is close to 1 (100% transmission through the atmosphere), whereas at a zenith angle of 90° the transmittance drops to 0.007 (less than 1%) if the AOT is 5.

AOT and thus transmittance are generally wavelength dependent.

*Sky irradiance.* Because the radiation is scattered on its travel down through the atmosphere, a particular pixel will be irradiated by energy on the direct path in Fig. 2.5 and by energy scattered from atmospheric constituents. The path for the latter is undefined and diffuse, and is referred to as sky irradiance component 1. A pixel can also receive energy that has been reflected from surrounding pixels and then scattered downwards by the atmosphere. That is the sky irradiance component 2 shown in Fig. 2.5. We call the total sky irradiance at the pixel $E_D$.

*Path radiance.* Again, because of scattering, radiation can reach the sensor from adjacent pixels and also via diffuse scattering of the incoming radiation directly to the sensor by atmospheric constituents before it reaches the ground. Those two components constitute path radiance, which is denoted as $L_p$.

Having defined these mechanisms, we are now in the position to determine how the radiance measured by the sensor is affected by the presence of the atmosphere. First, the total irradiance at the earth's surface now becomes, instead of (2.2),

$$E_G = E(\lambda)T_\theta(\lambda)\cos\theta + E_D \,\mathrm{Wm}^{-2}$$

where, for simplicity, it has been assumed that the diffuse sky irradiance $E_D$ is not a function of wavelength in the waveband of interest. The radiance resulting from this total irradiance of the pixel is thus

$$L = \{E(\lambda)T_\theta(\lambda)\cos\theta + E_D\}R(\lambda)/\pi \,\mathrm{Wm}^{-2}\mathrm{sr}^{-1}$$

Emerging from the atmosphere the total radiance detected by the sensor is composed of that term, reduced by atmospheric transmittance on the upward path, plus the path radiance $L_p$, to give

$$L = T_\phi(\lambda)\{E(\lambda)T_\theta(\lambda)\cos\theta + E_D\}R(\lambda)/\pi + L_p \,\mathrm{Wm}^{-2}\mathrm{sr}^{-1} \tag{2.5}$$

This equation gives the relationship between the radiance measured at the sensor $L$ and the reflectance of the surface material $R(\lambda)$ in a given waveband, assuming all the other quantities can be modelled or measured. Sometimes the path radiance term is written as $L_p = E(\lambda)R_A/\pi$ in which $R_A$ is called the reflectance of the atmosphere.[6] If the diffuse sky irradiance term $E_D$ is neglected (2.5) becomes

$$L(\lambda) = T(\lambda)R(\lambda)E(\lambda)\cos\theta/\pi + L_p(\lambda) \,\mathrm{Wm}^{-2}\mathrm{sr}^{-1} \tag{2.6}$$

---

[6] See *ACORN 4.0 Users Guide, Stand Alone Version*, Analytical Imaging and Geophysics LLC, Boulder, Colorado, 2002.

$E(\lambda)$ is the available solar irradiance in the channel of interest, $T(\lambda)$ is the transmittance of the total atmospheric path, $L(\lambda)$ is the radiance at the detector and $L_p(\lambda)$ is the path radiance. Equation (2.6) is used in many atmospheric correction procedures (see Sect. 2.9).

Image data products are expressed in digital numbers on a scale set by the radiometric resolution of the sensor[7]; 8 bit data is in the range 0–255, 10 bit data is in the range 0–1023 and 12 bit data is in the range 0–4095. The relationship between the detected radiance and the corresponding digital number ($DN$) in the image product can be expressed

$$L = \kappa DN + L_{min}\ \mathrm{Wm}^{-2}\mathrm{sr}^{-1} \tag{2.7a}$$

in which the sensor gain term $\kappa$ is

$$\kappa = (L_{max} - L_{min})/DN_{max}\ \mathrm{Wm}^{-2}\mathrm{sr}^{-1} \text{per digital value} \tag{2.7b}$$

$DN_{max}$ is the highest possible digital count for the sensor. Values for $L_{max}, L_{min}$ and $DN_{max}$ in each waveband are usually available from the sensor operator, allowing the digital data to be expressed in radiance. This is a necessary step before the correction of atmospheric errors.

## 2.7  Effect of the Atmosphere on Remote Sensing Imagery

The result of atmospheric distortion of the signal recorded by a sensor depends, to an extent, on the spectral resolution of the instrument. We consider broad waveband systems first, such Landsat ETM+ and SPOT HRG.

One effect of scattering is that fine detail in image data will be obscured. Consequently, in applications where the analyst depends on the limit of sensor resolution, it is important to take steps to correct for atmospheric effects.

It is important also to consider the effects of the atmosphere on systems with wide fields of view in which there will be an appreciable difference in atmospheric path length between nadir and the extremities of the swath. That will be significant with satellite missions such as NOAA.

Finally, and perhaps most importantly, because both Rayleigh and Mie scattering are wavelength dependent, the effects of the atmosphere will be different in the different wavebands of a given sensor system. In the case of the Landsat Thematic Mapper the visible blue band (0.45–0.52 µm) can be affected appreciably by comparison to the middle infrared band (1.55–1.75 µm). That leads to a loss in calibration over the set of brightness values for a particular pixel.

---

[7] See Appendix B.

In high spectral resolution systems, such as hyperspectral sensors, the effect of the atmosphere is complicated by the presence of the absorption lines superimposed by water vapour and other atmospheric constituents. We examine suitable correction mechanisms in Sect. 2.9.

## 2.8   Correcting Atmospheric Effects in Broad Waveband Systems

Correcting images from broad waveband sensors (typically multispectral) to remove as much as possible the degrading effects of the atmosphere requires modelling of the scattering and gross absorption processes to see how they determine the transmittances of the signal paths, and the components of sky irradiance and path radiance. When those quantities are available, they can be used in (2.5) or (2.6), and (2.7) to relate the digital numbers DN for the pixels in each band of data to the true reflectance R of the surface being imaged. An instructive example of how this can be done is given by Forster[8] for the case of Landsat MSS data; he also gives source material and tables to assist in the computations.

Forster considers the case of a Landsat 2 MSS image in the wavelength range 0.8–1.1 μm (near infrared; then called band 7) acquired at Sydney, Australia on 14th December 1980 at 9:05 am local time. At the time of overpass, the atmospheric conditions were.

| | |
|---|---|
| Temperature | 29 °C |
| relative humidity | 24% measured at 30 m above sea level |
| atmospheric pressure | 1004 mbar |
| visibility | 65 km |

Based on the equivalent mass of water vapour in the atmosphere (computed from temperature and humidity measurements) the absorbing effect of water molecules was computed. That is the only molecular absorption mechanism considered significant over the broad waveband involved. The measured value for visibility was used to estimate the effect of Mie scattering. That was combined with the known effect of Rayleigh scattering at that wavelength to give the total normal optical thickness of the atmosphere. Its value for this example is $\tau = 0.15$. Using this, with a solar zenith angle of 38° (at overpass) and a nadir viewing satellite we find from (2.4) that

$$T_\theta = 0.827$$

$$T_\phi = 0.861$$

[8] B. C. Forster, Derivation of atmospheric correction procedures for Landsat MSS with particular reference to urban data. *Int. J. Remote Sensing*, vol. 5. 1984, no. 5, pp. 799–817.

In the waveband of interest Forster shows that the solar irradiance at the earth's surface in the absence of an atmosphere is $E_0 = 256$ Wm$^{-2}$. He further computes the total global irradiance at the earth's surface as 186.6 Wm$^{-2}$. Noting in (2.5) that the term in brackets is the global irradiance, and using the relevant values of $T_\theta$ and $\cos \theta$, this gives the total diffuse sky irradiance as 19.6 Wm$^{-2}$—i.e., about 10% of the global irradiance for this example.

Using correction algorithms given by Turner and Spencer,[9] which account for Rayleigh and Mie scattering and atmospheric absorption, Forster computes the path radiance for this example as

$$L_p = 0.62\,\mathrm{Wm}^{-2}\mathrm{sr}^{-1}$$

so that (2.5) becomes for band 7

$$L_7 = 0.274 \times 186.6R_7 + 0.62\,\mathrm{Wm}^{-2}\mathrm{sr}^{-1}$$
$$\text{i.e.,} \quad L_7 = 51.5R_7 + 0.62\,\mathrm{Wm}^{-2}\mathrm{sr}^{-1} \tag{2.8}$$

At the time of overpass, it was established that for the band 7 sensor on Landsat 2 $L_{max} = 39.1$ Wm$^{-2}$sr$^{-1}$ and $L_{min} = 1.1$ Wm$^{-2}$sr$^{-1}$, while $DN_{max} = 63$ (6 bit data) so that, from (2.7b)

$$\kappa = 0.603\,\mathrm{Wm}^{-2}\mathrm{sr}^{-1}\text{per digital value}$$

From (2.7a) we thus have

$$L_7 = 0.603DN_7 + 1.1\,\mathrm{Wm}^{-2}\mathrm{sr}^{-1}$$

which, when combined with (2.8), gives the corrected reflectance in band 7 as

$$R_7 = 0.0118DN_7 + 0.0094$$

or, as a percentage, $R_7 = 1.18DN_7 + 0.94\%$

Similar calculations for the visible red band (band 5) give

$$R_5 = 0.44DN_5 + 0.5\%$$

---

[9] R. E. Turner and M. M. Spencer, Atmospheric model for the correction of spacecraft data, *Proc. 8th Int. Symposium on Remote Sensing of the Environment*, Ann Arbor, Michigan, 1972, pp. 895–934.

Methods such as this have now been operationalized in several large systems for multispectral satellite image production and distribution such as FORCE,[10] LaSRC[11] and MACCS.[12]

## 2.9  Correcting Atmospheric Effects in Narrow Waveband Systems

Correcting image data from narrow waveband sensors (hyperspectral) requires careful modelling of the differential absorption characteristics of atmospheric constituents and their scattering effects, and compensating for the solar curve as discussed in Sect. 2.5. The high spectral resolution means that fine atmospheric absorption features will be detected and may be confused with those of the ground cover type being imaged if not removed.

Correction in these cases consists of the following steps:

1. Conversion of raw recorded *DN* values to *radiance*, unless the product supplied is already in radiance form.
2. Compensating for the shape of the solar spectrum as outlined in Sect. 2.5. The measured radiances are divided by solar irradiances above the atmosphere to obtain the *apparent* reflectances of the surface.
3. Compensating for atmospheric gaseous transmittances, and molecular and aerosol scattering, by determining the aerosol optical thickness (AOT). Simulating these atmospheric effects allows the *apparent* reflectances to be converted to *scaled* surface reflectances.
4. Converting scaled surface reflectances to *real* surface reflectances after considering any topographic effects. If topographic data is not available, real reflectance is taken to be identical to scaled reflectance under the assumption that the surfaces of interest are Lambertian; that is the assumption made by many correction procedures although more recent approaches do allow for other surface behaviours.

We need now to consider how the third step can be performed. To do so we use the simplified expression of (2.6), which requires information on the absorptive and scattering properties of significant atmospheric constituents. Absorption enters via the transmittance $T(\lambda)$ and scattering via both the transmittance $T(\lambda)$ and the path radiance term $L_p$.

---

[10] D. Frantz, A. Roder, M. Stellmes and J. Hill, An operational radiometric Landsat pre-processing framework for large-area time series applications, *IEEE Transactions on Geoscience and Remote Sensing*, vol. 54, no. 7 2016, pp. 3928–3943.

[11] E. Vermote, J-C Roger, B. Franch and S. V. Skakun, LaSRC (Land Surface Reflectance Code): Overview, application and validation using MODIS, VIIRS, Landsat and Sentinel-2 data, *Proc. International Geoscience and Remote Sensing Symposium*, Valencia, July 2018, pp. 8173–8176.

[12] B. Petrucci, M. Huc, T. Feuvrier, C. Ruffel, O. Hagolie, V. Lonjou and C. Dejardins, MACCS: Multi-Mission Atmospheric Correction and Cloud Screening tool for high-frequency revisit data processing, *Proc. SPIE 9643, Image and Signal Processing for Remote Sensing XXI, 964307*, October 2015, https://doi.org/10.1117/12.2194797.

Over the years several data bases listing specific absorption characteristics of atmospheric gaseous components have been compiled. The most extensive, and that which is used by several atmospheric correction models in remote sensing, is HITRAN. Although its heritage can be traced back to 1973, successive refinements have led to an extensive compilation of the effects of a great number of important and less significant atmospheric constituents[13]; detailed information on the use of HITRAN and its development is available separately.[14] An on-line version of HITRAN is available and users can carry out their own analysis of the data supplied, via the HITRAN Application Programming Interface, HAPI.

Not all of the molecular constituents covered in HITRAN are significant when correcting high spectral resolution remote sensing imagery. The most important in the range relevant to optical remote sensing, 0.4–2.5 μm, are $H_2O$, $CO_2$, $O_3$, $N_2O$, $CO$, $CH_4$ and $O_2$.[15] Their transmission characteristics (referred to as transmission spectra) are illustrated in Fig. 2.7.

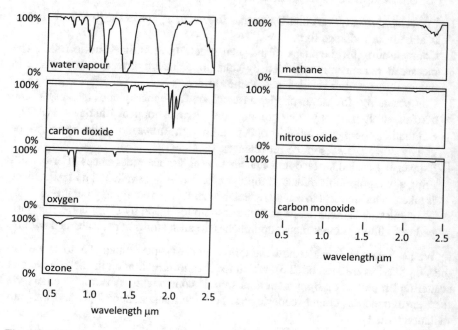

**Fig. 2.7** Indicative transmission spectra of the seven most significant atmospheric constituents[16]; the water vapour curve is for a tropical atmosphere

---

[13] L. S. Rothman and 42 others, The HITRAN 2008 molecular spectroscopic database, *J. Quantitative Spectroscopy and Radiative Transfer*, vol. 110, 2009, pp. 533–572.

[14] See www.cfa.harvard.edu/HITRAN/ accessed April 2021.

[15] B. C. Gao, K. B. Heidebrecht and A. F. H. Goetz, Derivation of scaled surface reflectance from AVIRIS data, *Remote Sensing of Environment*, vol. 44, 1993, pp. 165–178.

[16] Adapted from Figs. 2 and 3 of B. Gao, K. B. Heidebrecht and A. F. H. Goetz, *ibid*; used with permission of Elsevier.

Apart from ozone, which varies with latitude and season, but which can be modelled as a constant effect for a given image, all of $CO_2$, $N_2O$, $CO$, $CH_4$, $O_2$ can be considered relatively constant from image to image, and their absorption characteristics modelled[17] and used to correct for their absorbing effects on hyperspectral imagery.

Correction for the effects of water vapour is more complex because water in the atmosphere changes with humidity and can vary across a scene. Ideally it would be good to estimate the water vapour in the atmospheric path for each individual pixel so that each pixel can have its reflectivity corrected for water vapour absorption and scattering separately. Fortunately, with fine spectral resolution systems, that turns out to be possible through examining the resonant water absorption dips evident in Fig. 2.7 and reproduced in Fig. 2.8 with further relevant information added.

The depths of the minima at wavelengths of 0.94 and 1.14 μm depend on the amount of water vapour in the atmospheric path (column) for the relevant pixel. We can assess the quantity of water in the column by comparing the depths of either of those minima (as averages of a set of bands around those wavelengths) with the 100% transmission level shown; 100% transmission occurs in the water vapour windows near 0.865, 1.025 and 1.23 μm, so bands, or averages over sets of bands, near those wavelengths can be used to provide the 100% reference levels. Once the depth of a water absorption minimum has been estimated, usually by taking the ratio of the radiance at the minimum to the average radiance of the 100% transmission bands either side, a model is used to generate the water content in the path from the sun to the sensor, via the respective pixel. That allows the corresponding transmission coefficient to be derived and the path radiance contributed by the atmospheric water content to be determined.

Several packages are available that implement radiometric correction based on the processes just described. One of the earliest was ATREM (Atmosphere Removal) developed at the University of Colorado.[18] It accounts for the seven atmospheric constituents noted above, using the ratio technique of Fig. 2.8 to correct for atmospheric water vapour. Atmospheric scattering is incorporated using the 5S and 6S radiative transfer codes.[19] A version of 6S is now available that

---

[17] Many correction methodologies use the narrow band transmittance model in W. Malkmus, Random Lorentz band model with exponential-tailed S line intensity distribution function, *J. Optical Society of America*, vol. 57, 1967, pp. 323–329.

[18] *Atmosphere Removal Program (ATREM), Version 3.1 Users Guide*, Centre for the Study of Earth from Space, University of Colorado, 1999.

[19] D. Tanre, C. Deroo, P. Duhaut, M. Herman, J. J. Morchrette, J. Perbos and P. Y. Deschamps, *Simulation of the Satellite Signal in the Solar Spectrum (5S) Users Guide*, Laboratoire d'Optique Atmospherique, Universitat S. T. de Lille, 1986, E. F. Vermote, D. Tanre, J. L. Deuze, M. Herman, J-J Morc and J-J Morcretee, Second simulation of the satellite signal in the solar spectrum, 6S: An overview, *IEEE Transactions on Geoscience and Remote Sensing*, vol. 35, no. 3 1997, pp. 675–686, and https://salsa.umd.edu/6spage.html (the 6S users site).

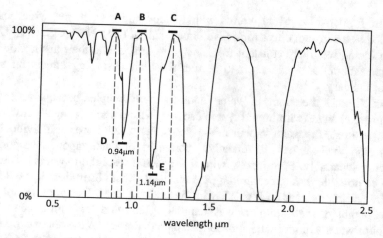

**Fig. 2.8** Using absorption features in the water spectrum of Fig. 2.7 to estimate atmospheric water content; the average signal at D is divided by the averages over the ranges A and B, and the average signal at E is divided by the averages of the ranges B and C to give two measurements for estimating water content

accounts for polarization effects and surfaces other than Lambertian.[20] The vegetation spectrum for a pixel that has been atmospherically corrected using ATREM is shown in Fig. 2.9.

ATREM is no longer available. Current correction programs tend to be based on, and are refinements of, MODTRAN4.[21] MODTRAN4, unlike ATREM allows the sky irradiance term 2 and path radiance term 2 in Fig. 2.5 to be incorporated into the atmospheric correction process. Two other approaches that build on MODTRAN4 are ACORN[22] (Atmospheric Correction Now) and FLAASH[23] (Fast Line-of-Sight Atmospheric Analysis of Hyperspectral Cubes). A comparison of

---

[20] S. Y. Kotchenova, E. F. Vermote, R. Matarrese, and F. J. Klemm, Jr, Validation of a vector version of the 6S radiative transfer code for atmospheric correction of satellite data, Part 1: path radiance, *Applied Optics*, Vol. 45, Issue 26, 2006, pp. 6762–6774 and S. Y. Kotchenova and E. F. Vermote, Validation of a vector version of the 6S radiative transfer code for atmospheric correction of satellite data, Part 2: Homogeneous, Lambertian and anisotropic surfaces, *Applied Optics*, Vol. 46, Issue 20, 2007, pp. 4455–4464.

[21] A. Berk, G. P. Anderson, L. S. Bernstein, P. K. Acharya, H. Dothe, M. W. Matthew, S. M. Adler-Golden, J. H. Chetwynd, Jr., S. C. Richtsmeier, B. Pukall, C. L. Allred, L. S. Jeong, and M. L. Hoke, MODTRAN4 Radiative Transfer Modeling for Atmospheric Correction, *Proc. SPIE Optical Stereoscopic Techniques and Instrumentation for Atmospheric and Space Research III*, vol. 3756, July 1999.

[22] *ACORN 4.0 Users Guide, Stand Alone Version*, loc. cit.

[23] S. M. Alder-Golden, M. W. Matthew, L. S. Bernstein, R. Y. Levine, A. Berk, S. C. Richtsmeier, P. K. Acharya, G. P. Anderson, G. Felde, J. Gardner, M. Hike, L. S. Jeong, B. Pukall, J. Mello, A. Ratkowski and H. H. Burke, Atmospheric correction for short wave spectral imagery based on MODTRAN4, *Proc. SPIE Imaging Spectrometry*, vol. 3753, 1999, pp. 61–69.

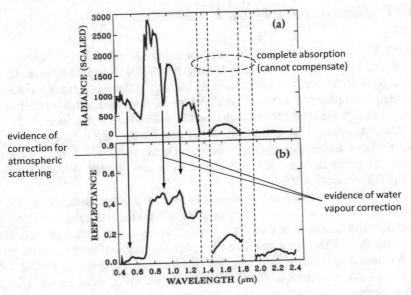

**Fig. 2.9** Correction of the raw spectrum of a vegetation pixel in which key features are evident; underlying diagram reprinted, with permission from Elsevier, from B. C. Gao, K. B. Heidebrecht and A. F. H. Goetz, Derivation of scaled surface reflectance from AVIRIS data, *Remote Sensing of Environment*, vol. 44, 1993, pp. 165–178

these packages will be found in Kruse,[24] while Gao et al.[25] review developments in correction algorithms and indicate where improvements are required.

## 2.10  Empirical, Data Driven Methods for Atmospheric Correction

Several approximate techniques are available for atmospheric correction that depend directly on measurements on the recorded image data. These are important when detailed data on the atmosphere, and particularly the water vapour content, are not available. The most common are considered here.

[24] F. A. Kruse, Comparison of ATREM, ACORN and FLAASH atmospheric corrections using low-altitude AVIRIS data of Boulder, CO, *Proc. 13th JPL Airborne Geoscience Workshop*, Pasadena, CA, 2004.

[25] B. C. Gao, M. J. Montes, C. O. Davis and A. F. H. Goetz, Atmospheric correction algorithms for hyperspectral remote sensing data of land and oceans, *Remote Sensing of Environment*, Supplement 1, Imaging Spectroscopy Special Issue, vol. 113, 2009, pp. S17–S24.

## 2.10.1  Haze Removal by Dark Subtraction

Frequently, detailed correction for the scattering and absorbing effects of the atmosphere is not required in broad waveband systems. Neither can detailed correction be implemented when the necessary ancillary information, such as visibility and relative humidity, is not readily available. In those cases, if the effect of the atmosphere is judged to be a problem, approximate correction can be carried out in the following manner. Effectively, it just corrects for the path radiance term $L_p$ in (2.5); some commercial image processing software systems use this method to account for path radiance before other procedures are applied to compensate for atmospheric absorption effects.

It makes the assumption that each band of data for a given scene should contain some pixels at or close to zero brightness value but that atmospheric effects, and especially path radiance, has added a constant level to each pixel in each band. Consequently, if histograms are taken of the bands (graphs of the number of pixels as a function of brightness value) the lowest significant occupied brightness value will be non-zero as shown in Fig. 2.10. Also, because path radiance varies as $\lambda^{-\alpha}$ (with $\alpha$ between 0 and 4 depending upon the extent of Mie scattering) the lowest occupied brightness value will be further from the origin for the shorter wavelengths, as depicted. Approximate correction requires, first, identifying the amount by which each histogram is apparently shifted in brightness from the origin and then subtracting that amount from each pixel brightness in that band.

It is clear that the effect of atmospheric scattering as implied in the histograms of Fig. 2.10 is to lift the overall brightness value of an image in each band. In the case of a colour composite product (see Sect. 3.2.1) this will appear as a whitish-bluish haze. Following correction in the manner just described—often called *dark subtraction*—the haze will be removed and the dynamic range of image intensity will be improved. Consequently, this approach is also frequently referred to as *haze removal*.

## 2.10.2  The Flat Field Method

The Flat Field method[26] depends on locating a large, spectrally uniform area in an image, such as sand or clouds, (a "spectrally" flat field) and computing its average radiance spectrum. It is assumed that the recorded shape and absorption features present in that spectrum are caused by solar and atmospheric effects since, in their absence, the spectrum should be flat. The reflectance of each image pixel is then

---

[26] D. A. Roberts, Y. Yamaguchi and R. J. P. Lyon, Comparison of various techniques for calibration of AIS data, *Proc. 2nd AIS Workshop*, JPL Publication 86–35, Jet Propulsion Laboratory, Pasadena CA, 1986.

**Fig. 2.10** Illustrating the
effect of path radiance
resulting from atmospheric
scattering

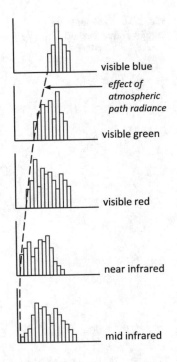

visible blue

*effect of
atmospheric
path radiance*

visible green

visible red

near infrared

mid infrared

corrected by dividing the spectrum of the pixel by the average radiance spectrum of
the flat field.

## 2.10.3  The Empirical Line Method

In this approach[27] two spectrally uniform targets in the image, one dark and one
bright, are identified; their actual reflectances are then determined by field or laboratory measurements. The radiance spectra for each target are extracted from the
image and then mapped to the actual reflectances using linear regression techniques.
The gain and offset so derived for each band are then applied to *all* pixels in the
image to calculate their reflectances, as illustrated in Fig. 2.11. While this is an
appealing technique and the computational load is manageable, it does require field
or laboratory reflectance data to be available.

---

[27] D. A. Roberts, Y. Yamaguchi and R. J. P. Lyon, Calibration of Airborne Imaging Spectrometer
data to percent reflectance using field spectral measurements, *Proc. 19th Int. Symposium on
Remote Sensing of Environment,* Ann Arbor, Michigan, 21–25 October 1986.

**Fig. 2.11** Illustrating the
empirical line method

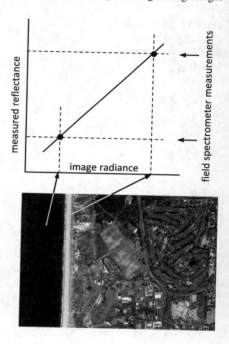

## 2.10.4  Log Residuals

This method is based on an assumed linear relationship between radiance and
reflectance of the form[28]

$$x_{i,n} = I_n S_i R_{i,n} \tag{2.9}$$

where $x_{i,n}$ $(i = 1\ldots K, n = 1\ldots N)$ is the radiance for pixel $i$ in waveband $n$ and $R_{i,n}$
is the reflectance to be found. $S_i$ accounts for the effect of topography, different for
each pixel but assumed to be constant for all wavelengths. $I_n$ accounts for wavelength dependent illumination, including the solar curve and atmospheric transmittance, which is assumed to be independent of pixel. $K$ and $N$ are the total
number of the pixels in the image and the total number of bands, respectively.

For reasons which will become clear shortly divide the measured radiance $x_{i,n}$ by
its geometric mean over the wavebands and its geometric mean over the pixels.
Here we denote the geometric mean of a quantity $x$ with respect to the index $n$ by
$\mathcal{G}_n(x)$ so that we have

---

[28] A. A. Green and M. D. Craig, Analysis of Airborne Imaging Spectrometer data with logarithmic
residuals, *Proc. 1st AIS Workshop*, JPL Publication 85–41, Jet Propulsion Laboratory,
Pasadena CA, 8–10 April 1985, pp. 111–119.

$$z_{i,n} = \frac{x_{i,n}}{\mathcal{G}_n(x_{i,n})\mathcal{G}_i(x_{i,n})} \tag{2.10}$$

Substituting from (2.9), this becomes

$$z_{i,n} = \frac{I_n S_i R_{i,n}}{\mathcal{G}_n(I_n)S_i \mathcal{G}_n(R_{i,n})I_n \mathcal{G}_i(S_i)\mathcal{G}_i(R_{i,n})}$$

i.e.,   $$z_{i,n} = \frac{R_{i,n}}{\mathcal{G}_n(I_n)\mathcal{G}_i(S_i)\mathcal{G}_n(R_{i,n})\mathcal{G}_i(R_{i,n})} \tag{2.11}$$

Now $\mathcal{G}_n(I_n)$ is independent of pixel and thus is a constant in (2.11); likewise, $\mathcal{G}_i(S_i)$ is independent of band and is also a constant. Therefore, to within a multiplicative constant, $z_{i,n}$ defined on the basis of measured radiance in (2.10) is an expression involving surface reflectance which is of the same form as that involving measured radiance and is independent of both the illumination conditions and the effect of topography. Accordingly, if we used (2.10) then the result can be considered to be a scaled reflectance.

We now take the logarithm of (2.10), to give

$$\log z_{i,n} = \log x_{i,n} - \log \mathcal{G}_n(x_{i,n}) - \log \mathcal{G}_i(x_{i,n})$$

i.e.,   $$\log z_{i,n} = \log x_{i,n} - \frac{1}{N}\sum_{n=1}^{N}\log x_{i,n} - \frac{1}{K}\sum_{i=1}^{K}\log x_{i,n}$$

This is the expression used for the log residuals method. It produces the logarithm of an expression equivalent to scaled reflectance and is thus independent of topographic and illumination effects.

## 2.11   Sources of Geometric Distortion

There are potentially many more sources of geometric distortion in images than radiometric distortion, and their effects can be quite severe. Some are more important with aircraft and drone platforms whereas others are a greater problem for satellites. They can be related to a number of factors, including

- the rotation of the earth during image acquisition
- variations in platform altitude, attitude and velocity
- the wide field of view of some sensors
- the curvature of the earth
- the finite scan rate of some sensors
- sensor non-idealities.

**Fig. 2.12** The display grid
used to build up an image
from the digital data stream of
pixels generated by a sensor

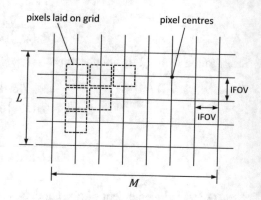

In the following sections we discuss the nature of the distortions that arise from these effects and means by which they can be compensated.

To appreciate why geometric distortion occurs and its manifestation in imagery, it is important to envisage how an image is formed from sequential lines of image data. If one imagines that a particular sensor records $L$ lines of $M$ pixels each then it would be natural to form the image by laying the $L$ lines down successively one under the other. If the IFOV of the sensor has an aspect ratio of unity—i.e., the pixels are the same size along and across the scan—then this is the same as arranging the pixels for display on a square grid, such as that shown in Fig. 2.12. The grid intersections are the pixel positions and the spacing between the grid points is determined by the sensor's IFOV.

## 2.12  The Effect of Earth Rotation

Sensors that record one line of data at a time across the image swath will incur distortion in the recorded image product as a result of the rotation of the earth during the finite time required to record a full scene. During the frame (or scene) acquisition time the earth rotates from west to east so that a pixel imaged at the end of the frame would have been further to the west when recording started. Therefore, if the lines of pixels recorded were arranged for display in the manner of Fig. 2.12 the later lines would be erroneously displaced to the east in terms of the terrain they represent. To give the pixels their correct positions relative to the ground it is necessary to offset the bottom of the image to the west by the amount by which the ground has moved during image acquisition, with all intervening lines displaced proportionately as depicted in Fig. 2.13. The amount the image has to be skewed to the west at the end of the frame depends on the relative velocities of the satellite and earth, and the length of the image frame recorded.

An example is presented here for Landsat 7. The angular velocity of the satellite is $\omega_o = 1.059\,\text{mrad s}^{-1}$ so that a nominal $L = 185\,\text{km}$ frame on the ground is scanned in

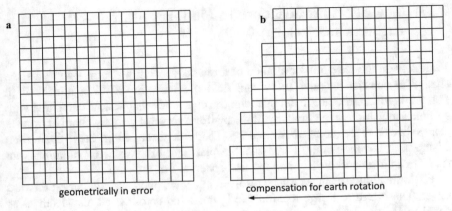

**Fig. 2.13** Effect of earth rotation on image geometry when data is acquired as scan lines: **a** image constructed according to Fig. 2.12 in which the pixels are arranged on a square grid, **b** offset of successive groups of lines to the west to correct for earth rotation during image acquisition

$$t_s = \frac{L}{\omega_o r_e} = 27.4\,\text{s}$$

where $r_e$ is the radius of the earth (6.37816 Mm). The surface velocity of the earth is

$$v_e = \omega_e r_e \cos \varphi$$

in which $\varphi$ is latitude and $\omega_e$ is the earth rotational velocity of 72.72 µ rad s$^{-1}$. At Sydney, Australia $\varphi = 33.8°$ so that

$$v_e = 385.4\,\text{ms}^{-1}$$

During the frame acquisition time the surface of the earth at Sydney moves to the east by

$$\Delta x_e = v_e t_s = 10.55\,\text{km}$$

This is 6% of the 185 km frame, which is quite severe and certainly noticeable. Since the satellite does not pass directly north–south, this figure has to be adjusted by the path inclination angle. At Sydney for Landsat 7 this is approximately 11° so that the effective sideways movement of the earth is actually

$$\Delta x = \Delta x_e \cos 11° = 10.34\,\text{km}$$

If steps are not taken to correct for the effect of earth rotation during Landsat 7 image acquisition the image will exhibit about 6% skew distortion to the east.

## 2.13    The Effect of Variations in Platform Altitude, Attitude and Velocity

Variations in the elevation or altitude of a remote sensing platform lead to a scale change at constant angular IFOV and field of view; the effect is illustrated in Fig. 2.14a for an increase in altitude with travel at a rate that is slow compared with a frame acquisition time. Similarly, if the platform forward velocity changes, a scale change occurs in the along-track direction. That is depicted in Fig. 2.14b again for a change that occurs slowly. For a satellite platform, orbit velocity variations can result from orbit eccentricity and the non-sphericity of the earth.

Platform attitude changes can be resolved into yaw, pitch and roll during forward travel. These lead to image rotation, along track and across track displacement as noted in Fig. 2.14c–e. The effects in the recorded imagery can be understood by again referring to Fig. 2.12 while looking at the diagrams in Fig. 2.14. For example, while Fig. 2.14a shows that the field of view of a sensor broadens with rising platform height, mapping the recorded pixels onto the grid of Fig. 2.12 will lead to an apparent compression of detail compared with that at lower altitudes.

Attitude variations in aircraft remote sensing systems can be quite significant owing to the effects of atmospheric turbulence. Those variations can occur over a short time, leading to localised distortions in aircraft scanner imagery.[29] Aircraft roll can be partially compensated in the data stream. That is made possible by having a data window that defines the swath width; the window is made smaller than the complete scan of data over the sensor field of view. A gyroscope mounted on the sensor is then used to move the position of the data window along the total scan line as the aircraft rolls. Pitch and yaw are generally not corrected unless the sensor is mounted on a three-axis stabilized platform.

While these variations can be described mathematically, at least in principle, a knowledge of the platform ephemeris is needed for their magnitudes to be computed.

## 2.14    The Effect of Sensor Field of View: Panoramic Distortion

For scanners used on spacecraft and aircraft remote sensing platforms the angular IFOV is constant. As a result, the effective pixel size on the ground is larger at the extremities of the scan than at nadir, as illustrated in Fig. 2.15. If the IFOV is $\beta$ and the pixel dimension at nadir is $p$ then its dimension in the scan direction at a scan angle $\theta$ as shown is

---

[29] For an extreme example see Fig. 3.1 in G. Camps-Valls and L. Bruzonne, eds., *Kernel Methods for Remote Sensing Data Analysis*, John Wiley & Sons, Chichester UK, 2009.

**Fig. 2.14** Effect of platform
position and attitude
variations on the region of the
earth being imaged for
variations that are slow
compared with image
acquisition

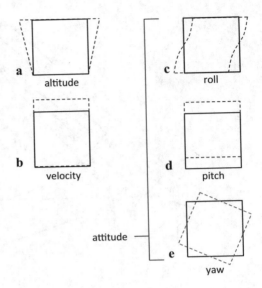

$$p_\theta = \beta h \sec^2 \theta = p \sec^2 \theta \qquad (2.12)$$

where $h$ is altitude. Its dimension across the scan line is $p \sec \theta$. For small values of $\theta$ distortion in pixel size is negligible. For Landsat 7 the largest value of $\theta$ is approximately $7.5°$ so that $p_\theta = 1.02p$. The effect can be quite severe for systems with larger fields of view, such as MODIS and aircraft scanners. For an aircraft scanner with FOV = $80°$ the distortion in pixel size along the scan line is $p_\theta = 1.70p$, i.e., the region on the ground measured at the extremities of the scan is 70% larger laterally than the region sensed at nadir. When the image data is arranged to form an image, as in Fig. 2.12, the pixels are all written as the same pixel size on a display device. While the displayed pixels are equal across the scan line the equivalent ground areas covered are not. This gives a compression of the image data towards its edges.

There is a second, related distortion introduced with wide field of view systems concerned with pixel position across the scan line. The scanner records pixels at constant angular increments and these are displayed on a grid of uniform centres, as in Fig. 2.12. However, the spacings of the effective pixels on the ground increase with scan angle. For example, if the pixels are recorded at an angular separation equal to the IFOV of the sensor then at nadir the pixels centres are spaced $p$ apart. At a scan angle $\theta$ the pixel centres will be spaced $p \sec^2 \theta$ apart as can be found from Fig. 2.15. By placing the pixels on a uniform display grid, the image will suffer an across track compression. Again, the effect for small angular field of view systems will be negligible in terms of the relative spacing of adjacent pixels. However, when the effect is aggregated to determine the location of a pixel at the swath edge relative to nadir the error can be significant. This can be determined by computing the arc SN in Fig. 2.15, S being the position to which the pixel at T

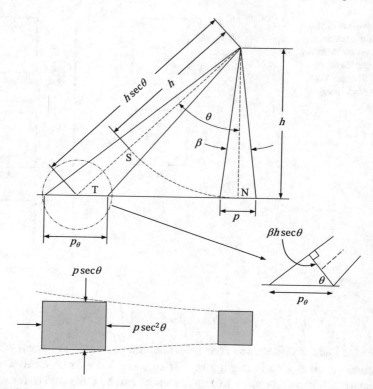

**Fig. 2.15**  Effect of scan angle on pixel size at constant angular instantaneous field of view

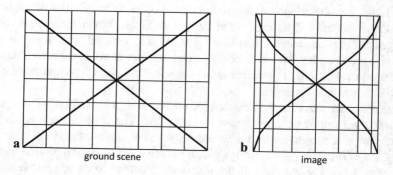

**Fig. 2.16**  Along scan line compression incurred in constant IFOV and constant scan rate sensors, leading to S bend distortion

would appear to be moved if the data is arrayed uniformly. It can be shown readily that SN/TN = $\theta/\tan\theta$ that being the degree of across track scale distortion. In the case of Landsat 7 $(\theta/\tan\theta)_{max} = 0.9936$. This indicates that a pixel at the swath edge (92.5 km from the sub-nadir point) will be 314 m out of position along the scan line compared with the ground, if the pixel at nadir is in its correct location.

These panoramic effects lead to an interesting distortion in the geometry of large field of view systems. To see this, consider the uniform mesh shown in Fig. 2.16a which represents a region being imaged. The cells in the grid could be considered to be large fields on the ground. Because of the compression caused by displaying equal-sized pixels on a uniform grid as discussed above, the uniform mesh will appear as shown in Fig. 2.16b. Recall that image pixels are recorded with a constant IFOV and at a constant angular sampling rate. The number of pixels recorded over the outer grid cells in the scan direction will be smaller therefore than over the fields near nadir. In the along-track direction there is no variation of pixel spacing or density with scan angle as this is established by the forward motion of the platform, although pixels near the swath edges will contain some information in common owing to the overlapping IFOV.

Linear features such as roads at an angle to the scan direction shown in Fig. 2.16 will appear bent in the displayed image because of the across scan compression effect. Owing to the change in shape, the distortion is frequently referred to as *S-bend distortion* and can be a common problem with aircraft line scanners. Clearly, not only linear features are affected; rather all of the detail near the swath edges is distorted.

## 2.15 The Effect of Earth Curvature

Aircraft and drone scanning systems, because of their low altitude and thus small absolute swath width, are usually not affected by earth curvature. Neither are small FOV spacecraft such as Landsat and SPOT, again because of the narrowness of their swaths. However wide swath width spaceborne imaging systems are affected. For MODIS, with a swath width of 2330 km and an altitude of 705 km, it can be shown that the deviation of the earth's surface from a plane amounts to less than 1% over the swath, which seems insignificant. However, it is the *inclination* of the earth's surface over the swath that causes the greater effect. At the edges of the swath the area of the earth's surface viewed at a given angular IFOV is larger than if the curvature of the earth were ignored; that exaggerates the panoramic effect treated in the previous section. The increase in pixel size can be computed by reference to the geometry of Fig. 2.17. The pixel dimension in the across track direction normal to the direction of the sensor is $\beta[h + r_e(1 - \cos\phi)]\sec\theta$ as shown. The effective pixel size on the inclined earth's surface is then

$$p_c = \beta[h + r_e(1 - \cos\phi)]\sec\theta\sec(\theta + \phi) \qquad (2.13)$$

where $\beta h$ is the pixel size at nadir and $\phi$ is the angle subtended at the centre of the earth. Note that this expression reduces to (2.12) as $\phi \to 0$, i.e., if earth curvature is negligible.

Using the NOAA satellite as an example $\theta = 54°$ at the edge of the swath and $\phi = 12°$. Equation (2.12) shows that the effective pixel size in the along scan

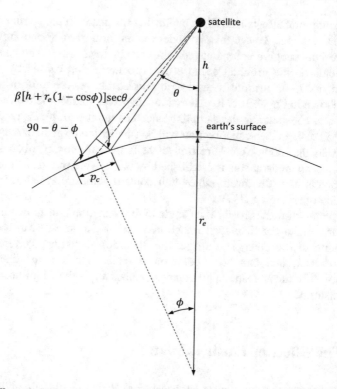

**Fig. 2.17** Effect of earth curvature on the size of a pixel in the across track direction

direction is 2.89 times larger than that at nadir when earth curvature is ignored but, from (2.13), is 4.94 times that at nadir when the effect of earth curvature is included. Thus, earth curvature introduces a significant additional compressive distortion in the image data acquired by satellites such as NOAA when an image is constructed on the uniform grid of Fig. 2.12. The effect of earth curvature in the along-track direction is negligible.

## 2.16  Geometric Distortion Caused by Instrumentation Characteristics

Depending on its style of operation, the sensor used for image acquisition can also introduce geometric distortion into the recorded image data. Here we look at three typical sources of distortion encountered with instruments that build up an image by scanning across the flight line, such as with Landsat, AVHRR, MODIS and some aircraft scanners.

### 2.16.1 Sensor Scan Nonlinearities

Line scanners that make use of rotating mirrors, such as the MODIS and AVHRR, have a scan rate across the swath that is constant, to the extent that the scan motor speed is constant. Systems that use an oscillating mirror, such as the Landsat Thematic Mapper, incur some nonlinearity near the swath edges in their scan angle versus time characteristic owing to the need for the mirror to slow down and change direction. That will lead to a displacement distortion of the recorded pixel data in the along-track direction.[30]

### 2.16.2 Finite Scan Time Distortion

Mechanical line scanners require a finite time to scan across the swath. During this time the satellite is moving forward, skewing the recorded image data in the along-track direction. As an illustration of the magnitude of the effect, the time required to record one Landsat MSS scan line of data is 33 ms. In this time the satellite travels forward by 213 m at its equivalent ground velocity of 6.461 $kms^{-1}$. The end of the scan line is advanced by that amount compared with its start. The Landsat Thematic Mapper compensates for this error source by using a scan skew corrector mirror.

### 2.16.3 Aspect Ratio Distortion

The aspect ratio of an image—its scale vertically compared with its scale horizontally—can be distorted by over-sampling or under-sampling across a scan line. In other words, samples of surface reflectance are taken at a rate not commensurate with the IFOV of the sensor. The most notable example of this in the past occurred with the Landsat Multispectral Scanner. By design, samples were taken across a scan line "too quickly" compared with the IFOV. That led to pixels having 56 m centres even though sampled with an IFOV of 79 m. Consequently, the effective pixel size in MSS imagery is 79 m × 56 m rather than square. As a result, if the pixels recorded by the Multispectral Scanner are displayed on the square grid of Fig. 2.12 the image will be too wide for its height when related to the corresponding region on the ground. The magnitude of the distortion is 79/56 = 1.411 which is quite severe and must be corrected for most applications. Similar distortion

---

[30] For Landsat multispectral scanner products this can lead to a maximum displacement in pixel position compared with a perfectly linear scan of about 395 m; see P. Anuta, Geometric correction of ERTS-1 digital MSS data, *Information Note 103073*, Laboratory for Applications of Remote Sensing, Purdue University West Lafayette, Indiana, 1973.

can occur with aircraft scanners if the velocity of the aircraft is not matched to the scanning rate of the sensor. Either under-scanning or over-scanning can occur distorting the along-track scale of the image.

## 2.17  Correction of Geometric Distortion

There are two techniques that can be used to correct the various types of geometric distortion present in digital image data. One is to model the nature and magnitude of the sources of distortion and use the model to establish correction formulas. That approach is effective when the types of distortion are well characterised, such as that caused by earth rotation. The second method depends on establishing mathematical relationships between the addresses of pixels in an image and the corresponding coordinates of those points on the ground (via a map).[31] Those relationships can be used to correct image geometry irrespective of the analyst's knowledge of the sources and types of distortion. This approach will be treated first since it is the most commonly used and, as a technique, is independent of the platform used for data acquisition. Correction by mathematical modelling is discussed later. Note that each band of image data has to be corrected. Since it can usually be assumed that the bands are well registered to each other, steps taken to correct one band, can be used on all remaining bands.

## 2.18  Use of Mapping Functions for Image Correction

An assumption made in this procedure is that there is available a map of the region covered by the image, which is correct geometrically. We then define two Cartesian coordinate systems as shown in Fig. 2.18. One describes the location of points in the map $(x, y)$ and the other defines the location of pixels in the image $(u, v)$. Suppose that the two coordinate systems can be related via a pair of mapping functions, such that

$$u = f(x, y) \tag{2.14a}$$

$$v = g(x, y) \tag{2.14b}$$

If these functions were known, then we could locate a point in the image knowing its position on the map. In principle, the reverse is also true. With this knowledge

---

[31] For a comprehensive treatment of image correction and registration see J. Le Moigne, N. S. Netanyahu and R. D. Eastman, eds., *Image Registration for Remote Sensing*, Cambridge University Press, Cambridge, 2011.

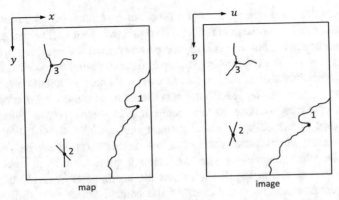

**Fig. 2.18** Map and image coordinate systems, and the concept of ground control points

we could build up a geometrically correct version of the image in the following manner. First, we define a grid over the map to act as the grid of pixel centres for the corrected image. This grid is parallel to, or could be, the map coordinate grid described by latitudes and longitudes, UTM coordinates, and so on. For simplicity we will refer to that grid as the *display grid*; by definition it is geometrically correct. We then move over the display grid pixel centre by pixel centre and use the mapping functions above to find the pixel in the image corresponding to each display grid position. Those pixels are then placed on the display grid. At the conclusion of the process, we have a geometrically correct image built up on the display grid using the original image as a source of pixels.

While the process is straightforward there are some practical matters that we must address. First, we do not know the explicit form of the mapping functions in (2.14). Secondly, even if we did, for a given display grid location they may not point exactly to a pixel in the image. In such a case some form of interpolation will be required.

## 2.18.1   Mapping Polynomials and the Use of Ground Control Points

Since explicit forms for the mapping functions in (2.14) are not known they are usually approximated by polynomials of first, second or third degree. In the case of second degree (or order)

$$u = a_0 + a_1 x + a_2 y + a_3 xy + a_4 x^2 + a_5 y^2 \tag{2.15a}$$

$$v = b_0 + b_1 x + b_2 y + b_3 xy + b_4 x^2 + b_5 y^2 \tag{2.15b}$$

Sometimes orders higher than three are used but care must be taken to avoid the introduction of errors worse than those to be corrected. That will be discussed later.

If the coefficients $a_i$ and $b_i$ in (2.15) were known then the mapping polynomials could be used to relate any point in the map to its corresponding point in the image, as in the discussion above. At present, however, those coefficients are unknown. Values can be estimated by identifying sets of features on the map that can also be identified on the image. Those features, referred to as *ground control points* (GCPs) or just *control points* (CPs), are well-defined and spatially small. They could be road intersections, street corners, airport runway intersections, sharp bends in rivers, prominent coastline features and the like. Enough are chosen, as pairs on the map and image as depicted in Fig. 2.18, so that the coefficients in (2.15) can be estimated by substituting the coordinates of the control points into those mapping polynomials to yield sets of equations in $a_i$ and $b_i$.

Equations (2.15) show that the minimum number of control points required for second order polynomial mapping is six. Likewise, a minimum of three is required for first order mapping and ten for third order mapping. In practice, significantly more than those minimums are chosen and the coefficients are evaluated using least squares estimation. In that manner any control points that contain significant positional errors, either on the map or in the image, will not have an undue influence on the estimated polynomial coefficients.

## 2.18.2 Building a Geometrically Correct Image

Having specified the mapping polynomials completely by the use of ground control points, the next step is to find points in the image that correspond to each location in display grid. The spacing of that grid is chosen according to the pixel size required in the corrected image and need not be the same as that of the original, geometrically distorted version. For the moment suppose that the points located in the image correspond exactly to image pixel centres, even though that rarely happens in practice. Then those pixels are simply transferred to the appropriate locations on the display grid to build up the rectified image. That is the case illustrated in Fig. 2.19.

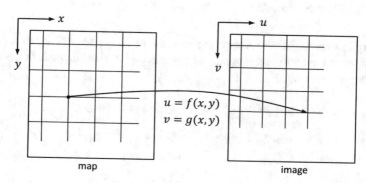

**Fig. 2.19** Using mapping functions to locate points in the image corresponding to particular display grid (map) positions

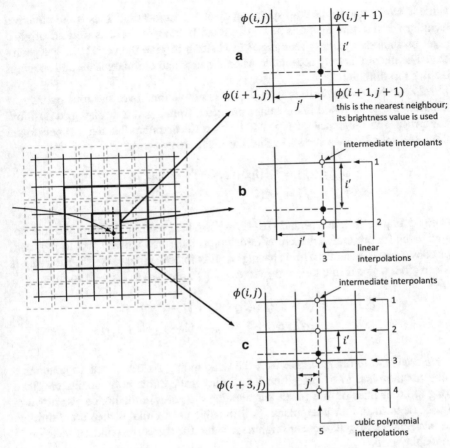

**Fig. 2.20** Determining a pixel brightness value for the display grid by **a** nearest neighbour resampling, **b** bilinear interpolation and **c** cubic convolution interpolation; $i,j$ etc. are discrete values of $u, v$

## 2.18.3 Resampling and the Need for Interpolation

As is to be expected, grid centres from the map-defined display grid will not usually project to exact pixel centre locations in the image. This is indicated in Fig. 2.20; some decision now has to be made about what pixel brightness value should be chosen for placement on the new grid.[32] Three principal techniques are used for this purpose.

*Nearest neighbour resampling* simply chooses the actual pixel that has its centre nearest to the point located in the image, as illustrated in Fig. 2.20a. That pixel

---

[32] In some treatments this is referred to as a radiometric transformation; see Le Moigne et al., loc. cit.

value is transferred to the corresponding display grid location. This is the preferred technique if the new image is to be classified because it consists only of original pixel brightnesses, simply rearranged in position to give the correct image geometry. The method is only acceptable when the new and old pixel sizes and spacings are not too different.

*Bilinear interpolation* uses three linear interpolations over the four pixels surrounding the point found in the image projected from a given display grid position. The process is illustrated in Fig. 2.20b. Two linear interpolations are performed along the scan lines as shown to find the interpolants

$$\phi(i,j+j') = j'\phi(i,j+1) + (1-j')\phi(i,j)$$
$$\phi(i+1,j+j') = j'\phi(i+1,j+1) + (1-j')\phi(i+1,j)$$

where $\phi$ is pixel brightness and $(i+i',j+j')$ is the position at which the interpolated value for brightness is required. Position is measured with respect to $(i,j)$ and, for convenience, assumes a grid spacing of unity in both directions. The final step is to carry out a linear interpolation over $\phi(i,j+j')$ and $\phi(i+1,j+j')$ to give

$$\phi(i+i',j+j') = (1-i')\{j'\,\phi(i,j+1) + (1-j')\phi(i,j)\} \\ + i'\{j'\phi(i+1,j+1) + (1-j')\phi(i+1,j)\} \tag{2.16}$$

*Cubic convolution interpolation* uses the surrounding sixteen pixels to generate a value for the pixel to be placed on the display grid. Cubic polynomials are fitted along the four lines of four pixels surrounding the point in the image, as shown in Fig. 2.20c to form four interpolants. A fifth cubic polynomial is then fitted through the interpolants to synthesise a brightness value for the corresponding location in the display grid.

The actual form of polynomial that is used for the interpolation is derived from considerations in sampling theory and by constructing a continuous function (an interpolant) from a set of samples.[33] The algorithm that is generally used to perform cubic convolution interpolation is[34]

$$\phi(I,j+1+j') = j'\{j'(j'[\phi(I,j+3) - \phi(I,j+2) + \phi(I,j+1) - \phi(I,j)] \\ + [\phi(I,j+2) - \phi(I,j+3) - 2\phi(I,j+1) + 2\phi(I,j)]) \\ + [\phi(I,j+2) - \phi(I,j)]\} + \phi(I,j+1) \tag{2.17a}$$

[33] An excellent treatment of the problem has been given by S. Shlien, Geometric correction, registration and resampling of Landsat imagery, *Canadian J. Remote Sensing*, vol. 5, 1979, pp. 74–89. He discusses several possible cubic polynomials that could be used for the interpolation process and demonstrates that the interpolation is a convolution operation.

[34] Based on the choice of interpolation polynomial in T. G. Moik, *Digital Processing of Remotely Sensed Images*, NASA, Washington, 1980.

with $I = i + n, n = 0, 1, 2, 3$ for the four lines of pixels surrounding the point for which the value is required. Note in Fig. 2.20c we have, for convenience, redefined the address of the $i, j$ pixel to be one line and column earlier. The four values from (2.17a) are interpolated vertically to give the estimate required

$$
\begin{aligned}
\phi(i+1+i',j+1+j') = i'\{i'(i'[\phi(i+3,j+1+j') - \phi(i+2,j+1+j') \\
+ \phi(i+1,j+1+j') - \phi(i,j+1+j')] \\
+ [\phi(i+2,j+1+j') - \phi(i+3,j+1+j') \\
- 2\phi(i+1,j+1+j') + 2\phi(i,j+1+j')]) \\
+ [\phi(i+2,j+1+j') - \phi(i,j+1+j')]\} \\
+ \phi(i+1,j+1+j')
\end{aligned}
\tag{2.17b}
$$

Cubic convolution interpolation, or resampling, yields a rectified image that is generally smooth in appearance and is used if the final product is to be analysed visually. However, since it gives pixels on the display grid with brightness values that are interpolated from the original data, it is not recommended if classification is to follow since the new brightness values may be slightly different from the actual radiances measured by the sensors.

The three interpolation methods just treated are not the only choices, although they are the most common; any effective two dimensional interpolation procedure could be used.[35]

## 2.18.4  *The Choice of Control Points*

When rectifying an image, enough well-defined control point pairs must be chosen to ensure that accurate mapping polynomials are generated. Care must also be given to the locations of the points. A general rule is that there should be a distribution of control points around the edges of the image to be corrected, with a scattering of points over the body of the image. That is necessary to ensure that the mapping polynomials are well-behaved over the scene. This concept can be illustrated by considering an example from curve fitting. While the nature of the problem is different, the undesirable effects that can be generated are similar. A set of data points is illustrated in Fig. 2.21 through which first order (linear), second order and third order curves are shown. As the order increases the curves pass closer to the actual points. If it is presumed that the data would have existed for larger values of $x$, with much the same trend as apparent in the points plotted, then the linear fit will extrapolate moderately acceptably. By contrast, the cubic curve can deviate markedly from the trend when used as an extrapolator. This is essentially true in

---

[35] For other methods see Le Moigne et al., loc. cit.

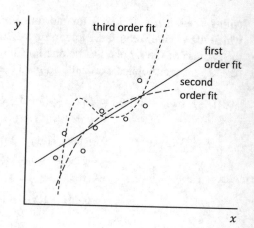

**Fig. 2.21** Illustration from curve fitting to emphasise the potentially poor behaviour of higher order polynomials when used to extrapolate

geometric correction of image data: while higher order polynomials may be more accurate in the vicinity of the control points, they can lead to significant errors, and thus distortions, for regions of an image outside the range of the control points. That will be seen in the example of Fig. 2.24.

## 2.18.5  *Example of Registration to a Map Grid*

To illustrate the techniques treated above a small segment of a Landsat MSS image of Sydney, Australia was registered to a map of the region.

It is important that the map has a scale not too different from the scale at which the image data is considered useful. Otherwise, the control point pairs may be difficult to establish. In this case a map at 1:250,000 scale was used. The relevant segment is shown reproduced in Fig. 2.22, along with the portion of image to be registered. Comparison of the two demonstrates the geometric distortion of the image. Eleven control points pairs were chosen for the registration, with the coordinates shown in Table 2.1.

Second order mapping polynomials were generated from the set of control points. To test their effectiveness in transferring pixels from the raw image grid to the map display grid, the UTM coordinates of the control points can be computed from their pixel coordinates in the image. They are compared with the actual UTM coordinates and the differences (residuals) calculated in both directions.[36] The root mean square of all the residuals is then computed in both directions (easting and northing) as shown in Table 2.1, giving an overall impression of the accuracy of the mapping process. In this case the control points lead to an average positional error of 56 m in easting and 63 m in northing, which is smaller than a pixel size in equivalent ground metres and thus would be considered acceptable.

---

[36] This registration exercise was carried out using the Dipix Systems Ltd R-STREAM Software.

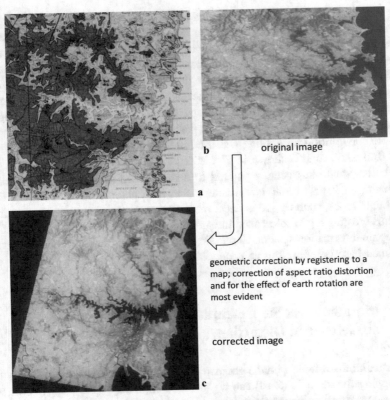

original image

geometric correction by registering to a
map; correction of aspect ratio distortion
and for the effect of earth rotation are
most evident

corrected image

**Fig. 2.22** **a** Map, **b** segment of a Landsat MSS image exhibiting geometric distortion, and **c** the result of registering the image to the map, using second order mapping polynomials and cubic convolution resampling to correct the geometry

**Table 2.1** Control points used in the image to map registration example

| GCP | Image pixel | Image line | Map easting actual | Map easting estimate | Map easting residual | Map northing actual | Map northing estimate | Map northing residual |
|---|---|---|---|---|---|---|---|---|
| 1 | 1909 | 1473 | 432,279 | 432,230.1 | 49.4 | 836,471 | 836,410.1 | 60.7 |
| 2 | 1950 | 1625 | 431,288 | 431,418.0 | −130.1 | 822,844 | 822,901.4 | −56.9 |
| 3 | 1951 | 1747 | 428,981 | 428,867.9 | 112.6 | 812,515 | 812,418.2 | 96.8 |
| 4 | 1959 | 1851 | 427,164 | 427,196.9 | −33.2 | 803,313 | 803,359.4 | −46.7 |
| 5 | 1797 | 1847 | 417,151 | 417,170.3 | −18.9 | 805,816 | 805,759.3 | 57.1 |
| 6 | 1496 | 1862 | 397,860 | 397,871.6 | −11.2 | 808,128 | 808,187.2 | −59.6 |
| 7 | 1555 | 1705 | 404,964 | 404,925.8 | 38.6 | 821,084 | 820,962.6 | 121.6 |
| 8 | 1599 | 1548 | 411,149 | 411,138.5 | 10.5 | 833,796 | 833,857.3 | −61.1 |
| 9 | 1675 | 1584 | 415,057 | 415,129.0 | −72.4 | 829,871 | 829,851.1 | 19.8 |
| 10 | 1829 | 1713 | 422,019 | 421,986.6 | 32.7 | 816,836 | 816,884.5 | −48.1 |
| 11 | 1823 | 1625 | 423,530 | 423,507.8 | 22.0 | 824,422 | 824,504.8 | −83.2 |

Standard error in easting = 55.92 m
Standard error in northing = 63.06 m

At this stage the table can be inspected to see if any individual control point has residuals that are unacceptably high. That could be the result of poor placement; if so, the control point coordinates would be re-entered and the polynomials recalculated. If re-entering the control point leaves the residuals unchanged it may be that there is significant local distortion in that particular region of the image. A choice has to be made then as to whether the control point should be used to give a degree of local correction, that might also influence the remainder of the image, or whether it should be removed and leave that part of the image in error.

Cubic convolution resampling was used in this illustration to generate the image on a 50 m × 50 m grid shown in Fig. 2.22c.

Once an image has been registered to a map coordinate system its pixels are addressable in terms of map coordinates rather than pixel and line numbers. Other spatial data types, such as geophysical measurements, can also be registered to the map thus creating a geo-referenced integrated spatial data base of the type used in a geographic information system. Expressing image pixel addresses in terms of a map coordinate base is referred to as *geocoding*.

## 2.19 Mathematical Representation and Correction of Geometric Distortion

If a particular distortion in image geometry can be represented mathematically then the mapping functions in (2.14) can be specified explicitly. That removes the need to choose arbitrary polynomials as in (2.15) and to use control points to determine the polynomial coefficients. In this section some of the more common distortions are treated from this point of view. Rather than commence with expressions that relate image coordinates $(u, v)$ to map coordinates $(x, y)$ it is simpler conceptually to start the other way around, i.e., to model what the true (map) positions of pixels should be, given their positions in an image. This expression can then be inverted, if required, to allow the image to be resampled onto the map grid.

### 2.19.1 Aspect Ratio Correction

The easiest source of geometric error to model is the distortion in aspect caused when the sampling rate across a scan line does not precisely match the IFOV of the sensor. A typical example is that caused by the 56 m ground spacing of the 79 m × 79 m pixels in the Landsat multispectral scanner. As noted in Sect. 2.16.3 that leads to an image that is too wide for its height by a factor of 1.411. Consequently, to produce a geometrically correct image either the vertical dimension has to be expanded by this amount or the horizontal dimension must be compressed. We consider the former. That requires the pixel axis horizontally to be left unchanged (i.e., $x = u$), but the axis vertically to be scaled (i.e., $y = 1.411v$). These can be expressed conveniently in matrix notation as

$$\begin{bmatrix} x \\ y \end{bmatrix} = \begin{bmatrix} 1 & 0 \\ 0 & 1.411 \end{bmatrix} \begin{bmatrix} u \\ v \end{bmatrix} \tag{2.18}$$

One way of implementing this correction would be to add extra lines of pixel data to expand the vertical scale which, for the multispectral scanner, could be done by duplicating about four lines in every ten. Alternatively, and more precisely, (2.18) can be inverted to give

$$\begin{bmatrix} u \\ v \end{bmatrix} = \begin{bmatrix} 1 & 0 \\ 0 & 0.709 \end{bmatrix} \begin{bmatrix} x \\ y \end{bmatrix} \tag{2.19}$$

As with the techniques of the previous section, a display grid with coordinates $(x, y)$ is defined over the map and (2.19) is used to find the corresponding location in the image $(u, v)$. The interpolation techniques of Sect. 2.18.3 are then used to generate brightness values for the display grid pixels.

## 2.19.2 Earth Rotation Skew Correction

To correct for the effect of earth rotation it is necessary to implement a shift of pixels to the left with the degree of shift dependent on the particular line of pixels, measured with respect to the top of the image. Their line addresses $(v)$ are not affected. Using the results of Sect. 2.12 the corrections are implemented by

$$\begin{bmatrix} x \\ y \end{bmatrix} = \begin{bmatrix} 1 & \alpha \\ 0 & 1 \end{bmatrix} \begin{bmatrix} u \\ v \end{bmatrix}$$

with $\alpha = -0.056$ for Sydney, Australia. Again, this can be implemented in an approximate sense for multispectral scanner data by making one pixel shift to the left every 17 lines of image data measured down from the top, or alternatively the expression can be inverted to give

$$\begin{bmatrix} u \\ v \end{bmatrix} = \begin{bmatrix} 1 & -\alpha \\ 0 & 1 \end{bmatrix} \begin{bmatrix} x \\ y \end{bmatrix} = \begin{bmatrix} 1 & 0.056 \\ 0 & 1 \end{bmatrix} \begin{bmatrix} x \\ y \end{bmatrix} \tag{2.20}$$

which again is used with the interpolation procedures from Sect. 2.18.3 to generate display grid pixels.

### 2.19.3 *Image Orientation to North–South*

Although not strictly a geometric distortion it is inconvenient to have an image that is correct for most major effects but is not oriented vertically in a north–south direction. It will be recalled for example that Landsat orbits are inclined to the north–south line by about 9°, dependent on latitude. To rotate an image by an angle $\zeta$ in the counter- or anticlockwise direction (as required in the case of Landsat) it is easily shown that[37]

$$\begin{bmatrix} x \\ y \end{bmatrix} = \begin{bmatrix} \cos\zeta & \sin\zeta \\ -\sin\zeta & \cos\zeta \end{bmatrix} \begin{bmatrix} u \\ v \end{bmatrix}$$

so that

$$\begin{bmatrix} u \\ v \end{bmatrix} = \begin{bmatrix} \cos\zeta & -\sin\zeta \\ \sin\zeta & \cos\zeta \end{bmatrix} \begin{bmatrix} x \\ y \end{bmatrix} \tag{2.21}$$

### 2.19.4 *Correcting Panoramic Effects*

The discussion in Sect. 2.14 notes the pixel positional error that results from scanning with a fixed IFOV at a constant angular rate. In terms of map and image coordinates the distortion can be described by

$$\begin{bmatrix} x \\ y \end{bmatrix} = \begin{bmatrix} \tan\theta/\theta & 0 \\ 0 & 1 \end{bmatrix} \begin{bmatrix} u \\ v \end{bmatrix}$$

where $\theta$ is the instantaneous scan angle, which in turn is related to $x$ or $u$ by $x = h\tan\theta$, $u = h\theta$, where $h$ is altitude. Consequently, resampling can be carried out according to

$$\begin{bmatrix} u \\ v \end{bmatrix} = \begin{bmatrix} \theta\cot\theta & 0 \\ 0 & 1 \end{bmatrix} \begin{bmatrix} x \\ y \end{bmatrix} = \begin{bmatrix} (h/x)\tan^{-1}(x/h) & 0 \\ 0 & 1 \end{bmatrix} \begin{bmatrix} x \\ y \end{bmatrix} \tag{2.22}$$

### 2.19.5 *Combining the Corrections*

Any exercise in image correction usually requires several distortions to be rectified. Using the techniques in Sect. 2.18 it is assumed that all sources are rectified simultaneously. When employing mathematical modelling, a correction matrix has

---

[37] K. R. Castleman, *Digital Image Processing*, 2nd ed., Prentice Hall, N.J., 1996.

to be devised for each separate source considered important, as in the preceding sub-sections, and the set of matrices combined. For example, if the aspect ratio of a Landsat MSS image is corrected first, followed by correction for the effect of earth rotation, the following single linear transformation can be established for resampling.

$$\begin{bmatrix} x \\ y \end{bmatrix} = \begin{bmatrix} 1 & \alpha \\ 0 & 1 \end{bmatrix} \begin{bmatrix} 1 & 0 \\ 0 & 1.411 \end{bmatrix} \begin{bmatrix} u \\ v \end{bmatrix}$$

$$= \begin{bmatrix} 1 & 1.411\alpha \\ 0 & 1.411 \end{bmatrix} \begin{bmatrix} u \\ v \end{bmatrix}$$

which, for $\alpha = -0.056$ at Sydney, gives

$$\begin{bmatrix} u \\ v \end{bmatrix} = \begin{bmatrix} 1 & 0.056 \\ 0 & 0.079 \end{bmatrix} \begin{bmatrix} x \\ y \end{bmatrix}$$

## 2.20   Image to Image Registration

Many applications in remote sensing require two or more scenes of the same geographical region, acquired at different times or by different sensors, to be processed together. Such a situation arises, for example, when changes are of interest in which case registered images allow a pixel-by-pixel comparison to be made, or when data is to be fused to help interpretation.

Two images can be registered to each other by registering each to a map coordinate base separately in the manner demonstrated in Sect. 2.18. Alternatively, and particularly if georeferencing is not important, one image can be chosen as a *master*, or reference, to which the other, known as the *slave*, is registered. Again, the techniques of Sect. 2.18 are used. However, the coordinates $(x, y)$ are now the pixel coordinates in the master image rather than the map coordinates. As before $(u, v)$ are the coordinates of the image to be registered (the slave). A benefit in image to image registration is that only one registration step is required, by comparison to two if both are taken back to a map base. Also, spatial correlation algorithms can be used to assist in accurate co-location of control point pairs, as discussed in Sect. 2.20.1.

### 2.20.1   Refining the Localisation of Control Points

In many applications control points are chosen manually in both the master and slave images. More recently, the trend has been towards automated techniques for control point identification[38] in order to minimise the need for analyst intervention. The method outlined in this section was developed initially as an aid to localising control points manually but is indicative of some of the procedures developed for automatic control point identification.

Usually, we want to check the chosen control point pairs to make sure that they are spatially matched as well as possible. Several means are available for assessing their correspondence, most of which involve selecting a rectangular sample, or window, of pixels surrounding the designated control point in the slave image, and moving and checking it against the master image, as illustrated in Fig. 2.23.

Because of the spatial properties of the pair of images near the control points, the best match should occur when the slave window is located exactly over its counterpart region in the master, in which case the master location of the control point is identified. Obviously, it is not necessary to move the slave window over the complete master image since the user knows approximately where the control point should occur in the master. It is only necessary to specify a search region in the neighbourhood of the approximate location.

Control point matching procedures of this sort are usually called sequential similarity detection algorithms (SSDA).[39] There are several means by which the match between the window and search region can be computed. Classically, the two could be correlated.[40] If we denote a pixel in the master image as $R(i,j)$, and a pixel in the slave image as $S(i,j)$ then their correlation over a window of pixels, defined about the control point in the master image, can be expressed

$$m(i,j) = \sum_m \sum_n R(m+i, n+j)S(m,n) \triangleq R \ast S$$

where $m, n$ are the pixel coordinates referred to the window. Frequently, this expression is squared and normalised by the autocorrelations over the master and slave to give the similarity measure:

$$SSDA = \frac{(R \ast S)^2}{(R \ast R)(S \ast S)} \tag{2.23}$$

---

[38] See Le Moigne et al., loc. cit. and T. T. Nguyen, Optimal ground control points for geometric correction using genetic algorithm with global accuracy, *European J. Remote Sensing*, vol. 47, no. 1, 2015, pp. 101–120 (which contains a review of automated techniques).

[39] See D. I. Barnea and H. F. Silverman, A class of algorithms for fast digital image registration, *IEEE Transactions on Computers*, vol. C-21, no. 2, February 1972, pp. 179–186, and R. Bernstein, Image Geometry and Registration, in R. N. Colwell, ed., *Manual of Remote Sensing*, 2nd ed., Chap. 21, American Society of Photogrammetry, Falls Church, Virginia, 1983.

[40] See P. E. Anuta, Spatial registration of multispectral and multitemporal digital imagery using fast Fourier transform techniques, *IEEE Transactions on Geoscience Electronics*, vol. GE-8, no. 4, October 1970, pp. 353–368.

**Fig. 2.23** Precisely locating control point pairs by using a window of pixels from the slave image to compare against the master image over a specified search region

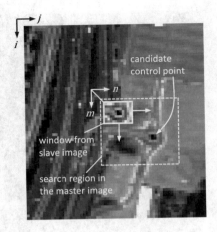

which is in the range (0,1). When the window is best matched to the control point in the master image this measure will be high; for a poor mismatch it will be low.

The operation in (2.23) is computationally demanding so simpler matching processes are used in practice. One is based on accumulating the absolute differences in brightness between the master and slave pixels in the window. In principle, when a match is achieved the accumulated difference should be a minimum. This measure is expressed

$$m(i,j) = \sum_m \sum_n |R(m+i, n+j) - S(m,n)| \qquad (2.24)$$

Clearly, measures such as those in (2.23) and (2.24) work well only when the contrast differences over the pixels in the vicinity of the control point are not too different in the master and slave images. If there are significant differences in the distributions of brightness values between the two images, owing to seasonal or noise effects for example, then SSDA techniques will suffer. Alternative procedures include those based on Fourier transforms and the concept of mutual information.[41] Most commercial image processing systems, however, still rely on matching processes such as that in (2.24).

### 2.20.2  Example of Image to Image Registration

To illustrate image to image registration, and also to see clearly the effect of control point distribution and the significance of the order of the mapping polynomials used in registration, two segments of Landsat Multispectral Scanner infrared image data

---

[41] See Le Moigne et al., loc. cit.

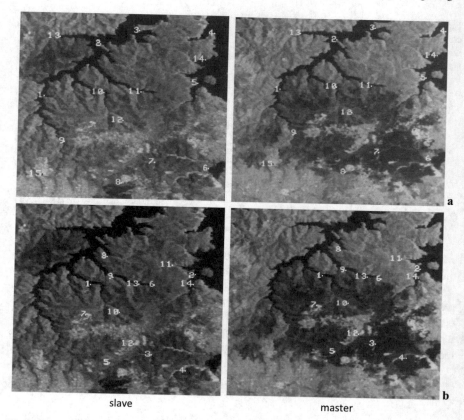

slave
master

**Fig. 2.24** Control points used in the image to image registration example: **a** good distribution **b** poor distribution

from the northern suburbs of Sydney were chosen. One was acquired on 29th December 1979 and was used as the master. The other was acquired on 14th December 1980 and was used as the slave image. Both are shown in Fig. 2.24; careful inspection shows the differences in image geometry.

Two sets of control points were chosen. In one, the points are distributed as nearly as possible in a uniform manner around the edge of the image segment as shown in Fig. 2.24a, with some points located across the centre of the image. This set would be expected to give reasonable registration of the images. The second set of control points was chosen injudiciously, closely grouped around one particular region, to illustrate the resampling errors that can occur. They are shown in Fig. 2.24b. In both cases the control point pairs were co-located with the assistance of the sequential similarity detection algorithm in (2.24). This worked well particularly for those control points around the coastal and river regions where the similarity between the images is unmistakable. To minimise tidal influences on the location of control points, those on water boundaries were chosen as near as possible to be on headlands, and never at the ends of inlets.

**Fig. 2.25  a** Registration of
1980 image (green) with 1979
image (red) using the control
points of Fig. 2.24a, and third
order mapping polynomials
**b** registration of 1980 image
(green) with 1979 image
(red) using the control points
of Fig. 2.24b, and third order
mapping polynomials
**c** registration of 1980 image
(green) with 1979 image
(red) using the control points
of Fig. 2.24b, and first order
mapping polynomials

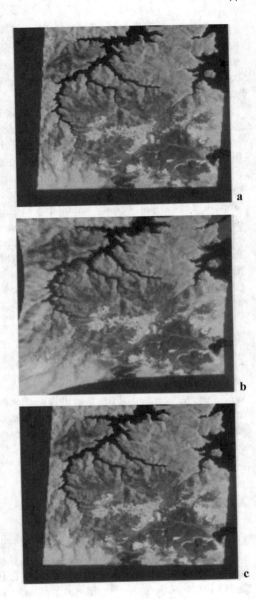

For both sets of control points third order mapping polynomials were used, along with cubic convolution resampling. As expected, the first set of points led to an acceptable registration of the images whereas the second set gave a good registration in the immediate neighbourhood of the control points but beyond that neighbourhood it produced gross distortion.

The adequacy of the registration process can be assessed visually if the master and resampled slave images are superimposed in different colours. Figures 2.25a and b show the master image in red with the resampled slave image superimposed in green. Where good registration has been achieved the result is yellow, with the exception of regions of gross dissimilarity in pixel brightness—in this case associated with fire burns. Misregistration shows quite graphically as a red-green separation. This is particularly noticeable in Fig. 2.25b where the poor extrapolation obtained with third order mapping is demonstrated.

The exercise using the poor set of control points in Fig. 2.25b was repeated. However, this time first order mapping polynomials were used. While they will not remove non-linear differences between the images, and will give poorer matches at the control points themselves, they are well behaved in extrapolation beyond the vicinity of the control points and lead to an acceptable registration as seen in Fig. 2.25c.

## 2.21    Other Image Geometry Operations

While the techniques of the previous sections have been devised for treating errors in image geometry, and for registering sets of images, they can also be used for performing intentional changes to image geometry. Image rotation and scale changing are chosen here as illustrations.

### 2.21.1    Image Rotation

Rotation of an image by an angle about the pixel grid can be useful for a number of applications. Most often it is used to align the pixel grid, and thus the image, to a north–south orientation as treated in Sect. 2.19.3. However, the transformation in (2.21) is perfectly general and can be used to rotate an image in an anticlockwise sense by any specified angle $\zeta$.

### 2.21.2    Scale Changing and Zooming

The scales of an image in both the vertical and horizontal directions can be altered by the transformation

$$\begin{bmatrix} x \\ y \end{bmatrix} = \begin{bmatrix} a & 0 \\ 0 & b \end{bmatrix} \begin{bmatrix} u \\ v \end{bmatrix}$$

where $a$ and $b$ are the desired scaling factors. To resample the scaled image onto the display grid we use the inverse operation to locate pixel positions in the original image corresponding to each display grid position, viz.

$$\begin{bmatrix} u \\ v \end{bmatrix} = \begin{bmatrix} 1/a & 0 \\ 0 & 1/b \end{bmatrix} \begin{bmatrix} x \\ y \end{bmatrix}$$

Interpolation is used to establish the actual pixel brightness values to use, since $u, v$ will not normally fall on exact pixel locations.

Frequently $a = b$ so that the image is simply magnified. This is called *zooming*. If the nearest neighbour interpolation procedure is used in the resampling process the zoom implemented is said to occur by *pixel replication* and the image will look progressively blocky for larger zoom factors. If cubic convolution interpolation is used there will be a change in magnification but the image will not take on the blocky appearance. Often this process is called *interpolative zoom*.

## 2.22   Bibliography on Correcting and Registering Images

A good general discussion on the effects of the atmosphere on the passage of radiation in the range of wavelengths important to optical remote sensing will be found in

P.N. Slater, *Remote Sensing: Optics and Optical Systems*, Addison-Wesley, Reading, Mass., 1980.

An introduction to the radiometric distortion problem facing high spectral resolution imagery, such as that produced by imaging spectrometers, is given, along with correction techniques, in

B.C. Gao, K. B. Heidebrecht and A.F.H. Goetz, Derivation of scaled surface reflectance from AVIRIS data, *Remote Sensing of Environment*, vol. 44, 1993, pp. 165–178

B. C. Gao, M.J. Montes, C.O. Davis and A.F.H. Goetz, Atmospheric correction algorithms for hyperspectral remote sensing data of land and oceans, *Remote Sensing of Environment*, Supplement 1, Imaging Spectroscopy Special Issue, vol. 113, 2009, pp. S17–S24

A very helpful comparison of the more common correction procedures will be found in

F.A. Kruse, Comparison of ATREM, ACORN and FLAASH atmospheric corrections using low-altitude AVIRIS data of Boulder, Colorado, *Proc. 13th JPL Airborne Geoscience Workshop*, Pasadena, CA, 2004

Techniques for correcting errors in image geometry are discussed in many standard image processing treatments in remote sensing. More recently, the range of operational procedures and new research directions in geometric correction and image registration are covered in

J. Le Moigne, N. S. Netanyahu and R. D. Eastman, eds., *Image Registration for Remote Sensing*, Cambridge University Press, Cambridge, 2011, and

J. Le Moigne, Introduction to remote sensing image registration, *Proc. IEEE 2017 Geoscience and Remote Sensing Symposium*, 23–28 July 2017, Fort Worth, Texas, pp. 2565–2568.

T.T. Nguyen, Optimal ground control points for geometric correction using genetic algorithm with global accuracy, *European J. Remote Sensing*, vol. 47, no. 1, 2015, pp. 101–120

A treatment of experimental fully automated image registration, including a review of available techniques, is given in

H. Gonçalves, L. Corte-Real and J.A. Gonçalves, Automatic image registration through image segmentation and SIFT, *IEEE Transactions on Geoscience and Remote Sensing*, vol. 49, no. 7, July 2011, pp. 2589–2600

while the use of wavelet-based approaches will be found in

J.M. Murphy, K. Le Moigne and D.J. Harding, Automatic image registration of multi-modal remotely sensed data, *IEEE Transactions on Geoscience and Remote Sensing*, vol. 54, no. 3, 2016, pp. 1685–1704.

Some earlier treatments should not be overlooked as they provide good insight into the problem of correcting geometry and are still relevant. They include

S. Shlien, Geometric correction, registration and resampling of Landsat imagery, *Canadian J. Remote Sensing*, vol. 5, 1979, pp. 74–89, and

F. Orti, Optimal distribution of control points to minimise Landsat registration errors, *Photogrammetric Engineering and Remote Sensing*, vol. 47, 1980, pp. 101–110.

Finally, books on computer graphics also contain very good material on image geometry correction and transformation, perhaps one of the most notable being

J.F. Hughes, A. van Dam, M. McGuire, D.F. Sklar, J.D. Foley, S.K. Feiner and K. Akeley, *Computer Graphics: Principles and Practice*, 3rd ed., Addison-Wesley, Boston, 2014.

## 2.23   Problems

2.1  (a)   Consider a region on the ground consisting of a square grid. For simplicity suppose the grid lines are 79 m in width and the grid spacing is 790 m. Sketch how the region would appear in Landsat multispectral scanner imagery before any geometric correction has been applied.

Include only the effects of earth rotation and the 56 m horizontal spacing of the 79 m × 79 m ground resolution elements.

(b) Develop a pair of linear mapping polynomials that will correct the image in (a). Assume the "lines" on the ground have a brightness of 100 and that the background brightness is 20. Resample onto a 50 m grid and use nearest neighbour interpolation. You will not want to compute all the resampled pixels unless a computer is used for the exercise. Instead, simply consider some significant pixels in the resampling to illustrate the accuracy of the geometric correction.

2.2 A sample of pixels from each of three cover types present in the Landsat MSS scene of Sydney, Australia, acquired on 14th December 1980 is given in Table 2.2a. Only the brightnesses (digital numbers) in the visible red band (0.6–0.7 μm) and the second of the infrared bands (0.8–1.1 μm) are given. For this image Forster[42] has computed the following relations between reflectance ($R$) and digital number ($DN$), where the subscript 7 refers to the infrared data and the subscript 5 refers to the visible red data:

$$R_5 = 0.44DN_5 + 0.5$$

$$R_7 = 1.18DN_7 + 0.9$$

Table 2.2 b shows samples of MSS digital numbers for a second scene of Sydney recorded on 8th June 1980. For this image Forster has determined

$$R_5 = 3.64DN_5 - 1.6$$

$$R_7 = 1.52DN_7 - 2.6$$

Compute the mean digital count value for each cover type in each scene and plot these, along with bars at ±1 standard deviation, in a spectral domain that has the infrared values on the ordinate and the visible red values along the abscissa. Now produce the same plots after converting the data to reflectances. Comment on the effect that correction of the raw digital numbers to reflectance data, in which atmospheric effects have been removed, has on the apparent separation of the three cover types in the spectral domain.

2.3 Aircraft line scanners frequently use a rotating mirror that sweeps out lines of data at right angles to the fuselage to acquire imagery. In the absence of a cross wind, scanning will be orthogonal to the aircraft ground track. Often scanning is carried out in the presence of a cross wind. The aircraft fuselage then maintains an angle to the ground track so that scanning is no longer orthogonal to the effective forward motion, leading to a distortion referred to as *crabbing*. Discuss the nature of this distortion when the image pixels are

---

[42] See B. C. Forster, loc. cit.

**Table 2.2** Digital numbers for a set of pixels from three cover types; note band 5 covers the wavelength range 0.6–0.7 μm and band 7 covers the range 0.8–1.1 μm

(a) Landsat MSS image of Sydney 14th December 1980

| Water | | Vegetation | | Soil | |
|---|---|---|---|---|---|
| Band 5 | Band 7 | Band 5 | Band 7 | Band 5 | Band 7 |
| 20 | 11 | 60 | 142 | 74 | 66 |
| 23 | 7 | 53 | 130 | 103 | 82 |
| 21 | 8 | 63 | 140 | 98 | 78 |
| 21 | 7 | 52 | 126 | 111 | 86 |
| 22 | 7 | 34 | 92 | 84 | 67 |
| 19 | 3 | 38 | 120 | 76 | 67 |
| 17 | 1 | 38 | 151 | 72 | 67 |
| 20 | 4 | 38 | 111 | 98 | 71 |
| 24 | 8 | 31 | 81 | 99 | 80 |
| 19 | 4 | 50 | 158 | 108 | 71 |

(b) Landsat MSS image of Sydney 8th June 1980

| Water | | Vegetation | | Soil | |
|---|---|---|---|---|---|
| Band 5 | Band 7 | Band 5 | Band 7 | Band 5 | Band 7 |
| 11 | 2 | 19 | 41 | 43 | 27 |
| 13 | 5 | 24 | 45 | 43 | 34 |
| 13 | 2 | 20 | 44 | 40 | 30 |
| 11 | 1 | 22 | 30 | 27 | 19 |
| 9 | 1 | 15 | 22 | 34 | 23 |
| 14 | 4 | 14 | 26 | 36 | 26 |
| 13 | 4 | 21 | 27 | 34 | 27 |
| 15 | 5 | 17 | 38 | 70 | 50 |
| 12 | 4 | 24 | 37 | 37 | 30 |
| 15 | 4 | 20 | 27 | 44 | 30 |

displayed on a square grid. Remember to take account of the finite time required to scan across a line of pixels.

Push broom scanners are also used on aircraft. What is the nature of the geometric distortion incurred with those sensors in the presence of a cross wind? Complete frames of pixels can also be captured from aircraft and drone platforms, effectively using digital cameras. What geometric distortion would be incurred with such a sensor in the presence of a cross wind?

2.4 Compute the skew distortion resulting from earth rotation in the case of Landsat 7 and SPOT.

2.5 For a particular application suppose it was necessary to apply geometric correction procedures to an image prior to classification (see Chap. 3 for an overview of classification). What interpolation technique would you prefer to use in the resampling process? Why?

2.6 Destriping Landsat multispectral scanner images is often performed by computing six modulo-6 line histograms and then either (i) matching all six to a standard histogram or (ii) choosing one of the six as a reference and matching the other five to it. Which method is to be preferred if the image is to be analysed by photointerpretation or by classification?

2.7 In a particular problem you have to register five images to a map. Would you register each image to the map separately, register one image to the map and then the other four images to that one, or image 1 to the map, image 2 to image 1, image 3 to image 2 etc.?

2.8 (This requires a background in digital signal processing and sampling theory—see Chap. 7). Remote sensing digital images are uniform two dimensional samples of the ground scene. One line of image data is a regular sequence of samples. The spatial frequency spectrum of a line of data will therefore be periodic as depicted in Fig. 2.26; the data can be recovered by low pass filtering the spectrum, using the ideal filter indicated in the figure. Multiplication of the spectrum by this ideal filter is equivalent to convolving the original line of samples by the inverse Fourier transform of the filter function. From the theory of the Fourier transform, the inverse of the filter function is

$$s(x) = \frac{2d}{\pi} \frac{\sin x}{x}$$

with $x = \zeta/2d$ in which $\zeta$ is a spatial variable along lines of data, and $d$ is the inter-pixel spacing. $s(x)$ is known generally as an interpolating function. Determine some cubic polynomial approximations to this function. These could be determined from a simple Taylor series expansion or could be derived from cubic splines.[43]

2.9 A scanner has been designed for aircraft operation. It has a field of view (FOV) of $\pm 35°$ about nadir and an instantaneous field of view (IFOV) of 2 mrad. The sensor is designed to operate at a flying height of 1000 m.

(i) Determine the pixel size, in metres, at nadir.
(ii) Determine the pixel size at the edge of a swath compared with that at nadir.
(iii) Discuss the nature of the distortion in image geometry encountered if the pixels across a scan line are displayed on uniform pixel centres.

2.10 Determine the maximum angle of the field of view (FOV) for an airborne optical sensor with a constant instantaneous field of view (IFOV), so that the pixel dimension along the scan line at the extremes is less than 1.5 times that at nadir (ignore the earth curvature effect).

2.11 Consider the panoramic along scan line distortion of an airborne optical remote sensing system with a constant instantaneous field of view (IFOV); sketch the image formed for the ground scene shown in Fig. 2.27 and explain why it appears as you have sketched it.

---

[43] For some examples see Shlien, loc. cit.

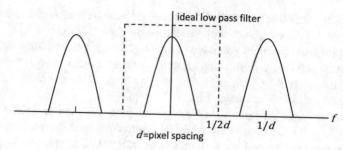

**Fig. 2.26** Idealised spatial frequency spectrum of line samples (pixels)

**Fig. 2.27** Ground scene

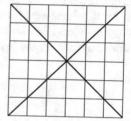

2.12  Compare the flat field, empirical line and log residuals methods for radio-metric correction from the point of view of ease of use compared with the nature of the result.

2.13  Following the development in Sect. 2.19 find a single matrix that describes the transformation from map coordinates to image coordinates for each of the following instruments, taking into account only earth rotation and orbital inclination at the equator.

SPOT HRG
AQUA MODIS
Ikonos.

2.14  Plot a graph of pixel size across the swath, as a function of pixel size at nadir, for look angles out to 70°; locate on the graph individual results for SPOT HRG, SPOT Vegetation, TERRA MODIS and EO-1 ALI.

2.15  Suppose a geostationary satellite carries a simple imager in the form of a digital camera. If the camera's field of view were such that it imaged a square area on the earth's surface with 5 km sides, discuss any distortions that might be present in the image and how they would appear in the final image product.

2.16   Is the wavelength dependence of atmospheric scattering strongest when
  (a) the atmosphere is foggy,
  (b) the atmosphere is clear, or
  (c) when there is a light haze with scattered clouds?

2.17   In the range of wavelengths commonly used in remote sensing, is the most significant absorbing constituent in the atmosphere
  (a) ozone,
  (b) carbon dioxide, or
  (c) water vapour?

2.18   Choose the correct answer below. One of the problems with nearest neighbour resampling is that
  (a) It can take a long time to compute, compared with other resampling methods
  (b) It can produce a blocky (pixelated) final product if the output scale is very different from the scale of the recorded image data
  (c) It requires more control points.

2.19   If you think the distribution of control point is poor is it best to
  (a) use a high degree mapping polynomial, such as a cubic,
  (b) use a first degree mapping polynomial, or
  (c) use any order (degree) since the degree of the polynomial is irrelevant?

2.20   In general, would you expect a smoother looking geometrically corrected image if you used
  (a) nearest neighbour resampling,
  (b) linear interpolation re-sampling, or
  (c) cubic convolution resampling?

# Chapter 3
# Interpreting Images

**Abstract** An introduction to the various methods for analysing remote sensing images is provided as the basis for the detailed treatments given in later chapters. Means for the creation of colour image products are described and used as the basis for photointerpretation—image interpretation by a skilled human analyst. The concept of machine assisted interpretation, referred to as quantitative analysis or classification, is presented as a means for producing a map of labels for image pixels, showing what they represent on the ground. Such a map is called a thematic map. The ideas of a pixel vector and spectral space are presented as the fundamental description of image data that underpins the mathematical models developed later. Statistical and geometric approaches to classification are summarised and the important distinction between information (on the ground) and spectral classes (in the data) is emphasised.

## 3.1  Introduction

With few exceptions the reason we record images of the earth in various wavebands is so that we can build up a picture of features on the surface. Sometimes we are interested in particular scientific goals but, even then, our objectives are largely satisfied if we can create a map of what is seen on the surface from the remotely sensed data available.[1] The principal focus of this book is on methods for analysing digital imagery and for creating maps from that analysis.

There are two broad approaches to image interpretation. One depends entirely on the skills of a human analyst—a so-called *photointerpreter*. The other involves computer assisted methods for analysis, in which various machine learning algorithms are used to automate what would otherwise be an impossibly tedious task. In this chapter we present an overview of the analytical methods used when

---

[1] In some cases, near sub-surface features can be seen in surface expressions in optical data. With radar, if the surface material is particularly dry, it is sometimes possible to image several metres under the surface.

© The Author(s), under exclusive license to Springer Nature Switzerland AG 2022
J. A. Richards, *Remote Sensing Digital Image Analysis*,
https://doi.org/10.1007/978-3-030-82327-6_3

interpreting imagery; this provides the context for the remainder of the book. We commence with an overview of photointerpretation and then move on to machine assisted analysis.

Although much of what is to be presented is regularly applied to radar imagery as much as to optical image data, the peculiarities of radar image data means that special analytical processes have been devised. They will be found in texts on radar remote sensing.[2]

## 3.2   Photointerpretation

A skilled photointerpreter extracts information from image data by visual inspection of an image product composed from the data. The analyst generally notes large-scale features and, in principle, is not concerned with the spatial and radiometric digitisations present. Spatial, spectral and temporal cues are used to guide the analysis, including the spatial properties of shape, size, orientation and texture. Roads, coastlines, river systems, fracture patterns and lineaments are usually readily identified by their spatial properties. Temporal cues are given by changes in a particular object or cover type from one date to another and assist in discriminating, for example, deciduous or ephemeral vegetation from perennial types. Spectral clues are based on the analyst's knowledge of, and experience with, the spectral reflectance characteristics of typical cover types including, if relevant, their radar scattering properties, and how those characteristics are sampled by the sensor on the platform used to acquire the image data.

Because photointerpretation is carried out by a human analyst it generally works at a scale much larger than the individual pixel in an image. It is a good approach for spatial assessment in general but is poor if the requirements of a particular exercise demand accurate quantitative estimates of the areas of particular cover types. It is also poor if the information required depends on detail in the spectral and radiometric properties of a particular image. By contrast, because humans reason at a higher level than computers, it is relatively straightforward for a photointerpreter to make decisions about context, proximity, shape and size, characteristics which challenge machine learning. It is in applications requiring those types of decision that photointerpretation is the preferred method for analysis.

---

[2] See J.A. Richards, *Remote Sensing with Imaging Radar*, Springer, Berlin, 2009 and F.T. Ulaby and D.G. Long, *Microwave Radar and Radiometric Remote Sensing*, University of Michigan Press, Ann Arbor, 2014.

## 3.2.1 *Forms of Imagery for Photointerpretation*

In order to carry out photointerpretation an image product has to be available, either in hard copy form or on a display device. That product can be a black and white image of an individual band or can be a colour image created from sets of bands. In the early days of remote sensing creating a colour image product presented little challenge because the number of bands of data recorded was not many more than the three primary colours of red, green, blue needed to form a display. With sensors now recording regularly more than 10 or so bands, and in the case of imaging spectrometers generating of the order of 100 bands, serious decisions have to be made about which bands to use when creating a colour product. In Chap. 6 we will address this problem by seeking to transform the recorded bands into a new compressed format that makes better use of the colour primaries for display. Here, however, we will focus on the simple task of selecting a set of the originally-recorded bands to create a colour product.

Essentially the task at hand is to choose three of the recorded bands and display them using the red, green and blue primary colours. It is conventional to order the chosen bands by wavelength in the same sequence as the colour primaries.[3] In other words, the shortest of the selected wavebands is displayed as blue and the longest is displayed as red.

Two simple considerations come to mind when seeking to select the wavebands to use. One is to create a colour product that is as natural in its colour as possible to that of the landscape being imaged. To do so entails choosing a band recorded in the blue part of the spectrum to display as blue, a band recorded in the green part of the spectrum to display as green, and a band recorded in the red part of the spectrum to display as red.

The other approach is to choose a set of wavebands that give better visual discrimination among the cover types of interest. When we look at spectral reflectance characteristics such as those shown in Fig. 1.11, it is clear that the red part of the spectrum will provide good discrimination among vegetated and bare cover types, while the infrared regime will give good separation from water and is also good for discriminating among vegetation types and condition. A popular colour product over many decades, therefore, has been one in which a green band has been displayed as blue, a red band has been displayed as green, and a near infrared band has been displayed as red. That has gone by several names, the most common of which is *colour infrared*. It is a product in which good healthy vegetation appears as bright red in the display. Of course, with other applications in mind, particularly in geology, choosing bands in the middle or thermal infrared ranges may be more appropriate. In those cases, user expertise will guide the choice of bands to display, but the principle of displaying the chosen bands in wavelength order is still maintained.

---

[3] In order of increasing wavelength, the additive colour primaries are blue, green and red.

Figure 3.1 shows a set of alternative displays for a portion of a HyVista HyMap image recorded over Perth in Western Australia. Similar displays can be created from mixed data sets such as that depicted in Fig. 1.15. For example, two of the display colours might be used to show bands of optical imagery with the third used to overlay synthetic aperture radar data.

When viewing image products like those shown in Fig. 3.1 it is important to know that the original bands of data have usually been enhanced in contrast before the colour composite image has been formed. As a result, it is sometimes difficult to associate accurately the colours observed with the spectral reflectance characteristics shown in Fig. 1.11. For example, in the colour infrared image a region of soil should appear reddish if the infrared band is displayed as red. Yet in Fig. 3.1 sparse vegetation and soils appear as blue-green. That is almost always the case in that type of imagery and is a direct result of each of the individual bands being expanded in contrast to cover the full range of brightness available before colour composition. Why is that necessary? Sensors are designed so that they can respond to features on the ground that have brightness values ranging from black (extreme shadows) to white (clouds, sand and snow). That means that the more common cover types such as vegetation and soils have about mid-range brightnesses and would thus appear dull if displayed as recorded. Consequently, the brightness values are stretched out over the available brightness range before display, using the techniques we cover in Chap. 4. If the bands were not contrast enhanced beforehand, the colour composite image would have a general reddish appearance for both vegetation and soil.

It is easy to see why the colour relativity is affected by changing the spread of brightness values in the individual bands before they are composed into the colour product. The simple illustration in Fig. 3.2 provides the explanation. A skilled photointerpreter takes this type of information into account when interpreting the colours seen in the image data. Not infrequently, the photointerpreter will also have available black and white images of significant bands so that the contrast differences within a band over different cover types can be taken into account during analysis.

### 3.2.2 Computer Enhancement of Imagery for Photointerpretation

While expanding the brightness range in an image is often performed to make a colour product potentially more attractive as illustrated in Fig. 3.2, a range of other image enhancement techniques can be applied to imagery to assist the photointerpretation task, as discussed in Chap. 4. New types of imagery can also be created by applying mathematical transformations to the original data set. In addition, image data can be processed geometrically, in which noise is smoothed or reduced, and features of particular significance, such as lines and edges, are enhanced. Those geometric processing methods are covered in Chap. 5 while Chaps. 6 and 7 cover image transformations.

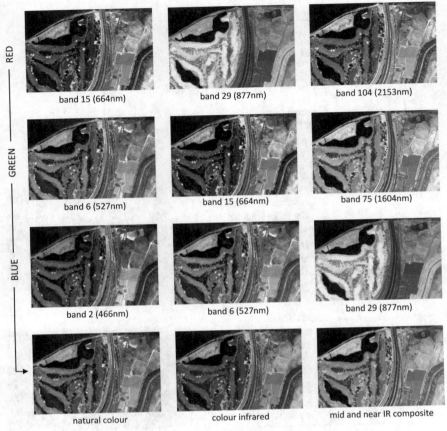

**Fig. 3.1** Colour image products formed by different combinations of recorded bands, in the sequence from top to bottom displayed respectively as red, green and blue: the image (which has north to the right) was recorded by the HyMap sensor over the city of Perth, Western Australia and shows how different band combinations highlight cover type variations

## 3.3 Quantitative Analysis: From Data to Labels

In order to allow a comparison with photointerpretation it is of value to consider briefly the fundamental nature of classification before a more detailed discussion of computer assisted interpretation is presented. Essentially, classification is a mapping from the spectral measurements acquired by a remote sensing instrument to a label for each pixel that identifies it with what's on the ground. Sometimes, several labels for a given pixel are generated, with varying degrees of likelihood, and sometimes mixtures of labels are given for each pixel. Those alternative cases will become evident later in this book. For the present, however, we will focus on obtaining a single name for a pixel in terms of known ground cover types.

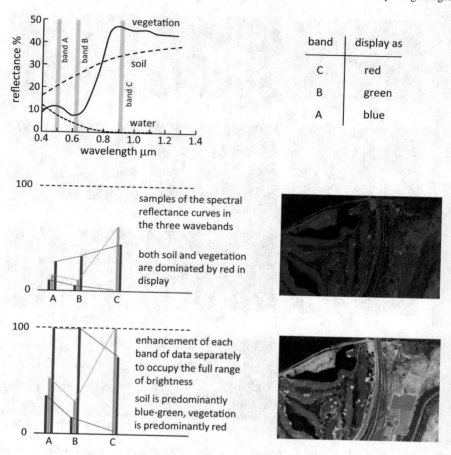

**Fig. 3.2** The impact of enhancing the contrasts of the bands individually before forming a colour image product

Figure 3.3 summarises the process of classification. Starting with the set of measurements, a computer processing algorithm is used to provide a unique label, or theme, for all the pixels in the image. Once complete, the operation has produced a map of themes on the ground from the recorded image data. The map is called a *thematic map* and the process of generating it is called *thematic mapping*. Once they have all been labelled, it is possible to count the pixels of a given cover type and note their geographic distributions. Knowing the size of a pixel in equivalent ground metres allows accurate estimates of the area of each cover type in the image to be produced. Because we are able to quantify the cover types in this manner, and because the procedures we use are inherently numerical and statistical, classification is often referred to as *quantitative analysis*.

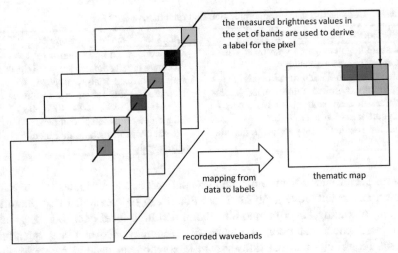

**Fig. 3.3**  Classification as a mapping from measurement or spectral space to a set of labels

## 3.4   Comparing Quantitative Analysis and Photointerpretation

We are now in the position to do a meaningful comparison of photointerpretation and classification, as the two principal means by which image understanding is carried out. Photointerpretation is effective for global assessment of geometric characteristics and for the general appraisal of ground cover types. It is, however, impracticable to apply at the pixel level unless only a handful of pixels is of interest. As a result, it is of limited value for determining accurate estimates of the area of an image corresponding to a particular ground cover type, such as the hectarage of a crop. Further, since photointerpretation is based on the ability of the human analyst to assimilate the data, only three or so of the complete set of spectral components of an image can easily be used. Yet there are of the order of 10–100 bands available in modern remote sensing image data sets. It is not that all of these would necessarily be needed in the identification of a pixel. However, should all, or a large subset, require consideration, analysis by photointerpretation is clearly limited. By comparison, if a machine can be used for analysis, as outlined in the previous section, it can work at the individual pixel level. Also, even though we have yet to consider specific algorithms for classification, we can presume from a knowledge of machine assisted computation in general, that it should be possible to devise approaches that handle as many bands as necessary to obtain an effective label for a pixel.

There is another point of difference between the ability of the photointerpreter and that of a machine. The latter can exploit the full radiometric resolution available in the image data. By comparison, a human's ability to discriminate levels of grey is limited to about 16, which again restricts the nature of the analysis able to be performed by a photointerpreter.

**Table 3.1** Comparison of photointerpretation and quantitative analysis

| Photointerpretation (human analyst) | Quantitative analysis (computer) |
|---|---|
| On a scale large compared with pixel size | Can work at the individual pixel level |
| Less accurate area estimates | Accurate area estimates are possible |
| Limited ability to handle many bands | Full multi-band analysis is possible |
| Can use only a limited number of brightness values in each band (about 16) | Can use the full radiometric resolution available (256, 1024, 4096, etc.) |
| Shape determination is easy | Shape determination is complex |
| Spatial information is easy to use in general | Spatial decision making in general is challenging |

Table 3.1 provides a more detailed comparison of the attributes of photointerpretation and quantitative analysis. From this it can be concluded that photointerpretation, involving direct human interaction and high-level decisions, is good for spatial assessment but poor in quantitative accuracy. By contrast, quantitative analysis, requiring some but little human interaction, in general has poor spatial reasoning ability but high quantitative accuracy. Its poor spatial properties come from the relative difficulty with which decisions about shape, size, orientation and, to a lesser extent, texture can be made using standard sequential computing techniques.

The interesting thing about the comparison in Table 3.1 is that each approach has its own strengths and, in several ways, they are complementary. In practice it is common to find both approaches employed when carrying out image analysis. As we will see shortly, photointerpretation is often an essential companion step to quantitative analysis because to make machine-assisted approaches work effectively some knowledge from the analyst has to be fed into the algorithms used.

## 3.5  The Fundamentals of Quantitative Analysis

### 3.5.1  Pixel Vectors and Spectral Space

We now look at the manner in which machine-assisted classification of remote sensing image data can be performed. Recall that the data recorded consists of a large number of pixels, with each pixel characterised by up to several hundred spectral measurements. If there is a sufficiently large number of fine bandwidth samples available, it is possible to reconstruct the reflectance spectrum for a pixel as seen by the sensor. Figure 3.4 shows a typical vegetation spectrum recorded by the HyMap imaging spectrometer. Provided such a spectrum has been corrected for the effects of the atmospheric path between the sun, the earth and the sensor, and the shape of the solar emission curve, then a skilled spectroscopist should, in principle, be able to identify the cover type, and its properties, from the measurements. While that approach is technically feasible, more often than not a pixel spectrum is identified by reference to a library of previously recorded spectra. We will have more to say about spectroscopic and library searching techniques in Chap. 11.

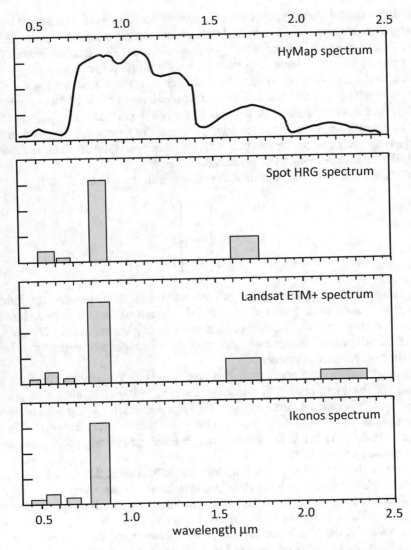

**Fig. 3.4** Typical single pixel vegetation spectrum recorded by the HyVista HyMap imaging spectrometer compared with the spectra for the same cover type that would be recorded by a number of other instruments; the latter have been estimated from the HyMap spectrum for illustration

Although the hyperspectral data sets provided by imaging spectrometers allow scientific methods of interpretation, often the smaller number of spectral measurements per pixel from many sensors makes that approach not feasible. The remaining spectra in Fig. 3.4 illustrate just how selectively the spectrum is sampled with some instruments. Nevertheless, while they do not fully replicate the spectrum, it is clear that the number and placement of the spectral samples should still be sufficient to permit some form of identification of the cover type represented by the pixel. What we want to do now is devise an automated analytical approach that works with sets of samples and which, when required, can be extended to work with the large number of spectral samples recorded by imaging spectrometers.

The first thing we have to do is to decide on a model for describing the data. Because each pixel is characterised by a set of measurements a useful summary tool is to collect those measurements together into a column called a *pixel vector* which has as many elements as there are measurements. By describing the pixel in this manner, we will, later on, be able to use the very powerful field of vector and matrix analysis when developing classification procedures.[4]

We write the pixel vector with square brackets in the column form:

$$\mathbf{x} = \begin{bmatrix} x_1 \\ x_2 \\ \vdots \\ x_N \end{bmatrix}$$

The elements listed in the column are the numerical measurements (brightness values) in each of bands 1 through to $N$, and the overall vector is represented by the lower-case character in bold. Because we will be using concepts from the field of mathematical pattern recognition and machine learning, the vector $\mathbf{x}$ is also sometimes called a *pattern vector*.

To help visualise the concepts that follow it is of value now to introduce the concept of the *spectral space* or spectral domain. In the terminology of pattern recognition, it is called a *pattern space*. This is a coordinate system with as many dimensions as there are measurements in the pixel vector. A particular pixel in an image will plot as a point in the spectral space according to its brightness along each of the coordinate directions.

Although the material to follow both here and in the rest of this book is designed to handle pixel vectors with as many measurements as necessary, it is helpful visually to restrict ourselves to just two measurements at this stage so that the spectral domain has only two coordinates. Figure 3.5 illustrates the idea of using measurements in the visible red portion of the spectrum and the near infrared.

As observed, sets of pixel vectors for different cover types appear in different regions of the spectral domain. Immediately, we can see that an effective way of labelling pixels as belonging to different cover types is to assess in what part of the spectral space they lie. Note that the different cover types will only be differentiated in the spectral domain if the wavebands of the sensor have been chosen to provide discrimination among the cover types of interest. The measurements chosen in Fig. 3.5 do provide separation between what we will now call *classes* of data: the vegetation class, the water class, and the soil class. Because this is the information

---

[4] See Appendix C for a summary of the essential elements of vector and matrix algebra.

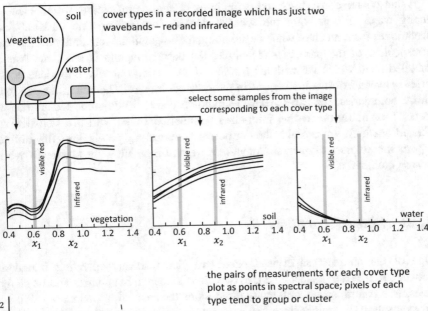

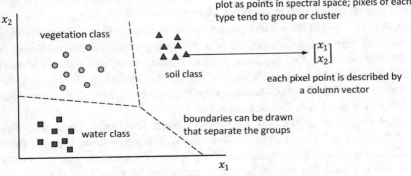

**Fig. 3.5** Pixels in a spectral space with coordinates that correspond to the spectral measurements made by a sensor; provided those measurements are well located spectrally the pixel points corresponding to different cover types will be separated in the spectral space, even though there will be natural variability within the spectral responses for each cover type, as illustrated

we are interested in obtaining from the remotely sensed data we commonly refer to classes of this type as *information classes*.

Suppose we knew beforehand the information classes associated with a small group of pixels in an image. We could plot the corresponding remote sensing measurements in the spectral domain to help identify where each of those classes is located, as shown in Fig. 3.5. We could then draw lines between the classes to break up the spectral domain into regions that could have information class labels attached to them. Having done that, we could take an unknown pixel and plot it in the spectral domain according to its measurements. We then label it as belonging to one of the available information classes, by reason of where it falls compared with the class boundaries.

What we have just described is the basis of *supervised classification*. There are many ways of separating the spectral domain into information classes. Those techniques form much of what we are going to look at in the remainder of this book. Irrespective of the particular technique, the basic principle is the same: we use labelled data, which we will call *training data*, to find out where to place boundaries between information classes in the spectral domain. Thereafter, having found those boundaries, we can label any unknown pixel. While some techniques will depend explicitly on finding inter-class boundaries, others will use statistical and related methods to achieve the purpose of separating pixels into the different information classes of interest. Sometimes we will even allow the classes to overlap across boundaries.

### 3.5.2  Linear Classifiers

One of the simplest supervised classifiers places linear separating boundaries between the classes, as just noted. In two dimensions the boundaries will be straight lines. In a pattern space with many dimensions the separating boundaries will be a generalisation of straight lines and surfaces; those higher order surfaces are called *hyperplanes*. A very straightforward method for finding appropriate hyperplanes is to use the training data to find the mean position (mean vector) of the pixels in each class, and then find those hyperplanes that are the perpendicular bisectors of the lines between the mean vectors. Such a classifier, which is treated in Chap 8, is referred to as the minimum distance classifier.

The field of pattern recognition essentially commenced using linear classifier theory of that nature.[5] Linear classification also forms the basis of the two most common machine learning approaches of the past two decades: *the support vector machine* (SVM), and the *neural network*, including its popular derivative the *convolutional neural network*. These are treated in Chap. 8.

One of the powerful features of the support vector machine approach is the ability to introduce data transformations that effectively turn the linear separating hyperplanes into more flexible, and thus more powerful, *hypercurves*.

The neural network can implement many decision boundaries that are piecewise linear in nature, allowing much more flexible class separation.

### 3.5.3  Statistical Classifiers

The original supervised classification procedure widely used in remote sensing since the 1970s is based on the assumption that the distribution of pixels in a given

---

[5] See N.J. Nilsson, *Learning Machines*, McGraw-Hill, N.Y., 1965.

class or group can be described by a probability distribution in spectral space. The probability distribution most often used is the multidimensional normal distribution; the technique is called *maximum likelihood classification* because a pixel previously unseen by the classifier algorithm is placed in the class for which the probability is the highest of all the classes.

When using a normal distribution model for each class the dispersion of pixels in the spectral space is described by their mean position and their multidimensional variance, or standard deviation. That is not unreasonable since it would be expected that most pixels in a distinct cluster or class would lay towards the centre and would decrease in likelihood for positions away from the class centre where the pixels are less typical.

It is important to recognise that the choice of the multidimensional normal, or Gaussian, distribution does not rest on the fact that the classes are actually normally distributed in nature; we will have more to say about that in Sect. 3.6 following. Instead, the reason we use the normal distribution as a class model is that its properties are well known for any dimensionality, its parameters are easily estimated, and it is robust in the sense that the accuracy of prediction when producing a thematic map is not overly sensitive to violations of the assumption that the classes are normal.

A two-dimensional spectral space with the classes modelled as normal distributions is shown in Fig. 3.6. The decision boundaries shown, which are the equivalent to the straight-line decision boundaries in Fig. 3.5, represent those points in the spectral space where a pixel has equal chance of belonging to either of two classes. Those boundaries partition the space into regions associated with each class; because of the mathematical form of the normal distribution, the boundaries are multidimensional quadratic functions.

In Chap. 8 we will look in detail at the mathematical form of the multidimensional normal distribution. Here it is sufficient to use the shorthand notation:

$$p(\mathbf{x}|\omega_i) \sim \mathcal{N}(\mathbf{m}, \mathbf{C})$$

which says that the probability of finding a pixel from class $\omega_i$ at the position $\mathbf{x}$ in the spectral domain is given by the value of a normal distribution which is described by a mean vector position $\mathbf{m}$ and whose spread is described by the covariance matrix $\mathbf{C}$. For the data sketched in Fig. 3.6 there are three such normal distributions, one for each of the classes. Therefore, there will be three different sets of the pair of parameters $\mathbf{m}$ and $\mathbf{C}$.

The multidimensional normal distribution is completely specified by its mean vector and covariance matrix. As a result, if the mean vectors and covariance matrices are known for all classes then it is possible to compute the set of probabilities that describe the relative likelihoods of a pixel at a particular location in spectral space as belonging to each of those classes. A pixel is allocated to the class for which the probability is highest. Before that can be done $\mathbf{m}$ and $\mathbf{C}$ have to be estimated for each class from representative sets of pixels—i.e., training sets of

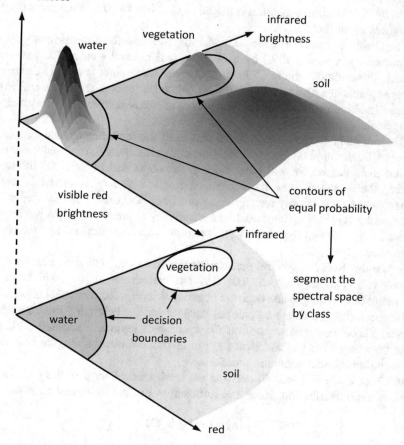

**Fig. 3.6** Two-dimensional spectral space with the classes represented by Gaussian probability distributions; note the unusual distribution of the soil class in this illustration, a problem that would be resolved by the application of thresholds (see Sect. 8.3.5)

pixels that the analyst knows belong to each of the classes of interest. Estimation of **m** and **C** from a training set is referred to as supervised learning. Based on this statistical approach, supervised classification therefore consists of three broad steps:

- A set of training pixels is selected for each class. That could be done using information from ground surveys, aerial photography, black and white and colour hard copy products of the actual image data, topographic maps, or any other relevant source of reference data.
- The mean vector and covariance matrix for each class are estimated from the training data. That completes the learning phase.

- The classification phase follows in which the relative likelihoods for each pixel in the image are computed and the pixel labelled according to the highest likelihood.

Since the approach to classification considered here involves estimation of the parameter sets **m** and **C**, the method goes under the general name of a *parametric* procedure. Other methods such as linear classifiers, support vector machines and neural networks are often referred to as *non-parametric* because there are no parameters that require estimation, despite the fact that they still require the estimation of certain constants during training.

## 3.6  Sub-classes and Spectral Classes

There is a significant assumption in the manner by which we have represented the clusters of pixels from given cover types in Figs. 3.5 and 3.6. We have shown the pixels from a given class as belonging to a single group or cluster. In practice, that is often not the case. Because of differences in soil types, vegetation condition, water turbidity, and similar, the information classes of one particular type often consist of, or can be conveniently represented as, collections of sub-classes. Also, when using the normal distribution model of the previous section, we have to face the realisation that sets of training pixels rarely fall into groups that can be well-represented by single normal distributions. Instead, the pixels are distributed in a fashion which is often significantly not normal. To apply a classifier based on a multidimensional normal model successfully we have to represent the pixels from a given information class as a set of normal distributions—again, effectively sub-classes—as illustrated in Fig. 3.7. If we assume that the subclasses are identifiable as individual groupings, or as representative partitions of the spectral space, we call them *spectral classes* to differentiate them from the information classes that are the ground cover type labels known to, and sought by, the analyst. Sometimes they are called *data classes*.

For most real image data, it is difficult to identify very many *distinct* information or spectral classes. To illustrate this point, Fig. 3.8 shows a two-dimensional spectral plot of the pixel points from the portion of imagery shown. In this diagram, which is called a *scatter plot* in general, the near infrared versus visible red data values of the pixels are seen mostly to form a continuum, with just a few groups that might be easily be ascribed to identifiable cover types. In the main, however, even though there are quite a few individual classes present in the image itself, those classes do not show up as distinct groups in the two-dimensional scatter plot.

To achieve a successful classification of the image data using any of the methods covered later in this book, it is important to recognise that the data space often is of the form of a continuum. The trick to making classification work well is to ensure that the data space is segmented in such a way that the properties of the chosen classifier algorithm are properly matched to the data. For linear classifiers we need to ensure that the separating boundaries appropriately segment the continuum; for

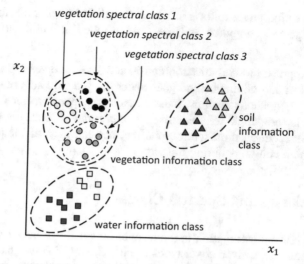

**Fig. 3.7** Modelling information classes by sets of spectral or sub-classes

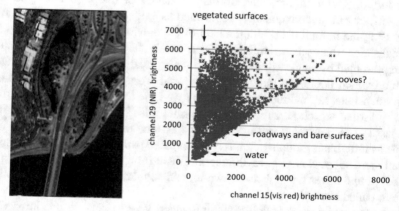

**Fig. 3.8** Scatter of pixel points in a near infrared versus red spectral subspace in which some individual classes are evident but in which much of the domain appears as a continuum; the continuum has to be represented by sets of sub-classes or spectral classes in order to achieve good classification results

statistical classifiers we need to ensure that the portions of the continuum corresponding to a given information class are resolved into an appropriate set of Gaussian spectral classes. Essentially, the spectral classes are the viable groups into which the pixels can be resolved in order best to match the properties of the classification algorithm being used. We will have a lot to say about spectral classes in Chap. 11 when we consider overall methodologies for performing classification and thematic mapping.

## 3.7   Unsupervised Classification

The supervised approach to classification outlined in the previous sections is not the only manner in which thematic maps can be created. There is a class of algorithm called *unsupervised* that also finds widespread use in the analysis of remote sensing image data.

Unsupervised classification is a method by which pixels are assigned to spectral (or information) classes without the user having any prior knowledge of the existence or names of those classes. It is most often performed using clustering methods, which are the topic of Chap. 9. Those procedures can be used to determine the number and location of the spectral classes into which the data naturally falls, and then to find the spectral class for each pixel in the image. The output from such an analysis is generally a map of symbols—a cluster map—that depict the class memberships of the pixels without the analyst yet knowing what those symbols represent, apart from the fact that pixels of a given symbol fall into the same spectral class or cluster. The statistics of the spectral classes are also generally available from the application of a clustering algorithm. Once the cluster map is available the analyst identifies the classes or clusters by associating a sample of pixels from the cluster map with the available reference data, which could include other maps and information from ground visits.

Clustering procedures are generally computationally expensive, yet they are central to the analysis of remote sensing imagery. While the information classes for a particular exercise are known, the analyst is usually totally unaware of the spectral classes, or subclasses, beforehand. Unsupervised classification is often useful for determining a spectral class decomposition of the data prior to detailed analysis by the methods of supervised classification.

## 3.8   Bibliography on Interpreting Images

The field of image analysis has a rich history, covering many fields. The text which has become a standard treatment is

R.O. Duda, P.E. Hart and D.G. Stork, *Pattern Classification*, 2nd ed., John Wiley & Sons, N.Y., 2001.

There are many treatments now available that have a remote sensing focus, including

R.A. Schowengerdt, *Remote Sensing: Models and Methods for Image Processing*, 3rd ed., Academic, Burlington, Mass., 2006.

B. Tso and P.M. Mather, *Classification Methods for Remotely Sensed Data*, 2nd ed., CRC Press, Taylor and Francis Group, Boca Raton, Florida, 2016.

J.R. Jensen, *Introductory Digital Image Processing: A Remote Sensing Perspective*, 4th ed., Prentice-Hall, Upper Saddle River, N.J, 2015.

J.R. Shott, *Remote Sensing: The Image Chain Approach*, 2nd Ed., Oxford UP, N.Y., 2007.

Even though this book is concerned with image processing and analysis it is important not to overlook the principles of the application domain of remote sensing in which that work is located. It makes little sense processing data blind, without some appreciation of the requirements of a given application and an understanding of the fundamental principles of remote sensing. Accordingly, the following books might also be consulted. Although the first is now a little dated in some respects, it still has one of the best chapters on the spectral reflectance characteristics of ground cover types and contains good overview discussions on the objectives and execution of an image analysis exercise.

P.H. Swain and S.M. Davis, eds, *Remote Sensing: The Quantitative Approach*, McGraw-Hill, N.Y., 1978.

T. Lillesand, R.W. Kiefer and J. Chipman, *Remote Sensing and Image Interpretation*, 7th ed., J. Wiley and Sons, N.Y., 2015.

J.B. Campbell and R.H. Wynne, *Introduction to Remote Sensing*, 5th ed., Guildford, N.Y., 2011.

F.F. Sabins and J.M. Ellis, *Remote Sensing: Principles and Interpretation and Applications*, 4th ed., Waveland, Long Grove, IL., 2020.

The following is an introductory treatment which also covers geographic information systems:

J.G. Liu and P.J. Mason, *Image Processing and GIS for Remote Sensing: Techniques and Applications*, Wiley-Blackwell, N.J., 2016.

## 3.9   Problems

3.1  For each of the following applications would photointerpretation or quantitative analysis be the most appropriate analytical technique? Where necessary, assume spectral discrimination is possible.

- creating maps of land use
- mapping the movement of floods
- determining the area of crops
- mapping lithology in geology
- structural mapping in geology
- assessing forest condition
- mapping drainage patterns
- creating bathymetric charts.

3.2 Can contrast enhancing an image beforehand improve its discrimination for machine analysis? Could it impair machine analysis by classification methods?

3.3 Prepare a table comparing the attributes of supervised and unsupervised classification. You may wish to consider problems with collecting training data, the cost of processing, the extent of analyst interaction and the determination of spectral classes.

3.4 A problem with using probability models to describe classes in spectral space is that atypical pixels can be erroneously classified. For example, a pixel with low red brightness might be wrongly classified as soil even though it is more reasonably vegetation. This is a result of the positions of the decision boundaries as seen in Fig. 3.6. Suggest a means by which this situation can be avoided (see Sect. 8.3.5).

3.5 The collection of brightness values for a pixel in a given image data set is called a vector. Each of the components of the vector can take on a discrete number of brightness values determined by the radiometric resolution of the sensor. If the radiometric resolution is 8 bits the number of brightness values is 256. If the radiometric resolution is 10 bits the number of brightness values is 1024. How many distinct pixel vectors are possible with SPOT HRG, Ikonos and Landsat ETM+ data?
It is estimated that the human visual system can discriminate about 20,000 colours. Comment on the radiometric handling capability of a computer compared to colour discrimination by a human analyst.

3.6 Information classes are resolved into spectral classes prior to classification. In the case of the multidimensional normal distribution, those spectral classes are individual Gaussian models. Why are more complex statistical distributions not employed to overcome the need to establish individual, normally distributed spectral classes?

3.7 A very simple sensor that might be used to discriminate between water and non-water could consist of a single infrared band with 1 bit radiometric resolution. A low response indicates water and a high response indicates non-water. What would the spectral space look like? Suppose the sensor now had 4 bit radiometric resolution. Again, describe the spectral space but in this case noting the need to position the boundary between water and non-water optimally within the limits of the available radiometric resolution. How might you determine that boundary?

3.8 Plot the pixels from Table 2.2a in Question 2.2 in a coordinate space with band 5 brightness horizontally and band 7 brightness vertically. Do they naturally separate into three classes? Find the two-dimensional mean vectors for each class and find the perpendicular bisectors of lines drawn between each pair. Show that, as a set, those bisectors partition the coordinate space by class. Repeat the exercise for the data in Table 2.2b. Is the separation now poorer? If so, why? Note that June in the southern hemisphere is mid-winter and December is mid-summer. To assess separability you may wish to mark plus and minus one standard deviation about each mean in both coordinates.

3.9   In Problem 3.8 assume that the reflectance of water does not change between dates and that the difference in the water mean values is the result of solar illumination changes with season. Using the water mean for the summer image as a standard, find a simple scale change that will adjust the mean of the water class for the winter image to that of the summer image. Then apply that scale change to the other two winter classes and plot all classes (three for each of summer and winter) on the same coordinates. Interpret the changes observed in the mean positions of the vegetation and soil classes.

3.10  Thinking carefully about the spectral reflectance curves of vegetation, soil and water, where would pixels of each of those cover types appear in a two-dimensional spectral space in which the brightness in a near infrared band is plotted vertically, while the brightness in a visible red band is plotted horizontally? Why are there parts of that spectral space which are always empty? Where would deep shadows appear in a spectral space?

# Chapter 4
# Radiometric Enhancement of Images

**Abstract** Techniques for altering the brightness and contrast of images, generically called contrast modification, are presented. These are based on methods for modifying the brightness values of individual pixels within given bands of data, and are seen to be valuable in improving the visual quality of an image. The image histogram is described and used as the basis of many of the contrast modification procedures covered. It is seen that the image histogram can be used also for matching the contrasts (brightness distributions) of two geographically adjacent images so they can be joined side-by-side in a mosaic without there being a significant change in contrast across the common boundary. Examples are given of the procedures covered.

## 4.1 Introduction

### 4.1.1 Point Operations and Look Up Tables

Image analysis by photointerpretation is often made easier when the radiometric nature of an image is enhanced to improve its visual characteristics. Specific differences in vegetation and soil type, for example, may be brought out by increasing the contrast of an image. Highlighting subtle differences in brightness value by applying contrast modification, or by assigning different colours to different brightness ranges in the method known as colour density slicing, will often reveal features not otherwise easily seen.

It is the purpose of this chapter to present a variety of radiometric modification procedures that are regularly used with remote sensing image data. The methods treated are characterised by the common feature that a new brightness value for a pixel is generated only from its existing value. Neighbouring pixels have no influence, as they do in the geometric enhancement procedures that are the subject of Chap. 5. Consequently, radiometric enhancement techniques are sometimes referred to as point or pixel-specific operations.

© The Author(s), under exclusive license to Springer Nature Switzerland AG 2022     107
J. A. Richards, *Remote Sensing Digital Image Analysis*,
https://doi.org/10.1007/978-3-030-82327-6_4

All of the techniques to be covered in this chapter can be represented either as a graph or a table that expresses the relationship between the old and new brightness values. In tabular form this is referred to as a look up table (LUT).

### 4.1.2   Scalar and Vector Images

Two particular image types require consideration when treating image enhancement. The first is referred to as a *scalar image*, in which each pixel has only a single brightness value associated with it. Such is the case for a simple black and white image. The second is a *vector image*, in which each pixel is represented by a vector of brightness values, which might be the blue, green and red components of the pixel in a colour scene or, for a remote sensing image, would be the various spectral response components for the pixel. Most image enhancement techniques relate to scalar images or the scalar components of vector imagery. That is the case with all techniques covered in this chapter. Enhancement methods that relate particularly to vector imagery tend to be transformation oriented. They are treated in Chap. 6.

## 4.2   The Image Histogram

Consider a typical remote sensing image, composed of pixels spatially and in which each pixel is quantised radiometrically into discrete brightness levels. If pixels with the same brightness value are counted, a graph of the number of pixels at a given brightness, versus brightness value, can be constructed. That is referred to as the *histogram* of the image. The tonal or radiometric quality of an image can be assessed by inspecting its histogram, as illustrated in Fig. 4.1. An image which makes good use of the available range of brightness values has a histogram with occupied bins (brightness values) over the full range, but without significantly large values at either extreme. The vertical scale of a histogram is sometimes called frequency (of the occurrence of specific brightness values), or proportion, or occurrence.

An image has a unique histogram, but the reverse is not true in general since a histogram contains only radiometric and no spatial information. A point of some importance is that the histogram can be viewed as a discrete probability distribution since the relative height of a particular bar or occurrence, normalised by the total number of pixels in the image, indicates the chance of finding a pixel with that particular brightness value somewhere in the image.

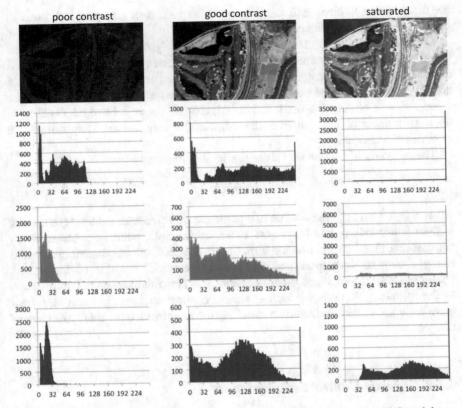

**Fig. 4.1** Some images and the histograms of their three colour components; from left to right: an image with low contrast and brightness, in which the histograms occupy a limited set of brightness values, the same image with a good range of brightness and contrast and for which the histograms show good use of the available range, and an overly contrasting image in which detail is lost and the histograms show saturation at the upper extremes; although shown in colour to link them to the image components, most histograms would be displayed in black and white

## 4.3   Contrast Modification

### 4.3.1   Histogram Modification Rule

Suppose we have available a digital image with poor contrast, such as that with the histograms on the left-hand side of Fig. 4.1. We want to improve its contrast to obtain an image with histograms that have a good spread of occurrences over the available brightness range, resembling that in the centre of Fig. 4.1. An operation called *contrast modification or stretching* is required; it is applied to the individual image components. Often the degree of stretching desired is apparent. For example, an original histogram may occupy brightness values between 40 and 75 and we might want to expand that range to the maximum possible, say 0–255. Even though the modification is obvious it is necessary to express it in mathematical terms in order to transfer the problem to a computer.

Contrast modification is just a mapping of brightness values, in that the brightness value (the abscissa) of a particular histogram occurrence is respecified more favourably. The magnitudes of the occurrences themselves though are not altered, although in some cases some may be mapped to the same new brightness value and will be superimposed. In general, though, the new histogram will have the same number of occurrences as the old, with the same values. They will just be at different locations.

The mapping of brightness values can be described by

$$y = f(x) \tag{4.1}$$

where $x$ is the old brightness value of a particular occurrence in the histogram and $y$ is the corresponding new brightness value. In principle, what we want to do in contrast modification is to find the form of $y = f(x)$ that will implement the desired changes in pixel brightness and thus in the perceived contrast of the image. Sometimes that is simple; on other occasions $y = f(x)$ might be quite complicated. In the following sections we look at simple contrast changes first and then treat more complex situations, including matching the brightness value ranges of pairs of images.

### 4.3.2   Linear Contrast Modification

The most common contrast modification operation is that in which the new and old brightness values of the pixels in an image are related in a linear fashion, so that (4.1) can be expressed

$$y = f(x) = ax + b$$

A simple numerical example of linear contrast modification is shown in Fig. 4.2; the look-up table is included in the figure. In practice this would be used in software to produce the new image. That is done by reading the original brightness values of the pixels one at a time, substituting those brightnesses into the left-hand side of the table and then reading the new brightness values for the pixels from the corresponding entries on the right-hand side of the table.

It is important to note in digital image handling that the new brightness values, just as the old, must be discrete integers and cover usually the same range. That may require some rounding to integer form of the new values calculated from the mapping function $y = f(x)$. A further point to note in the example of Fig. 4.2 is that the look-up table is valid only for the range of inputs from 2 to 4. Beyond that, output brightness values would be generated that lay outside the range valid for this example. In practice, linear contrast stretching is generally implemented as the

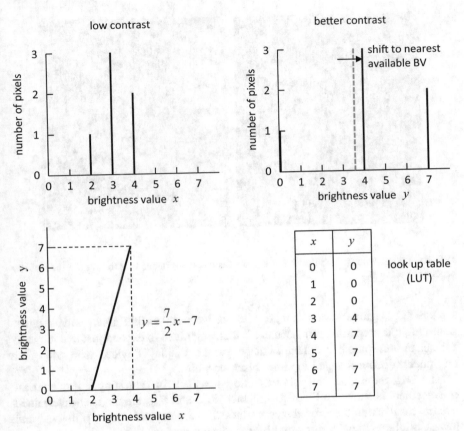

**Fig. 4.2** Simple example of linear contrast modification; the look up table (LUT) created from the brightness value mapping function is used to re-position the occurrences in the histogram

saturating linear contrast enhancement technique in Sect. 4.3.3 in which, for this example, the outputs are set to 0 and 7 for input brightnesses less than 2 and greater than 4 respectively.

An image with poor contrast that has been radiometrically enhanced by linear contrast stretching is shown in Fig. 4.3.

## 4.3.3   Saturating Linear Contrast Enhancement

Frequently a better image product is generated when a degree of saturation is created at the black and white ends of the histogram while applying linear contrast enhancement. Such is the case, for example, if the darker regions in an image belong to the same ground cover type within which small radiometric variations are of no interest. Alternatively, a particular region of interest in an image may occupy

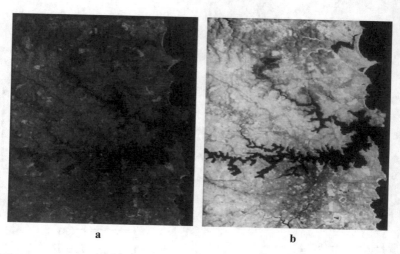

a                                                      b

**Fig. 4.3** Linear contrast modification of the raw (as recorded) image in **a** to produce the visually better product in **b**

a restricted brightness value range, such as water in visible green wavelengths; saturating linear contrast enhancement is then employed to expand just that range to the maximum possible dynamic range of the display device, with all other brightnesses being mapped to either black or white.

The brightness value mapping function $y = f(x)$ for saturating linear contrast enhancement is shown in Fig. 4.4, in which $B_{max}$ and $B_{min}$ are the user-determined maximum and minimum brightness values that are expanded to the highest and lowest brightness levels supported by the display device.

## 4.3.4   Automatic Contrast Enhancement

Most remote sensing image data is too low in brightness and poor in contrast to give an acceptable image product if displayed directly in raw form as recorded by a sensor. That is a result of the need to have the dynamic range of satellite and aircraft sensors so adjusted that a variety of cover types over many images can be detected without leading to saturation of the detectors, or without useful signals being lost in noise. As a consequence, a typical single image will contain a restricted set of brightnesses. Image display systems frequently implement an automatic contrast stretch on the raw data in order to give a product with good contrast. Typically, the automatic enhancement procedure is a saturating linear stretch. The cut-off and saturation limits $B_{min}$ and $B_{max}$ are chosen by determining the mean brightness of the raw data and its standard deviation, and then making $B_{min}$ equal to the mean less three standard deviations and $B_{max}$ equal to the mean plus three standard deviations.

**Fig. 4.4** Saturating linear
brightness value mapping
function

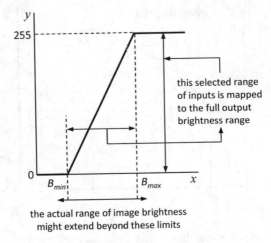

this selected range
of inputs is mapped
to the full output
brightness range

the actual range of image brightness
might extend beyond these limits

### 4.3.5   *Logarithmic and Exponential Contrast Enhancement*

Logarithmic and exponential mappings of brightness value between the original and
modified images are useful processes for enhancing dark and light features
respectively. The mapping functions are shown in Fig. 4.5, along with their
mathematical expressions. It is particularly important with these that the output
values be scaled to lie within the range of the device used to display the product and
that they be rounded to allowed, discrete values.

### 4.3.6   *Piecewise Linear Contrast Modification*

A particularly useful and flexible contrast modification procedure is the piecewise
linear mapping function shown in Fig. 4.6, which is characterised by a set of
user-specified break points. Generally, the user can also specify the number of
break points. This method has particular value in implementing some of the contrast
matching procedures in Sects. 4.4 and 4.5 following. It is a generalisation of the
saturating linear contrast stretch of Sect. 4.3.3.

## 4.4   Histogram Equalisation

### 4.4.1   *Use of the Cumulative Histogram*

The previous sections have addressed the task of simple expansion (or contraction)
of the histogram of an image. In many situations, however, it is desirable to modify

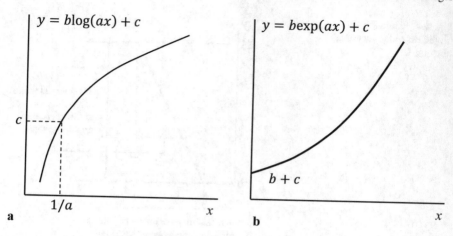

**Fig. 4.5** **a** Logarithmic and **b** exponential brightness mapping functions that respectively enhance low brightness value and high brightness value pixels

**Fig. 4.6** Piecewise linear brightness value modification function defined by a set of user-specified break points that commence at 0, 0 and increase monotonically to finish at $L$–1, $L$–1

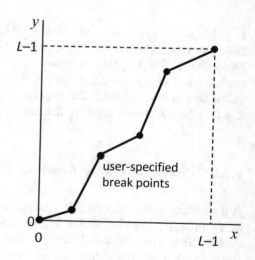

contrast so that an image histogram matches a preconceived shape, other than a simple mathematical modification of the original version. An important modified shape is the uniform histogram in which, in principle, each occurrence has the same height—i.e., each is equally likely. Such a histogram is associated with an image that utilises the available brightness levels equally and thus should give a display in which there is good representation of detail at all brightness values. In practice a perfectly uniform histogram cannot be achieved for digital image data. Nevertheless, the procedure following produces a histogram that is quasi-uniform on the average.

The method for producing a uniform histogram is known as *histogram equalisation*. It is useful, in developing the method to achieve this, if we regard the histograms as continuous curves as depicted in Fig. 4.7.[1] In this $h_i(x)$ represents the original image histogram (the "input" to the modification process) and $h_o(y)$ represents the histogram of the image after it has had its contrast modified (the "output" from the modification process). These functions are similar to probability density functions in statistics, in that the number of pixels covered by a brightness range $\delta x$ about brightness value $x$ in the original histogram is $h_i(x)\delta x$, and similarly for the modified histogram.

In Fig. 4.7 the number of pixels included the range $\delta y$ in the modified histogram must, by definition in the diagram, be equal to the number of pixels included in the range $\delta x$ in the original histogram. Given that $h_i(x)$ and $h_o(y)$ are density functions and that their values don't vary much over the ranges $\delta x$ and $\delta y$, this requires

$$h_o(y)\delta y = h_i(x)\delta x$$

so that in the limit for a small range of brightness values, i.e., $\delta x, \delta y \to 0$, we have

$$h_o(y) = h_i(x)\frac{dx}{dy} \tag{4.2}$$

We can use the last expression in two ways. First, if we know the original (input) histogram—which is usually always the case—and the function $y = f(x)$, we can determine the resulting (output) histogram. Alternatively, if we know the original histogram and the shape of the output histogram we want—"flat" in the case of contrast equalisation—then we can use (4.2) to help us find the $y = f(x)$ that will generate that result. Our interest here is in the second approach.

Since $y = f(x)$, and thus $x = f^{-1}(y)$, (4.2) can be expressed

$$h_o(y) = h_i\{f^{-1}(y)\}\frac{df^{-1}(y)}{dy}$$

which is a mathematical expression for the modified histogram in general.[2,3] To develop the brightness value modification procedure for contrast equalisation in particular, it is convenient to re-express (4.2) as

---

[1] This figure is adapted from K.R. Castleman, *Digital Image Processing*, 2nd ed., Prentice Hall, N. J., 1996.

[2] This requires the inverse $x = f^{-1}(y)$ to exist. For the contrast modification procedures used in remote sensing that is usually always the case. Should an inverse not exist, for example if $y = f(x)$ is not monotonic, Castleman, loc. cit., recommends that the original brightness value range $x$ be treated as a set of contiguous sub-ranges within each of which $y = f(x)$ is monotonic.

[3] If we apply this expression to the brightness value modification function for linear contrast enhancement, $y = f(x) = ax + b$ then $x = \frac{y-b}{a}$ so that $h_o(y) = \frac{1}{a}h\left(\frac{y-b}{a}\right)$. Relative to the original histogram, the modified version is shifted because of the effect of $b$, is spread or compressed depending on whether $a$ is greater or less than 1, and is modified in amplitude. The last effect only relates to the continuous function and cannot happen with discrete brightness value data.

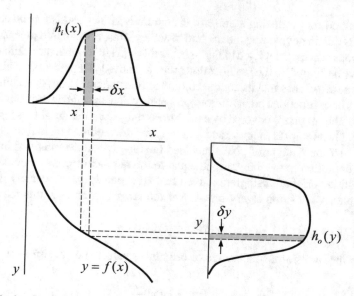

**Fig. 4.7** Setting the mathematical basis for histogram equalisation; the same numbers of pixels are represented in the two shaded areas

$$\frac{dy}{dx} = \frac{h_i(x)}{h_o(y)}$$

For a uniform histogram $h_o(y)$, and thus $1/h_o(y)$, should be constant—i.e., independent of $y$. This is a mathematical idealisation for real data, and rarely will we achieve a totally flat modified histogram in practice, as the examples in the following will show. However, making this assumption mathematically will generate for us the process we need to adopt for equalisation; thus

$$\frac{dy}{dx} = \text{constant} \times h_i(x)$$

so that

$$dy = \text{constant} \times h_i(x)dx$$

giving by integration

$$y = \text{constant} \int h_i(x)dx$$

How should we interpret the integral on the right-hand side of this last expression? In effect it is the continuous version of a *cumulative histogram* which, in discrete form, is a graph of the number of pixels below a given brightness value as a function

**Fig. 4.8** A simple histogram
and the corresponding
cumulative version

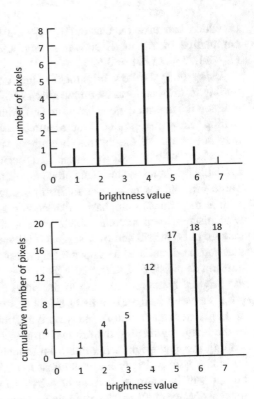

of brightness value, as shown in Fig. 4.8. The cumulative histogram is computed by summing the occurrence bars in the ordinary histogram from left to right.

If we call the cumulative histogram $C(x)$, then

$$y = \text{constant} \times C(x)$$

is the brightness value modification formula for histogram (contrast) equalisation. How do we find the value of the "constant"? We note, first, that the range of values of $y$ is required to be 0 to $L - 1$ to match the $L$ brightness values available in the image. Secondly, note that the maximum value of $C(x)$ is $N$, the total number of pixels in the image, as seen in Fig. 4.8. Thus, the constant needs to be $(L - 1)/N$ in order to generate the correct range for $y$. In summary, the brightness value mapping function that gives contrast equalisation is

$$y = \frac{L - 1}{N} C(x) \tag{4.3}$$

This equation is, in effect, a look-up table that can be used to move histogram occurrence bars to new brightness value locations to create the equalised product.

To illustrate the concept, consider the need to "flatten" the simple histogram shown in Fig. 4.9a. This corresponds to a hypothetical image with 24 pixels, each of which can take on one of 16 possible brightness values. The corresponding cumulative histogram is shown in Fig. 4.9b, and the scaling factor in (4.3) is $(L-1)/N = 15/24 = 0.625$.

Using (4.3) the new brightness value location of a histogram bar is given by finding its original location on the abscissa of the cumulative histogram ($x$) and then reading its unscaled new location ($y$) from the ordinate. Multiplication by the scaling factor then produces the required new value. It is likely, however, that this may not be one of the discrete brightness values available (for the output display device) in which case the associated occurrence bar is moved to the nearest available brightness value. This procedure is summarised for the example at hand in Table 4.1, and the new, quasi-uniform histogram is given in Fig. 4.9c.

It is important to emphasise that additional brightness values cannot be created with discrete data nor can pixels from a single brightness value in an original histogram be distributed over several brightness values in the modified version. All that can be done is to re-map the brightness values to give a histogram that is as uniform as possible. Occasionally that entails some neighbouring occurrences from the original histogram moving to the same new location and thus being superimposed, as seen in the example at brightness value 1.

In practice, the look up table created in Table 4.1 would be applied to every pixel in the image by feeding the original brightness value for a pixel into the table and reading the new brightness value from the table. Figure 4.10 shows an example of an image with a simple linear contrast modification compared to the same image, but in which contrast modification by histogram equalisation has been implemented. Many contrast changing techniques only give perceived improvement of detail on some image types and sometimes require all components of a colour composite image to be so processed before the change is noticeable.

It is not necessary to retain the same number of distinct brightness values in an equalised histogram as in the original. Sometimes it is desirable to have a smaller output set and thereby produce a histogram with (fewer) occurrences that are closer in height than would otherwise be the case. That can implemented by redefining $L$ in (4.3) to be the new total number of bars. Repeating the example of Table 4.1 and Fig. 4.9 for the case of $L = 8$ (rather than 16) gives the look up table of Table 4.2. Such a strategy would be an appropriate one to adopt when using an output device with a small number of brightness values (grey levels).

The maximum value of the cumulative histogram in (4.3) will be $N$, the total number of pixels in the image. The divisor $N$ in (4.3) has the effect then of normalising the cumulative histogram to unity. Multiplication by $L - 1$ as shown means the magnitude of the cumulative histogram goes from 0 to $L - 1$, as does its argument, and is therefore directly in the form of the look up table.

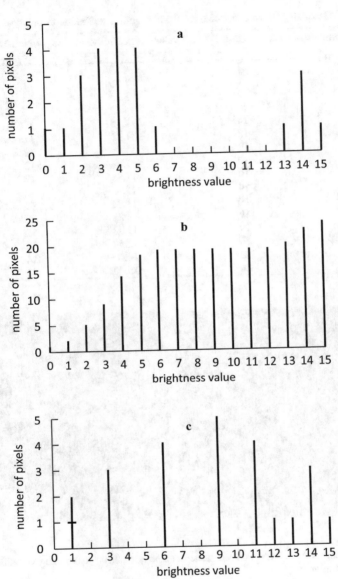

**Fig. 4.9** Histogram equalisation: **a** original histogram, **b** cumulative histogram used to produce the LUT in Table 4.1, and **c** the resulting quasi-uniform histogram

**Table 4.1** Generating the look up table for the histogram equalisation example of Fig. 4.9

| Original brightness value | Unscaled new value | Scaled new value | Nearest allowable brightness value |
|---|---|---|---|
| 0 | 1 | 0.63 | 1 |
| 1 | 2 | 1.25 | 1 |
| 2 | 5 | 3.13 | 3 |
| 3 | 9 | 5.63 | 6 |
| 4 | 14 | 8.75 | 9 |
| 5 | 18 | 11.25 | 11 |
| 6 | 19 | 11.88 | 12 |
| 7 | 19 | 11.88 | 12 |
| 8 | 19 | 11.88 | 12 |
| 9 | 19 | 11.88 | 12 |
| 10 | 19 | 11.88 | 12 |
| 11 | 19 | 11.88 | 12 |
| 12 | 19 | 11.88 | 12 |
| 13 | 20 | 12.50 | 13 |
| 14 | 23 | 14.40 | 14 |
| 15 | 24 | 15.00 | 15 |

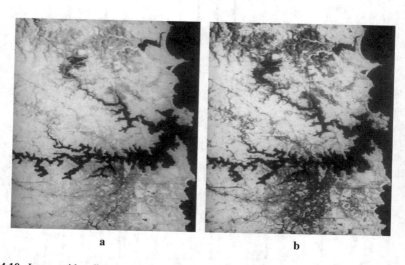

a                                              b

**Fig. 4.10** Image with a linear contrast stretch compared with the same image modified by histogram equalisation

## 4.4.2  Anomalies in Histogram Equalisation

Images containing large homogeneous regions will give rise to histograms with high occurrences at the corresponding brightness values. A particular example is a near infrared image with a large expanse of water. Because histogram equalisation creates a histogram that is uniform on the average, the equalised version of such an image will have poor contrast and little detail—quite the opposite to what is

**Table 4.2** The look up table for histogram equalisation using 8 output brightness values from 16 input brightness levels

| Original brightness value | Unscaled new value | Scaled new value | Nearest allowable brightness value |
|---|---|---|---|
| 0 | 1 | 0.29 | 0 |
| 1 | 2 | 0.58 | 1 |
| 2 | 5 | 1.46 | 1 |
| 3 | 9 | 2.63 | 3 |
| 4 | 14 | 4.08 | 4 |
| 5 | 18 | 5.25 | 5 |
| 6 | 19 | 5.54 | 6 |
| 7 | 19 | 5.54 | 6 |
| 8 | 19 | 5.54 | 6 |
| 9 | 19 | 5.54 | 6 |
| 10 | 19 | 5.54 | 6 |
| 11 | 19 | 5.54 | 6 |
| 12 | 19 | 5.54 | 6 |
| 13 | 20 | 5.83 | 6 |
| 14 | 23 | 6.70 | 7 |
| 15 | 24 | 7.00 | 7 |

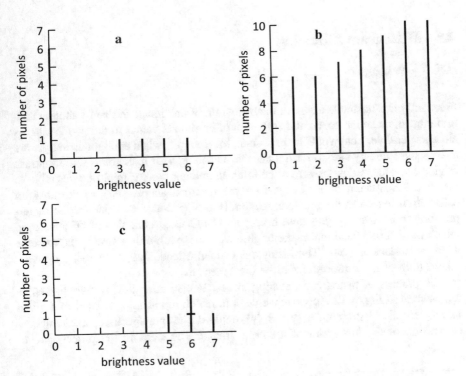

**Fig. 4.11** Illustration of anomalous histogram equalisation caused by very large occurrences in the original histogram: **a** original with a large number of pixels at 0, **b** cumulative histogram of the original, and **c** equalised histogram

intended. The reason for this can be seen in the simple illustration of Fig. 4.11. The cumulative histogram used as the look-up table for the enhancement is dominated by the large bar at brightness value 0. The resulting image would be mostly grey and white with little discrimination within the grey levels.

A similar situation happens when the automatic contrast enhancement procedure of Sect. 4.3.4 is applied to images with large regions of constant brightness. That can generate highly contrasting images on colour display systems; an acceptable display may require some manual adjustment of contrast, taking due regard of the abnormally large histogram bars.

To avoid the anomaly in histogram equalisation caused with the types of image discussed it is necessary to reduce the significance of the dominating occurrences. That can be done simply by arbitrarily reducing their size when constructing the look up table, remembering to take account of that in the scale factor in (4.3). Another approach is to produce the cumulative histogram and thus look-up table on a subset of the image that does not include any, or any substantial portion, of the dominating region. Yet another solution[4] is based on accumulating the histogram over "buckets" of brightness value—once a bucket is full to a pre-specified level, a new bucket is started.

## 4.5  Histogram Matching

### 4.5.1  Principle

Frequently it is desirable to match the histogram of one image to that of another and, in doing so, make the apparent distribution of brightness values in the two images as close as possible. That would be important, for example, when contiguous images are to be joined side by side to form a mosaic. Matching their histograms will minimise brightness value variations across the joint. In another case, it might be desirable to match the histogram of an image to a pre-specified shape, such as the uniform distribution treated in the previous section. It is often found of value in photointerpretation to have an image whose histogram is a Gaussian function of brightness, in which most pixels have mid-range brightness values with only a few in the extreme white and black regions. The histogram matching technique, now to be derived, allows both of those procedures to be implemented.

The process of histogram matching is best looked at as having two stages, as represented in Fig. 4.12. Suppose we want to match the histogram $h_i(x)$ of a given image, to the histogram $h_o(y)$; $h_o(y)$ could be a pre-specified mathematical expression or the histogram of a second image. The steps in the process are to

---

[4] See A. Hogan, A piecewise linear contrast stretch algorithm suitable for batch Landsat image processing. *Proc. 2nd Australasian Conference on Remote Sensing*, Canberra, Australia, 1981, pp. 6.4.1–6.4.4.

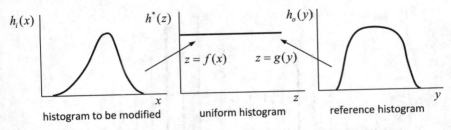

**Fig. 4.12** The stages in histogram matching, using the uniform histogram as a bridge

equalise the histogram $h_i(x)$, by the methods of the previous section, to obtain an intermediate histogram $h^*(z)$, which is then transformed to the desired shape $h_o(y)$.

If $z = f(x)$ is the transformation that flattens $h_i(x)$ to produce $h^*(z)$ and $z = g(y)$ is the operation that would flatten the reference histogram $h_o(y)$ then the overall mapping of brightness values required to produce $h_o(y)$ from $h_i(x)$ is

$$y = g^{-1}(z) \text{ with } z = f(x), \text{ or } y = g^{-1}\{f(x)\} \tag{4.4}$$

If the number of pixels and brightness values in $h_i(x)$ and $h_o(y)$ are the same, then the $(L-1)/N$ scaling factor in (4.3) will cancel in (4.4) and can be ignored in establishing the look up table that implements the contrast matching process. If the number of pixels is different, say $N_1$ in the image to be modified and $N_2$ in the reference image, and the number of brightness levels $L$ is the same in both images, then a scaling factor of $N_2/N_1$ will be included in (4.4). Scaling in (4.4) however is not a consideration if the cumulative histograms are normalised to some value such as unity, or as a percentage of the total number of pixels in an image.

## 4.5.2 Image to Image Contrast Matching

Figure 4.13 illustrates the steps implicit in (4.4) when matching source and reference histograms. In this section the reference histogram is that of a second image. The procedure is to use the cumulative histogram of the source image to obtain new brightness values in the manner of the previous section. We commence by reading the ordinate values corresponding to original brightness values entered on the abscissa of the source cumulative histogram. These ordinate values are then entered into the ordinate of the cumulative reference histogram and the final brightness values (for the occurrence bars of the source histogram) are read from the abscissa of the cumulative reference histogram; in this stage we see the cumulative reference histogram being used in reverse as indicated by the $g^{-1}$ operation in (4.4). The look up table for this example is shown in Table 4.3. Again, note that some of the new brightness values produced may not be in the available range; as before, they are adjusted to the nearest available value.

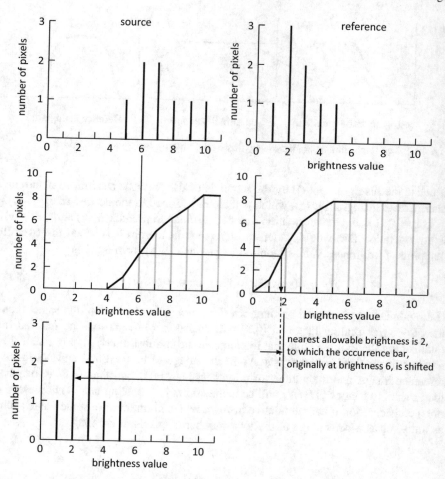

**Fig. 4.13** An example of the steps in histogram matching

An example using a pair of contiguous image segments is shown in Fig. 4.14. Because of seasonal differences the original contrasts are quite different. Using the cumulative histograms in (4.4) creates a good matching of image contrasts. Such a process, as noted earlier, is an essential step in producing a mosaic of separate contiguous images. Another step is to ensure geometric integrity of the join. That is done using the geometric registration procedures of Sect. 2.20.

**Table 4.3** Generating the look up table for contrast matching

| Source histogram brightness value $x$ | Intermediate (equalised) value $z$ | Brightness value matched to reference $y$ | Nearest allowable brightness value |
|---|---|---|---|
| 0 | 0 | 0 | 0 |
| 1 | 0 | 0 | 0 |
| 2 | 0 | 0 | 0 |
| 3 | 0 | 0 | 0 |
| 4 | 0 | 0 | 0 |
| 5 | 1 | 1 | 1 |
| 6 | 3 | 1.8 | 2 |
| 7 | 5 | 2.6 | 3 |
| 8 | 6 | 3 | 3 |
| 9 | 7 | 4 | 4 |
| 10 | 8 | 8 | 5 |
| 11 | 8 | 8 | 5 |

**Fig. 4.14 a** Contiguous Landsat multispectral scanner images showing contrast differences resulting from seasonal effects; the left-hand scene was recorded in autumn while the right-hand scene was recorded in summer, **b** the same image pair but in which the contrast of the autumn scene has been matched to that of the summer scene

### 4.5.3  Matching to a Mathematical Reference

In some applications it is of value to pre-specify the desired shape of an image histogram to give a modified image with a particular distribution of brightness values. To implement this, we take an existing image histogram and modify it according to the procedures of Sect. 4.5.1. However, the reference is now a mathematical function that describes the desired shape.

A particular example is to match an image histogram to a Gaussian or normal shape. That is referred to as applying a *Gaussian stretch* to an image. It yields a new image with few black and white regions and in which most detail is contained in the mid-grey range. Here the reference histogram is the normal distribution. Since a cumulative version of the reference is to be used, the cumulative normal distribution is required. To use that distribution in the contrast matching situation, either its ordinate has to be adjusted to the total number of pixels in the image to be modified or both cumulative histograms must have the same vertical scale. Further, the abscissa of the cumulative normal distribution needs to be scaled to match the maximum allowable brightness range in the image. That requires consideration to be given to the number of standard deviations of the Gaussian distribution to be contained in the total brightness value range, having in mind that the Gaussian function is continuous to $\pm\infty$. The mean of the distribution is placed usually at the mid-point of the brightness scale and the standard deviation is chosen such that the extreme black and white regions are three standard deviations from the mean. A simple illustration is shown in Fig. 4.15.

## 4.6  Density Slicing

### 4.6.1  Black and White Density Slicing

A point operation often performed with remote sensing image data is to map ranges of brightness value to particular shades of grey. In that way the overall discrete number of brightness values used in the image is reduced and some detail is lost. However, the effect of noise can be reduced and the image becomes segmented, or sometimes contoured, into sections of similar grey level, in which each segment is represented by a user-specified brightness. The technique is known as density slicing and finds value, for example, in highlighting bathymetry in images of water when penetration is acceptable. When used to segment a scalar image into significant regions of interest it is acting as a simple one-dimensional parallelepiped classifier (see Sect. 8.6).

The brightness value mapping function for density slicing is illustrated in Fig. 4.16. The thresholds in such a function are entered by the user. An image in which the technique has been used to highlight bathymetry is shown in Fig. 4.18b. Here differences in Landsat multispectral scanner visible imagery, with brightness values too low to be discriminated by eye, have been remapped to new grey levels to make detail apparent.

**Fig. 4.15** Modification of a histogram to a pseudo-Gaussian shape: **a** original histogram, which is the same as that in Fig. 4.9a, **b** cumulative normal, and **c** the histogram matched to the Gaussian reference, which also requires use of Fig. 4.9b

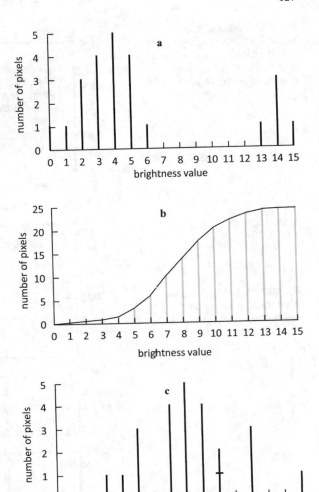

## 4.6.2  *Colour Density Slicing and Pseudocolouring*

A simple, yet lucid, extension of black and white density slicing is to use colours to highlight brightness value ranges, rather than simple grey levels. That is known as colour density slicing. Provided the colours are chosen suitably, it can allow fine detail to be made immediately apparent. It is a particularly simple operation to implement by establishing three brightness value mapping functions in the manner indicated in Fig. 4.17. Here one function is applied to each of the colour primaries used in the display device. An example of the use of colour density slicing, again for bathymetric purposes, is given in Fig. 4.18c. The technique is also used to give a colour rendition to black and white imagery. It is then usually called

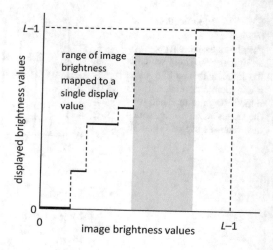

**Fig. 4.16** Example of a brightness value mapping function for black and white density slicing in which the transitions are user-specified

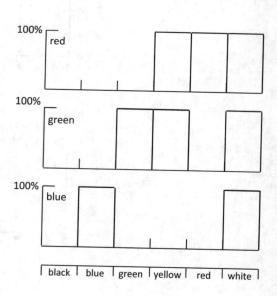

**Fig. 4.17** A look up table for colour density slicing, in this case into six strong colours

*pseudocolouring.* Where possible this uses as many distinct hues as there are brightness values in the image. In this way the contours introduced by density slicing are avoided. It is of value in perception if the hues used are graded continuously. For example, starting with black, moving through dark blue, mid blue, light blue, dark green, etc. and then moving to oranges and reds will give a much more acceptable coloured product than one in which the hues are chosen arbitrarily.

**Fig. 4.18** Illustration of
**b** black and white and
**c** colour density slicing to
enhance water detail; the
original image **a** is a visible
green plus near infrared
Landsat multispectral scanner
composite that was smoothed
to reduce sensor line striping

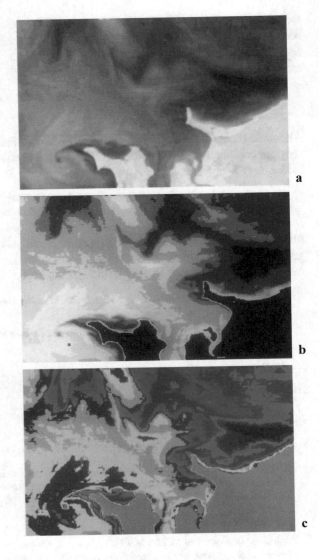

## 4.7  Bibliography on Radiometric Enhancement of Images

There are many books available that cover the essentials of radiometric enhancement and modification, including

K.R. Castleman, *Digital Image Processing*, 2nd ed., Prentice Hall, N.J., 1996.

R.C. Gonzalez and R.E. Woods, *Digital Image Processing*, 4th ed. Pearson Prentice-Hall, Upper Saddle River, N.J., 2018.

Gonzalez and Woods was first published in 1977 as Gonzalez and Wintz, and has been through several editions (and changes of author) since, each time revising and adding new material. There is a companion volume, showing how the various techniques can be implemented in Matlab®:

> R.C. Gonzalez, R.E. Woods and S.L. Eddins, *Digital Image Processing Using Matlab*®, 3rd ed., Gatesmark, Knoxville, 2020.

Although their motivation and perspectives are different from those in remote sensing, computer graphics texts often contain good coverage of image processing techniques. See, for example

> J.F. Hughes, A. van Dam, M. McGuire, D.F. Sklar, J.D. Foley, S.K. Feiner and K. Akeley, *Computer Graphics: Principles and Practice*, 3rd ed., Addison-Wesley, Boston, 2014.

Many books that treat digital processing and analysis in remote sensing more generally also contain good treatments on radiometric enhancement techniques, most of which have gone to later editions, including

> R.A. Schowengerdt, *Remote Sensing: Models and Methods for Image Processing*, 3rd ed., Academic, Burlington, Mass., 2006.

> J.R. Jensen, *Introductory Digital Image Processing: A Remote Sensing Perspective*, 4th ed., Prentice-Hall, Upper Saddle River, N.J, 2015.

One of the first treatments of digital image processing in the remote sensing context, now more of historical value, is

> J.G. Moik, *Digital Processing of Remotely Sensed Images*, NASA, Washington, 1980.

For examples of histogram equalisation and Gaussian contrast stretching see

> A. Schwartz, New techniques for digital image enhancement, *Proc. Caltech/JPL Conference on Image Processing Technology, Data Sources and Software for Commercial and Scientific Applications*, Pasadena, California, 3–5 Nov. 1976, pp. 2.1–2.12, and

> J.M. Soha, A.R. Gillespie, M.J. Abrams and D.P. Madura, Computer techniques for geological applications, *Proc. Caltech/JPL Conference on Image Processing Technology, Data Sources and Software for Commercial and Scientific Applications*, Pasadena, California, 3–5 Nov. 1976, pp. 4.1–4.21.

In order to enhance spatial detail a multicycle version of contrast enhancement can be used, in which the brightness value mapping function of (4.1) is cyclic. With this approach several sub-ranges of image brightness are each mapped to the full range of output brightness. The method is attributable to the following report

> P.S. Chavez, G.L. Berlin and W.B. Mitchell, *Computer Enhancement Techniques of Landsat MSS Digital Images for Land Use/Land Cover Assessment*. US Geological Survey, Flagstaff, Arizona, 1979.

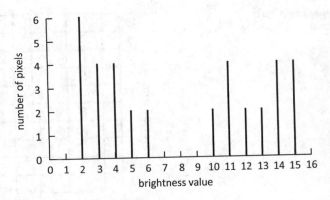

**Fig. 4.19** Histogram to be matched to a Gaussian

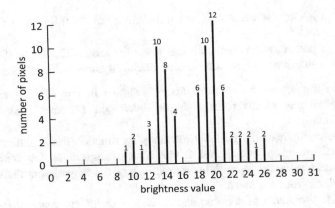

**Fig. 4.20** Histogram

## 4.8 Problems

4.1 One popular type of histogram modification is to match the histogram of an image to a Gaussian or normal function. Suppose a raw image has the histogram indicated in Fig. 4.19. Produce the look-up table that describes how the brightness values of the image should be changed if the histogram is to be mapped, as nearly as possible, to a Gaussian histogram with a mean of 8 and a standard deviation of 2 brightness values. Note that the sum of counts in the Gaussian reference histogram must be the same as that in the raw data histogram, or both should be normalised to unity.

4.2 The histogram of a particular image is shown in Fig. 4.20. Produce the modified version that results from:

**Fig. 4.21** Two dimensional image histogram

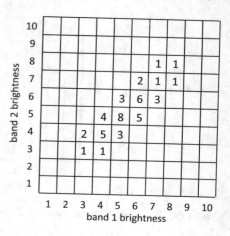

(i) a linear contrast stretch which makes use of the full range of brightness values

(ii) a piecewise linear stretch that maps the range (12, 23) to (0, 31) and

(iii) histogram equalisation (i.e., producing a quasi-uniform histogram).

4.3 The histogram for a two-band image is shown in Fig. 4.21. Determine the histogram that results from a simple linear contrast stretch on each band individually.

4.4 Derive mathematically the contrast mapping function that equalises the contrast of an image which has a Gaussian histogram at the centre of the brightness value range, with the extremities of the range being three standard deviations from the mean.

4.5 What is the shape of the cumulative histogram of an image that has been contrast (histogram) equalised? Can this be used as a figure of merit in histogram equalisation?

4.6 Clouds and large regions of clear, deep water frequently give histograms for near infrared imagery that have large high brightness value or large low brightness value occurrence bars respectively. Sketch typical histograms of these types. Using the material of Sect. 4.4 show how these histograms would be equalised and comment on the likely undesirable appearance of the corresponding contrast enhanced images. Show that the situation can be rectified somewhat by artificially limiting the large bars to values not greatly different from the heights of other bars in the histogram, provided the accompanying cumulative histograms are normalised to correspond to the correct number of pixels in the image.

4.7 Two images are to be joined side by side to form a mosaic for a particular application. To give the new, combined image a uniform appearance it is decided that the range and distribution of brightness levels in the first image should be made to match those of the second image, before they are joined. This is to be carried out by matching the histogram of image 1 to that of image

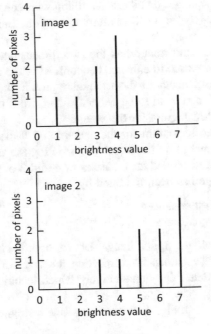

**Fig. 4.22** Histograms of images to be contrast matched

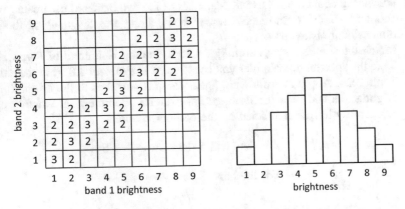

**Fig. 4.23** Two dimensional image histogram, and a triangular reference histogram

2 (the reference). The original histograms are shown in Fig. 4.22. Produce a look up table that can be used to transform the pixel brightness values of image 1 in order to match the histograms as nearly as possible. Use the look-up table to modify the histogram of image 1 and comment on the degree to which contrast matching has been achieved. Now repeat the exercise with the histogram of image 1 chosen as the reference.

4.8 Describe the advantages or otherwise in applying contrast enhancement if an image is to be analysed by (i) photointerpretation, and (ii) quantitative computer methods.

4.9 A particular two band image has the two dimensional histogram shown in Fig. 4.23. It is proposed to enhance the contrast of the image by matching the histograms in each band to the triangular profile shown. Produce look-up tables to enable each band to be enhanced and, from these, produce the new two-dimensional histogram for the image.

4.10 Plot the equalised histogram for the example of Table 4.2. Compare it with Fig. 4.9 and comment on the effect of restricting the range of output brightnesses. Repeat the exercise for the cases of 4 and 2 output brightness values.

4.11 Suppose an image has been modified by

   (i)  linear contrast enhancement
   (ii) histogram equalisation.

You have available the digital image data for both the original image and the contrast modified versions. By inspecting the data (or histograms) describe how you might determine quantitatively which technique was used in each case.

4.12 Repeat the example in Fig. 4.13 but reverse the histograms used as the source and reference.

4.13 Examine the two dimensional histogram of Fig. 4.21. Noting the clustering of the occurrences close to the diagonal through the histogram can you say something about the relative appearances of the two bands of the image. What about if all the occurrences were exactly along the diagonal of the two dimensional histogram?

4.14 Match the source histogram in Fig. 4.15 to the top histogram of Fig. 4.22.

4.15 Take the spectral domain plot you constructed in Problem 3.8 from the data in Table 2.2b. Apply the following linear contrast stretches to the original pixel brightnesses and redo the Problem 3.8 exercise. Comment on whether class separability has been affected by the contrast changes.

For the band 5 data   $y = 2x + 5$

For the band 7 data   $y = 3x + 4$

# Chapter 5
# Geometric Processing and Enhancement: Image Domain Techniques

**Abstract** Methods by which the geometric properties of an image can be modified are covered. These include reducing noise, enhancing edges or sharpening image detail, all of which are illustrated by examples. Those operations are shown to depend upon processing a neighbourhood of pixels about a central pixel of interest; this is identified as the spatial convolution operation. Most commonly, spatial convolution and thus the operations of smoothing and sharpening are implemented by the use of templates, kernels or filters defined over a neighbourhood about the pixel being processed. It is shown that the desired results are obtained by running the template over an image, pixel by pixel, and executing a defined operation between the template entries and pixel brightness values. Means are also presented for describing geometric properties such as texture and spatial correlation. Image morphological analysis is covered as a further example of template-based processing, in which image objects can be refined.

## 5.1 Introduction

This chapter presents methods that allow us to analyse or modify the geometric properties of an image. Our attention, first, is on techniques for filtering images to remove noise or to enhance geometric detail. We will then look at means by which we can characterise geometric properties like texture, and processes that allow us to analyse objects and shapes that appear in imagery.

Our methods here are called image domain techniques because the results are generated by working on the pixel brightness values directly. An alternative approach is based on the spatial frequency domain, in which the images are transformed first using Fourier and wavelet methods. Those are the subject of Chap. 7.

© The Author(s), under exclusive license to Springer Nature Switzerland AG 2022
J. A. Richards, *Remote Sensing Digital Image Analysis*,
https://doi.org/10.1007/978-3-030-82327-6_5

## 5.2   Neighbourhood Operations in Image Filtering

In contrast to the point operations used for radiometric modification of image data, techniques for geometric processing are characterised by operations over local neighbourhoods of pixels, as illustrated in Fig. 5.1. The result of a neighbourhood operation is still a modified brightness value for the single pixel at the centre of the neighbourhood; however, the new value is determined by the brightnesses of all the local neighbours rather than just the original brightness value of the central pixel alone.

The neighbourhood shown in Fig. 5.1 is square and of dimension $3 \times 3$. It is defined by a *template* or window that is laid over the image. In practice the template can be any shape and size and, as we shall see, that partly defines the outcome of the resulting operation; usually though it has an odd number of cells horizontally and vertically so that it has a natural centre to place over the image pixel being processed.

While the template has been introduced as a means for defining the image neighbourhood of interest it most often has numbers associated with each of its cells as seen in Fig. 5.2; those numbers contribute to the outcome of the operation. The result is most easily expressed in the $(m,n)$ coordinate system of the template. The new brightness value for the pixel $(i,j)$ at the centre of the template is

$$r(i,j) = \sum_m \sum_n \phi(m,n)t(m,n) \tag{5.1}$$

in which $\phi(m,n)$ are original pixel brightness values addressed according to their positions under the template; $t(m,n)$ is the corresponding template entry with its centre located at $(i,j)$. The origin for the template coordinates is at the upper left-hand corner. Sometimes the template entries collectively are referred to as its *kernel* and the template itself is sometimes called a *mask or filter*. Different kernels will implement different geometric modifications of the image. The coordinate system we have used here to address image pixels also has its origin at the top left-hand corner, consistent with referring to the first row as the "first line of pixels" as normally displayed.

If the template is filled entirely with zeros except for an entry of "1" at its centre, then application of (5.1) will leave the image unchanged. While this may seem to be a strange and unproductive operation it will feature in compound template-based operations designed to achieve specific image processing outcomes, as we will see later.

Equation (5.1) shows the template operating algebraically on the image pixels. Non-algebraic operations are also possible as we will see later in this chapter. We could write (5.1) more generally as

$$r(i,j) = \mathcal{T}_{m,n}\{\phi(i,j)\} \tag{5.2}$$

in which the operator $\mathcal{T}_{m,n}$ represents how the template processes the brightness values of the pixels covered by the template and centred on $(i,j)$. It could be an operator whose kernel selects the maximum value of the pixels or calculates their median value for example.

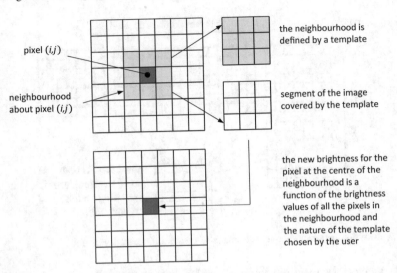

**Fig. 5.1** Neighbourhood processing defined by a template operator

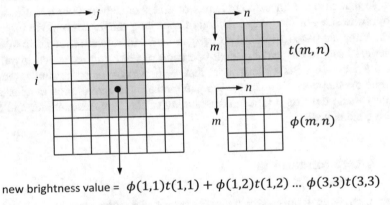

new brightness value = $\phi(1,1)t(1,1) + \phi(1,2)t(1,2) \ldots \phi(3,3)t(3,3)$

**Fig. 5.2** Demonstrating the operation of a numerical template of dimension $3 \times 3$ and the coordinate convention used, based on the template

  To produce a geometrically modified image the template is run over the original image line by line and column by column, centred on each pixel in turn, as illustrated in Fig. 5.3, creating new brightness values for the central pixels. Note that the border pixels cannot, in principle, be modified because they do not have a full set of neighbours. Often, they are left unprocessed and removed because there are usually sufficiently many pixels in the original image that the lost edges are not a problem. Sometimes artificial borders are created outside the image to allow the actual borders to be processed. They are used in the generation of new edge pixel values but are not themselves replaced by a template response. The external artificial borders can be made up by replicating the actual border pixels. A theoretically more

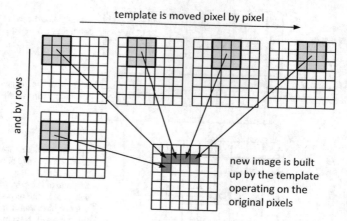

**Fig. 5.3** Moving the template over the image to create a processed version

correct method for handling the borders derives from sampling theory, which we treat in Chap. 7. That considers an image as though it were just one cycle within an infinitely repeating sequence of images in both the horizontal and vertical directions. In that case the lines and columns of pixels outside the borders are those on the opposite sides of the image. Although appealing this is rarely done in practice.

The template-based approach to the geometric modification of an image has a theoretical basis in the mathematical operation of *convolution*, which is treated in Sect. 5.8 following. Before looking at that we first examine some simple templates and see how they modify imagery. We will then look at convolution in a little detail so that we can develop a fuller understanding of the template method, and how it can be generalised.

## 5.3  Image Smoothing

Images often contain noise, which usually shows as random variations of the brightnesses of the pixels about the actual values associated with real features and objects in the scene. The noise can originate in the sensors that were used to acquire the data, from any communications channels used to carry the information, and from quantisation noise introduced when the raw signal data was digitised. One of the most common geometric processing operations carried out with imagery is to smooth the data to reduce the effect of noise. In this section we look at two approaches for smoothing.

### 5.3.1  *Mean Value Smoothing*

If the template has dimensions $M \times N$ and each of its entries has the value $1/MN$ then (5.1) becomes

$$r(i,j) = \frac{1}{MN} \sum_m \sum_n \phi(m,n) \tag{5.3}$$

which is the mean value of the pixels covered by the template. The new brightness for the pixel at the centre of the template is then the average brightness value in the neighbourhood of that pixel.

Figure 5.4 shows the result of running a $1 \times 3$ smoothing template over a single line of image data. The data contains an edge between bright (left-hand) and dark (right-hand) regions. Either side of the boundary the noise fluctuations have been reduced by the smoothing template. However, the edge has been smeared out over several pixels. That degradation can be avoided if a threshold is applied to the smoothing operation such that if the new brightness value is significantly different from its old value then the old value is used. That is implemented by the following algorithm. Let

$$\rho(i,j) = \frac{1}{MN} \sum_m \sum_n \phi(m,n) \tag{5.4a}$$

then $\quad r(i,j) = \rho(i,j) \quad \text{if } |\phi(i,j) - \rho(i,j)| < T$

$$\qquad\qquad = \phi(i,j) \quad \text{otherwise} \tag{5.4b}$$

where $T$ is a user-specified threshold, which could be determined from a knowledge or estimate of the signal to noise ratio of the image. Figure 5.4 shows how the use of such a threshold can preserve boundaries and edges.

Templates of any shape and size can be used. Larger templates give more smoothing and greater loss of fine detail. We describe fine detail in terms of *high spatial frequencies* (see Chap. 7). If the detail changes rapidly across or down the scene it is said to be high frequency, whereas low spatial frequency features vary slowly across or down an image. Horizontal templates will smooth horizontal noise and detail but leave detail in the vertical direction unaffected. In Fig. 5.5 the results of applying several different smoothing templates to the same image are shown to illustrate these points. Sometimes smoothing with a simple averaging template is called *box car filtering*.

## 5.3.2  *Median Filtering*

An alternative to applying thresholding for avoiding edge deterioration when smoothing is to use a median filter. In this case the kernel of the operator in (5.2) is designed to find the median brightness of the pixels covered by the template, which is then used as the new brightness for the central pixel. Whereas the mean of a set of numbers is their average, the median is that number which sits numerically in the middle of the set. For example, the median of the set 4, 6, 3, 7, 9, 2, 1, 8, 8 is 6, whereas its mean is 5.3. Figure 5.6 shows the effect of applying a median filter to

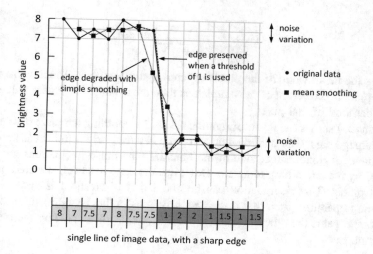

**Fig. 5.4** Illustration of $1 \times 3$ smoothing along a single line of image data showing the degradation of an edge, and its preservation if a threshold is used in the smoothing operation

the data of Fig. 5.4 compared with simple mean value smoothing. It can be seen that most of the boundary is preserved with the median operator.

An application for which median filtering is well suited is the removal of impulse like noise. That is because pixels corresponding to noise spikes are atypical among their neighbours and will be replaced by the most typical pixel in the neighbourhood. Figure 5.7 shows the value of median filtering on an image that contains impulsive black and white noise (sometimes called salt and pepper noise).

### 5.3.3  Modal Filtering

Another form of smoothing filter replaces the brightness of the central pixel by that most commonly occurring among the pixels covered by the template. That is referred to as the *mode* of the set and is illustrated in Fig. 5.8.

## 5.4  Sharpening and Edge Detection

The opposite to smoothing, in which high spatial frequency detail is reduced, is image sharpening in which detail, including edges and lines, is enhanced. Two techniques are in common use and are treated in the following. While the procedures to be covered sharpen all high spatial frequency detail, the examples used are based on edge detection and enhancement. We also consider edge detection explicitly in Sect. 5.5.

**Fig. 5.5** Mean value smoothing; in the last case only horizontal detail has been averaged

original

3x3 smoothed

5x5 smoothed

3x1 smoothed

## 5.4.1 Spatial Gradient Methods

There are several implementations of this approach, but all depend on calculating the local gradient in the brightness of an image in a pair of orthogonal directions. Let $\nabla_1$ and $\nabla_2$ be measures of how the brightness changes in those directions, at right angles to each other. We define the magnitude of the change as

$$|\nabla| = \sqrt{\nabla_1^2 + \nabla_2^2} \qquad (5.5a)$$

and its direction as

$$\angle\nabla = \tan^{-1}\{\nabla_2/\nabla_1\} \qquad (5.5b)$$

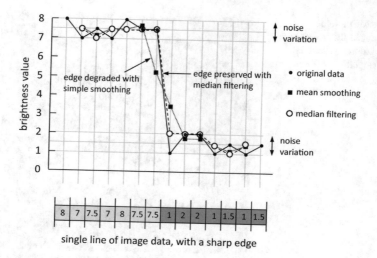

**Fig. 5.6** Demonstration of the value of median filtering for preserving edges while smoothing noise, compared with mean value smoothing

**Fig. 5.7** Median filtering of an image to reduce impulsive noise

**Fig. 5.8** Choosing the
(dominant) mode within the
template

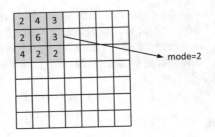

The direction will not concern us any further here since it does not feature in sharpening operations. It is of interest when finding contours in imagery and for deriving slope and aspect models.

Application of (5.5) will extract the high spatial frequency detail. It is usual to add that detail back to the original data so that the high frequencies are enhanced and the image appears sharper. Usually, a weighted combination is employed along the lines of

$$sharpened\ image = original\ image + c \times high\ spatial\ frequency\ detail$$

where $c$ is a constant that controls the degree of sharpening.

Different spatial gradient methods for sharpening are distinguished by how they estimate the orthogonal gradients $\nabla_1$ and $\nabla_2$. We now look at the most commonly adopted definitions.

### 5.4.1.1 The Roberts Operator

If the estimates of $\nabla_1$ and $\nabla_2$ are produced from the differences between the brightnesses of pixels diagonally separated, i.e.

$$\nabla_1 = \phi(i,j) - \phi(i+1,j+1) \tag{5.6a}$$

$$\nabla_2 = \phi(i+1,j) - \phi(i,j+1) \tag{5.6b}$$

then the Roberts Operator is generated from (5.5a). In principle, it computes the gradient at the point $(i + \frac{1}{2}, j + \frac{1}{2})$, although the result is generally associated with the pixel at $(i,j)$. The application of the Roberts Operator to the artificial image in Fig. 5.9a, which has sharp horizontal and vertical transitions in brightness, is shown in Fig. 5.9b.

### 5.4.1.2 The Sobel Operator

Perhaps a better spatial gradient estimator than the Roberts Operator is the Sobel Operator which generates $\nabla_1$ and $\nabla_2$ for detecting edges in the horizontal and vertical directions centred on the pixel at $(i,j)$. While it derives most of its weight from the pixels immediately above, below and to the sides of that at $(i,j)$, it is also sensitive to the pixels diagonally about the pixel of interest. The individual orthogonal components of gradient are

$$\nabla_1 = \{\phi(i+1,j-1)+2\phi(i+1,j)+\phi(i+1,j+1)\}$$
$$- \{\phi(i-1,j-1)+2\phi(i-1,j)+\phi(i-1,j+1)\} \quad (5.7a)$$

and
$$\nabla_2 = \{\phi(i-1,j+1)+2\phi(i,j+1)+\phi(i+1,j+1)\}$$
$$- \{\phi(i-1,j-1)+2\phi(i,j-1)+\phi(i+1,j-1)\} \quad (5.7b)$$

which are equivalent to applying the $3 \times 3$ templates.

$$\nabla_1 \equiv \begin{array}{|c|c|c|} \hline -1 & -2 & -1 \\ \hline 0 & 0 & 0 \\ \hline 1 & 2 & 1 \\ \hline \end{array} \qquad \nabla_2 \equiv \begin{array}{|c|c|c|} \hline -1 & 0 & 1 \\ \hline -2 & 0 & 2 \\ \hline -1 & 0 & 1 \\ \hline \end{array}$$

These templates are separately run over the image in the manner of Fig. 5.3 and the results combined in (5.5a).

Applying the Sobel Operator to the image of Fig. 5.9a generates the result shown in Fig. 5.9c. Again, a threshold would be specified to highlight major transitions in brightness. In this case the boundaries in the image are highlighted by two rows of pixels, one either side of the boundary.

### 5.4.1.3   The Prewitt Operator

If the weightings of 2 for the pixels directly above and below, and to either side of, the central pixel in the calculations of (5.7) are changed to 1 then the result is the Prewitt Operator, which has the template equivalents.

$$\nabla_1 \equiv \begin{array}{|c|c|c|} \hline -1 & -1 & -1 \\ \hline 0 & 0 & 0 \\ \hline 1 & 1 & 1 \\ \hline \end{array} \qquad \nabla_2 \equiv \begin{array}{|c|c|c|} \hline -1 & 0 & 1 \\ \hline -1 & 0 & 1 \\ \hline -1 & 0 & 1 \\ \hline \end{array}$$

Again, the templates are applied to the image and the results combined in (5.5a).

Note that for the Roberts, Sobel and Prewitt operators the template entries sum to zero. That is a feature of all operators that seek to highlight high spatial frequency detail such as lines and edges, including the Laplacian operator below. The reason

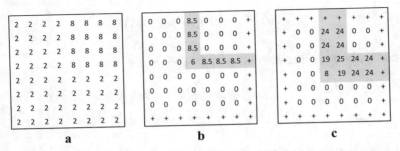

**Fig. 5.9  a** Artificial image with vertical and horizontal edges, **b** application of the Roberts Operator and **c** application of the Sobel Operator; the + entries represent indeterminate responses; since the procedure finds a local gradient, edges and other features can be identified by selecting a threshold above which they are kept and below which they are suppressed

is that there should be zero response when those templates are applied to areas of constant image brightness. In contrast, the entries in the templates used for smoothing encountered earlier sum to a non-zero value—unity in the case of mean value smoothing so that original image brightness values are preserved in homogeneous regions.

### 5.4.1.4   The Laplacian Operator

Examination of (5.6) shows that its two components are in fact the classical definition of the first derivative of the image brightness value function $\phi(i,j)$, albeit in the diagonal directions. It is, of course, possible to look at the first derivatives as incremental differences in the horizontal and vertical directions:

$$\nabla_1^{i,j} = \phi(i,j+1) - \phi(i,j) \tag{5.8a}$$

$$\nabla_2^{i,j} = \phi(i+1,j) - \phi(i,j) \tag{5.8b}$$

in which the superscripts have been added to indicate the pixel address *from which* the difference is computed.

The Laplacian operator[1] is based on estimating the *second* derivatives in the horizontal and vertical directions and summing them. The second derivative is the derivative of the first derivative—it is a measure of how rapidly the first derivative changes. Consider the horizontal direction first. The first derivative from the $j$th to the $(j+1)$th pixel is given by (5.8a). In a similar way the first derivative (the gradient) from the $(j-1)$th to the $j$th pixel is

---

[1] R. C. Gonzalez and R. E. Woods, *Digital Image Processing*, 4th ed. Pearson Prentice-Hall, Upper Saddle River, N.J., 2018.

$$\nabla_1^{i,j-1} = \phi(i,j) - \phi(i,j-1)$$

The change in those first derivatives is the difference

$$\nabla_1^{i,j} - \nabla_1^{i,j-1} = \phi(i,j+1) - \phi(i,j) - \phi(i,j) + \phi(i,j-1)$$
$$= \phi(i,j+1) - 2\phi(i,j) + \phi(i,j-1) \tag{5.9a}$$

which is the second derivative at the point $i,j$. Similarly, in the vertical direction the change from the $i$th to the $(i+1)$th pixel, less that from the $(i-1)$th to the $i$th pixel, i.e., the vertical second derivative at the point $i,j$, is

$$\nabla_2^{i,j} - \nabla_2^{i-1,j} = \phi(i+1,j) - 2\phi(i,j) + \phi(i-1,j) \tag{5.9b}$$

The Laplacian operator is the sum of (5.9a) and (5.9b) and given the symbol

$$\nabla^2\phi(i,j) = \phi(i-1,j) + \phi(i,j-1) - 4\phi(i,j) + \phi(i,j+1) + \phi(i+1,j) \tag{5.10}$$

which is equivalent to the template.

$$\nabla^2 \equiv
\begin{array}{|c|c|c|}
\hline
0 & +1 & 0 \\
\hline
+1 & -4 & +1 \\
\hline
0 & +1 & 0 \\
\hline
\end{array}
\quad \text{or} \quad
\begin{array}{|c|c|c|}
\hline
0 & -1 & 0 \\
\hline
-1 & +4 & -1 \\
\hline
0 & -1 & 0 \\
\hline
\end{array}$$

The second version, with the signs reversed, is often encountered in practice, particularly in the context of unsharp masking treated in the next section.

Sometimes the second derivatives in the two diagonal directions are added as well to give the templates.

$$\nabla^2 \equiv
\begin{array}{|c|c|c|}
\hline
+1 & +1 & +1 \\
\hline
+1 & -8 & +1 \\
\hline
+1 & +1 & +1 \\
\hline
\end{array}
\quad \text{or} \quad
\begin{array}{|c|c|c|}
\hline
-1 & -1 & -1 \\
\hline
-1 & +8 & -1 \\
\hline
-1 & -1 & -1 \\
\hline
\end{array}$$

## 5.4.2 Subtractive Smoothing (Unsharp Masking)

As seen in Figs. 5.4 and 5.5 a smoothed image retains low spatial frequency detail but has its high spatial frequency features such as edges and lines attenuated (unless thresholding has been employed to preserve sharp transitions). If a smoothed image were subtracted from the original, we would be left, therefore, with just the high spatial frequency detail—largely the edges and lines. That is illustrated in Fig. 5.10, using the data of Fig. 5.4. If the high frequency detail so detected is then added back to the original data, the result is an image in which the higher spatial frequencies, including edges and lines, are enhanced, as seen in Fig. 5.10c.

The difference operation can result in negative values as seen in Fig. 5.10b. Provided the result is not for display that is not a problem. If display is required, a fixed brightness value offset can be added to all pixels and the results rescaled to the display brightness range so that mid grey represents no difference. Positive differences will be brighter than the mid-range and negative differences will be darker. The same scaling approach is adopted for the final result in which the difference image has been added back to the original. Again, the result is scaled to fit within the allowed brightnesses of the display device.

Figure 5.11 shows the technique applied to each of 3 bands of a Landsat Multispectral Scanner colour composite image. As seen, the sharpened image has clearer high spatial frequency detail, although there is also a tendency for noise to be enhanced.

The technique commonly goes by the name of *unsharp masking*; the high frequency image that results from subtracting a smoothed version from the original is sometimes referred to as a mask. Although it apparently requires three steps (smoothing, subtraction, adding to the original) it can in fact be implemented by a single template that combines those steps into a single operation, as the template arithmetic shown in Fig. 5.12 demonstrates. Sharpening of a colour image product can also be performed using the pan sharpening procedure treated in Sect. 6.8.

## 5.5 Edge Detection

The operators defined in Sect. 5.4 effectively detect edges, although as we have seen they will enhance high spatial frequency detail in general. If the gradient estimators $\nabla_1$ and $\nabla_2$ are kept separate and not combined in the magnitude operation of (5.5a) they will individually detect edges in the vertical and horizontal directions. That can be seen by looking at the structures of the templates for the Roberts, Sobel and Prewitt operators.

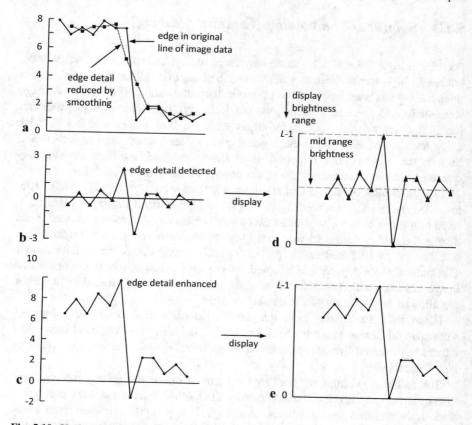

**Fig. 5.10** Unsharp masking: **a** original line of image data and smoothed version, **b** detected high spatial frequency detail, **c** sharpened version and **d, e** mapping the results of **b, c** to the allowed brightness range of a display device

It is possible to build templates with kernels that detect edges in diagonal directions as well. Suitable Roberts and Sobel diagonal edge detectors are.

Roberts

$\nabla_{45°} \equiv$

| +1 | +1 | 0 |
|----|----|----|
| +1 | 0 | -1 |
| 0 | -1 | -1 |

$\nabla_{-45°} \equiv$

| 0 | +1 | +1 |
|----|----|----|
| -1 | 0 | +1 |
| -1 | -1 | 0 |

Sobel

$\nabla_{45°} \equiv$

| +2 | +1 | 0 |
|----|----|----|
| +1 | 0 | -1 |
| 0 | -1 | -2 |

$\nabla_{-45°} \equiv$

| 0 | +1 | +2 |
|----|----|----|
| -1 | 0 | +1 |
| -2 | -1 | 0 |

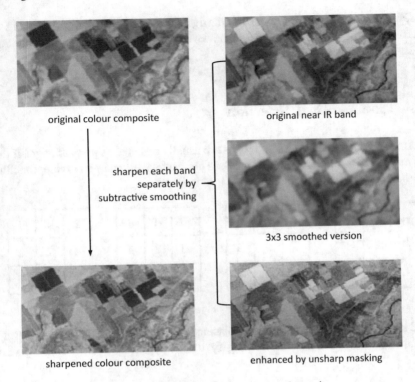

original colour composite

original near IR band

sharpen each band
separately by
subtractive smoothing

3x3 smoothed version

sharpened colour composite

enhanced by unsharp masking

**Fig. 5.11** Subtractive smoothing (unsharp masking) for image sharpening

this template leaves the
image unchanged, apart
from loss of borders

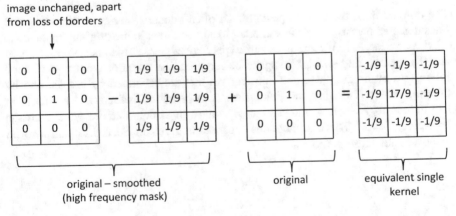

| 0 | 0 | 0 |
|---|---|---|
| 0 | 1 | 0 |
| 0 | 0 | 0 |

$-$

| 1/9 | 1/9 | 1/9 |
|---|---|---|
| 1/9 | 1/9 | 1/9 |
| 1/9 | 1/9 | 1/9 |

$+$

| 0 | 0 | 0 |
|---|---|---|
| 0 | 1 | 0 |
| 0 | 0 | 0 |

$=$

| -1/9 | -1/9 | -1/9 |
|---|---|---|
| -1/9 | 17/9 | -1/9 |
| -1/9 | -1/9 | -1/9 |

original – smoothed
(high frequency mask)

original

equivalent single
kernel

**Fig. 5.12** The steps in unsharp masking, and a single equivalent template

Other, more sophisticated forms of edge detector are available including the Marr-Hildreth and Canny operators[2]; however, they are not usually encountered in remote sensing applications.

## 5.6   Line and Spot Detection

Linear features such as rivers and roads are usually detected as pairs of edges if they are more than a pixel in width. If they are almost exactly one pixel wide, then they can be extracted with the templates.

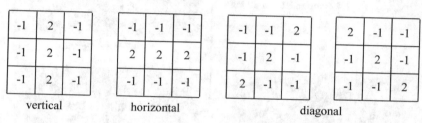

vertical       horizontal       diagonal

Examination of the form of the Laplacian operator in Sect. 5.4.1.4 suggests it is suited to the detection of spots in imagery, characteristically about a single pixel in size.

## 5.7   Thinning and Linking

The outputs from the above operators can often contains breaks in the edges and lines detected because of noise and other local variations in the original image data. Also, as we have seen, some lines and edges may be thicker than necessary. Should an edge or line map of high integrity be required, the operator outputs may need tidying up by linking edge and line features that are separated by breaks, and by thinning others. Thinning and linking are not operations encountered regularly in remote sensing. Good treatments can be found in standard, more general image processing texts.[3] The morphological operations given in Sect. 5.11 can also be employed for these purposes.

---

[2] Gonzalez and Woods, loc. cit.

[3] See Gonzalez and Woods, loc. cit., and K. R. Castleman, *Digital Image Processing,* 2nd ed., Prentice Hall, N.J., 1996.

## 5.8   Geometric Processing as a Convolution Operation

While the template method for smoothing and sharpening presented above is intuitive, it does in fact have a theoretical basis. In this section we develop the more formal convolution approach to geometric processing. Having done so we will understand more clearly the origins of the methods just treated and will be able to devise still more sophisticated templates. The convolution approach also allows a direct comparison of these image domain techniques with those based on the Fourier transformation that operate in the spatial frequency domain.

Convolution is a process that occurs surprisingly often in physical systems including optical imaging devices, imaging radar, and the transmission of signals through electronic circuits. We introduce the concept here for a signal passing though some unspecified system, and then generalise it to the case of images.

Suppose we have a function of time $f(t)$. If it is passed through some system,[4] such as an amplifier, a telephone network, a link to a satellite in space or similar, as shown in Fig. 5.13, then the signal $y(t)$ that emerges from the system is given by the *convolution integral*

$$y(t) = \int_{-\infty}^{\infty} f(\tau)h(t - \tau)d\tau \triangleq f(t) \circledast h(t) \qquad (5.11)$$

in which $h(t)$ describes the properties of the system through which the signal passes. It is sometimes called its *transfer function* but is more properly called its *impulse response*. The convolution operation also features strongly in the handling of digital signals and images as seen in the material of Sects. 7.6 and 7.10.

Note that there is a dummy variable $\tau$ in (5.11) which describes the various functions inside the integral and disappears once the integration is performed. Also, note that we adopt the symbol $\circledast$ to represent the convolution of two functions. Detailed treatments of the concept of convolution in the context of processing images will be found in Castleman[5] and Gonzalez and Woods.[6]

Equation (5.11) is the convolution integral in just one dimension—time. For images we have two independent variables—the spatial coordinates that describe pixel location. Consequently, we need a two dimensional version of the convolution integral.

Even though the pixels in a remotely sensed image are described by discrete image coordinates, for the moment assume that any point in a scene can be described by the pair of continuous variables x and y and that the scene properties are represented by the brightness function $\phi(x, y)$. Now suppose we image or view

---

[4] This assumes the system is linear, which is almost always the case for those we encounter in image processing. A linear system is one in which adding two inputs gives a response which is the sum of the individual responses.

[5] Castleman, loc cit.

[6] Gonzalez and Woods, loc. cit.

**Fig. 5.13** System model for demonstrating the convolution operation

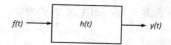

that scene through some form of filter or lens system that has the two dimensional impulse response $k(x, y)$. Then the resulting image is given by the two dimensional convolution

$$r(x, y) = \int\limits_{-\infty}^{\infty} \int\limits_{-\infty}^{\infty} \phi(u, v)k(x - u, y - v)dudv \qquad (5.12)$$

$k(x, y)$ is the impulse response of the system that operates on the image; we have given it the symbol $k$ to emphasise that it is a kernel, similar to the templates used earlier. It is also often referred to as the *system function*. If the system represents an imaging device, then $k(x, y)$ would be the device's point spread function; (5.12) then describes the degradation of the properties of the scene observed in an image. Alternatively, if $k(x, y)$ is one of the kernels specified earlier[7] for geometrically processing an existing image, then (5.12) is a mathematical specification of the sliding template operation illustrated in Fig. 5.3; in this case $\phi(x, y)$ and $r(x, y)$ represent the original and processed versions of the image respectively.

For digital image data $r(x, y)$, $\phi(x, y)$ and $k(x, y)$ are discrete rather than continuous and have limited ranges in the two coordinate directions, not extending to $\pm\infty$ as implied by the integrals in (5.12). It is necessary therefore to modify (5.12) so it can be applied to digital imagery. The integrals will be replaced by discrete sums and the continuous functions by their digital counterparts. If we let $(i, j)$ be the discrete versions of $(x, y)$ and $(m, n)$ be discrete values of the integration variables $(u, v)$ then (5.12) is written

$$r(i, j) = \sum_m \sum_n \phi(m, n)k(i - m, j - n) \qquad (5.13)$$

The sums are taken over all values of the dummy variables $m, n$. Strictly, the ranges of $m, n$ are the same as the ranges of $i, j$ and they have the same origin in the manner in which (5.13) is expressed. Similarly, the template is, in principle, the same size as the image and its origin is the same as the image origin.[8]

To see how (5.13) would be used in practice it is necessary to interpret the sequence of operations it incorporates. The negative signs on $(m, n)$ in (5.13) imply

---

[7] With one small modification, seen in the following paragraphs.

[8] There is a further subtlety that only becomes apparent when sampling theory is understood (see Chap. 7). When a continuous image is sampled to convert it to digital form it effectively becomes just one cycle in each dimension of an infinite periodic repetition of samples; the same is the case for the template. In practice, we often ignore that property. It is however important when considering the interaction with a template that is comparable in size to the image.

reflections through the origin in each of the $m$ and $n$ axes. That is equivalent to a rotation of the system function $k(i,j)$ by 180° before it is used in (5.13). We call the rotated function $t(m-i, n-j)$.

Next, (5.13) says that the value of $r(i,j)$ is given by multiplying, for all values of $m, n$, the image and the rotated system function and then summing the result. That gives the new value for the pixel at the specific location $(i,j)$—in other words just for one pixel in the new image. To get the modified values for all pixels we need to adopt in turn all values of $(i,j)$ in (5.13), which has the effect of relocating the template to each $(i,j)$, pixel by pixel, as in Fig. 5.3.

The pixel and template origins in (5.13) are in their respective upper left-hand corners. It is more convenient in practice, when we restrict the templates or kernels to a small neighbourhood about $(i,j)$, to address pixel and template entries by coordinates which have their origin at the upper left-hand corner of the finite sized template. That allows (5.13), with the kernel $k(i-m, j-n)$ replaced by the rotated system function $t(m-i, n-j)$, to be re-expressed in simple form as (5.1).

The templates of the previous sections, and in (5.1) in particular, are equivalent to the rotated version of the system function $k(m, n)$. Consequently, any geometric processing operation that can be modelled by convolution can also be expressed in template form. For example, if the point spread function of an imaging device is known then an equivalent template can be derived for computing what the image of a scene will look like, noting that the 180° rotation is important if the system function is not symmetric.

Templates for altering the geometric properties of an image can be chosen intuitively, as with smoothing in Sect. 5.3, or can be designed using a knowledge of filtering in the spatial frequency domain, based on the Fourier transformation discussed in Chap. 7.

## 5.9 Image Domain Techniques Compared with Using the Fourier Transform

Most geometric processing operations can be implemented using either the image domain procedures of this chapter or the Fourier transformation to be treated in Chap. 7. Which option to adopt depends on several factors, such as user familiarity and processing limitations. The Fourier transform method is more flexible and allows a much greater range of processing operations to be applied. Another consideration relates to processing time. This matter is pursued here in order to indicate, from a cost viewpoint, when one method might be chosen in favour of the other.

The Fourier transform spatial frequency domain process and the template approach both consist only of sets of multiplications and additions. No other mathematical operations are involved. It is sufficient, therefore, to make a time comparison based upon the number of multiplications and additions necessary to achieve a result. Here we will ignore the additions since they are generally faster than multiplications in most cases and also since they are comparable in number to the multiplications involved. For an image of $K \times K$ pixels, and a template of size $M \times N$, the total number of multiplications necessary to evaluate (5.1) for every image pixel (ignoring any difficulties with the edges of the image) is

$$N_C = MNK^2$$

From the material presented in Sect. 7.9 it can be seen that the number of (complex) multiplications required in the frequency domain approach is

$$N_F = 2K^2 \log_2 K + K^2$$

A processing time comparison is, therefore, given by

$$\frac{N_C}{N_F} = MN/(2\log_2 K + 1)$$

When this figure is below unity it is more economical to use the template operator approach. Otherwise, the Fourier transform method is more cost effective. That does not take into account program overheads, such as the bit shuffling required in the frequency domain approach and the added cost of complex multiplications; however, it is a reasonable starting point in choosing between the methods.

Table 5.1 shows values of $N_c/N_p$ for various image and template sizes, from which it is seen that, provided a $3 \times 3$ template will implement the operation required, it is always more cost-effective than processing based on the Fourier transformation. Similarly, a rectangular $3 \times 5$ template is more cost effective for practical image sizes. The spatial frequency domain technique is seen to be economical if very large templates are needed, although only marginally so for large images. Note, however, that the frequency domain method is able to implement processes not possible (or at least not viable) with template operators. Removal of periodic noise is one example. That is particularly simple in the spatial frequency domain but requires unduly complex templates or even nonlinear operators (such as median filtering) in the image domain. Nevertheless, the template approach is a popular one since often $3 \times 3$ and $5 \times 5$ templates are sufficient in many cases.

## 5.10 Geometric Properties of Images

The grey level histograms treated in Chap. 4 summarise the radiometric properties of images. We now look at a number of measures that characterise geometric structure. Image geometry is effectively characterised by the inter-relationships between pixels at different locations. For example, sets of pixels in a given neighbourhood may describe an object, such as a field or roadway, while repeating patterns of pixels define any texture-like qualities in a scene. We now look at both of those concepts.

**Table 5.1** Time comparison of geometric processing using templates compared with the Fourier transformation approach; this is based upon a comparison of multiplications (real or complex)

| Image size | Template size 3 x 3 | 3 x 5 | 5 x 5 | 5 x 7 | 7 x 7 |
|---|---|---|---|---|---|
| 128 x 128 | 0.60 | 1.00 | 1.67 | 2.33 | 3.27 |
| 256 x 256 | 0.53 | 0.88 | 1.47 | 2.06 | 2.88 |
| 512 x 512 | 0.47 | 0.79 | 1.32 | 1.84 | 2.58 |
| 1024 x 1024 | 0.43 | 0.71 | 1.19 | 1.67 | 2.33 |
| 2048 x 2048 | 0.39 | 0.65 | 1.09 | 1.52 | 2.13 |
| 4096 x 4096 | 0.36 | 0.60 | 1.00 | 1.40 | 1.96 |

## 5.10.1  Measuring Geometric Properties

When thinking about the geometric nature of an image a logical consideration is how related adjacent pixels might be. In agricultural regions, for example, it is highly likely that neighbouring pixels will be of the same ground cover type and thus will be similar in brightness in each of the recorded data channels. Also, road and river systems consist of sets of connected pixels of comparable brightness. We are then led to think about means by which we might describe the spatial relationships between pixels. In Chap. 8, when we look at image classification, we will do that using a set of conditional probabilities that describe the likelihoods of neighbouring pixels being from the same ground cover class. Here we are interested in measures that apply to the pixel brightness values in a single channel of image data.

A simple and obvious measure is to compare the brightnesses of pixels by taking their differences. Although logically we might look at adjacent pixels there is also value in comparing pixels further apart. Among other things that will lead us to definitions for texture. If we use the symbol $k \in \{i, j\}$ to represent either the row or column index of a pixel, then the brightness difference of two pixels from the same channel, spaced $h$ apart along a given row or down a given column, is

$$\phi(k) - \phi(k+h)$$

Usually, single pixel measures are not that helpful. Instead, we might be interested in the difference between *all* pixels in the image and their neighbours $h$ away as some sort of measure of similarity or correlation over a given separation. To do so we could average the brightness differences, but of course that risks the cancellation of positive and negative differences, so we average the squared distances instead, to get

$$\text{var} = \frac{1}{K} \sum_{k=1}^{K} \{\phi(k) - \phi(k+h)\}^2 \tag{5.14}$$

which will be recognised as a variance-like measure—in fact it is the variance in pixel brightness in the direction of interest; $K$ is the number of brightness pairs chosen for the computation. Often that might be all the available pixel pairs in that direction.

If we vary the separation $h$ then we can construct a graph that shows the variance (essentially how different the pairs of pixels are on the average) as a function of separation. The graph is called a *variogram*. Sometimes half the variance is plotted; we then have a *semivariogram*. As the semivariance increases the correlation or similarity of pixels on the average decreases. Conversely, highly correlated pixel pairs will exhibit small semivariance and will plot low on a semivariogram. If there is spatial periodicity in the landscape the semivariogram will reflect that behaviour.

Several properties can be derived from the semivariogram, which are best illustrated on the idealised form shown in Fig. 5.14; they include the *sill* (its asymptotic maximum value, if it exists), the *nuggett variance* (the extrapolated point of intersection with the ordinate), sometimes taken to indicate the noise properties of the image since it represents variance that is not related to the spatial properties of the scene, and the *range*, which is the lag or separation at which the sill is reached.

The image of Fig. 5.15 is a Landsat ETM+ image of a region surrounding Canberra, Australia. Figure 5.16 plots horizontal semivariograms for the four regions indicated.[9] Note that grassland exhibits the least variance and thus the greatest correlation over the image, while the suburban region is least correlated.

### 5.10.2  Describing Texture

Many of the surfaces we observe about us in our day-to-day life are textured—that is they seem to have some form of quasi-repeating pattern that readily tells us whether the surface is moderately smooth, rough but nominally repetitive (such as a carpet) or rough and strongly repetitive, such as a piece of linen cloth. The same is the case for satellite imagery. Grassland, crops and forests all appear differently textured to our observation—that is they are composed of some form of natural scale that seems to repeat on the average. While we can describe textures qualitatively, and can certainly discriminate among them using our own eyes, quantitative characterisation of texture is not so simple. We have to start by finding a

---

[9] These results were produced by Associate Professor Xiuping Jia of the University of New South Wales.

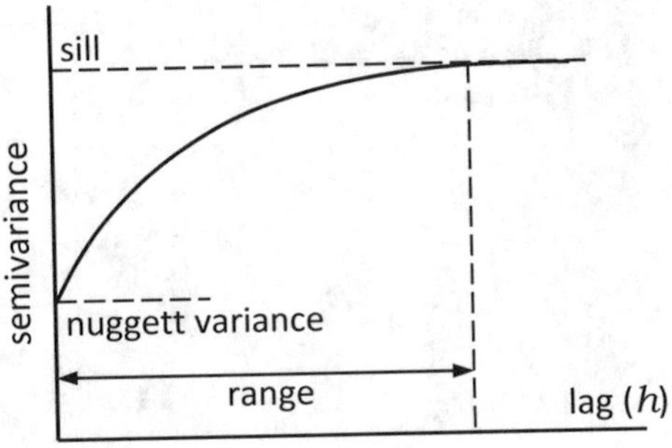

**Fig. 5.14** An idealised
semivariogram when there is
no spatial periodicity present

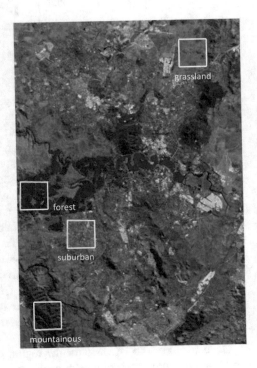

**Fig. 5.15** Portion of a
Landsat ETM+ image near
Canberra, Australia showing
four regions used to compute
geometric properties

measure that captures the spatial properties of a scene which reveal texture. Those
properties have to do with repetition, on the average.

A long-standing spatial measure is the grey level co-occurrence matrix (GLCM)
defined in the following way.[10] To make the development simple, imagine we want
to detect a component of texture just in the horizontal direction in a particular
region. To do that we could see how often two particular grey levels occur in that
direction in the selected region, separated by a specified distance. Let $g(\phi_1, \phi_2|h, \theta)$

---

[10] R. M. Haralick, Statistical and structural approaches to texture, *Proc. IEEE*, vol. 67, no. 5, May
1979.

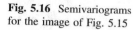

**Fig. 5.16** Semivariograms
for the image of Fig. 5.15

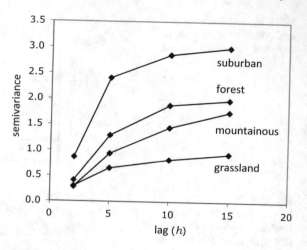

be the relative occurrence of pixels with grey levels $\phi_1$ and $\phi_2$ spaced $h$ apart in the
$\theta$ direction—for the moment chosen as the horizontal direction. Relative occurrence
is the number of times each grey level pair occurs, divided by the total possible
number of grey level pairs. The GLCM for the region of interest is the matrix of
those measurements over all grey level pairs; there will be as many GLCMs as there
are values of $h$ and $\theta$. If the pixels are described by $L$ possible brightness values
then the matrix will be $L \times L$. Given that $L$ can be quite large, the brightness value
range is often reduced by looking for co-occurring pairs of brightness value ranges.
Alternatively, the dynamic range of the data can be reduced (e.g., from 10 to 5 bit
data) before processing. We generally look for similar behaviour in other directions,
such as vertically and diagonally, in which case there would be four matrices for the
chosen values of $h$. Often $h$ is used as a variable to see whether texture exists on a
local or more regional scale. Sometimes the GLCMs computed for various values of
$\theta$ are kept separate to see whether the texture is orientation dependent; alternatively,
they can be averaged on the assumption that texture will not vary significantly with
orientation.

Once we have the GLCMs for the region of interest it is then usual to set up a
single metric computed from each matrix to use as a texture descriptor. A range of
measures is possible one of which is to describe the *entropy* of the information
contained in the GLCM, defined by

$$H(h, \theta) = -\sum_{\phi_1=1}^{L} \sum_{\phi_2=1}^{L} g(\phi_1, \phi_2 | h, \theta) \log\{g(\phi_1, \phi_2 | h, \theta)\} \qquad (5.15)$$

Entropy will be highest when all entries in the GLCM are equiprobable—when the
image is not obviously textured—and will be low when there is a large disparity in
the probabilities, as happens when significant texture is present. Another measure is

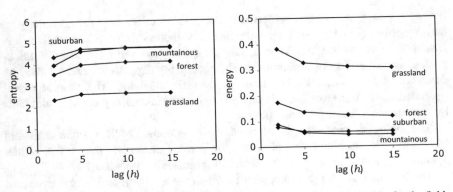

**Fig. 5.17** Entropy and energy computed from the grey level co-occurrence matrices for the fields in Fig. 5.15

*energy* which is the sum of the squared elements of the GLCM. It will be small when the GLCM elements are small, indicating low texture.

Figure 5.17 shows graphs[11] of entropy and energy for the four regions indicated in Fig. 5.15, within which just the horizontal GLCMs were computed for a range of values of lag, $h$. Those calculations used just the first ETM+ band—the visible blue, which was reduced in dynamic range to 5 bits before any calculations were performed. Two points are noteworthy. First, entropy increases, and energy decreases, with lag, indicating that the texture is falling away at larger spacing. Secondly the four cover types chosen—grass, forest, mountainous and suburban—are separable by their texture, with grassland exhibiting the strongest texture. The suburban and mountainous regions are seen to be low in texture by comparison and are comparable to each other for the range of scales chosen. Note that entropy and energy behave oppositely to each other as expected.

## 5.11   Morphological Analysis

While most interest in digital processing of remotely sensed imagery relates to enhancing radiometric and geometric properties, or to interpreting cover types using the mapping and labelling procedures to be treated in later chapters, there are occasions when we are interested in specific objects. Those objects will be defined by connected groups of pixels and could represent agricultural fields, river systems and road networks, or the building blocks of an urban scene.

Often those objects are not well defined. The image of an agricultural field may have less regular boundaries than is the case in reality because of system noise, or

---

[11] These results were produced by Associate Professor Xiuping Jia of the University of New South Wales.

limitations in the analytical procedures used to identify the field. Similarly, a homogeneous field may end up with incorrect inclusions of other cover types. Roads and river networks, likewise, might exhibit gaps and have variable thicknesses, again because of limitations in the processing algorithms, such as the line detectors of Sect. 5.6, and the nature of the data available. The morphological operations to be introduced in this section are helpful in cleaning up such problems by operating on the objects themselves.

Morphological processing is a template-based operation of the form described in (5.2). As with the template operations we looked at earlier, the choice of the operator $T_{m,n}$ establishes how the object in an image is modified.

The images that contain the objects in which we are interested in this treatment have binary brightness values for each pixel. Said another way, the background has one value and all the pixels that define the object have a different brightness value. Figure 5.18 contains such an object, which might be a field. Whether it consists of a single field or a long narrow field adjacent to a larger rectangular field is hard to discern because of the join between what could be two objects; that may be the result of the processing operation that led to the object(s) being identified in the image in the first place. Also shown in Fig. 5.18 is an elongated object that emulates a river which might have been extracted from an image using a sharpening template.

In morphological processing the template operation $T_{m,n}$ is defined in terms of a *structuring element* (SE), which is effectively a template in which the elements are present or not present, often represented respectively by template entries of 1 or 0. As with the geometric processing operations considered earlier, the SE is placed over each image pixel in turn, similar to the process shown in Fig. 5.3. For each location, the result of applying the structuring element to the image is a decision as to whether the pixel under the centre of the SE is a member of the object or not. That decision will depend on how the entries in the structuring element are used with respect to the object pixels being examined. This is a logical, rather than an algebraic, operation.

Note that there is no concept of generating a new brightness value for the pixel. Instead, the outcome of the operation is whether the pixel is a member of the set of

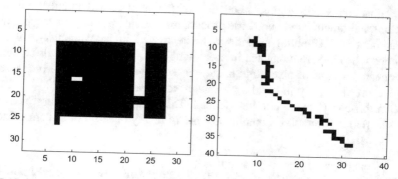

**Fig. 5.18** Two objects to be used to demonstrate morphological operations

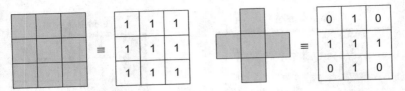

**Fig. 5.19** Examples of 3 × 3 structuring elements

pixels that define the object, or a member of the set of pixels that define the background. That leads us naturally to consider morphological processing in the language of set theory.

The structuring element can, in principle, be any shape or size. We will choose a 3 × 3 example to illustrate the essential points about morphological processing, but it will become obvious how to use other variants. Two 3 × 3 examples are shown in Fig. 5.19. Both can be represented as a box with binary elements. In image processing it is common to define the SE by rectangles and to indicate by 1 those cells which take part in the morphological processing and by 0 those cells which do not, as shown.

We will now introduce some specific processing operations.

### 5.11.1  Erosion

As its name implies this operation has the effect of eroding, and thus reducing the size of, an object. It has the particular advantage that it can help to reduce ragged edges. It is defined by deciding that a pixel is part of an object if the SE, when centred on that pixel, is completely enclosed by the object. As illustrated in Fig. 5.20, if any part of the SE is outside the object, even though centred on an object pixel, that pixel is removed from the object. Thus, the size of the object is reduced.

We now express this operation in the notation of set theory. Let $\mathbf{S}$ represent the set of members of the structuring element and $\mathbf{O}$ represent the set of pixels from which the object is composed. If the structuring element is centred on the image pixel at $(i,j)$ then we represent that by adding the pixel address as subscripts to the SE, viz. $\mathbf{S}_{i,j}$. If, at a particular location, the structuring element is completely inside the object then it can be said to be a sub-set of the object, which is written $\mathbf{S}_{i,j} \subseteq \mathbf{O}$. The eroded object is the set of pixels, with addresses $(i,j)$ that satisfy that subset condition. If we let $\mathbf{E}$ represent the set of pixels that describe the eroded object, then we express that object as

$$\mathbf{E} = \mathbf{O} \ominus \mathbf{S} = \{i,j | \mathbf{S}_{i,j} \subseteq \mathbf{O}\} \qquad (5.16)$$

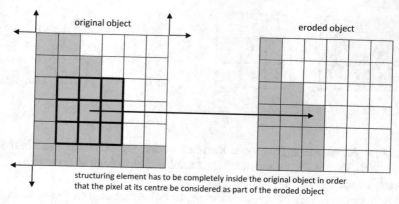

**Fig. 5.20** Illustrating the erosion operation; the image with the object in it is assumed to extend beyond the sample shown, with continuity in brightness value, as indicated by the corner arrows

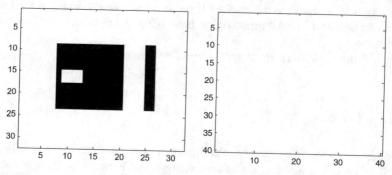

**Fig. 5.21** Erosion of the objects of Fig. 5.18 using a 3 × 3 structuring element

In words, this equation says that the eroded object is the set of points $(i,j)$ that, when used as the centre of the SE, satisfy the condition that the SE is a sub-set of (is completely enclosed within) the original object. The symbol $\ominus$ is used to represent erosion.[12] Figure 5.21 shows the effect of eroding the objects of Fig. 5.18 with a square 3 × 3 SE. As seen, the outcome is a greatly reduced object, with protrusions and thin linkages reduced. In the case of the river object, since its width is no greater than 3 pixels at any point, it is totally eroded away.

## 5.11.2  Dilation

Dilation has the opposite effect on an object to erosion; it has the tendency to grow the object's size and to fill in holes. It is defined by deciding that a pixel is part of an

---

[12] Castleman uses $\otimes$, which some other authors use for a thinning operation.

object if the SE, when centred on that pixel, partly overlaps the object. This is illustrated in Fig. 5.22, showing that the size of the object is expanded. How can we write this in set notation? We start by saying that the only SE centre locations that are not part of the dilated object are those for which the SE lies entirely outside the original object. If the set of SE members lies outside the set of object members, then their intersection will be the null set. In dilation we are happy to accept any location as part of a dilated object provided the SE has a non-null intersection with the original object. Using the same set notation as above, but calling the set of pixels that describe the dilated object **D**, then

$$\mathbf{D} = \mathbf{O} \oplus \mathbf{S} = \{i, j | \mathbf{S}_{i,j} \cap \mathbf{O} \neq \emptyset\} \tag{5.17}$$

in which $\emptyset$ is the null set. The symbol $\oplus$ is used to signify dilation. Figure 5.23 shows the effect of dilating the objects of Fig. 5.18 with a square $3 \times 3$ SE. As seen, the outcome is a greatly expanded object, with small holes closed up. The gaps in the river object are closed but at the expense of a broadened line overall.

### 5.11.3 Opening and Closing

Erosion and dilation are not the inverse of each other. In other words, the original objects cannot be recovered after dilation by applying an erosion operation, and vice-versa. However, the sequential application of erosion and dilation, or the sequential application of dilation and erosion, give interesting operations in their own right.

Erosion followed by dilation is referred to as *opening*, while dilation followed by erosion is called *closing*. The reasons for these names will be apparent from the

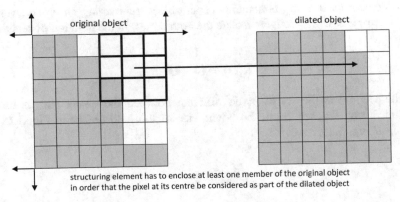

**Fig. 5.22** Illustrating the dilation operation; the image with the object in it is assumed to extend beyond the sample shown, as indicated by the corner arrows

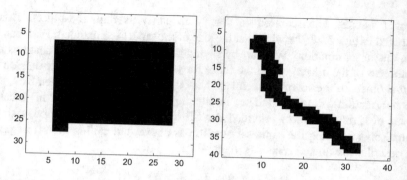

**Fig. 5.23** Dilation of the objects of Fig. 5.18 using a 3 × 3 structuring element

results of their application. Using these definitions, we can write the set of pixels (the object) that results from an opening operation, in which the bracketed operation is performed first, as

$$\mathbf{P} = \mathbf{O} \circ \mathbf{S} = (\mathbf{O} \ominus \mathbf{S}) \oplus \mathbf{S} \tag{5.18}$$

and the object that results from a closing operation as

$$\mathbf{C} = \mathbf{O} \cdot \mathbf{S} = (\mathbf{O} \oplus \mathbf{S}) \ominus \mathbf{S} \tag{5.19}$$

The results of applying these to the objects in Fig. 5.18 are shown in Fig. 5.24.

### 5.11.4 Boundary Extraction

Since erosion shrinks the boundaries of an object, subtracting an eroded version from its original will effectively isolate the boundaries. In the notation of sets this is written

$$B(\mathbf{O}) = \mathbf{O} - \mathbf{O} \ominus \mathbf{S}$$

in which $B(\mathbf{O})$ is the set of pixels different between the object and its eroded version. Using this approach, the boundaries of the field object in Fig. 5.18 are shown in Fig. 5.25.

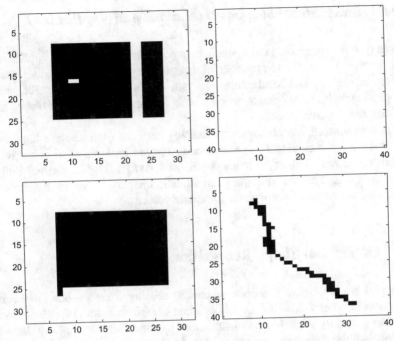

**Fig. 5.24** Result of applying opening (top) and closing (bottom) morphologic operations to the images of Fig. 5.18

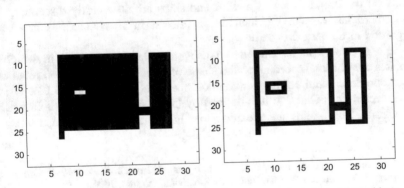

**Fig. 5.25** Boundary extraction by subtracting an eroded version from the original object

### 5.11.5   *Other Morphological Operations and Applications*

A range of other morphological operations is possible and finds wide application in fields in which visual interpretation of imagery is common, including medical imaging, astronomy and handwriting analysis.[13] It is also possible to set up morphological techniques to operate on grey scale images, as against the binary image data considered here.

In remote sensing, morphological processing has been extended to a high degree of sophistication, including its application in detailed analysis of high spatial and spectral resolution imagery.[14] It has been used successfully to extract building shapes from high spatial resolution panchromatic imagery,[15] and to classify urban land cover, where again shape is an important feature.[16]

## 5.12   Object and Shape Recognition

In the past the analysis of shapes in remote sensing imagery has not been as common as in other fields such as robotics and computer vision, often because the resolution generally available was insufficient to define shape with any degree of precision, and the algorithms did not exist to do object recognition easily. However, with pixels of the order of 1 m resolution and better now available, shapes such rectangular and circular pivotal irrigation fields in agriculture, urban features, and objects on the ground such as aircraft and ships are now easily discerned. As a result, there has been a big increase in interest in shape analysis and object recognition in the past decade or so.

Shape analysis can be carried out using template techniques, in which the templates are chosen according to the shape of interest. The operation required is one of correlation and not the convolution operation of (5.1) and (5.13). We will meet correlation in Chap. 7; it is defined by (5.13) but with additions in place of subtractions. That yields an operation that has a maximum response when the

---

[13] See Gonzalez and Woods, loc. cit., and R. Berry and J. Burnell, *Handbook of Astronomical Image Processing*, 2nd ed., William Bell, Inc., Richmond Virginia, 2006.

[14] See M. Pesaresi and J. A. Benediktsson, A new approach for the morphological segmentation of high resolution satellite imagery, *IEEE Transactions on Geoscience and Remote Sensing*, vol. 39, no. 2, February 2001, pp. 309–320, and P. Soille and M. Pesaresi, Advances in mathematical morphology applied to geosciences and remote sensing, *IEEE Transactions on Geoscience and Remote Sensing*, vol. 40, no. 9, September 2002, pp. 2042–2055.

[15] N. L. Gavankar and S. K. Chosh, Automatic building footprint extraction from high-resolution satellite image using mathematical morphology, *European Journal of Remote Sensing*, vol. 51, no. 1, 2018, pp. 182–193.

[16] L. T. Tsoeleng, J. Odini and P. Mhangara, A comparison of two morphological techniques in the classification of urban land cover, *Remote Sensing*, Vol. 12, 2020, https://doi.org/10.3390/rs12071089.

kernel matches as closely as possible the underlying segment of image data in shape, size and orientation. A difficulty with this simple approach, which as a consequence renders the technique of limited value in practice, is the need to match shape, size and orientation exactly.

However, when convolution forms the basis of neural networks, as we will see in Sect. 8.21, the whole field of object recognition and analysis in remote sensing opens up. Shape and object analysis is almost fundamental to the development and application of the convolutional neural network (CNN),[17] which is now used routinely in fields such as picture analysis and handwriting recognition.

Although the theory of CNNs is not covered until Chap. 8, examples of their application to object recognition in remote sensing can be seen in Liu et al.[18]

Other methods have been used in the past but are now largely of historical value. They include the adoption of shape factors, moments of area and Fourier transforms of shape boundaries. In each of these the shape must first be delineated from the rest of the image. That is achieved by histogram slicing (to separate objects from backgrounds), quantitative analysis and edge and line detection processes.[19]

## 5.13   Bibliography on Geometric Processing and Enhancement: Image Domain Techniques

Many of the books that cover radiometric enhancement of images also include good treatments of geometric processing. They include.

K.R. Castleman, *Digital Image Processing*, 2nd *ed., Prentice Hall, N.J., 1996.*

R.C. Gonzalez and R.E. Woods, *Digital Image Processing*, 4th ed. Pearson Prentice-Hall, Upper Saddle River, N.J., 2018.

R.A. Schowengerdt, *Remote Sensing: Models and Methods for Image Processing*, 3rd ed., Academic, Burlington, Mass, 2006.

J.R. Jensen, *Introductory Digital Image Processing: A Remote Sensing Perspective*, 4th ed., Prentice-Hall, Upper Saddle River, N.J, 2015.

The classic reference on texture analysis and processing is.

R.M. Haralick, Statistical and structural approaches to texture, *Proc. IEEE.*, vol. 67, no. 5., May 1979, pp. 786–802.

---

[17] I. Goodfellow, Y. Bengio, and A. Courville, *Deep Learning (Adaptive Computation and Machine Learning series)* The MIT Press, Cambridge Mass., 2016.

[18] Y. Liu, H-Y Cui, Z. Kuang and G-Q Li, Ship detection and classification on optical remote sensing images using deep learning, *4th Annual Conference on Information Technology and Applications*, ITM Web of Conferences, vol. 12, 05012, 2017.

[19] See S. Loncaric, A survey of shape analysis techniques, *Pattern Recognition*, vol. 31, no. 8 August 1988, pp. 983–1001.

For a comprehensive overview of recent advances in morphological processing applied to remote sensing images, and which includes a good reference list of salient contributions in morphology, see.

M. D. Mura, J.A. Benediktsson, J. Chanussot and L. Bruzzone, The evolution of the morphological profile: from panchromatic to hyperspectral images, in S. Prasad, L.M. Bruce and J. Chanussot, eds., *Optical Remote Sensing: Advances in Signal Processing and Exploitation Techniques*, Springer, Berlin, 2011.

Castleman, loc. cit., contains a short overview of common shape analysis methods. While the application of convolutional neural networks to shape and object recognition will be found in.

I. Goodfellow, Y. Bengio, and A Courville, *Deep Learning (Adaptive Computation and Machine Learning series)* The MIT Press, Cambridge Mass., 2016.

Many tutorials on object detection and recognition will be found through web searching.

## 5.14  Problems

5.1. The template entries for line and edge detection sum to zero whereas those for smoothing do not. Why?

5.2. Repeat the example of Fig. 5.10 but by using a $[1 \times 5]$ smoothing operation in part (a), rather than $[1 \times 3]$ smoothing.

5.3. Repeat the example of Fig. 5.10 but by using a $[1 \times 3]$ median filtering operation in part (a) rather than $[1 \times 3]$ mean value smoothing.

5.4. An alternative smoothing process to median and mean value filtering using template methods is modal filtering. Apply $[1 \times 3]$ and $[1 \times 5]$ modal filters to the image data of Fig. 5.6. Note differences in the results compared with mean value and median smoothing, particularly around the edges.

5.5. Suppose S is a template operation that implements smoothing, and O is the template operator that leaves an image unchanged. Then an edge enhanced image created by unsharp masking (Sect. 5.4.2) can be expressed

new image = O (old image) + O (old image) − S (old image)

Rewrite this expression to incorporate two user defined parameters A and B that will cause the formula to implement any of smoothing, edge detection or edge enhancement.

5.6. This requires vector algebra background. Show that template methods for line and edge detection can be expressed as the scalar product of a vector composed from the template entries and a vector formed from the neighbourhood of pixels currently covered by the template. Show how the angle between the template and pixel vectors can be used to assess the edge or line feature to which a current pixel most closely corresponds.

5.7. The following kernel is sometimes convolved with image data. What operation will it implement?

| 0 | -1 | 0 |
|---|----|---|
| -1 | +4 | -1 |
| 0 | -1 | 0 |

5.8. Consider the middle pixel shown in the figure below and calculate its new value if

  (i)   a [3 × 3] median filter is applied,

  (ii)  a [3 × 3] unsharp mask is applied,

  (iii) a [1 × 3] image smoothing template with a threshold of 2 is applied, and

  (iv) the Sobel operator is applied.

| 3 | 7 | 0 |
|---|---|---|
| 8 | 1 | 1 |
| 7 | 2 | 9 |

5.9. Image smoothing can be performed by template operators that implement averaging or median filtering. Compare those methods, particularly as they affect edges. Would you expect median filtering to be useful in edge enhancement using unsharp masking?

5.10. If a [3 × 3] smoothing template is applied to an image twice in succession how many neighbours will have played a part in modifying the brightness of a given pixel? Design a single template to achieve the same result in one pass.

5.11. The kernel function $k$ (,) in either (5.12) or (5.13) can be used to demonstrate the degrading effect of the point spread function (PSF) of an imaging sensor on a scene being recorded. If $\phi$ (,) is the ideal image of the scene and $k$ (,) is the instrument PSF, what form should $k$ (,) take in order that the instrument cause minimum degradation to the image data?

5.12. Inspection of Figs. 5.18, 5.21 and 5.23 demonstrates that erosion of an object is the same as dilation of the background for a binary image, and vice versa. Can you demonstrates that using set theory?

5.13. Repeat the examples of Figs. 5.20 and 5.22 using structuring elements with shapes: 1 × 3, 3 × 1, 5 × 3, 3 × 5.

# Chapter 6
# Spectral Domain Image Transforms

**Abstract** After looking at the value of band arithmetic, in which sums, differences and other operations are performed at the pixel level between the different bands of an image, focus shifts to the very important concept of principal components analysis. In preparation for this, the properties of the mean vector and covariance matrix are described and illustrated, along with the notion of band-to-band correlation. The principal components transform is seen as an operation which produces a new set of bands from the originally recorded set, in which the new bands are uncorrelated and in which detail, described by variance, is ranked-ordered by component. The first component contains most variance, while the last contains least. These properties are illustrated by example. Applications of the principal components transform are covered, including its value in detecting changes between images. The applications-specific Kauth-Thomas Tasseled cap transform is also developed. The chapter concludes with the transformation that converts the red, green, blue image description into a hue, saturation, intensity description, and with the operation of pan sharpening.

## 6.1 Introduction

Because of the multispectral or vector character of remote sensing image data we can devise transformations that generate new pixel vector descriptions from the old, and thereby create synthetic image components, or bands, as shown in Fig. 6.1. They represent an alternative description of the data. The new components are related to the old brightness values of the pixels in the original set of spectral bands via a mathematical operation, which is usually linear. Such a transformed image may make evident features that are not easily seen in the original data; alternatively, it might be possible to preserve most of the essential information content of the original image using a reduced number of the transformed dimensions. The last point has significance for displaying data in the three colour primaries, and for compressed transmission and storage of data.

© The Author(s), under exclusive license to Springer Nature Switzerland AG 2022    171
J. A. Richards, *Remote Sensing Digital Image Analysis*,
https://doi.org/10.1007/978-3-030-82327-6_6

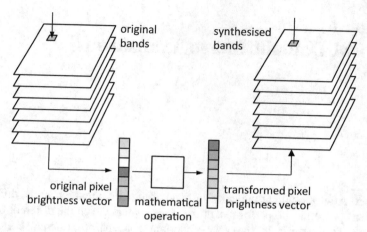

**Fig. 6.1** The principle of spectral domain transformation

It is also possible to devise pixel descriptors matched to particular applications. Vegetation indices are a good example. By combining the components of a pixel vector, a single descriptor can be generated that indicates the degree of vegetation present in that pixel. By doing so for all pixels, a vegetation index image is produced.

This chapter presents image transformations that are of value in enhancing and interpreting remote sensing imagery; some also find application in preconditioning image data prior to classification by the techniques of Chaps. 8 and 9. The methods covered here, which appeal directly to the vector nature of the image, include the principal components transformation and so-called band arithmetic. The latter includes the creation of ratio images and specialised indices. Special purpose transformations, such as the Kauth-Thomas tasseled cap transform, are also treated.

## 6.2  Image Arithmetic and Vegetation Indices

The simplest of all transformed images results from basic arithmetic operations among the pixel brightnesses in Fig. 6.1. Addition, subtraction and division (ratios) of the brightness values of two or more bands are the most common. While band products can be formed, they are rarely encountered.

Differences can be used to highlight regions of change between two images of the same area. That requires the images to be registered beforehand using the techniques of Chap. 2. The resultant difference image must be scaled to remove negative brightness values; normally that is done so that regions of no change appear mid-grey, while changes show as brighter or duller than mid-grey according to the sign of the difference.

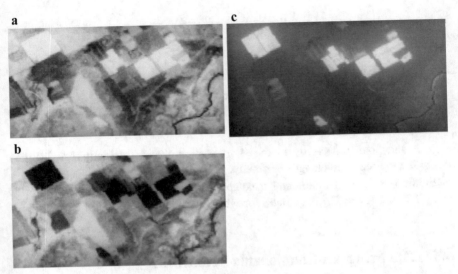

**Fig. 6.2 a** Near infrared and **b** visible red Landsat MSS images, along with **c** their ratio, which shows vegetated regions as bright, soils as mid to dark grey and water as black

Ratios of different spectral bands from the same image find use in reducing the effect of topography, as a vegetation index, and for enhancing subtle differences in the spectral reflectance characteristics for rocks and soils. As an illustration of the value of band ratios for providing a simple vegetation index image, Fig. 6.2 shows Landsat near infrared and red channel Multispectral Scanner images of an agricultural region, and their ratio. As seen, healthy vegetated areas are bright, soils are mid to dark grey, and water is black. Those shades are readily understood from an examination of the corresponding spectral reflectance curves.

More sophisticated vegetation indices can be created using composite band arithmetic operations.[1] The most widely encountered uses the ratio of the difference and sum of reflected infrared and visible red measurements. It is generally referred to as the Normalised Difference Vegetation Index (NDVI):

$$NDVI = \frac{\phi_{nir} - \phi_{red}}{\phi_{nir} + \phi_{red}}$$

in which $\phi_{nir}$ is the near infrared brightness of a pixel and $\phi_{red}$ is its visible red brightness. Two other indices sometimes used are the Normalised Difference Water Index (NDWI), which gives an indication of soil moisture

---

[1] For a comprehensive set of indices see http://www.harrisgeospatial.com/docs/VegetationIndices. html accessed March 2021.

$$NDWI = \frac{\phi_{green} - \phi_{nir}}{\phi_{green} + \phi_{nir}}$$

and the Land Surface Water Index (LSWI):

$$LSWI = \frac{\phi_{nir} - \phi_{swir}}{\phi_{nir} + \phi_{swir}}$$

$\phi_{swir}$ is the pixel reflectivity in the short wave infrared (typically 1.4–3.0 μm). $\phi_{green}$ is the green wavelength brightness of a pixel.

Figure 6.3 shows a continental-scale application of NDVI to drought monitoring using NOAA AVHRR image data, for which the pixel size is 1 km.


## 6.3  The Principal Components Transform

We can represent the pixels in a vector space with as many axes or dimensions as there are spectral components associated with each pixel (see Sect. 3.5.1 and Appendix C). In the case of Landsat ETM+ multispectral data it will have seven dimensions while for SPOT HRG data it will be four dimensional. For hyperspectral data, such as that from Hyperion, there may be several hundred axes. A particular pixel plots as a point in such a space with coordinates that correspond to its brightness values in the spectral components. We call the space, generically, a spectral space.[2] For simplicity the treatment to be developed here is based on a two dimensional space (say visible red and near infrared) since the diagrams are then easily understood and the mathematical examples are easily followed. The results to be derived, however, are perfectly general and apply to data of any dimensionality.


### 6.3.1  The Mean Vector and the Covariance Matrix

The positions of the pixel points in spectral space can be described mathematically by column vectors, the components of which are the individual spectral responses in each band. Strictly, they are vectors drawn from the origin to the pixel points as seen in Appendix C, but this concept is not used explicitly.

Consider a spectral space containing a large number of pixels as shown in Fig. 6.4, with each pixel described by its appropriate vector **x**. The mean position of

---

[2] It can have any number of dimensions; occasionally we will talk about hyperspectral space if the dimensionality is exceptionally high, although spectral and multispectral are even then still sufficient descriptors mathematically.

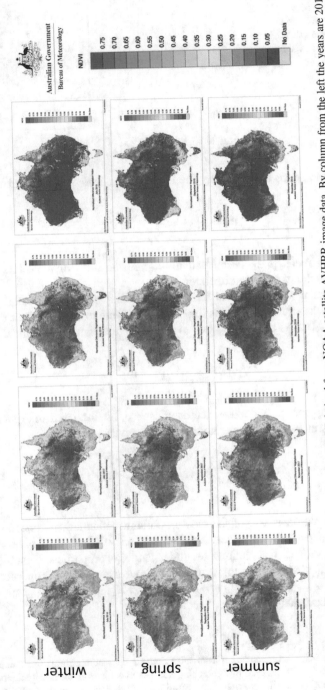

**Fig. 6.3** Monitoring the progression of drought using NDVI derived from NOAA satellite AVHRR image data. By column from the left the years are 2016, 2017, 2018 and 2019. *Used by permission of the Bureau of Meteorology and the Australian Government.* See www.bom.gov.au/climate/austmaps/about-ndvi-maps.shtml accessed 2021

**Fig. 6.4** A two dimensional spectral space showing individual pixel vectors and their average, or expected, position described by the mean vector **m**

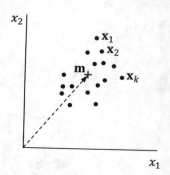

the pixels in the space is defined by the expected (or average) value of the pixel vector **x**

$$\mathbf{m} = \mathcal{E}\{\mathbf{x}\} = \frac{1}{K}\sum_{k=1}^{K}\mathbf{x}_k \tag{6.1}$$

where **m** is the mean pixel vector and the $\mathbf{x}_k$ are the individual vectors of total number $K$; $\mathcal{E}$ is the expectation operator.[3]

While the mean vector is useful for describing the average or expected position of the pixels in the vector space, it is important to be able to describe their scatter or spread as well. That is the role of the *covariance matrix* which is defined as

$$\mathbf{C}_x = \mathcal{E}\left\{(\mathbf{x} - \mathbf{m})(\mathbf{x} - \mathbf{m})^{\mathrm{T}}\right\} \tag{6.2}$$

in which the superscript T denotes vector transpose, as discussed in Appendix C. An unbiased estimate of the covariance matrix is given by

$$\mathbf{C}_x = \frac{1}{K-1}\sum_{k=1}^{K}(\mathbf{x}_k - \mathbf{m})(\mathbf{x}_k - \mathbf{m})^{\mathrm{T}} \tag{6.3}$$

The covariance matrix is one of the most important mathematical concepts in the analysis of multispectral remote sensing data, as a result of which it is of value to consider some sample calculations to understand its properties. First, there is another matrix that is helpful to know about when we examine the properties of the covariance matrix. It is called the *correlation matrix* $\mathbf{R}_x$; its elements $r_{ij}$ are related to those of the covariance matrix $c_{ij}$ according to

---

[3] Sometimes the principal components transform is derived by assuming that the mean of the distribution is zero. That is achieved by subtracting the mean vector from all pixel vectors before processing, an operation called *centering*. See Sect. 6.3.3 for the effect of an origin shift and the invariance of the transformation.

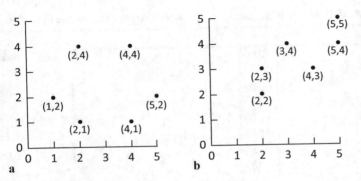

**Fig. 6.5** Two dimensional data sets **a** with no correlation between the components, and **b** with high correlation between the components

$$r_{ij} = c_{ij}/\sqrt{c_{ii}c_{jj}} \tag{6.4}$$

The elements of the correlation matrix describe explicitly the degree of mathematical correlation between the data dimensions, spectral measurements, or coordinates, indexed by the two subscripts.

Consider the two sets of pixel vectors shown in Fig. 6.5. They have different distribution patterns in spectral space—we will see shortly that the correlations between the coordinates in each case are different.

Table 6.1 shows a set of hand calculations for finding the covariance matrix of the set in Fig. 6.5a. As seen, the covariance matrix has zero off-diagonal entries and the correlation matrix, computed from the entries of the covariance matrix, also has zero off-diagonal entries. That indicates that there is no correlation between the two spectral components of the data set in Fig. 6.5a. A similar set of calculations for the data set of Fig. 6.5b gives

$$\mathbf{m} = \begin{bmatrix} 3.50 \\ 3.50 \end{bmatrix} \quad \mathbf{C}_x = \begin{bmatrix} 1.90 & 1.10 \\ 1.10 & 1.10 \end{bmatrix} \quad \mathbf{R}_x = \begin{bmatrix} 1.00 & 0.76 \\ 0.76 & 1.00 \end{bmatrix}$$

which shows the two spectral components to be 76% correlated. What does that mean? Fig. 6.6 shows what the images represented by the data in Fig. 6.5 might look like. For the data of Fig. 6.5b when one band is dark the other is generally dark, and vice versa. On the other hand, for the uncorrelated data of Fig. 6.5a, there is no near-correspondence in dark and light between the bands.

If the points in Fig. 6.5b happened all to lie on a straight line, then the corresponding images of both bands would be identical, and it would be possible to predict what the pixel in one band would look like knowing its appearance in the other band. The distribution of points in a fairly random, circular pattern as in Fig. 6.5a is typical of bands that exhibit little mutual correlation, whereas the elongate distribution of points in Fig. 6.5b is typical of the pixels in a correlated image.

**Table 6.1** Computation of the covariance and correlation matrices for the data of Fig. 6.5a

| $\mathbf{x}$ | $\mathbf{x} - \mathbf{m}$ | $(\mathbf{x} - \mathbf{m})(\mathbf{x} - \mathbf{m})^T$ | |
|---|---|---|---|
| $\begin{bmatrix} 1 \\ 2 \end{bmatrix}$ | $\begin{bmatrix} -2.00 \\ -0.33 \end{bmatrix}$ | $\begin{bmatrix} 4.00 & 0.66 \\ 0.66 & 0.11 \end{bmatrix}$ | |
| $\begin{bmatrix} 2 \\ 1 \end{bmatrix}$ | $\begin{bmatrix} -1.00 \\ -1.33 \end{bmatrix}$ | $\begin{bmatrix} 1.00 & 1.33 \\ 1.33 & 1.77 \end{bmatrix}$ | |
| $\begin{bmatrix} 4 \\ 1 \end{bmatrix}$ | $\begin{bmatrix} 1.00 \\ -1.33 \end{bmatrix}$ | $\begin{bmatrix} 1.00 & -1.33 \\ -1.33 & 1.77 \end{bmatrix}$ | |
| $\begin{bmatrix} 5 \\ 2 \end{bmatrix}$ | $\begin{bmatrix} 2.00 \\ -0.33 \end{bmatrix}$ | $\begin{bmatrix} 4.00 & -0.66 \\ -0.66 & 0.11 \end{bmatrix}$ | |
| $\begin{bmatrix} 4 \\ 4 \end{bmatrix}$ | $\begin{bmatrix} 1.00 \\ 1.67 \end{bmatrix}$ | $\begin{bmatrix} 1.00 & 1.67 \\ 1.67 & 2.79 \end{bmatrix}$ | |
| $\begin{bmatrix} 2 \\ 4 \end{bmatrix}$ | $\begin{bmatrix} -1.00 \\ 1.67 \end{bmatrix}$ | $\begin{bmatrix} 1.00 & -1.67 \\ -1.67 & 2.79 \end{bmatrix}$ | |

From which $\quad \mathbf{C}_x = \begin{bmatrix} 2.40 & 0 \\ 0 & 1.87 \end{bmatrix} \quad$ and $\quad \mathbf{R}_x = \begin{bmatrix} 1.00 & 0 \\ 0 & 1.00 \end{bmatrix}$

The mean vector for this data set is $\mathbf{m} = \begin{bmatrix} 3.00 \\ 2.33 \end{bmatrix}$

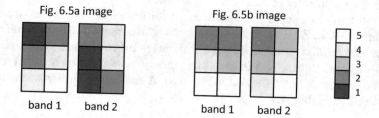

**Fig. 6.6** Six-pixel images that might correspond to the data of Fig. 6.5; shades of grey have been used to represent numerical values when moving clockwise around the data points of Fig. 6.5

Both the covariance and correlation matrices are symmetric and will be diagonal if there is no correlation between the image components. Although this example has been based on two dimensional data, the symmetry of the matrices holds in general, as does the special case of diagonal correlation and covariance matrices if all bands are uncorrelated.

We are now in the position to understand the basis of the principal components transform, one of the most frequently encountered image domain transformations in remote sensing image analysis.

## 6.3.2  A Zero Correlation, Rotational Transform

It is fundamental to the development of the principal components transformation to ask whether there is a new coordinate system in spectral space in which the data can be represented without correlation; in other words, so that the covariance matrix in the new coordinates is diagonal. For two dimensions such a new coordinate system is depicted in Fig. 6.7. If the vectors describing the pixel points are represented as $\mathbf{y}$ in the new coordinates then we want to find a linear transformation $\mathbf{G}$ of the original $\mathbf{x}$ coordinates, such that

$$\mathbf{y} = \mathbf{Gx} = \mathbf{D}^T\mathbf{x} \tag{6.5}$$

subject to the condition (constraint) that the covariance matrix of the pixel data in the $\mathbf{y}$ coordinates is diagonal.[4] Expressing $\mathbf{G}$ as $\mathbf{D}^T$ will allow a simpler comparison of principal components with other transformation operations to be treated later. Equation (6.5) is a vector–matrix shorthand method for saying that each component of $\mathbf{y}$ is a linear combination of all of the elements of $\mathbf{x}$; the weighting coefficients are the elements of the matrix $\mathbf{G}$.

In the $\mathbf{y}$ coordinate space the covariance matrix of the data is, by definition,

$$\mathbf{C}_y = \mathcal{E}\left\{ (\mathbf{y} - \mathbf{m}_y)(\mathbf{y} - \mathbf{m}_y)^T \right\}$$

in which $\mathbf{m}_y$ is the mean vector (mean position of the data) expressed in $\mathbf{y}$ coordinates. It is readily shown that[5]

$$\mathbf{m}_y = \mathcal{E}\{\mathbf{y}\} = \mathcal{E}\{\mathbf{D}^T\mathbf{x}\} = \mathbf{D}^T\mathcal{E}\{\mathbf{x}\} = \mathbf{D}^T\mathbf{m}_x$$

where $\mathbf{m}_x$ is the mean vector in the original $\mathbf{x}$ coordinate system. Thus, the covariance matrix in the new $\mathbf{y}$ coordinates becomes

$$\mathbf{C}_y = \mathcal{E}\left\{ (\mathbf{D}^T\mathbf{x} - \mathbf{D}^T\mathbf{m}_x)(\mathbf{D}^T\mathbf{x} - \mathbf{D}^T\mathbf{m}_x)^T \right\}$$

which can be written[6]

$$\mathbf{C}_y = \mathbf{D}^T\mathcal{E}\left\{ (\mathbf{x} - \mathbf{m}_x)(\mathbf{x} - \mathbf{m}_x)^T \right\}\mathbf{D}$$

---

[4] See Appendix C for a summary of relevant material in vector and matrix algebra.

[5] Since $\mathbf{D}^T$ is a matrix of constants it can be taken outside the expectation operator.

[6] Since $[\mathbf{A}\varsigma]^T = \varsigma^T\mathbf{A}^T$, which is called the *reverse order law of matrices*. Note also that $[\mathbf{A}\varsigma]^{-1} = \varsigma^{-1}\mathbf{A}^{-1}$.

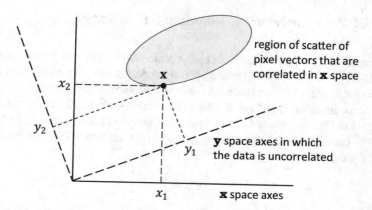

**Fig. 6.7** Illustration of how data that is correlated in one coordinate space (**x**) can be uncorrelated in a new, rotated coordinate space (**y**)

$$\text{i.e.}\quad \mathbf{C}_y = \mathbf{D}^{\mathrm{T}}\mathbf{C}_x\mathbf{D} = \mathbf{G}\mathbf{C}_x\mathbf{G}^{\mathrm{T}} \tag{6.6}$$

in which $\mathbf{C}_x$ is the covariance matrix of the data in the **x** coordinate space.

It is now important to recall that we are looking for a **y** coordinate space in which the pixel data exhibits no correlation. That requires $\mathbf{C}_y$ to be a diagonal matrix, as demonstrated in the simple two dimensional example of the previous section. Thus, from matrix algebra, (6.6) will be recognised as the diagonal form of the original covariance matrix $\mathbf{C}_x$, which requires $\mathbf{C}_y$ to be a diagonal matrix of the eigenvalues of $\mathbf{C}_x$ and **D** to be a matrix of the eigenvectors of $\mathbf{C}_x$.

We write the **y** space covariance matrix as

$$\mathbf{C}_y = \begin{bmatrix} \lambda_1 & 0 & & & \\ 0 & \lambda_2 & & & \\ & & \cdot & & \\ & & & \lambda_{N-1} & 0 \\ & & & 0 & \lambda_N \end{bmatrix}$$

where $N$ is the dimensionality of the data (the number of spectral bands) and the $\lambda_n$ are the eigenvalues of $\mathbf{C}_x$. The matrix is arranged diagonally such that $\lambda_1 > \lambda_2 \ldots > \lambda_N$. The elements of a diagonal covariance matrix are the simple variances along each coordinate direction. Thus, the eigenvalues of $\mathbf{C}_x$ will be the variances along each of the new coordinates in the **y** space. In view of the rank ordering of the eigenvalues, the data will show greatest spread along the first **y** coordinate. Its second greatest dispersion will be along the second new coordinate, which corresponds to the second largest eigenvalue, and so on.

The transformation described by (6.5), subject to (6.6) with a diagonal transformed covariance matrix, is called the *Principal Components Transform*. It is also known as the *Karhunen-Loève* or *Hotelling Transform*.

To summarise, the steps involved in the principal components transformation are:

- Generate the covariance matrix of the data in the coordinates of the original spectral measurements
- Find the eigenvalues and eigenvectors of the covariance matrix
- Rank order the eigenvalues to identify the first, second and subsequent principal axes
- Use the matrix of eigenvectors in (6.5) to generate the new brightness values for the pixels in each of the principal axes.

To demonstrate these points further we return to the example of Fig. 6.5b. Recall that the **x** space covariance matrix of the that highly correlated data is

$$\mathbf{C}_x = \begin{bmatrix} 1.90 & 1.10 \\ 1.10 & 1.10 \end{bmatrix}$$

To find its eigenvalues we have to solve the characteristic equation

$$|\mathbf{C}_x - \lambda \mathbf{I}| = 0$$

in which **I** is the identify matrix, and the vertical bars signify the determinant of the enclosed matrix. Substituting for the elements of $\mathbf{C}_x$ this gives

$$\begin{vmatrix} 1.90 - \lambda & 1.10 \\ 1.10 & 1.10 - \lambda \end{vmatrix} = 0$$

which, when evaluated, leads in this second order case to the quadratic equation in the unknown eigenvalue $\lambda$

$$\lambda^2 - 3.00\lambda + 0.88 = 0$$

On solving, this gives $\lambda = 2.67$ or $0.33$ as the two eigenvalues ($\lambda_1$ and $\lambda_2$ respectively) of the covariance matrix $\mathbf{C}_x$. As a handy check on the analysis, it may be noted that the sum of the eigenvalues is equal to the trace of the covariance matrix, which is simply the sum of its diagonal elements. Having found the eigenvalues we now know that the covariance matrix in the **y** coordinate system is

$$\mathbf{C}_y = \begin{bmatrix} 2.67 & 0 \\ 0 & 0.33 \end{bmatrix}$$

Thus, the variance of the data (loosely, its scatter) in the $y_1$ coordinate direction, which is referred to as the *first principal component*, is 2.67 and its variance along the $y_2$ direction, the *second principal component*, is 0.33. Using these values we can say that the first principal component accounts in this case for $2.67/(2.67 + 0.33) \equiv$ 89% of the total variance of the data.

We can now proceed to find the actual principal components transformation matrix $\mathbf{G}$ in (6.5). Since $\mathbf{D}$ is the matrix of eigenvectors of $\mathbf{C}_x$, then $\mathbf{G}$ is the transposed matrix of eigenvectors.

There are two eigenvectors, one for each of the eigenvalues. We commence by seeking that which corresponds to $\lambda_1 = 2.67$, in which the subscript 1 has been added so that we can associate the eigenvalue with its eigenvector. The eigenvector $\mathbf{g}_1$ is a solution to the vector equation

$$[\mathbf{C}_x - \lambda_1 \mathbf{I}]\mathbf{g}_1 = 0$$

with $\mathbf{g}_1 = \begin{bmatrix} g_{11} \\ g_{12} \end{bmatrix} = \mathbf{d}_1^T$.

Substituting for $\mathbf{C}_x$ and $\lambda_1$ gives the pair of equations

$$-0.77g_{11} + 1.10g_{21} = 0$$
$$1.10g_{11} - 1.57g_{21} = 0$$

both of which give

$$g_{11} = 1.43g_{21} \tag{6.7}$$

The equations are not independent because they are a homogeneous set.[7] At this stage either $g_{11}$ or $g_{21}$ would be chosen arbitrarily; however, there is a property we haven't yet used that allows us to determine unique values for $g_{11}$ and $g_{21}$. The covariance matrices we encounter in remote sensing are always real and symmetric. That ensures the eigenvalues are real and that the eigenvectors can be normalised.[8] From this last property we have another equation we can use to tie down values for $g_{11}$ and $g_{21}$, viz. that the vector magnitude is unity:

$$g_{11}^2 + g_{21}^2 = 1 \tag{6.8}$$

From (6.7) and (6.8) we find

$$\mathbf{g}_1 = \begin{bmatrix} 0.82 \\ 0.57 \end{bmatrix}$$

[7] Although these simultaneous equations are homogeneous (both equate to zero) the solution, as we have seen, is non-trivial. That is because the determinant of the coefficient matrix is zero, a property we used earlier in finding the eigenvalues.

[8] See C.M. Bishop, *Pattern Recognition and Machine Learning*, Springer Science + Business Media, N.Y., 2006.

In a similar manner it can be shown that the eigenvector corresponding to $\lambda_2 = 0.33$ is

$$\mathbf{g}_2 = \begin{bmatrix} -0.57 \\ 0.82 \end{bmatrix}$$

The required principal components transformation matrix is, therefore,

$$\mathbf{G} = \mathbf{D}^{\mathrm{T}} = \begin{bmatrix} 0.82 & -0.57 \\ 0.57 & 0.82 \end{bmatrix}^{\mathrm{T}} = \begin{bmatrix} 0.82 & 0.57 \\ -0.57 & 0.82 \end{bmatrix}$$

We now consider how to use these results. First, the individual eigenvectors $\mathbf{g}_1$ and $\mathbf{g}_2$ define the principal component axes in the original coordinate space. They are shown in Fig. 6.8, expressed in terms of a pair of unit vectors $\mathbf{e}_1$ and $\mathbf{e}_2$ that point along the original coordinate directions. It is evident that the data is uncorrelated in the new axes and that the new axes are a rotation of the original coordinates. This holds with data of any dimensionality. For this reason, the principal components transform is called a *rotational* transform.

Now consider the application of the transformation matrix $\mathbf{G}$ to find the brightness values of the pixels in the principal axes. For this example (6.5) is

$$\begin{bmatrix} y_1 \\ y_2 \end{bmatrix} = \begin{bmatrix} 0.82 & 0.57 \\ -0.57 & 0.82 \end{bmatrix} \begin{bmatrix} x_1 \\ x_2 \end{bmatrix} \tag{6.9}$$

For the pixels given in Fig. 6.5b

$$\mathbf{x} = \begin{bmatrix} 2 \\ 2 \end{bmatrix}, \begin{bmatrix} 4 \\ 3 \end{bmatrix}, \begin{bmatrix} 5 \\ 4 \end{bmatrix}, \begin{bmatrix} 5 \\ 5 \end{bmatrix}, \begin{bmatrix} 3 \\ 4 \end{bmatrix}, \begin{bmatrix} 2 \\ 3 \end{bmatrix}$$

we find

$$\mathbf{y} = \begin{bmatrix} 2.78 \\ 0.50 \end{bmatrix}, \begin{bmatrix} 4.99 \\ 0.18 \end{bmatrix}, \begin{bmatrix} 6.38 \\ 0.43 \end{bmatrix}, \begin{bmatrix} 6.95 \\ 1.25 \end{bmatrix}, \begin{bmatrix} 4.74 \\ 1.57 \end{bmatrix}, \begin{bmatrix} 3.35 \\ 1.32 \end{bmatrix}$$

The pixels plotted in $\mathbf{y}$ space are shown in Fig. 6.9. Several important points can be noted. First, the data exhibits no obvious correlation in the new coordinates, meaning that there is no correlation between the two new axes for this data. Secondly, most of the spread is in the direction of the first principal component; we could interpret this to mean that, for this example, the first principal component contains most of the information in the image.[9] Finally, if the pair of principal component images is generated using the $y_1$ and $y_2$ component brightness values for

[9] This needs to be interpreted carefully. Since the principal components transform was generated using the global covariance matrix for the data, comments like this mean "on the average." As will be seen in practice, later principal components may contain small, yet informative detail.

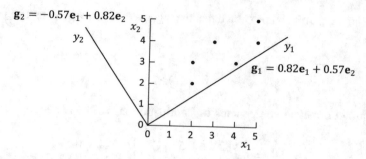

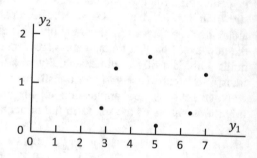

**Fig. 6.8** Principal component axes for the data of Fig. 6.5b; $e_1$ and $e_2$ are directional unit vectors in the $x_1$ and $x_2$ directions respectively

**Fig. 6.9** Pixel data in the principal components space, in which the correlation has been removed

the pixels, then the first principal component will show a high degree of contrast while the second will have lower contrast by comparison. Its brightness values will be spread over a smaller part of the available range so that it appears to lack the detail of the first component. This trend continues for the higher order components. The brightness value range of the higher order components is generally so small that they look noisy because of the limited number of discrete brightness values used to represent them in a display system.

## 6.3.3 The Effect of an Origin Shift

It will be evident that some principal component pixel brightness values could be negative owing to the fact that the transformation is a simple axis rotation. Clearly a combination of positive and negative brightnesses cannot be displayed. Nor can negative brightness pixels be ignored because their appearance relative to other pixels serves to define detail. In practice, the problem with negative values is handled by shifting the origin of the principal components space so that all components have positive, and thus displayable, brightnesses. That has no effect on the properties of the transformation as can be seen by inserting an origin shift term in the definition of the covariance matrix in the principal component space.

Let $\mathbf{y}' = \mathbf{y} - \mathbf{y}_0$ in which $\mathbf{y}_0$ is the position of a new origin. In the new $\mathbf{y}'$ coordinates

$$\mathbf{C}_{y'} = \mathcal{E}\left\{ (\mathbf{y}' - \mathbf{m}_{y'})(\mathbf{y}' - \mathbf{m}_{y'})^{\mathrm{T}} \right\}$$

Now $\mathbf{m}_{y'} = \mathbf{m}_y - \mathbf{y}_0$ so that $\mathbf{y}' - \mathbf{m}_{y'} = \mathbf{y} - \mathbf{y}_0 - \mathbf{m}_y + \mathbf{y}_0 = \mathbf{y} - \mathbf{m}_y$. Thus $\mathbf{C}_{y'} = \mathbf{C}_y$. Therefore, the origin shift has no influence on the covariance (and thus correlation) of the data in the principal components axes and can be used for convenience in displaying principal component images.

### 6.3.4 *Example and Some Practical Considerations*

The material presented in Sect. 6.3.2 provides the background theory for the principal components transform. By working through the numerical example in detail the importance of eigenanalysis of the covariance matrix has been seen. When using principal components analysis in practice the user is generally not involved in that level of detail. Rather only three steps are necessary. They are, first, the assembling of the covariance matrix of the image that is to be transformed, according to (6.5). Normally, software will be available for that step, often in conjunction with the need to generate signatures for classification as described in Chap. 8. The second step is to determine the eigenvalues and eigenvectors of the covariance matrix. At this stage the eigenvalues can be used to assess the distribution of data variance over the respective components. A rapid fall off in the sizes of the eigenvalues indicates that the original band description of the image data exhibits a high degree of correlation and that the results from the transformation step to follow will be significantly different from the original bands.[10]

The final step is to calculate the principal components using the eigenvectors of the covariance matrix as the weighting coefficients. As seen in (6.5) and as demonstrated in (6.9), the principal component brightness values for a pixel are the weighted sums of its brightnesses in the original bands; the weights are the elements of the eigenvectors of the covariance matrix. The first eigenvector produces the first principal component from the original data, the second eigenvector gives rise to the second component and so on.

Figure 6.10 shows two examples of principal components drawn from the same image data set.[11] These are band subsets that exhibit differing degrees of

---

[10] If there were no correlation among the original bands for a given data set, then the principal components transform will generate components that are identical with the original bands. That can be demonstrated easily by taking the ideal case in which the covariance matrix for the original data is the identity matrix.

[11] A HyVista Hymap 124 channel hyperspectral image over Perth, Western Australia; MultiSpec was used for the principal components processing.

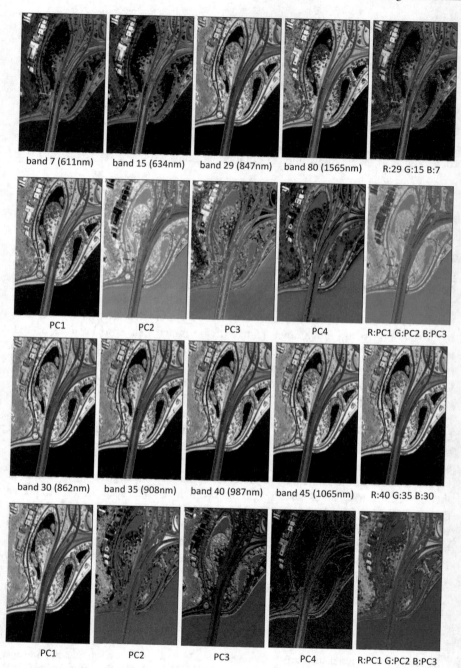

band 7 (611nm)    band 15 (634nm)    band 29 (847nm)    band 80 (1565nm)    R:29 G:15 B:7

PC1              PC2              PC3              PC4              R:PC1 G:PC2 B:PC3

band 30 (862nm)   band 35 (908nm)   band 40 (987nm)   band 45 (1065nm)   R:40 G:35 B:30

PC1              PC2              PC3              PC4              R:PC1 G:PC2 B:PC3

**Fig. 6.10** First row: four bands with low correlation and the colour composite formed from the first three; second row: the four principal components generated from the bands in the first row and a colour composite formed from the first three; third row: four bands with high correlation and a colour composite formed from the first three; fourth row: the four principal components generated from the third row and the colour composite formed from the first three

**Table 6.2** Eigenvalues and eigenvectors for the two image data sets of Fig. 6.10; the data is 16 bit

| | Data set with lower correlation | | | | | | Data set with higher correlation | | | | | | |
|---|---|---|---|---|---|---|---|---|---|---|---|---|---|
| | Eigenvalues | | Eigenvectors by row | | | | Eigenvalues | | Eigenvectors by row | | | | |
| | ×1000 | % | | | | | ×1000 | % | | | | | |
| 1 | 4495 | 87.51 | 0.22 | 0.26 | 0.72 | 0.61 | 10,822 | 99.79 | 0.47 | 0.49 | 0.52 | 0.52 | |
| 2 | 539 | 10.49 | −0.43 | −0.58 | 0.61 | −0.32 | 16 | 0.15 | −0.81 | 0.03 | 0.14 | 0.57 | |
| 3 | 95 | 1.84 | 0.55 | 0.31 | 0.32 | −0.71 | 5 | 0.05 | 0.25 | −0.87 | 0.26 | 0.33 | |
| 4 | 8 | 0.16 | 0.68 | −0.70 | −0.10 | 0.17 | 2 | 0.02 | 0.25 | 0.04 | −0.80 | 0.54 | |
| Covariance | 0.34 | | | | | | 2.44 | | | | | | |
| matrices | 0.40 | 0.49 | | | | | 2.49 | 2.56 | | | | | |
| (×10⁶) | 0.57 | 0.65 | 2.54 | | | | 2.68 | 2.75 | 2.96 | | | | |
| | 0.63 | 0.78 | 1.84 | 1.76 | | | 2.64 | 2.71 | 2.92 | 2.89 | | | |
| Correlation | 1.00 | | | | | | 1.00 | | | | | | |
| matrices | 0.97 | 1.00 | | | | | 1.00 | 1.00 | | | | | |
| | 0.61 | 0.58 | 1.00 | | | | 1.00 | 1.00 | 1.00 | | | | |
| | 0.81 | 0.84 | 0.87 | 1.00 | | | 0.99 | 1.00 | 1.00 | 1.00 | | | |

correlation. In the first row, four bands are chosen that have lower correlation; the principal components of those bands are shown in the second row. Each of the individual band images and the individual principal component images has been contrast stretched to allow detail to be seen better. At the end of each row is a colour composite image formed from the channels indicated. Although the principal components image is brightly coloured it does not seem to provide much added detail. For comparison, the four bands shown in the third row are more highly correlated; note how similar they are and how the colour composite product is almost a grey level image, with only tiny touches of colour. In this case, the principal components, shown in the fourth row, exhibit a greater drop in variance, with a corresponding increase in discretisation noise. The colour product is brightly coloured, and highlights detail not seen in the original bands, particularly among the buildings.

Table 6.2 shows the eigenvalues for the two examples, from which the compressive property of the transformation can be seen. Also shown are the eigenvectors so that each component can be seen to be a weighted combination of the original bands. The covariance and correlation matrices are also included although, for clarity, only to three significant figures. Because they are symmetric only the lower triangles are shown.

## 6.3.5   Application of Principal Components in Image Enhancement and Display

When displaying remotely sensed image data only three bands can be mapped to the three colour primaries of the display device. For imagery with more than three bands the user must choose the most appropriate subset of three to use. A less ad hoc means for colour assignment is based on performing a principal components transform and assigning the first three components to the red, green and blue colour primaries. Examination of a typical set of principal component images, such as those seen in Fig. 6.10 for the highly correlated data set, reveals that there is little detail in the fourth (and later) components so that, in general, they could be ignored without prejudicing the ability to extract meaningful information from the scene visually.

A difficulty with principal components colour display, however, is that there is no longer a one-to-one mapping between sensor wavelengths and colours. Rather, each colour now represents a linear combination of spectral components, making photointerpretation difficult for many applications. An exception might be in exploration geology where structural differences may be enhanced in principal components imagery, there sometimes being little interest in the meanings of the actual colours.

At this point we might ask why principal components imagery is often more colourful than the original image data. In preparation for answering that question consider the three dimensional colour space shown in Fig. 6.11. That represents how data would be displayed in terms of the three additive primary colours. It also shows how combinations of the fully saturated colour primaries create: yellow through the addition of red and green; magenta by the addition of red and blue; and cyan through the addition of blue and green. Note also that white results from the addition of the fully saturated three colour primaries, black is the result of zero values of red, green and blue, and shades of grey occur along the three dimensional diagonal shown.

When creating remote sensing imagery, it is sometimes the case that the three bands chosen for display are moderately correlated. Plotting the pixel points in a three dimensional space corresponding to those bands will give a scatter of points about the three dimensional diagonal that represents 100% correlation. When those three bands are allocated to the colour primaries for display, the pixels points similarly group about the three dimensional colour space diagonal. The colour product then appears to be poorly coloured, with many grey tones. That is depicted in Fig. 6.12a, in which the scatter of pixel points is shown as an ellipsoid. If the data is subject to simple linear contrast stretching of each band, in an attempt to improve its appearance, the general shape of the data ellipsoid will not change, the correlation remains and the colour of the final product, while brighter, will not have vastly improved tones. Any brightly coloured parts of an image will generally be the result of image segments (cover types) that have less correlation among the chosen bands. Some images will appear more colourful, especially if the correlations are low, but even though the colour primaries (red, green and blue) might show richly in some imagery, rarely will the fully saturated combinations of yellow, cyan and magenta appear.

If the bands chosen for display are totally uncorrelated then the data will scatter in a roughly spherical fashion as shown in Fig. 6.12b. Contrast stretching can push the data out into all corners of the data space, including those corresponding to yellow, cyan and magenta, resulting in an image that makes good use of all

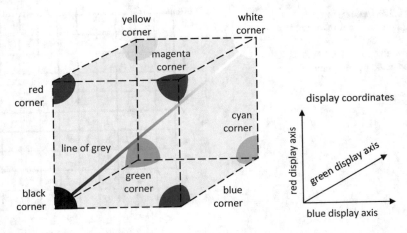

**Fig. 6.11** The colour space in which three bands of remotely sensed image data are displayed; fully correlated data points would fall along the line of grey

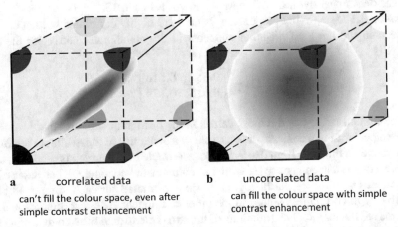

**a**        correlated data

can't fill the colour space, even after simple contrast enhancement

**b**        uncorrelated data

can fill the colour space with simple contrast enhancement

**Fig. 6.12** Showing the use of the colour space made when displaying **a** correlated and **b** uncorrelated image data

available colours; that is the reason why principal components imagery, derived from an original highly correlated set of bands, is much more colourful than the original image display.

## 6.3.6 The Taylor Method of Contrast Enhancement

Based on the observations of the previous section on filling the colour space with uncorrelated image data, an interesting contrast stretching procedure can be

**Fig. 6.13** Histogram for a hypothetical two dimensional image which has strongly correlated bands; the numbers in the cells are the pixel counts

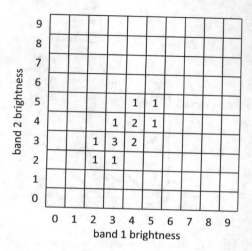

developed using the principal components transformation.[12] Consider a two dimensional image with the histogram shown in Fig. 6.13.

As observed, the two components (bands) are highly correlated, which can also be seen in the large off diagonal elements in the covariance matrix for the data, which is

$$\mathbf{C}_x = \begin{bmatrix} 0.885 & 0.616 \\ 0.616 & 0.879 \end{bmatrix} \tag{6.10}$$

The limited range of brightness values in the histogram suggests there would be value in performing a contrast stretch to allow a better use of the available brightness value range, in this case from 0 to 9. Suppose a simple linear stretch is used. It is usual to apply that enhancement to each image component separately. That requires the equivalent one dimensional histogram to be constructed for each component, irrespective of the brightness values in the other component. The one dimensional histograms are the marginal distributions of the two dimensional histogram of Fig. 6.13 and are shown in Fig. 6.14a. When contrast stretched the versions in Fig. 6.14b result. The equivalent two dimensional histogram resulting from the separate stretches in each band is shown in Fig. 6.15. Note that the correlation between the bands is still present and that if component 1 is displayed as red and component 2 as green no highly saturated reds or greens will be evident in the enhanced image although, for this two dimensional example, brighter yellows will appear.

---

[12] This was demonstrated initially in M.M. Taylor, Principal components display of ERTS-1 imagery, *Third Earth Resources Technology Satellite-1 Symposium*, NASA SP-351, 1973, pp. 1877–1897, and later in J.M. Soha and A.A. Schwartz, Multispectral histogram normalization contrast enhancement. *Proc. 5th Canadian Symposium on Remote Sensing*, 1978, pp. 86–93. A recent, general treatment will be found in N.A. Campbell, The decorrelation stretch transformation, *Int. J. Remote Sensing*, vol. 17, 1996, pp. 1939–1949.

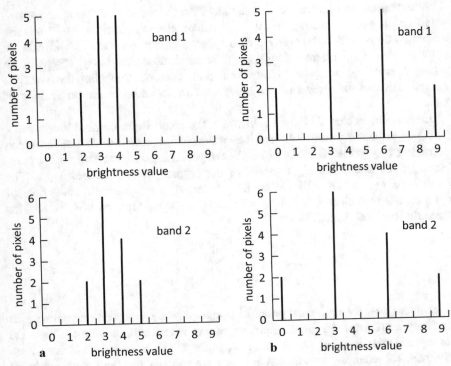

**Fig. 6.14  a** Individual histograms for the image with the two dimensional histogram of Fig. 6.13, and **b** individual histograms after a simple linear contrast stretch

**Fig. 6.15** Two dimensional result of the simple contrast stretch of each band of the image with histogram in Fig. 6.13

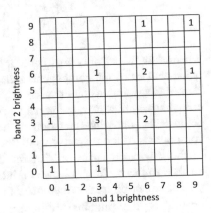

As noted in the previous section the situation for three dimensional correlated data is even worse, showing no saturated colour primaries and no bright mixture colours. The procedure recommended by Taylor overcomes this, as will now be demonstrated, allowing the colour space to be filled more completely. It is based on taking a principal components transformation, contrast stretching the principal components and then inverting the transformation to return to the original set of bands.

Let $\mathbf{x}$ be the vector of brightness values of the pixels in the original image and $\mathbf{y}$ be the corresponding vector of intensities after principal components transformation according to $\mathbf{y} = \mathbf{Gx}$. $\mathbf{G}$ is the principal components transformation matrix, composed of the transposed eigenvectors of the original covariance matrix $\mathbf{C}_x$. The covariance matrix which describes the scatter of pixel points in the principal components vector space is a diagonal matrix of the eigenvalues of $\mathbf{G}$ which, for three dimensional data, is of the form

$$\mathbf{C}_y = \begin{bmatrix} \lambda_1 & 0 & 0 \\ 0 & \lambda_2 & 0 \\ 0 & 0 & \lambda_3 \end{bmatrix}$$

Suppose now the individual principal components are enhanced in contrast such that they each cover the full range of brightness values and, in addition, have the same variances; in other words, the histograms of the principal components are matched, for example, to a Gaussian histogram that has the same variance in all dimensions. The new covariance matrix will be of the form

$$\mathbf{C}_y' = \begin{bmatrix} \sigma^2 & 0 & 0 \\ 0 & \sigma^2 & 0 \\ 0 & 0 & \sigma^2 \end{bmatrix} = \sigma^2 \mathbf{I} \tag{6.11}$$

where $\mathbf{I}$ is the identity matrix. Since the principal components are uncorrelated, enhancement of the components independently yields an image with good utilisation of the available colour space, with all hues technically possible. The axes in the colour space, however, are principal components and are not as desirable for image analysis by photointerpretation as having a colour space based on the original components of the image. It would be of value if the image data could be returned to the original $\mathbf{x}$ space so that the display colours represent the original image bands. Let the contrast enhanced principal components be represented by the vector $\mathbf{y}'$. These can be transformed back to the original axes for the image by using the inverse of the principal components transformation matrix $\mathbf{G}^{-1}$. Since $\mathbf{G}$ is orthogonal its inverse is its transpose, which is readily available, so that the modified pixel vector in the original coordinate space (i.e., the spectral measurement space) is given from $\mathbf{x}' = \mathbf{G}^T \mathbf{y}'$.

The new covariance matrix of the data back in the original image domain is, using (6.6),

$$\mathbf{C}'_x = \mathcal{E}\left\{[\mathbf{x}' - \mathcal{E}(\mathbf{x}')][\mathbf{x}' - \mathcal{E}(\mathbf{x}')]^T\right\} = \mathbf{G}^T \mathcal{E}\left\{[\mathbf{y}' - \mathcal{E}(\mathbf{y}')][\mathbf{y}' - \mathcal{E}(\mathbf{y}')]^T\right\}\mathbf{G} = \mathbf{G}^T \mathbf{C}'_y \mathbf{G}$$

In view of (6.11) this gives

$$\mathbf{C}'_x = \mathbf{G}^T \sigma^2 \mathbf{I} \mathbf{G} = \sigma^2 \mathbf{I}.$$

Thus, the covariance matrix of the enhanced principal components data is preserved on transformation back to the original image space. No correlation is introduced, and the data shows good utilisation of the colour space while using the original image data components.[13]

In practice, one problem encountered with the Taylor procedure is the discretisation noise introduced into the final results by the contrast enhanced third principal component. If the brightness values were continuous, rather than discrete, that would not occur. However, because image analysis software treats image data in integer format, the rounding of intermediate results to integer form produces the noise. One possible remedy is to low pass filter the noisy components before the inverse transform is carried out.

### 6.3.7 Use of Principal Components for Image Compression

Recall that the principal components transformation generates images of decreasing variance, and essentially compresses most of the image spectral detail into the first few component images. It is therefore suited to generating a reduced representation of image data for storage or transmission. In such a situation, it is only the uppermost significant components that are retained to represent the image. The information content lost is indicated by the sum of the eigenvalues of the components ignored compared with the total sum; this represents the mean square error of the approximation. Since the eigenvalues are ranked, the compression is optimal in terms of mean square error.

If the original image is to be restored, either on reception through a communications channel or on retrieval from memory, then the inverse of the transformation matrix is used to reconstruct the image from the reduced set of components. Since the matrix is orthogonal its inverse is its transpose. This technique is known as bandwidth compression in the field of telecommunications.

---

[13] See Problem 6.10 for a completion of this example.

## 6.3.8  The Principal Components Transform in Change Detection Applications

The principal components transformation is a redundancy reduction technique that generates data that is uncorrelated in the principal axes. It can also be used as a data transform to enhance regions of localised change in multitemporal image data.[14] That is a direct result of the high correlation that exist between images for regions that don't change significantly, and the relatively low correlations associated with regions that change. Provided the major portion of the variance in a multitemporal image data set is associated with constant cover types, regions of localised change will be enhanced in the higher principal components generated from the multi-temporal data. Regions of localised urban development and floods can be enhanced in this manner. In this section we demonstrate the technique using multitemporal imagery that shows fire damage between dates.

As might be expected, the major effect of fire damage on healthy green vegetation shows up in the near infrared region of the spectrum. When burnt, the strong infrared response of vegetation drops sharply. Conversely, when fire damaged vegetation regenerates, the near infrared response increases. To illustrate the value of principal components for highlighting changes associated with fire events consider near infrared data from two dates, one prior to a fire and the other afterwards.

We can construct a two-date scatter diagram as shown in Fig. 6.16. Pixels that correspond to cover types that remain essentially constant between dates, apart from normal seasonal differences, distribute in an elongated fashion as shown, corresponding to vegetation, water and soils. Cover types that change between dates will appear as major departures from that general trend. Pixels that were vegetated in the first date and burnt in the second lead to the off-diagonal group which has a low date 2 near infrared response as shown. Similarly, pixels that were bare (burnt) in the first date and revegetated in the time before the second acquisition, also cluster, but in the opposite direction.

Principal components analysis will lead to the axes shown in the figure. Variations in the near infrared response associated with the localised fire changes project into both component axes, but the effect is masked in the first component by the large range of brightness values of the near-constant cover types. By comparison, the change effect dominates the second principal component because the cover

---

[14] See G.R. Byrne and P.F. Crapper, An example of the detection of changes between successive Landsat images by numerical methods in an urban area, *Proc. 1st Australasian Conf. on Remote Sensing (Landsat '79)*, Sydney, 1979; G.R. Byrne, P.F. Crapper and K.K. Mayo, Monitoring land cover change by principal components analysis of multitemporal Landsat data, *Remote Sensing of Environment*, vol. 10, 1980, pp. 175–184; J.A. Richards, Thematic mapping from multitemporal image data using the principal components transformation, *Remote Sensing of Environment*, vol. 16, 1984, pp. 35–46; S.E. Ingebritsen and R.J.P. Lyon, Principal components analysis of multi-temporal image pairs, *Int. J. Remote Sensing*, vol. 6, 1985, pp. 687–696; T. Fung and E. Le Drew, Application of principal components analysis to change detection, *Photogrammetric Engineering and Remote Sensing*, vol. 53, 1987, pp. 1649–1658.

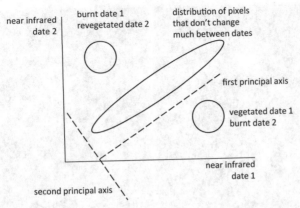

**Fig. 6.16** A two-date near infrared scatter diagram showing the likely distribution of pixels that remain largely constant in brightness between dates and those that change with fire events

types that were unchanged between dates project to a small range of brightnesses in the second component. The same general situation occurs when all available bands of image data are used, and all principal components are generated. Several of the higher order components may show local change information.

Figure 6.17 shows portions of two near infrared images recorded by Landsat in 1979 and 1980, over the northern suburbs of the city of Sydney, Australia. In the first there is a large fire scar evident as the result of a severe bush fire that occurred just prior to image acquisition. That region is revegetating in the second image, but two new burns are evident from fires that occurred earlier in 1980. While the figure shows only the near infrared images, four bands were available for each date, corresponding the visible green and red bands and the two near infrared bands of the Landsat Multispectral Scanner instrument. After the two images were registered, an 8 band multispectral, multitemporal image data set was available for use in the study.[15]

The principal components of the 8 band data set were generated, the first four of which are shown in Fig. 6.18. The remainder do not show any features of significance to this analysis. The first component is tantamount to a total brightness image, whereas the later components highlight changes. It is the second, third and fourth components that are most striking in relation to the fire features of interest. Pixels that have essentially the same cover type in both dates e.g., vegetation and vegetation, fire burn and fire burn, show as mid grey in the second, third and fourth components. Those that have changed, either as vegetation to fire burn or as fire burn to vegetation, show as darker or brighter than mid grey, depending on the component.

[15] See details in J.A. Richards, Thematic mapping from multitemporal image data using the principal components transformation, *Remote Sensing of Environment*, vol. 16, 1984, pp. 35–46.

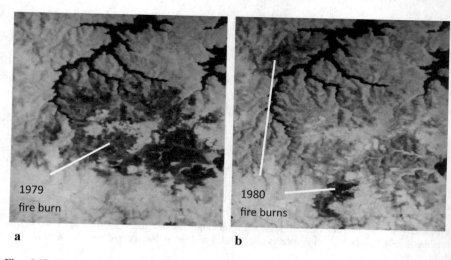

**Fig. 6.17** Landsat MSS infrared images recorded in **a** 1979 and **b** 1980, showing fire and revegetation events

Those effects are easily verified by substituting typical spectral reflectance characteristics into the equations that generate the components. Each component is a linear combination of the original eight bands of data, where the weighting coefficients are the components of the corresponding eigenvector of the $8 \times 8$ covariance matrix. Those eigenvectors, along with their eigenvalues, which are the variances of the components, are shown in Table 6.3.

When interpreting the fourth component it is necessary to account for a sign change introduced by the software[16] that generated the set of principal components. The second principal component image expresses the 1979 fire burn as lighter than average tone, while the third principal component highlights the two fire burns. The 1979 burn region shows as darker than average whereas that for 1980 shows as slightly lighter than average. In the fourth component the 1980 fire burn shows as darker than average with the 1979 scar not evident. What can be seen, however, is revegetation in 1980 from the 1979 fire. That shows as lighter regions. A particular example is revegetation in two stream beds on the right-hand side of the image a little over halfway down.

A colour-composite image formed by displaying the second principal component as red, the third as green, and the fourth as blue is given in Fig. 6.19. That shows the area that was vegetated in 1979 but burnt in 1980 as lime green; regions from the 1979 burn that remain without vegetation or have only a light vegetation cover in 1980 show as bright red; revegetated regions in 1980 from the 1979 fire display as bright blue/purple whereas the vegetated, urban, and water backgrounds that remained essentially unchanged between dates show as dark green/grey.

---

[16] Dipix Aries II.

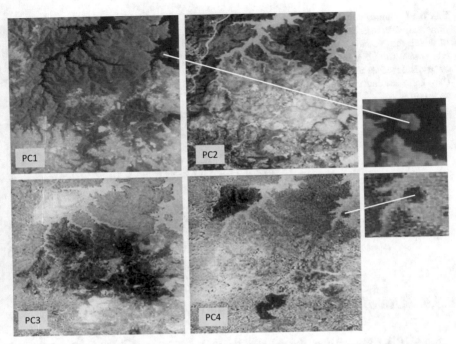

**Fig. 6.18** The first four principal component images from the 1979, 1980 multitemporal Landsat multispectral scanner data set; the expanded portions demonstrate the high noise of the fourth component compared with the low noise of the first; underlying diagram from J.A. Richards, Thematic mapping from multitemporal image data using the principal components transformation, *Remote Sensing of Environment*, vol. 16, 1984, pp. 35–46 used with permission of Elsevier

**Table 6.3** Eigenvalues and eigenvectors of the 8 band multitemporal covariance matrix; the eigenvector elements weight the original bands when generating the principal components

| Component | Eigenvalue | Eigenvector elements (by row) | | | | | | | |
|---|---|---|---|---|---|---|---|---|---|
| 1 | 1884 | 0.14 | 0.21 | 0.38 | 0.38 | 0.15 | 0.30 | 0.53 | 0.50 |
| 2 | 236 | 0.24 | 0.32 | −0.21 | −0.45 | 0.36 | 0.63 | 0.06 | −0.25 |
| 3 | 119 | 0.24 | 0.21 | 0.49 | 0.46 | 0.07 | 0.08 | −0.40 | −0.53 |
| 4 | 19 | −0.51 | −0.58 | −0.03 | 0.27 | 0.13 | 0.55 | −0.04 | −0.12 |
| 5 | 6 | 0.37 | −0.50 | 0.07 | −0.04 | 0.38 | −0.30 | 0.49 | −0.37 |
| 6 | 5 | 0.44 | −0.14 | −0.54 | 0.41 | 0.31 | 0.00 | −0.37 | 0.32 |
| 7 | 4 | −0.17 | 0.35 | −0.52 | 0.45 | −0.19 | −0.05 | 0.42 | −0.39 |
| 8 | 3 | 0.50 | −0.29 | −0.04 | −0.02 | −0.74 | 0.34 | 0.08 | −0.04 |

**Fig. 6.19** Colour composite
principal components image
in which the second principal
component from Fig. 6.18 has
been displayed as red, the
third as green and the fourth
as blue

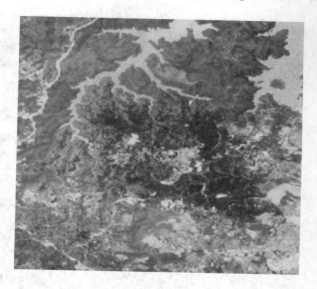

## 6.3.9   Use of Principal Components for Feature Reduction

In Sect. 10.3.1 we consider the value of the principal components transformation as
a tool for reducing the number of features that need to be processed when classifying image data. Because of its compressive properties the essence of the technique is to apply a principal components transformation and retain only the high
variance bands for subsequent quantitative analysis. Provided the classes of interest
are discriminable in the principal axes then the technique is suitable. However,
noting that principal components analysis is based on the global covariance matrix,
which is insensitive to the class composition of the data, the procedure is often
unsatisfactory particularly, as we have seen in Sect. 6.3.8, some spatially small
classes are emphasised in later components.

## 6.4   The Noise Adjusted Principal Components Transform

In the examples of Figs. 6.10 and 6.18 it is apparent that noise present in the
original image has been concentrated in the later principal components. Ordinarily
that is what would be expected: the components would become progressively
noisier as their eigenvalues decrease. In practice that is not always the case. It is

sometimes found that earlier components are noisier than those with the smallest eigenvalues. The noise adjusted transformation overcomes that problem.[17] Let

$$y = Gx = D^T x \qquad (6.12)$$

be a transformation that will achieve what we want. As with the principal components transformation, if $C_x$ is the covariance of the data in the original (as recorded) coordinate system, the diagonal covariance matrix after transformation will be

$$C_y = D^T C_x D \qquad (6.13)$$

To find the transformation matrix $D^T$ that will order the noise by component we start by defining the *noise fraction*

$$n = \frac{v^n}{v} \qquad (6.14)$$

where $v^n$ is the noise variance along a particular axis and $v$ is the total variance along that axis, consisting of the sum of the signal (wanted) variance and the noise variance. We assume that the signal and noise are uncorrelated. The total noise variance over all bands in the recorded data can be expressed as a noise covariance matrix $C_x^n$ so that after transformation according to (6.12) the noise covariance matrix will be

$$C_y^n = D^T C_x^n D \qquad (6.15)$$

This last equation is a matrix-based summary of the set of equations

$$v^n = d^T C_x^n d$$

in which $v^n$ is the noise variance along the axis $g = d^T$. From (6.13) the total signal plus noise variance along that axis is

$$v = d^T C_x d$$

so that, from (6.14), the noise fraction along that axis is

$$n = \frac{d^T C_x^n d}{d^T C_x d} \qquad (6.16)$$

[17] J. B. Lee, A. S. Woodyatt, and M. Berman, Enhancement of high spectral resolution remote-sensing data by a noise-adjusted principal components transform, *IEEE Transactions on Geoscience and Remote Sensing*, vol. 28, no. 3, May 1990, pp. 295–304.

We now want to find the coordinate direction $\mathbf{g} = \mathbf{d}^{\mathrm{T}}$ that minimises $n$. To do that we equate to zero the first derivative of $n$ with respect to $\mathbf{d}$. Thus[18]

$$\frac{\partial n}{\partial \mathbf{d}} = 2\mathbf{C}_x^n \mathbf{d} \{\mathbf{d}^{\mathrm{T}} \mathbf{C}_x \mathbf{d}\}^{-1} - 2\mathbf{C}_x \mathbf{d} \{\mathbf{d}^{\mathrm{T}} \mathbf{C}_x \mathbf{d}\}^{-2} \{\mathbf{d}^{\mathrm{T}} \mathbf{C}_x^n \mathbf{d}\} = 0$$

which, after simplification, gives

$$\mathbf{C}_x^n \mathbf{d} - \mathbf{C}_x \mathbf{d} \frac{\mathbf{d}^{\mathrm{T}} \mathbf{C}_x^n \mathbf{d}}{\mathbf{d}^{\mathrm{T}} \mathbf{C}_x \mathbf{d}} = 0$$

i.e. $(\mathbf{C}_x^n - \mathbf{C}_x n)\mathbf{d} = 0$

so that

$$(\mathbf{C}_x^n \mathbf{C}_x^{-1} - n\mathbf{I})\mathbf{d} = 0 \tag{6.17}$$

We now recognise the $n$ as the eigenvalues of $\mathbf{C}_x^n \mathbf{C}_x^{-1}$ and the $\mathbf{d}$ as the corresponding eigenvectors. If we rank the eigenvalues in *increasing* order then the image components will be ranked from that with the lowest noise variance to that with the highest, as required.

Suppose the noise covariance can be transformed to the identity matrix $\mathbf{I}$, by the technique we develop below. Then (6.17) becomes

$$(\mathbf{C}_x^{-1} - n\mathbf{I})\mathbf{d} = 0$$

and, from (6.16), $n = v^{-1}$, so that after multiplying throughout by $\mathbf{C}_x$ the last expression becomes

$$(\mathbf{C}_x - v\mathbf{I})\mathbf{d} = 0$$

which is the standard eigenvalue equation for the principal components transform, in which $v$ is explicitly the *image* variance associated with the relevant principal component, as before. This leads to a simple way to apply the noise adjusted principal components transform. First, we transform the original data so that the noise covariance matrix becomes the identity matrix. We then apply the standard principal components procedure.

The only outstanding step is to know how to transform the original data so that its noise covariance becomes the identity matrix. That is achieved in the following manner. From Appendix C we see that the diagonal form of the noise covariance matrix is

$$\Lambda = \mathbf{E}^{-1} \mathbf{C}_x^n \mathbf{E}$$

---

[18] Note $\frac{\partial}{\partial \mathbf{x}} \{\mathbf{x}^{\mathrm{T}} \mathbf{A} \mathbf{x}\} = 2\mathbf{A}\mathbf{x}$.

in which $\mathbf{\Lambda}$ is the diagonal matrix of its eigenvalues and $\mathbf{E}$ is the matrix of its eigenvectors. If we pre-multiply the last expression by $(\mathbf{\Lambda}^{-0.5})^{\mathrm{T}}$ and post-multiply it by $\mathbf{\Lambda}^{-0.5}$ we have

$$\mathbf{I} = (\mathbf{\Lambda}^{-0.5})^{\mathrm{T}} \mathbf{E}^{-1} \mathbf{C}_x^n \mathbf{E} \mathbf{\Lambda}^{-0.5}$$

Defining $\mathbf{F} = \mathbf{E}\mathbf{\Lambda}^{-0.5}$, and noting that $\mathbf{E}$ is orthogonal, the last expression becomes

$$\mathbf{I} = \mathbf{F}^{\mathrm{T}} \mathbf{C}_x^n \mathbf{F}$$

We thus recognise $\mathbf{y} = \mathbf{F}^{\mathrm{T}}\mathbf{x}$ as the transformation of the original data which yields a new set in which the noise covariance matrix is unity. Provided this transformation is carried out first, the standard principal components transform can be applied.

The procedure just outlined requires an estimate of the noise covariance matrix for the original image data. There are several means by which this estimate might be found.[19] Many are based on examining an image in segments thought to represent relatively pure, homogeneous regions on the ground. For those segments, an estimate of the noise is generated from the result of having subtracted a smoothed version of the image from the original.

## 6.5   The Kauth-Thomas Tasseled Cap Transform

The principal components transform yields a new coordinate description of remote sensing image data by establishing the diagonal form of the global covariance matrix. The new coordinates are linear combinations of the original spectral bands. Other linear transformations are possible. One is the procedure referred to as canonical analysis, treated in Chap. 10. Others are application-specific in that the new axes in which the data are described have been devised to maximise information of importance to particular needs. The "tasseled cap" transform[20] is an agriculture-specific transformation designed to highlight the most important, spectrally observable phenomena of crop development, in such a way that discrimination among crops, and crops from other vegetative cover is maximised. Its basis lies in the behaviour of crop trajectories with time in infrared versus visible

---

[19] See S.I. Olsen, Estimation of noise in images: an evaluation, *Graphical Models and Image Processing*, vol. 55, 1993, pp. 319–323.

[20] See R.J. Kauth and G.S. Thomas, The tasselled cap—a graphic description of the spectral-temporal development of agricultural crops as seen by Landsat. *Proc. LARS 1976 Symposium on Machine Processing of Remotely Sensed Data*, Purdue University, 1976; E.P. Crist and R.J. Kauth, The tasseled cap de-mystified, *Photogrammetric Engineering and Remote Sensing*, vol. 52, 1986, pp. 81–86.

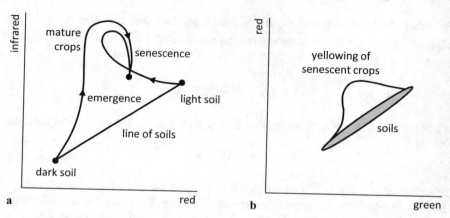

**Fig. 6.20** Spectral subspaces showing stages of crop development

red, and visible red versus visible green image subspaces. Consider the infrared/red space shown in Fig. 6.20a.

The first observation that can be made is that the variety of soil types on which specific crops might be planted appear as pixel points scattered about the diagonal of the infrared/red space. That is well-known and can be assessed by observing the spectral reflectance characteristics for soils.[21] Darker soils lie nearer the origin and lighter soils are at higher values in both bands. The actual slope of the line of soils will depend on global external factors such as atmospheric haze and soil moisture effects. If the transformation to be derived is to be used quantitatively those effects need to be modelled and the data calibrated or corrected beforehand.

Consider now the trajectories followed in the infrared/red subspace for crop pixels corresponding to growth on different soils as shown in the figure—in this case looking just at the extreme light and dark soil types. For both, the spectral response at planting is dominated by soil, as expected. As crops emerge the shadows cast over the soil dominate any green matter response. As a result, there is a darkening of the response of the lighter soil crop fields and possibly a slight darkening of fields on dark soil. When crops reach maturity their trajectories come together, implying closure of the crop canopy over the soil. The response is then dominated by the green biomass, being in a high infrared and low red region, as is well known.

When the crops senesce and turn yellow their trajectories remain together and move away from the green biomass point in the manner depicted in the diagram. However, whereas the development to maturity takes place almost totally in the same plane, the yellowing development moves out of that plane, as seen by how the trajectories develop in the red versus green subspace during senescence, as depicted

---

[21] See P.H. Swain and S.M. Davis, eds., *Remote Sensing: The Quantitative Approach*, McGraw-Hill, N.Y., 1978, Chap. 5.

in Fig. 6.20b. Should the crops then be harvested, the trajectories move back towards their original soil positions.

Having made those observations, the two diagrams of Fig. 6.20 can now be combined into a single three dimensional version in which the stages of the crop trajectories can be described according to the parts of a cap, with tassels, from which the name of the subsequent transformation is derived. That is shown in Fig. 6.21. The first point to note is that the line of soils used in Fig. 6.20a is now shown as a plane of soils. Its maximum spread is along the three dimensional diagonal as indicated; it has a scatter about this line consistent with the spread in red versus green shown in Fig. 6.20b. This plane of soils forms the brim and base of the cap. As crops develop on any soil type their trajectories converge essentially towards the crown of the cap at maturity, after which they fold over and continue to yellowing as indicated. Thereafter they break up to return ultimately to various soil positions, forming the tassels on the cap.

The behaviour observable in Fig. 6.21 suggests the development of a linear transformation that would be useful in crop discrimination. As with principal components analysis, this transformation should be based on orthogonal axes. However, the axis directions are chosen according to the behaviour seen in Fig. 6.21.

Three orthogonal directions of significance in agriculture can be identified in the new coordinate system. The first is the principal diagonal along which soils are distributed. That is chosen as the first axis in the tasseled cap transform. The development of green biomass as crops move towards maturity appears to occur orthogonal to the soil axis. That direction is chosen for the second axis, with the intention of providing a greenness indicator. Crop yellowing takes place in a

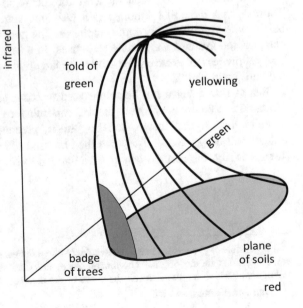

**Fig. 6.21** Crop trajectories in a green, red and infrared space, having the appearance of a tasseled cap

different plane from maturity. Consequently, choosing a third axis orthogonal to the soil line and greenness axis will give a yellowness measure. Finally, a fourth axis is required to account for data variance not substantially associated with differences in soil brightness, or vegetative greenness or yellowness if using four dimensional data such as Landsat MSS. Again, that needs to be orthogonal to the previous three. The transformation that produces the new description of the data may be expressed

$$\mathbf{u} = \mathbf{R}\mathbf{x} + \mathbf{c} \qquad\qquad (6.18)$$

where $\mathbf{x}$ is the original image data vector, and $\mathbf{u}$ is the vector of transformed brightness values. This has soil brightness as its first component, greenness as its second and yellowness as its third. They can be used as indices. $\mathbf{R}$ is the transformation matrix and $\mathbf{c}$ is a constant vector chosen to avoid negative values in $\mathbf{u}$. The transformation matrix $\mathbf{R}$ is the transposed matrix of column unit vectors along each of the transformed axes (compare with the principal components transformation matrix). For a given agricultural region the first unit vector can be chosen as the line of best fit through a set of soil classes. The subsequent unit vectors can then be generated by using a Gram Schmidt orthogonalization procedure in the directions required.[22] The transformation matrix generated by Kauth and Thomas for a Landsat Multispectral Scanner data set is[23]

$$\mathbf{R} = \begin{bmatrix} 0.433 & 0.632 & 0.586 & 0.264 \\ -0.290 & -0.562 & 0.600 & 0.491 \\ -0.829 & 0.522 & -0.039 & 0.194 \\ 0.223 & 0.012 & -0.543 & 0.810 \end{bmatrix}$$

From this it can be seen, at least for the region investigated by Kauth and Thomas, that the soil brightness is a weighted sum of the original four Landsat bands, with approximately equal emphases. The greenness measure is the difference between the infrared and visible responses. In a sense it is more a biomass index. The yellowness measure can be seen to be largely the difference between the visible red and green bands.

Just as new images can be synthesised to correspond to principal components, the same can be done with the Tasseled cap transform. By applying (6.18) to every pixel in the original image, soil brightness, greenness, yellowness and residual images can be produced. They can then be used to assess stages in crop development. Clearly, the method can also be applied to other sensors.

---

[22] It is possible to generate as many independent orthogonal vectors, and thus coordinates, as there are dimensions in the original coordinate space; the Gram-Schmidt process allows such an orthogonal basis to be found to suit the needs of the data set.

[23] Kauth and Thomas, loc. cit.

## 6.6 The Kernel Principal Components Transform

As the examples in Sect. 6.3.4 show the principal components transform produces the most striking results on data that exhibits a strong degree of correlation. This suggests that the data distributions on which the transformation has most impact are those which are elongated in the spectral domain. In such a situation the transformation often works well for feature reduction, because many classes will be distributed along the line of maximum correlation. However, Fig. 3.8 demonstrates that many of the classes of interest in remote sensing are distributed in a more general fashion. Although the principal components transformation can be limited as a feature reduction tool, as discussed in Sect. 6.3.9, if it can be made to discriminate among classes that are distributed in the fashion of Fig. 3.8, then its value for feature selection is significantly improved.

One way of achieving that would be to apply some form of axis transformation to the data before principal components analysis in order to redistribute the data points in a form more suited to generating a better outcome from the principal components transformation. We will now explore that approach; by exploiting an artifice known as the kernel trick we will see that we do not have to find the axis transform explicitly.[24]

Assume there is a transformation $\mathbf{z} = \Phi(\mathbf{x})$ that projects the data into a new set of coordinates in which the principal components transformation is better able to discriminate the classes of interest. We are imposing no particular constraints on the transformation $\Phi(\mathbf{x})$ at this stage. It could project the data into a space with dimensions higher than that of the original recorded measurements.

It helps our analysis if we assume that the original measurements $\mathbf{x}$ are centred. That implies we have assumed a mean position of zero, which can be achieved by subtracting the mean vector from each of the original measurement vectors. Likewise, we assume that the $\mathbf{z}$ are centred. The covariance matrix in the $\mathbf{z}$ coordinates is

$$\mathbf{C}_z = \mathcal{E}\{\mathbf{z}\mathbf{z}^T\} = \frac{1}{N} \sum_{j=1}^{N} \mathbf{z}_j \mathbf{z}_j^T = \frac{1}{N} \sum_{j=1}^{N} \Phi(\mathbf{x}_j)\Phi(\mathbf{x}_j)^T \qquad (6.19)$$

in which $N$ is the total number of data points. The equation in $\mathbf{z}$ space for finding the eigenvalues $\lambda_z$ and eigenvectors $\mathbf{v}_z$ of $\mathbf{C}_z$ is[25]

$$\lambda_z \mathbf{v}_z = \mathbf{C}_z \mathbf{v}_z$$

---

[24] The formative paper on kernel principal components is B. Schölkopf, A. Smola and K-R Müller, Kernel principal components analysis, in B. Schölkopf, C.J.C. Burges, and A.J. Smola, eds., *Advances in Kernel Methods–Support Vector Learning*, MIT Press, Cambridge, Mass., 1999.

[25] See (C.6).

For reasons which will become clear shortly, take the scalar product of both sides of this equation with $\Phi(\mathbf{x}_k), k = 1 \ldots N$. The scalar product of two vectors can be expressed in dot product or transposed form; here we choose the latter

$$\lambda_z \Phi(\mathbf{x}_k)^{\mathrm{T}} \mathbf{v}_z = \Phi(\mathbf{x}_k)^{\mathrm{T}} \mathbf{C}_z \mathbf{v}_z, \quad k = 1 \ldots N \tag{6.20}$$

Recall that the eigenvectors can be expressed as a linear combination of the coordinates of the $\mathbf{z}$ space; the set of expansion coefficients $\alpha_i$ below are what we normally refer to as the eigenvectors. Thus

$$\mathbf{v}_z = \sum_{i=1}^{N} \alpha_i \mathbf{z}_i = \sum_{i=1}^{N} \alpha_i \Phi(\mathbf{x}_i) \tag{6.21}$$

If we now substitute (6.21) and (6.19) into (6.20) we obtain

$$\lambda_z \Phi(\mathbf{x}_k)^{\mathrm{T}} \sum_{i=1}^{N} \alpha_i \Phi(\mathbf{x}_i) = \Phi(\mathbf{x}_k)^{\mathrm{T}} \frac{1}{N} \sum_{j=1}^{N} \Phi(\mathbf{x}_j) \Phi(\mathbf{x}_j)^{\mathrm{T}} \sum_{i=1}^{N} \alpha_i \Phi(\mathbf{x}_i), \quad k = 1 \ldots N$$

This can be written

$$N \lambda_z \Phi(\mathbf{x}_k)^{\mathrm{T}} \sum_{i=1}^{N} \alpha_i \Phi(\mathbf{x}_i) = \Phi(\mathbf{x}_k)^{\mathrm{T}} \sum_{i,j=1}^{N} \Phi(\mathbf{x}_j) \Phi(\mathbf{x}_j)^{\mathrm{T}} \alpha_i \Phi(\mathbf{x}_i), \quad k = 1 \ldots N$$

or

$$N \lambda_z \sum_{i=1}^{N} \alpha_i \Phi(\mathbf{x}_k)^{\mathrm{T}} \Phi(\mathbf{x}_i) = \sum_{i,j=1}^{N} \Phi(\mathbf{x}_k)^{\mathrm{T}} \Phi(\mathbf{x}_j) \Phi(\mathbf{x}_j)^{\mathrm{T}} \Phi(\mathbf{x}_i) \alpha_i, \quad k = 1 \ldots N$$

Note that this last equation is expressed in the form of three scalar products of the transformed vector $\mathbf{z} = \Phi(\mathbf{x})$. If we define the vector of coefficients $\alpha_i$ by $\boldsymbol{\alpha}$ then the last expression can be written compactly as

$$N \lambda_z \mathbf{K} \boldsymbol{\alpha} = \mathbf{K}^2 \boldsymbol{\alpha}$$

in which $\mathbf{K}$ is the matrix of all scalar products of the general form $\Phi(\mathbf{x}_i)^{\mathrm{T}} \Phi(\mathbf{x}_j), i, j = 1 \ldots N$. The last expression simplifies to

$$N \lambda_z \boldsymbol{\alpha} = \mathbf{K} \boldsymbol{\alpha} \tag{6.22}$$

so that $N\lambda_z$ is an eigenvalue of $\mathbf{K}$ and $\boldsymbol{\alpha}$ is the eigenvector. We require the eigenvalues $\mathbf{v}_z$ to be normalised, as with traditional principal components analysis, so that $\mathbf{v}_z^{\mathrm{T}}\mathbf{v}_z = 1$. From (6.21) that gives for the $k$th eigenvector[26]

$$\sum_{i,j=1}^{N} \alpha_i^k \alpha_j^k \Phi(\mathbf{x}_i)^{\mathrm{T}}\Phi(\mathbf{x}_j) = 1$$

which is equivalent to $(\boldsymbol{\alpha}^k)^{\mathrm{T}}\mathbf{K}\boldsymbol{\alpha}^k = 1$. From (6.22) this gives $(\boldsymbol{\alpha}^k)^{\mathrm{T}}N\lambda_z^k\boldsymbol{\alpha}^k = 1$ which requires the eigenvector $\boldsymbol{\alpha}^k$ to be divided by $\sqrt{N\lambda_z^k}$ to ensure that the eigenvectors $\mathbf{v}_z$ have unit magnitude.

How do we now create the actual principal components in $\mathbf{z}$ space? As in (6.9) we generate the pixel points by taking the scalar products of the eigenvectors of the covariance matrix and the original pixel vector. The equivalent operation in this analysis for the $k$th component is

$$k\text{th kernel PC} = (\mathbf{v}^k)^{\mathrm{T}}\Phi(\mathbf{x}) = \sum_{i=1}^{N} \alpha_i^k \Phi(\mathbf{x}_i)^{\mathrm{T}}\Phi(\mathbf{x}) \tag{6.23}$$

Thus, to compute the $k$th principal component in the transformed data space it is only necessary to know the scalar product $\Phi(\mathbf{x}_i)^{\mathrm{T}}\Phi(\mathbf{x})$ in (6.23) and not the actual form of the axis transform itself. We replace that scalar product by the so-called *kernel function* $k(\mathbf{x}_i, \mathbf{x})$. Once that is specified, we then only need the eigenvectors $\boldsymbol{\alpha}$ in order to apply (6.23). Those eigenvectors are found by solving (6.22) which requires the *kernel matrix* $\mathbf{K}$ to be found. Note that its elements are also defined entirely in terms of the scalar products and thus in terms of the chosen kernel function.

The kernel function $k(.,.)$ must satisfy certain conditions, as all the more commonly encountered kernels do (see Sect. 8.16).

If the data are not centred beforehand, the above analysis still holds provided the kernel matrix is modified to[27]

$$\mathbf{K} = \mathbf{K} - \mathbf{I}_N\mathbf{K} - \mathbf{K}\mathbf{I}_N + \mathbf{I}_N\mathbf{K}\mathbf{I}_N \tag{6.24}$$

in which $\mathbf{I}_N = \frac{1}{N}\mathbf{I}$.

---

[26] Note that the superscript here refers to the eigenvector index and is not a power.

[27] See Schölkopf et al., loc. cit.

The kernel based PCA has been used to improve feature selection[28] and to assist image registration.[29] It has also been used to enhance the PCA-based change detection procedure covered in Sect. 6.3.8.[30]

## 6.7  HSI Image Display

While not strictly a spectral domain transform in the sense of principal components analysis, the transform which creates a hue, saturation, intensity image from a blue, green, red counterpart is a useful operation for display and is employed as a step in the process of colour image sharpening (see Sect. 6.8).

The human visual system sees detail predominantly because of variations in brightness rather than changes in colour. That is why black and white photographs and black and white television are acceptable. Colour has the effect of adding to the visual experience but is not nearly as important as image intensity. As a consequence, in television systems more bandwidth (bit rate) is given to the transmission of intensity information than to the transmission of colour.

A useful model with which to describe colour is the colour wheel or colour solid, such as that shown in Fig. 6.22. It shows image brightness, or *intensity*, along the vertical axis and colour around a circular base plane. Two properties are required to describe colour in that plane. The most common are the *hue*, which represents the actual colour itself, and the *saturation*, which represents how vivid the colour is. A fully saturated red is very bright whereas an unsaturated red would be a washed-out pink colour. The hue and saturation description of colour are effectively the polar coordinates in the colour plane. Points in the colour plane can be located with other coordinate systems as well such, as the U, V axes used for the PAL colour television system.

The hue (H), saturation (S) and intensity (I) components of an image which has been displayed with red (R), green (G) and blue (B) components can be found in the following manner. First, it is convenient to define the normalised colours

---

[28] M. Fauvel, J. Chanussot and J.A. Benediktsson, Kernel principal component analysis for feature reduction in hyperspectral image analysis, *Proc. 7th Nordic Signal Processing Symposium 2006*, pp. 238–241, and M. Fauvel, J. Chanussot and J.A. Benediktsson, Kernel principal component analysis for the construction of the extended morphological profile, *Proc. Int. Geoscience and Remote Sensing Symposium, IGARSS2009*, Cape Town, 12–17 July 2009, pp. II843–II846.

[29] M. Ding, Z. Tian, Z. Jin, M. Xu and C. Cao, Registration using robust kernel principal components for object-based change detection, *Geoscience and Remote Sensing Letters*, vol. 7, no. 4, October 2010, pp. 761–765.

[30] A.A. Neilson and M.J. Canty, Kernel principal components analysis for change detection, *Proc. SPIE Image and Signal Processing for Remote Sensing XIV*, L. Bruzzone, C. Notarnicola and F. Posa, eds., vol. 7109, 2008, pp. 71090 T-1–71090 T-10.

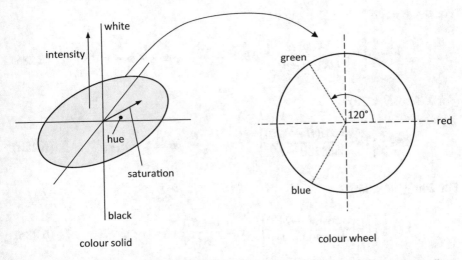

**Fig. 6.22** Colour representation in HSI coordinates; effectively this is a cylindrical coordinate system, whereas the RGB colour cube of Fig. 6.11 is a Cartesian colour space; the grey axis in Fig. 6.11 corresponds to the vertical intensity axis above

$$r = \frac{R}{R+G+B} \quad g = \frac{G}{R+G+B} \quad b = \frac{B}{R+G+B}$$

Intensity is the average of the red, green and blue signals:

$$I = \frac{R+G+B}{3} \tag{6.25}$$

Hue is given by

$$\text{for } b < g \qquad H = \cos^{-1} \frac{(r-g)+(r-b)}{2\sqrt{(r-g)^2 + (r-b)(g-b)}} \tag{6.26a}$$

$$\text{for } b > g \qquad H = 2\pi - H \tag{6.26b}$$

while saturation is defined by

$$S = 1 - 3\min(r,g,b) \tag{6.27}$$

Intensity is called luminance in the television industry and given the symbol Y. Note that on these definitions the range of hue is [0, 2π] and that of saturation is [0, 1]. When used for display, those ranges, as that for intensity, need to be scaled to the range of the display device. As seen in Fig. 6.22 the hue origin corresponds to red.

We can go from HSI to RGB with the following expressions. They depend on the sector of the colour wheel in which the colour point falls.

For $0 \leq H < 120°$

$$r = \frac{1}{3}\left[1 + \frac{S\cos H}{\cos(60 - H)}\right] \quad b = \frac{1-S}{3} \quad g = 1 - b - r \tag{6.28a}$$

For $120° \leq H < 240°$

$$g = \frac{1}{3}\left[1 + \frac{S\cos(H - 120)}{\cos(180 - H)}\right] \quad r = \frac{1-S}{3} \quad b = 1 - r - g \tag{6.28b}$$

For $240° \leq H < 360°$

$$b = \frac{1}{3}\left[1 + \frac{S\cos(H - 240)}{\cos(300 - H)}\right] \quad g = \frac{1-S}{3} \quad r = 1 - g - b \tag{6.28c}$$

## 6.8  Pan Sharpening

Having recognised in the previous section that it is possible to map from the red, green, blue colour space to a hue, saturation and intensity representation, and the fact that spatial detail is more easily discerned in intensity data, a simple method arises for sharpening the spatial detail in an image.

Many remote sensing instruments contain a higher resolution panchromatic band along with the set of multispectral bands. A straightforward way to produce a colour product with the spatial resolution of the panchromatic band is to choose three of the original multispectral bands from which a blue, green, red colour version is formed. A hue, saturation and intensity transformation is then carried out, following which the intensity channel is discarded and replaced by the panchromatic data. The hue and saturation images are then resampled to the spatial resolution of the panchromatic band. The new combination is transformed back to the red, green, blue colour space. This technique is known as *pan sharpening*. The intensity channel could of course be replaced by a co-registered higher resolution intensity product from any other source.

A variation on this approach uses the principal components transformation. This has the advantage that the original data can consist of more than just three bands, such as red, green and blue. In this method a principal components transformation is applied to the original multispectral image data set. The panchromatic band is contrast matched to the first principal component and is then used in place of that component. After resampling the remaining principal components to the new panchromatic band, an inverse principal components transformation is performed to regenerate the original set of bands, but with the spatial resolution of the panchromatic band.

Reviews of pan sharpening techniques will be found in Xiangchao et al.,[31] and Ehlers et al.[32] More recently the convolutional neural network has been applied to the sharpening problem.[33]

## 6.9   Bibliography on Spectral Domain Image Transforms

One of the earliest and most easily read treatments of the principal components transformation in a remote sensing context is the formative paper.

> S.K. Jensen and F.A. Waltz, Principal components analysis and canonical analysis in remote sensing, *Proc. American Photogrammetric Society 45th Annual Meeting*, 1979, pp. 337–348.

Full theoretical treatments can be found in many books on image processing and statistics, although often under the alternative titles of Karhunen-Loève or Hotelling transform. They include

> R.C. Gonzalez and R.E. Woods, *Digital Image Processing*, 4th ed. Pearson Prentice-Hall, Upper Saddle River, N.J., 2018.

Other books on remote sensing image analysis also normally contain extensive sections on the principal components transformation and related topics. The foundation paper for the kernel principal components transformation is

> B. Schölkopf, A. Smola and K-R Müller, Kernel principal components analysis, in B. Schölkopf, C. J. C. Burges and A. J. Smola, eds., *Advances in Kernel Methods–Support Vector Learning*, MIT Press, Cambridge, MA, 1999.

Kernel methods in general, with a remote sensing context, will be found in

> G. Camps-Valls and L. Bruzzone, eds., *Kernel Methods for Remote Sensing Image Analysis*, John Wiley & Sons, Chichester UK, 2009.

The use of the principal components transformation for change detection will be found in

---

[31] M. Xiangchao, S. Huanfeng, L. Huifang, Z. Liangpei and F. Randi, Review of the pansharpening methods for remote sensing images based on the idea of meta-analysis: practical discussion and challenges, *Information Fusion*, vol. 46, 2019, pp. 102–113.

[32] M. Ehlers, S. Klonus, P.J. Åstrand and P. Rosso, Multi-sensor image fusion for pansharpening in remote sensing, *International Journal of Image and Data Fusion*, vol.1, no.1, 2010, pp. 25–45.

[33] X. Li, F. Xu, X. Lyu, Y. Tong, Z. Chen, S. Li and D. Liu, A remote-sensing image pan-sharpening method based on multi-scale channel attenuation residual network, *IEEE Access*, vol. 8, 2020, pp. 27,163–27,177.

G.R. Byrne and P.F. Crapper, An example of the detection of changes between successive Landsat images by numerical methods in an urban area, *Proc. 1st Australasian Conf on Remote Sensing (Landsat'79)*, Sydney, 1979.

G.R. Byrne, P.F. Crapper and K.K. Mayo, Monitoring land cover change by principal components analysis of multitemporal Landsat data, *Remote Sensing of Environment*, vol. 10, 1980, pp. 175–184.

J.A. Richards, Thematic mapping from multitemporal image data using the principal components transformation, *Remote Sensing of Environment*, vol. 16, 1984, pp. 35–46.

S.E. Ingebritsen and R.J.P. Lyon, Principal components analysis of multitemporal image pairs, *Int. J. Remote Sensing*, vol. 6, 1985, pp. 687–696.

T. Fung and E. Le Drew, Application of principal components analysis to change detection, *Photogrammetric Engineering and Remote Sensing*, vol. 53, 1987, pp. 1649–1658.

The tasselled cap transformation is developed and refined in

R.J. Kauth and G.S. Thomas, The tasselled cap – a graphic description of the spectral-temporal development of agricultural crops as seen by Landsat. *Proc. LARS 1976 Symposium on Machine Processing of Remotely Sensed Data*, Purdue University, 1976.

E.P. Crist and R.J. Kauth, The tassled cap de-mystified, *Photogrammetric Engineering and Remote Sensing*, vol. 52, 1986, pp. 81–86.

Colour space transformations will be found in books on computer graphics and image processing and those that treat the transmission of colour in applications such as colour television. Easily read treatments that might be consulted include

K.R. Castleman, *Digital Image Processing*, 2nd ed., Prentice Hall, N.J., 1996, and

J.F. Hughes, A. van Dam, M. McGuire, D.F. Sklar, J.D. Foley, S.K. Feiner and K. Akeley, *Computer Graphics: Principles and Practice*, 3rd ed., Addison-Wesley, Boston, 2014.

Gonzalez and Woods (see above) has a detailed and excellent section on colour models and representation.


## 6.10  Problems


6.1  (a)  At a conference two research groups (A and B) presented papers on the value of the principal components transform for reducing the number of features required to represent image data. Group A showed very good results. Group B thought it was of little use. Both groups were using image data with only two spectral components. The covariance matrices for their respective images are:

$$\mathbf{C}_A = \begin{bmatrix} 5.4 & 4.5 \\ 4.5 & 6.1 \end{bmatrix} \quad \mathbf{C}_B = \begin{bmatrix} 28.0 & 4.2 \\ 4.2 & 16.4 \end{bmatrix}$$

Explain the points of view of both groups.

(b) If information content can be related directly to variance, indicate how much information is discarded if only the first principal component is retained by each group?

6.2 Suppose you have been asked to describe the principal components transformation to a non-specialist. Write a single paragraph summary of its essential features, using diagrams if you wish, but no mathematics.

6.3 (For those mathematically inclined). Demonstrate that the principal components transformation matrix developed in Sect. 6.3.2 is orthogonal.

6.4 Colour image products formed from principal components generally appear richer in colour than a colour composite product formed by combining three of the original bands of remote sensing image data. Why?

6.5 (a) The steps for computing principal component images may be summarised as:

- calculation of the image covariance matrix
- eigenanalysis of the covariance matrix
- computation of the principal components.

Assessments can be made in the first two steps as to the likely value in proceeding to compute the components. Describe what you would look for in each case.

(b) The covariance matrix need not be computed over the full image to produce a principal components transformation. Discuss the value of using training areas to define the portion of image data to be taken into account in compiling the covariance matrix.

6.6 Imagine you have two images from a sensor which has a single band in the range 0.9–1.1 μm. One image was taken before a flood occurred. The second shows the extent of flood inundation. Produce a sketch of what the two-date spectral space would look like if the image from the first date contained rich vegetation, sand and water and that in the second date contains the same cover types but with an expanded region of water. Demonstrate how a two dimensional principal components transform can be used to highlight the extent of flooding.

6.7 Repeat the exercise of 6.6 but for a region that was initially largely soil and on which a crop has been sown and reached maturity. Comment on the extent to which the crop cover type can fill the image if the technique is to work.

6.8 Describe the nature of the correlations between the pairs of axis variables (e.g., bands) in each of the cases in Fig. 6.23.

**Fig. 6.23** Examples of two dimensional correlations

6.9 The covariance matrix for an image recorded by a particular four channel sensor is shown below. Which band would you discard if you had to construct a colour composite display of the image by assigning the remaining three bands to each of the colour primaries?

$$\mathbf{C} = \begin{bmatrix} 35 & 10 & 10 & 5 \\ 10 & 20 & 12 & 2 \\ 10 & 12 & 40 & 30 \\ 5 & 2 & 30 & 30 \end{bmatrix}$$

6.10 Verify the elements of the covariance matrix in (6.10) using the data of Fig. 6.13. Although lengthy, you should now be able to complete the example of Sect. 6.3.6 by computing the principal components, linearly stretching the components, and transforming them back to the original coordinate system.

6.11 Perform a principal components transformation of the data shown in Fig. 4.21 and then produce a simple linear contrast stretch on each of the components separately. Compare the result to that from Problem 4.3.

6.12 Is the principal components transformation just a rotation of axes because

(a) the principal components values are linear combinations of the original brightness values in the image, or

(b) the eigenvalues decrease monotonically, or

(c) the principal components values are linear combinations of the original brightness values in the image, and the resulting covariance matrix is diagonal?

6.13 Consider an artificial landscape consisting of a black and white chequerboard pattern. Suppose a sensor with blue, green and red bands images that landscape. Which of the following statements is correct?

(a) The covariance matrix of the image data will be diagonal

(b) All the elements of the correlation matrix will be unity

(c) One of the eigenvalues of the covariance matrix will be unity and the other two will be zero

6.14  When using the principal components transformation for data compression in which the later principal components are discarded, is it possible to quantify the information loss?

(a) Not easily because we have no reliable metrics
(b) Yes, by looking at the distribution of the components in the first eigenvector
(c) Yes, by looking at the sum of the eigenvalues of the discarded bands compared with the sum of all eigenvalues

6.15  Is it practically conceivable that we can find an image of the natural landscape for which there is no correlation among bands recorded in the optical wavelengths?

# Chapter 7
# Spatial Domain Image Transforms

**Abstract** A number of special functions are introduced including the complex exponential, delta and Heaviside functions. The Fourier series and Fourier transforms are then covered in continuous and discrete forms, leading to the definition of the Fourier transform of an image, and how it can be evaluated. Convolution, including in two dimensions, is then introduced both as a basis for developing sampling theory and for understanding the theoretical origin of the spatial domain geometric processing techniques of Chap. 5. Examples are presented to consolidate the material developed, before moving on to consider an introduction to the wavelet transform.

## 7.1 Introduction

In the material presented in Chap. 5 we looked at a number of geometric processing operations that involve the spatial properties of an image. Averaging over adjacent groups of pixels to reduce noise and looking at local spatial gradients to enhance edge and line features are examples. There are, however, more sophisticated approaches for processing the spatial domain properties of an image. The most recognisable is probably the Fourier transformation which we consider in this chapter, allowing us to understand what are called the spatial frequency components of an image. Once we can use the Fourier transformation, we will see that it offers a powerful method for generating the sorts of operation we did with templates in Chap. 5.

There are other spatial transformation techniques as well. Some we will mention in passing that find application in image compression in the video and television industry. Other techniques, such as the wavelet transform, have emerged as important image processing tools in their own right, including in remote sensing. The wavelet transform is treated later in this chapter.

We commence with some necessary mathematical background that leads, in the first instance, to the principles of sampling theory. That is sampling, not in the statistical sense that we use when testing map accuracy in Chap. 11, but in the role

© The Author(s), under exclusive license to Springer Nature Switzerland AG 2022
J. A. Richards, *Remote Sensing Digital Image Analysis*,
https://doi.org/10.1007/978-3-030-82327-6_7

of digitising a continuous landscape to produce an image composed of discrete pixels. This material is based on two mathematical fields that the earth science reader may not have encountered in the past: the first is complex numbers and the second is integral calculus. We will work our way through that material as carefully as possible but for those readers not needing a background on transformation methods, this chapter can be passed over without affecting an understanding of the remainder of the book.

## 7.2  Special Functions

Three special functions are important in understanding the development of sampling theory and the transformations treated here. We will consider them as functions of time, because that is the way most are presented in classical texts, but they will be interpreted as functions of position, in either one or two dimensions as required, later in the chapter.

### 7.2.1  The Complex Exponential Function

Several of the functions we meet here involve *imaginary numbers* which arise when we try to take the square root of a negative number. The most basic is the square root of $-1$. Although that may seem to be an unusual concept, it is an enormously valuable mathematical construct in developing transformations. It is sufficient here to consider $\sqrt{-1}$ as a special symbol rather than try to understand the logical implications of taking the square root of something that is negative. It is given the symbol $j$ in the engineering literature but is represented by $i$ in the mathematical literature.

By definition, a complex exponential function[1] that is periodic with time is

$$g(t) = e^{j\omega t} \tag{7.1}$$

This is best looked at as a vector that rotates in an anticlockwise direction in a two dimensional plane (called an Argand diagram) described by real and imaginary number axes as shown in Fig. 7.1. The concept of an imaginary number is developed further below. If the exponent in (7.1) were negative the vector would rotate in the clockwise direction. As the vector rotates from its position at time zero its projection onto the axis of real numbers plots out the cosine function whereas its projection onto the axis of imaginary numbers plots out the sine function. One

---

[1] We use the symbol $g$ here for a general function, rather than the more usual $f$, to avoid confusion with the symbol universally used for frequency.

**Fig. 7.1** Showing how sinusoids can be generated from a rotating complex exponential function

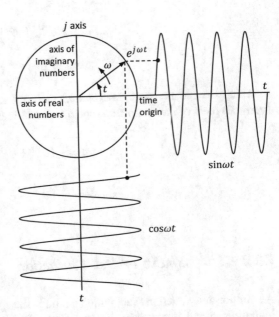

complete rotation of the vector, covering 360° or $2\pi$ radians, takes place in a time $t = T$ defined by $\omega T = 2\pi$. $T = 2\pi/\omega$ is called the *period* of the function, measured in seconds, and $\omega$ is called its *radian frequency*, with units of radians per second. For example, if the period is 10 ms, the radian frequency would be $200\pi = 628$ rad s$^{-1}$. Often, we describe frequency $f$ in hertz rather than as $\omega$ in radians per second. The two measures are related by

$$\omega = 2\pi f \tag{7.2}$$

The complex exponential expression in (7.1) can be written[2]

$$e^{j\omega t} = \cos \omega t + j \sin \omega t \tag{7.3a}$$

or, if the sign of the exponent is reversed,

$$e^{-j\omega t} = \cos \omega t - j \sin \omega t \tag{7.3b}$$

Both equations in (7.3) can be written in the form $g(t) = a(t) + jb(t)$ in which $a(t)$ is called the real part of $g(t)$ and $b(t)$ is called the imaginary part.[3]

The expressions in (7.3) can be demonstrated from Fig. 7.1 if the rotating vector is represented by two Cartesian coordinates with unit vectors $(1, j)$ in the horizontal and vertical directions respectively. In this way the symbol $j$ is nothing other than a vector that points in the vertical direction. From (7.3a) we can see that

---

[2] This is known as Euler's formula.

[3] If $a$, $b$ and $g$ are constants then $g$ is called a complex number.

$$\cos \omega t = \mathrm{Re}\{e^{j\omega t}\} \qquad (7.4\mathrm{a})$$

$$\sin \omega t = \mathrm{Im}\{e^{j\omega t}\} \qquad (7.4\mathrm{b})$$

in which Re and Im are operators that pick out the real (horizontal) or imaginary (vertical) parts of the complex exponential. Alternatively, from (7.3) we can see that

$$\cos \omega t = \frac{1}{2}\left(e^{j\omega t} + e^{-j\omega t}\right) \qquad (7.5\mathrm{a})$$

$$\sin \omega t = \frac{1}{2j}\left(e^{j\omega t} - e^{-j\omega t}\right) \qquad (7.5\mathrm{b})$$

## 7.2.2  The Impulse or Delta Function

An important function for understanding the properties of sampled signals, including digital image data, is the impulse function. It is also referred to as the Dirac delta function, or simply the delta function. It is spike-like, of infinite amplitude and infinitesimal duration. It cannot be defined explicitly. Instead, it is described by a limiting operation in the following manner.

Consider the rectangular pulse of duration $\alpha$ and amplitude $1/\alpha$ shown in Fig. 7.2. Note that the area under the curve is 1. If we let the value of $\alpha$ go to 0 then the function grows in amplitude to infinity and tends to an infinitesimal width, while its area stays the same. We define the delta function by that limiting operation. As a formal definition, the best we can do then is to say

$$\delta(t) = 0 \quad \text{for } t \neq 0 \qquad (7.6\mathrm{a})$$

and[4]

$$\int_{-\infty}^{\infty} \delta(t)dt = 1 \qquad (7.6\mathrm{b})$$

This turns out to be sufficient for most purposes in engineering and science.

Equation (7.6) defines a delta function at the origin; an impulse located at time $t_0$ is defined by

$$\delta(t - t_0) = 0 \quad \text{for } t \neq t_0 \qquad (7.6\mathrm{c})$$

---

[4] Recall that the integral of a function over its range is equal to the area under its curve.

**Fig. 7.2** A rectangular pulse
that approaches an impulse in
the limit as $\alpha \to 0$

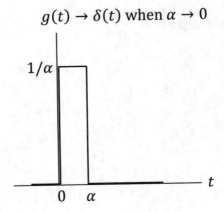

$$g(t) \to \delta(t) \text{ when } \alpha \to 0$$

and

$$\int_{-\infty}^{\infty} \delta(t - t_0)dt = 1 \qquad (7.6\text{d})$$

If we take the product of a delta function with another function the result, from (7.6c), is

$$\delta(t - t_0)g(t) = \delta(t - t_0)g(t_0) \qquad (7.7)$$

From this we see

$$\int_{-\infty}^{\infty} \delta(t - t_0)g(t)dt = \int_{-\infty}^{\infty} \delta(t - t_0)g(t_0)dt = g(t_0) \int_{-\infty}^{\infty} \delta(t - t_0)dt = g(t_0) \ (7.8)$$

This is known as the *sifting property* of the impulse because it picks (sifts) out the value of $g(t)$ at time $t_0$.

### 7.2.3   The Heaviside Step Function

The Heaviside or unit step function is shown in Fig. 7.3 and is defined by

$$\begin{aligned} u(t - t_0) &= 1 \quad \text{for } t \geq t_0 \\ &= 0 \quad \text{for } t < t_0 \end{aligned} \qquad (7.9)$$

**Fig. 7.3** The Heaviside (unit) step function

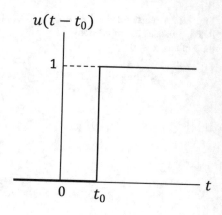

The step and delta functions are related by

$$\delta(t - t_0) = \frac{du(t - t_0)}{dt}$$

## 7.3 The Fourier Series

We now come to a very important concept in the analysis of functions and signals, which we will apply later to images. If a function of time is periodic, in the sense that it repeats itself with some regular interval such as the square waveform shown in Fig. 7.4, then it can be written as the sum of sinusoidal signals or the sum of complex exponential signals; that is called a Fourier series. We write a periodic function as $g(t) = g(t + T)$ where, again, $T$ is its period. In the terminology of the complex exponential the Fourier series of the function $g(t)$ is written

$$g(t) = \sum_{n=-\infty}^{\infty} G_n e^{jn\omega_0 t} \tag{7.10a}$$

in which $\omega_0 = 2\pi/T$, and $n$ is an integer. The coefficients $G_n$ tell us how much of each sinusoidal frequency component is present in the composition of $g(t)$. Notice however that there are coefficients with positive and negative indices corresponding to positive and negative frequency components. That can be understood by noting in (7.5) that the pure trigonometric functions are composed of exponentials with positive and negative exponents. The two-sided summation in (7.10a) recognises that property. One might ask why the trigonometric functions themselves are not used in developing the Fourier series. The fact is they can be and are; it is just that the exponential form is more convenient mathematically and has become the standard expression in engineering and science.

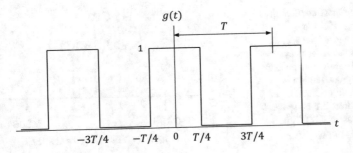

**Fig. 7.4** A square waveform with period $T$

The expansion coefficients $G_n$ are, in general, complex numbers. Finding their values is the most significant part of using the Fourier series. They are given by[5]

$$G_n = \frac{1}{T} \int_{-T/2}^{T/2} g(t)e^{-jn\omega_0 t} dt \qquad (7.10b)$$

To understand their importance, consider the Fourier series of the square waveform of Fig. 7.4. Over the range $(-T/2, T/2)$ covered by the integral the square waveform is zero except between $(-T/4, T/4)$ over which it is unity. Therefore Eq. (7.10b) becomes

$$G_n = \frac{1}{T} \int_{-T/4}^{T/4} e^{-jn\omega_0 t} dt = \frac{1}{n\pi} \sin \frac{n\pi}{2}$$

The first few values of this last expression for positive and negative values of $n$ are

| $n =$ | $-5$ | $-4$ | $-3$ | $-2$ | $-1$ | 0 | 1 | 2 | 3 | 4 | 5 |
|---|---|---|---|---|---|---|---|---|---|---|---|
| $G_n =$ | $1/5\pi$ | 0 | $-1/3\pi$ | 0 | $1/\pi$ | 0.5 | $1/\pi$ | 0 | $-1/3\pi$ | 0 | $1/5\pi$ |

These are shown plotted in Fig. 7.5 in two forms: in the second, they are represented by their amplitudes and phase angles. Any complex number can be written in either of two forms—the Cartesian and polar representations. Which one to use is dictated by the application at any given time. The Cartesian form $a + jb$ can be converted to the polar form $Re^{j\theta}$ where $R = \sqrt{a^2 + b^2}$ and $\theta = \tan^{-1}(b/a)$. In our work the polar form, in which the number has an amplitude $R$ and phase $\theta$, is most common.

[5] See L.A. Pipes, *Applied Mathematics for Engineers and Physicists*, 2nd ed., McGraw-Hill, N.Y., 1958 for an easily read treatment. See also L.A. Pipes and L.R. Harvill, *Applied Mathematics for Engineers and Physicists*, 3rd ed., Dover, N.Y., 2014.

**Fig. 7.5** Fourier coefficients (top), along with the amplitude and phase spectra of the square wave; the phase spectrum is shown as an odd function by convention although both odd and even representations are equivalent since phases of $+180°$ and $-180°$ both imply $-1$

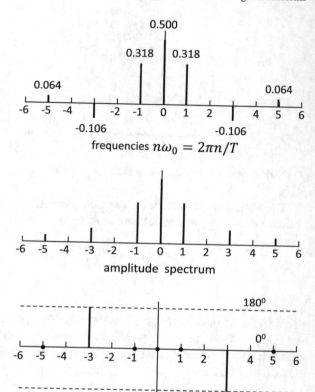

The graphs shown in Fig. 7.5 are frequently referred to as *spectra*, as indicated; generically the *set* of $G_n$ are also referred to as the spectrum of the function $g(t)$. Note that an angle (phase) of $\pm180$ is equivalent to $-1$ in complex numbers. These results tell us that the square wave of Fig. 7.4 can be made up from 0.500 parts of a constant, plus 0.318 parts of a pure sinusoid with the same fundamental frequency as the square wave, minus 0.106 parts of a sine wave with three times the frequency (said to be the third harmonic), plus 0.064 parts of a sine wave at five times the frequency (fifth harmonic), and so on. The square wave has no even harmonic components. Again, remember that the two components for the same frequency either side of the origin are the amplitudes of the positive and negative exponential components that constitute a sinusoid as in (7.5).

## 7.4 The Fourier Transform

The Fourier series above describes a periodic function as the sum of a discrete number of sinusoids, expressed in complex exponentials at integral multiples of the fundamental frequency. For functions that are non-periodic, called *aperiodic*, decomposition into sets of sinusoids is still possible. However, whereas the Fourier series yields a set of distinct and countable components, the spectrum of a non-periodic function can consist of an infinite set of sinusoids at every conceivable frequency. To find that composition we use the Fourier transform, or Fourier integral, defined by[6]

$$G(\omega) = \int_{-\infty}^{\infty} g(t)e^{-j\omega t}dt \qquad (7.11a)$$

This is the equivalent expression to (7.10b). The major difference is that there is no sense of period in this equation; in addition, the frequency term in the exponent is now a variable $\omega$ rather than a discrete set $\{n\omega_0\}$ as was the case for the Fourier series. Writing the transform as a function of the continuous frequency variable indicates the likelihood that it exists at all possible frequencies.

If we know the Fourier transform, or spectrum, of a function then the function can be reconstructed according to

$$g(t) = \frac{1}{2\pi} \int_{-\infty}^{\infty} G(\omega)e^{j\omega t}d\omega \qquad (7.11b)$$

This is the equivalent of the Fourier series expression of (7.10a).

To demonstrate the application of the Fourier transform, consider the single unit pulse shown in Fig. 7.6a. From (7.11a) the transform is seen to be

$$G(\omega) = \int_{-a}^{a} e^{-j\omega t}dt = 2a\frac{\sin a\omega}{a\omega}$$

which is shown plotted in Fig. 7.6b.[7] Again, note that the frequency axis accommodates both positive and negative frequencies, which together compose sinusoids. Note also that every possible frequency exists in the spectrum of the pulse, apart from a set where the spectrum crosses the frequency axis. The spectrum has negative as well as positive values. When negative, the phase spectrum has a value of $\pm 180°$.

An important Fourier transform is that of an impulse function. From the sifting property of the impulse and the fact that $e^0 = 1$, the transform is

---

[6] There are a number of conventions for defining the Fourier integral, largely to do with where the $2\pi$ is placed between (7.11a) and (7.11b).

[7] The function of the form $\frac{\sin x}{x}$, and shown in Fig. 7.6b, is called a sinc function.

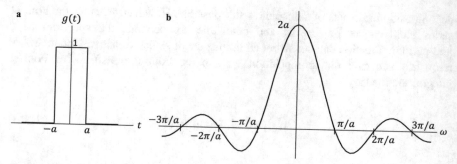

**Fig. 7.6** **a** Unit pulse and **b** its Fourier transform

$$G(\omega) = \int_{-a}^{a} \delta(t)e^{-j\omega t}dt = 1$$

This tells us that the impulse is composed of equal amounts of every possible frequency! It also suggests that functions which change rapidly in time will have large numbers of frequency components.

It is also interesting to consider the Fourier transform of a constant $c$. That can be shown to be

$$G(\omega) = \int_{-\infty}^{\infty} ce^{-j\omega t}dt = 2\pi c\delta(\omega)$$

In other words, as expected, the spectrum of a constant exists only at the origin in the frequency domain.[8] This expression can be derived based on the properties of the complex exponential; it can be verified by working backwards through (7.11b). While it might seem strange having the impulse function appear as a multiplier for the constant, we can safely interpret that as just a reminder that the constant exists at $\omega = 0$ and not as something that changes the amplitude of the constant.

The Fourier transform of a periodic function, normally expressed using the Fourier series, is important. We can obtain it by substituting (7.10a) into (7.11a) to give

$$G(\omega) = \int_{-\infty}^{\infty} \sum_{n=-\infty}^{\infty} G_n e^{jn\omega_0 t} e^{-j\omega t}dt$$

---

[8] If the transform at (7.11) had used the $2\pi$ as a denominator in (7.11a) then the $2\pi$ would not appear in the transform of a constant. If the transforms were defined in terms of frequency $f = \omega/2\pi$, the $2\pi$ factor would not appear at all. Some authors use that form.

$$\text{i.e.,} \quad G(\omega) = \sum_{n=-\infty}^{\infty} G_n \int_{-\infty}^{\infty} e^{j(n\omega_0 - \omega)t} dt$$

which becomes

$$G(\omega) = 2\pi \sum_{n=-\infty}^{\infty} G_n \delta(\omega - n\omega_0) \tag{7.12}$$

This last expression uses the property for the Fourier transform of a constant (in this case unity). It tells us that the only frequencies that can exist in the Fourier transform of a periodic function are those which are integral multiples of the fundamental frequency of the periodic waveform.

The last example is important because it tells us that we do not need to work with both Fourier series and Fourier transforms. Because the Fourier transform also applies to periodic functions, it is sufficient in the following to focus on the transform alone.

## 7.5 The Discrete Fourier Transform

Because our interest is in digital imagery, which is simply a two-dimensional version of the one-dimensional functions we have been considering up to now, it is important to move away from continuous functions of time (or any other independent variable) and instead look at the situation when the independent variable is discrete, and the dependent variable consists of a set of samples.

Suppose we have a set of $K$ samples over a time interval $T$ of the continuous function $g(t)$. Call these samples $\gamma(k), k = 0 \ldots K - 1$. The individual samples occur at the times $t_k = k\Delta t$ where $\Delta t$ is the spacing between samples. Note that $T = K\Delta t$. Consider now how (7.11a) needs to be modified to handle the set of samples rather than a continuous function of time. First, obviously the function itself is replaced by the samples. Secondly the integral over time is replaced by the sum over the samples and the infinitesimal time increment $dt$ in the integral is replaced by the sampling time increment $\Delta t$. The time variable $t$ is replaced by $k\Delta t = \frac{kT}{K}, k = 0 \ldots K - 1$. So far this gives as a discrete form of (7.11a)

$$G(\omega) = \Delta t \sum_{k=0}^{K-1} \gamma(k) e^{-j\omega k \Delta t}$$

We now have to consider how to treat the frequency variable $\omega$. We are developing this discrete form of the Fourier transformation so that it can handle digitised data and so that it can be processed by computer. Therefore, the frequency domain also has to be digitised by replacing $\omega$ by the frequency samples $\omega = r\Delta\omega, r = 0 \ldots K - 1$. We have deliberately chosen the number of samples in the frequency domain to be the same as the number in the time domain for convenience. What value now do we give to the increment in frequency $\Delta\omega$? That is not easily answered until we have treated sampling theory later in this chapter, so for the present note that it will be $2\pi/T$ and thus is directly related to the time over which the original function has been sampled. With this treatment of the frequency variable the last expression for the discrete version of the Fourier transform now becomes

$$G(r) = \Delta t \sum_{k=0}^{K-1} \gamma(k) e^{-\frac{j2\pi rk}{K}} \quad r = 0 \ldots K - 1$$

It is common to define

$$W = e^{-j2\pi/K} \tag{7.13}$$

so that the last expression is written

$$G(r) = \Delta t \sum_{k=0}^{K-1} \gamma(k) W^{rk} \quad r = 0 \ldots K - 1 \tag{7.14a}$$

Equation (7.14a) is known as the *discrete Fourier transform* (DFT). In a similar manner we can derive a *discrete inverse Fourier transform* (IDFT) to allow the original sequence $\gamma(k)$ to be reconstructed from the frequency samples $G(r)$. That is given by

$$\gamma(k) = \frac{1}{T} \sum_{r=0}^{K-1} G(r) W^{-rk} \quad k = 0 \ldots K - 1 \tag{7.14b}$$

If we substitute (7.14a) into (7.14b) we see they do in fact constitute a transform pair. To do so we need different indices for $k$; we will use $l$ instead of $k$ in (7.14b) so that

$$\gamma(l) = \frac{1}{T} \sum_{r=0}^{K-1} G(r) W^{-rl} = \frac{1}{T} \sum_{r=0}^{K-1} \Delta t \sum_{k=0}^{K-1} \gamma(k) W^{rk} W^{-rl}$$

i.e., $\quad \gamma(l) = \frac{1}{K} \sum_{K=0}^{K-1} \gamma(k) \sum_{r=0}^{K-1} W^{r(k-l)}$

From the properties of the complex exponential function the second sum is zero for $k \neq l$ and is $K$ when $k = l$, so that the right-hand side becomes $\gamma(l)$ as required. An interesting by-product of this analysis has been that the $\Delta t$ and $T$ divide to leave $K$, the number of samples. As a result, the transform pair in (7.14) can be written in the simpler form, used in software that computes the discrete Fourier transform:

$$G(r) = \sum_{k=0}^{K-1} \gamma(k) W^{rk} \quad r = 0 \ldots K-1 \tag{7.15a}$$

$$\gamma(k) = \frac{1}{K} \sum_{r=0}^{K-1} G(r) W^{-rk} \quad k = 0 \ldots K-1 \tag{7.15b}$$

These last two expressions are particularly simple. All they involve are the sets of samples to be transformed (or inverse transformed) and the complex constants $W$, which can be computed beforehand.

## 7.5.1   Properties of the Discrete Fourier Transform

Three properties of the DFT and IDFT are important.

*Linearity*

Both the DFT and IDFT are linear operations. Thus, if the set $G_1(r)$ is the DFT of the sequence $\gamma_1(k)$ and the set $G_2(r)$ is the DFT of the sequence $\gamma_2(k)$ then, for any constants, $a$ and $b$, $aG_1(r) + bG_2(r)$ is the DFT of $a\gamma_1(k) + b\gamma_2(k)$.

*Periodicity*

From (7.13) $W^{\pm mkK} = 1$ where $m$ and $k$ are integers, so that

$$G(r + mK) = \sum_{k=0}^{K-1} \gamma(k) W^{(r+mK)k} = G(r) \tag{7.16a}$$

Similarly,

$$\gamma(k + mK) = \frac{1}{K} \sum_{r=0}^{K-1} G(r) W^{-r(k+mK)} = \gamma(k) \tag{7.16b}$$

Thus, both the sequence of samples and the set of transformed samples are periodic with period $K$. This has two important implications. First, to generate the Fourier series of a periodic function it is only necessary to sample it over a single period. Secondly, the spectrum of an aperiodic function will be that of a periodic repetition of that function over the sampling duration—in other words it is made to look periodic by the limited time sampling. Therefore, for a limited time function such as the rectangular pulse shown in Fig. 7.6, it is necessary to sample the signal well beyond the range of arguments for which it is non-zero to ensure it looks approximately aperiodic.

*Symmetry*

Put $r' = K - r$ in (7.15a) to give

$$G(r') = \sum_{k=0}^{K-1} \gamma(k) W^{(K-r)k}$$

Since $W^{kK} = 1$ then $G(K - r) = G^*(r)$ where the * represents the complex conjugate. This implies that the amplitude spectrum is symmetric about $K/2$ and the phase spectrum is antisymmetric.

## 7.5.2 Computing the Discrete Fourier Transform

Evaluating the $K$ values of $G(r)$ from the $K$ values of $\gamma(k)$ in (7.15a) requires $K^2$ multiplications and $K^2$ additions, assuming that the values of $W^{rk}$ have been calculated beforehand. Since those numbers are complex, the multiplications and additions required to evaluate the Fourier transform are also complex. It is the multiplications that are the problem; complex multiplications require significant computing resources, so that transforms involving many samples can take significant time. Fortunately, a fast algorithm, called the *fast Fourier transform* (FFT), is available.[9] It only requires $\frac{K}{2} \log_2 K$ complex multiplications, which is a substantial reduction in computational demand. The implementation of the DFT in software uses the FFT algorithm. The only penalty in using this method is that the number of samples taken of the function to be transformed, and the number of samples in the transform, each have to be a power of two.

## 7.6 Convolution

### 7.6.1 The Convolution Integral

Before proceeding to look at the Fourier transform of an image it is of value to appreciate the operation called convolution. It was introduced in Sect. 5.8 in the context of geometric enhancement of imagery. We now look at it in more detail because of its importance in understanding both sampled data and the spatial processing of images. As with the development of the Fourier transform, we commence with its application to one dimensional, continuous functions of time. We will then modify it to apply to samples of functions. After having considered the Fourier transform of an image, we will look at the two-dimensional version of convolution.

---

[9] See E.O. Brigham, *The Fast Fourier Transform and its Applications*, Prentice-Hall, N. J., 1988.

Convolution is defined in terms of a pair of functions. For $g_1(t)$ and $g_2(t)$ the result is

$$y(t) = g_1(t) * g_2(t) = \int\limits_{-\infty}^{\infty} g_1(\tau)g_2(t-\tau)d\tau \qquad (7.17)$$

in which the symbol $*$ indicates convolution. The operation is commutative, i.e., $g_1(t) * g_2(t) = g_2(t) * g_1(t)$, a fact sometimes used when evaluating the integral.

We can understand the convolution operation if we break it down into the following four steps, which are illustrated in Fig. 7.7:

i. Folding: form $g_2(-\tau)$ by taking the mirror image of $g_2(\tau)$ about the vertical axis
ii. Shifting: form $g_2(t-\tau)$ by shifting $g_2(-\tau)$ by the variable amount $t$
iii. Multiplication: form the product $g_1(\tau)g_2(t-\tau)$
iv. Integration: compute the area under the product.

### 7.6.2 Convolution with an Impulse

Convolution of a function with an impulse is important in understanding sampling. The delta function sifting property in (7.8) gives

$$y(t) = g(t) * \delta(t-t_0) = \int\limits_{-\infty}^{\infty} g(\tau)\delta(t-\tau-t_0)d\tau = g(t-t_0)$$

Thus, the result is to shift the function to a new origin. Clearly, for $t_0 = 0$, $y(t) = g(t)$.

### 7.6.3 The Convolution Theorem

This theorem can be verified using the definitions of convolution and the Fourier transform.[10] It has two forms:

If $y(t) = g_1(t) * g_2(t)$        then $Y(\omega) = G_1(\omega)G_2(\omega)$      (7.18a)

If $Y(\omega) = G_1(\omega) * G_2(\omega)$     then $y(t) = 2\pi g_1(t)g_2(t)$      (7.18b)

---

[10] See A. Papoulis, *Circuits and Systems: a Modern Approach*, Holt-Saunders, Tokyo, 1980, and Brigham, loc. cit.

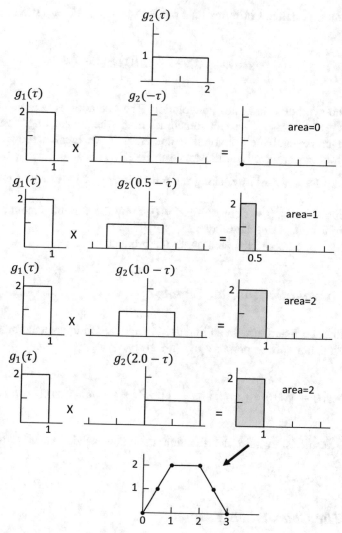

**Fig. 7.7** Graphical illustration of convolution

## 7.6.4   Discrete Convolution

Just as we can modify the Fourier transform formula to handle discrete samples rather than continuous functions, we can do the same with convolution. Suppose $\gamma_1(k)$ and $\gamma_2(k)$ are sampled versions of the functions $g_1(t)$ and $g_2(t)$ then (7.17) can be written in discrete form as

**Fig. 7.8** Discrete
convolution using two
sequences {5, −3, 7} and
{4, 6, − 5}; the second is
reversed and run past the first
as shown, generating the
aggregate sums {0, 20, 18,
−15, 57, −35, 0}

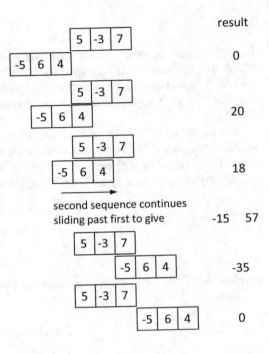

$$y(k) = \sum_{n} \gamma_1(n)\gamma_2(k - n) \quad \text{for all k} \tag{7.19}$$

The integral has been replaced by the sum operation. Strictly $dt$ should be replaced by the time increment between the samples, but it is usually left out in the discrete form. To evaluate (7.19) the set of samples $\gamma_2(k)$ is first reversed in order to form $\gamma_2(-n)$ and then slid past $\gamma_1(n)$ as shown in Fig. 7.8, taking products where samples coincide and then summing the results.

## 7.7 Sampling Theory

Discrete time functions, such as the sequence of samples that we considered when developing the discrete Fourier transform, and digital images, can be considered to be the result of sampling the corresponding continuous functions or scenes on a regular basis. The periodic sequence of impulses, spaced $\Delta t$ apart, sometimes called a *Dirac comb*,

$$\mathfrak{D}(t) = \sum_{k=-\infty}^{\infty} \delta(t - k\Delta t) \tag{7.20}$$

can be used to extract a uniform set of samples from a function $g(t)$ by forming the product

$$\mathfrak{D}(t)g(t) \tag{7.21}$$

From (7.7) this is seen to be a sequence of samples of value $g(kT)\delta(t - k\Delta t)$, which we represent by $\gamma(k)$. Despite the undefined magnitude of the delta function we will be content in this treatment to regard the product simply as a sample of the function $\gamma(t)$, so that (7.21) can be interpreted as a set of uniformly spaced samples of $\gamma(t)$.

It is important now to know the Fourier transform of the samples in (7.21). We will find that by calling on the convolution theorem in (7.18b), which requires the Fourier transforms of $g(t)$ and $\mathfrak{D}(t)$. Since $\mathfrak{D}(t)$ is periodic we can work via the Fourier series coefficient formula of (7.10b), which shows

$$\mathfrak{D}_n = \frac{1}{\Delta t} \int_{-\Delta t/2}^{\Delta t/2} \delta(t)e^{-jn\omega_0 t}dt = \frac{1}{\Delta t}$$

so that from (7.12) the Fourier transform of the periodic sequence of impulses in (7.20) is

$$\mathfrak{D}(\omega) = \frac{2\pi}{\Delta t} \sum_{n=-\infty}^{\infty} \delta(\omega - n\omega_s) \tag{7.22}$$

in which $\omega_s = 2\pi/\Delta t$. Thus, the Fourier transform of the periodic sequence of impulses spaced $\Delta t$ apart in the time domain is itself a periodic sequence of impulses in the frequency domain spaced apart $\frac{2\pi}{\Delta t}$ rad s$^{-1}$, or $\frac{1}{\Delta t}$ Hz.

Suppose $g(t)$ has the Fourier transform, or spectrum, $G(\omega)$; then from (7.18b) and (7.22) the spectrum of the *samples* of $g(t)$ represented by (7.21) is given by convolving $G(\omega)$ with (7.22). Convolution with an impulse shifts a function to a new origin centred on the impulse so that the outcome of this operation is a periodic repetition of the spectrum $G(\omega)$—the spectrum of the unsampled function—as shown in Fig. 7.9. The repetition period in the frequency domain is determined by the rate at which the time function is sampled —it is the inverse of the time domain sampling rate. Note that if the sampling rate is high then the repeated segments of the spectrum of the original function are well separated. If the sampling rate is low, then those segments are closer together.

Imagine that the frequency components of the original function are limited to frequencies below $B$ Hz (or $2\pi B$ rad s$^{-1}$). $B$ is called the bandwidth of $g(t)$. Not all real non-periodic functions have a limited bandwidth. The single pulse of Fig. 7.6 is an example. However, it suits our purpose here to assume that there is a limit $B$ to the frequency composition of those functions of interest to us. That allows us to introduce a particularly important concept in the sampling of signals and images. It is clear that if adjacent spectral segments in Fig. 7.9 are to remain separated then we require

**Fig. 7.9** Development of the Fourier transform of a sampled function:
**a** unsampled function and its spectrum, **b** periodic sequence of impulses and its spectrum, **c** sampled function and its spectrum; $\mathcal{F}$ is the Fourier transform operation, **d** sub-Nyquist rate sampling leading to overlap in the spectrum and thus aliasing distortion

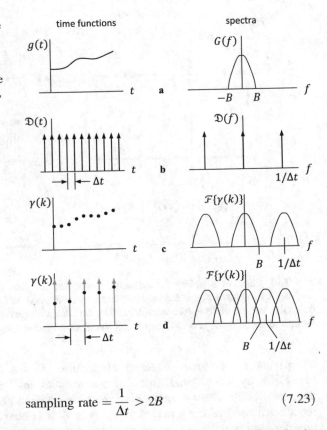

$$\text{sampling rate} = \frac{1}{\Delta t} > 2B \qquad (7.23)$$

In other words, the rate at which the function $g(t)$ is sampled must exceed twice its bandwidth. Should that not be the case then the segments of the spectrum will overlap as shown in Fig. 7.9d, causing a form of distortion called *aliasing*. The sampling rate of $2B$ in (7.23) is called the *Nyquist rate*. Equation (7.23) is often called the *sampling theorem*.

Return now to the concept of the discrete Fourier transform. On the basis of the sampling material just presented we know that the spectrum of a band limited sampled signal is a periodic repetition of the spectrum of the original unsampled signal. We now need to represent the spectrum by a set of samples so that we can handle the data digitally. Effectively that means sampling the spectrum shown in Fig. 7.10a by a periodic sequence of impulses (a sampling comb) in the frequency domain. For this purpose, consider an infinite periodic sequence of impulses in frequency spaced $\Delta f$ or $\Delta\omega/2\pi$ apart as seen in Fig. 7.10b. Using (7.20) and (7.22), but going from the frequency to the time domain, it can be shown that the inverse transform of the frequency domain sampling sequence is a set of impulses in the time domain, spaced $T = 1/\Delta f$ apart.

Multiplication of the spectrum of the sampled time function in Fig. 7.10a by the frequency domain sampling function in Fig. 7.10b produces the sampled spectrum of Fig. 7.10c. By the convolution theorem of (7.18a), that is equivalent to

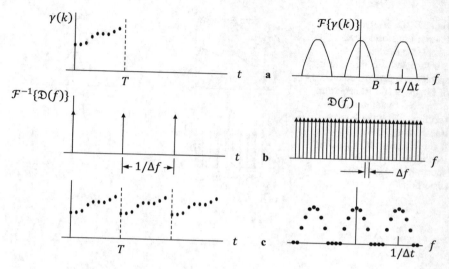

**Fig. 7.10** Effect of sampling the spectrum: **a** sampled function and its spectrum, **b** periodic sequence of impulses (right) used to sample the spectrum, and its time domain equivalent (left), **c** sampled version of the spectrum (right) and its time domain equivalent (left), which is a periodic version of the samples in **a**; $\mathcal{F}^{-1}$ represents the inverse Fourier transform

convolving the periodic sequence of impulses in the time domain shown in Fig. 7.10b by the original time function samples in Fig. 7.10a to produce the periodic repetition of the samples of the time function shown in Fig. 7.10c. The repetition period of the group of samples is $T$. It is convenient if the number of samples used to represent the spectrum is the same as the number of samples taken of the time function. Let that number be $K$, consistent with the development in Sect. 7.5. Since the time domain samples are spaced $\Delta t$ apart, the duration of sampling is $K\Delta t$. By the manner in which we have drawn Fig. 7.10, we have, for convenience, synchronised the total sampling period of the time function $T$ with the repetition period of the inverse Fourier transform of the frequency domain sampling comb in Fig. 7.10b. As a result, $\Delta f = 1/T = 1/K\Delta t$, the inverse of the sampling duration. Similarly, the total unambiguous bandwidth in the frequency domain is $K\Delta f = 1/\Delta t$ which covers just one segment of the spectrum.

## 7.8 The Discrete Fourier Transform of an Image

### 7.8.1 The Transform Equations

The previous sections have treated functions with a single independent variable. That could have been time, or position along a line of image data. We now turn our attention to functions with two independent variables to allow Fourier transforms of images to be computed and understood. Despite this implicit increase in complexity, we will find that we can take full advantage of the material of the previous sections to help in our understanding. Let

$$\phi(k,l)\, k,l = 0\ldots K - 1 \qquad (7.24)$$

be the brightness of a pixel at location $k,l$ in an image of $K \times K$ pixels. The set of image pixels is a digital sample of the scene recorded by the remote sensing imaging instrument. Therefore, the Fourier transform of the scene is given by the discrete Fourier transform of the set of pixel brightnesses. Building on the material of Sect. 7.5 it can be seen that the discrete Fourier transform of an image is given by

$$\Phi(r,s) = \sum_{k=0}^{K-1}\sum_{l=0}^{K-1} \phi(k,l) W^{rk} W^{sl} \qquad (7.25a)$$

An image can be reconstructed from its transform according to

$$\phi(k,l) = \frac{1}{K^2}\sum_{r=0}^{K-1}\sum_{s=0}^{K-1} \Phi(r,s) W^{-rk} W^{-sl} \qquad (7.25b)$$

### 7.8.2 *Evaluating the Fourier Transform of an Image*

We can rewrite (7.25a) as

$$\Phi(r,s) = \sum_{k=0}^{K-1} W^{rk} \sum_{l=0}^{K-1} \phi(k,l) W^{sl} \qquad (7.26)$$

The right-hand sum will be recognised as the one-dimensional Fourier transform of the $k$th row of pixels in the image, which we write as

$$\Phi(k,s) = \sum_{l=0}^{K-1} \phi(k,l) W^{sl} \quad k = 0\ldots K - 1 \qquad (7.27)$$

Thus, the first step is to Fourier transform each row of the image. We then replace the row by its transform. The transformed pixels are now addressed by the spatial frequency index $s$ across the row rather than the positional index $l$. Using (7.27) in (7.26) we have

$$\Phi(r,s) = \sum_{k=0}^{K-1} \Phi(k,s) W^{rk} \qquad (7.28)$$

which is the one-dimensional discrete Fourier transform of the $s$th column of the image, after the row transforms of (7.27) have been computed. Therefore, to compute the two dimensional Fourier transform of an image, it is only necessary to transform each row individually to generate an intermediate image, and then transform that result by column to yield the final transform. Both the row and column transformations would be carried out using the fast Fourier transform algorithm, which requires $K^2 \log_2 K$ complex multiplications to transform the complete image.

### 7.8.3  The Concept of Spatial Frequency

Entries in the Fourier transformed image $\Phi(r, s)$ represent the composition of the original image in terms of spatial frequency components, vertically and horizontally. Spatial frequency is the image analogue of the time frequency of signal. A sinusoidal function with a high frequency alternates rapidly, whereas a low-frequency function changes slowly with time. Similarly, an image with high spatial frequency in, say, the horizontal direction shows frequent changes of brightness with position horizontally. A picture of a crowd of people would be a particular example, whereas a head and shoulders view of a person reading the news on television is likely to be characterised mainly by low spatial frequencies. Typically, an image is composed of a collection of both horizontal and vertical components with different spatial frequencies of differing strengths. They are what the discrete Fourier transform describes.

The upper left-hand pixel in $\Phi(r, s)$—$\Phi(0, 0)$—is the average brightness value of the image. In engineering this would sometimes be called the DC value. That is the component of the spectrum with zero frequency variation in both directions. Thereafter, pixels in $\Phi(r, s)$ both horizontally and vertically represent components with frequencies that increment by $1/K$ where the original image is of size $K \times K$. In most cases we would know the scale of the image, in other words the distance on the ground covered by the $K$ pixels across the lines and down the columns. That allows us to define the spatial frequency increment in terms of metres$^{-1}$. For example, if an image covered 5 km $\times$ 5 km, then the spatial frequency increment in both directions is $2 \times 10^{-4}$ m$^{-1}$.

### 7.8.4  Displaying the DFT of an Image

In Fig. 7.9 we saw that the one dimensional discrete Fourier transformation of a function is periodic with period $K$. The same is true of the discrete Fourier transform of an image. The $K \times K$ pixels of $\Phi(r, s)$ can be viewed as one period of an infinitely periodic two-dimensional array in the manner depicted in Fig. 7.11. We also saw that the amplitude spectrum of the one dimensional DFT is symmetric about $K/2$. Similarly, $\Phi(r,s)$ is symmetric about its centre. Therefore, no new amplitude information is shown by displaying transformed pixels horizontally and vertically beyond $K/2$. Rather than ignore them (since their accompanying phase is important) the display is adjusted as shown in Fig. 7.11 to bring $\Phi(0,0)$ to the centre. In that manner the pixel at the centre of the Fourier transform array represents the image average brightness value. Pixels away from the centre represent the proportions of image components with increasing spatial frequency. That is the usual manner for presenting two-dimensional image transforms. Examples of spectra displayed in this manner are given in Fig. 7.12. To make visible those components with smaller amplitudes logarithmic scaling[11] is sometimes used

$$D(r,s) = \log\{|\Phi(r,s)|\} \tag{7.29}$$

## 7.9 Image Processing Using the Fourier Transform

Having a knowledge of the discrete Fourier transform of an image allows us to develop more general geometric processing operations than those treated in Chap. 5. In preparation for this, note that the high spatial frequency content of an image is associated with frequent changes of brightness with position. Edges, lines and some types of noise are examples of high spatial frequency data. In contrast, gradual changes of brightness with position account for the low frequency components of the spatial spectrum. Since ranges of spatial frequency are identifiable with regions in the spectrum, we can understand how the spectrum of an image can be altered to produce different geometric enhancements of the image itself. For example, if regions near the centre of the spectrum are removed, leaving behind only the high frequencies, and the image is then reconstructed from the modified spectrum, a version containing only edges and line-like features will be produced. On the other hand, if the high frequency components are removed, leaving behind only the region near the centre of the spectrum, the reconstructed image will appear smoothed, since edges, lines and other high-frequency detail will have been removed.

Modification of the two-dimensional discrete image spectrum in the manner just described can be expressed as the product of the image spectrum $\Phi(r,s)$ and a filter function $H(r,s)$ to generate the new spectrum:

$$Y(r,s) = H(r,s)\Phi(r,s) \tag{7.30}$$

---

[11] R.C. Gonzalez and R.E. Woods, *Digital Image Processing*, 4th ed. Pearson Prentice-Hall, Upper Saddle River, N.J., 2018.

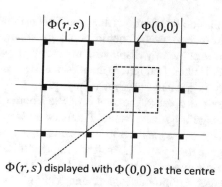

$\Phi(r, s)$ displayed with $\Phi(0,0)$ at the centre

**Fig. 7.11** Showing the periodic nature of the two dimensional discrete Fourier transform, indicating how an array centred on the average value $\Phi(0,0)$ is chosen for symmetric display purposes

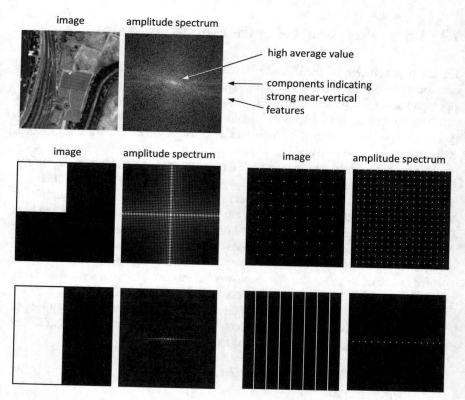

**Fig. 7.12** Examples of Fourier transforms for images: the top set is a visible red image with strong vertical features, and its spectrum, while the others are further demonstrations of the influence of geometry; in each case the logarithm of the amplitude spectrum is shown

To implement simple sharpening or smoothing as described above $H(r, s)$ would be set to 0 for those frequency components to be removed and 1 for those frequency components to be retained. Equation (7.30) also allows more complicated filtering operations to be carried out. For example, a specific band of spatial frequencies could be excluded if they corresponded to some form of periodic noise, such as the line striping sometimes observed with line scanner data. Also, the entries in $H(r, s)$ can be different from 0 or 1, allowing more sophisticated changes to the spectrum of an image. Figure 7.13 shows the effect of applying ideal (sharp cut-off) filters to the image segment in Fig. 7.12. The filter cut off values are shown superimposed over the log amplitude spectrum of the image by circles.

The low pass filtered images are those generated by retaining only those frequency components inside the circles, whereas the high pass filtered versions are made up from those spectral components outside the filter circles. Even though the filters are shown for convenience over the amplitude spectra they are applied to the full complex Fourier transform of the original image. Modification of the spatial (geometric) features of an image in the frequency domain in this manner involves three steps. First, the image has to be Fourier transformed to produce a spectrum. Secondly, the spectrum is modified according to (7.30). Finally, the image is reconstructed from the modified spectrum using the inverse discrete Fourier transform. Together, these three operations require $2K^2 \log_2 K + K^2$ multiplications.

## 7.10 Convolution in Two Dimensions

The convolution theorem for functions given in Sect. 7.6.3 has a two dimensional counterpart, again in two forms:

If $y(k, l) = \phi(k, l) \ast h(k, l)$      then $Y(r, s) = \Phi(r, s)H(r, s)$     (7.31a)

If $Y(r, s) = \Phi(r, s) \ast H(r, s)$      then $y(k, l) = \phi(k, l)h(k, l)$     (7.31b)

Unlike (7.18b) there is no $1/2\pi$ scaling factor here since the spatial frequency variables $r$, $s$ are equivalent to frequency $f$ in hertz and not frequency $\omega$ in radians per second.

The operation in (7.31) is the discrete version of convolution shown in (5.13) and described in Sect. 5.8. Equation (7.31a) shows that any of the geometric enhancement operations that can be implemented by modifying the image spectrum can also be carried out by performing a convolution between the image itself and the inverse Fourier transform of the filter function $H(r, s)$. Conversely, operations such as simple mean value filtering described in Sect. 5.3.1 can be implemented in the spatial frequency domain; that needs the Fourier transform of the template. The template has to have the same dimensions as the image but with values of 0 everywhere except for the set of pixels that are used to implement the prescribed operation.

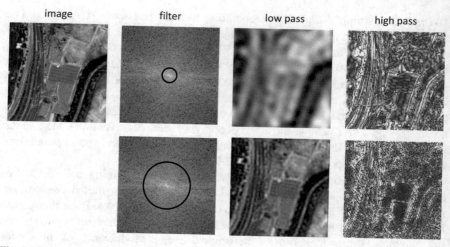

**Fig. 7.13** Examples of low pass and high pass spatial filtering based on the Fourier transform of the original image; in this case filters with sharp cut-offs have been used

## 7.11   Other Fourier Transforms

If (7.3b) is substituted in (7.11a) we have

$$G(\omega) = \int_{-\infty}^{\infty} g(t)(\cos \omega t - j \sin \omega t)dt$$

which can be separated into

$$G(\omega) = \int_{-\infty}^{\infty} g(t) \cos \omega t \, dt \tag{7.32a}$$

and

$$G(\omega) = \int_{-\infty}^{\infty} g(t) \sin \omega t \, dt \tag{7.32b}$$

the first of which is called a Fourier cosine transform, and the second of which is called a Fourier sine transform. They are applied to even and odd functions respectively, in which case the integrals usually go from 0 to ∞. There is a discrete version of the Fourier cosine transform, called the DCT or discrete cosine transform, which is given by discretising (7.32a) in the same manner we discretised

(7.11a) to generate the DFT. The DCT finds widespread application in video compression for the television industry.[12]

## 7.12   Leakage and Window Functions

In Sect. 7.7 we noted that a sampled function can be regarded as the unsampled version multiplied by an infinite periodic sequence of impulses. The spectrum of the set of samples produced is the spectrum of the original function convolved with the spectrum of the sequence of impulses; we saw that in Fig. 7.9.

In practice it is not possible to take an infinite number of samples of the function. Instead, sampling is commenced at a given time and terminated after some period $\tau$, as illustrated in Fig. 7.14. A finite set of samples can be represented by a long pulse of unit amplitude and duration $\tau$—*a sampling window*—that multiplies an infinite sequence of samples. The spectrum of the finite set of samples is, as a consequence, modified by being convolved by the spectrum of the sampling window, again as illustrated in Fig. 7.14. Since the sampling window is a rectangular pulse its Fourier transform is as shown in Fig. 7.6. Because the pulse is long compared with the sampling interval, the spectrum shown in Fig. 7.6 is compressed and looks like a finite amplitude impulse, thus approximating well the situation with an infinite sampling comb. However, when finite in length, its side lobes create problems during convolution with the spectrum of the sequence of samples, causing distortion, as depicted in Fig. 7.14.

To minimise that form of distortion, which is referred to as *leakage*, the rectangular sampling window is replaced by a function which avoids the sharp turn on and turn off with time that characterises the rectangular function. There are several candidates for these so-called *window functions*,[13] perhaps the most common of which is the raised cosine or Hanning window:

$$w(t) = 0.5 - 0.5 \cos\left(\frac{2\pi t}{\tau}\right) \tag{7.33}$$

This has smaller side lobes in its spectrum than the simple rectangular pulse and, as a result, leads to less leakage distortion.

If the function being sampled is periodic, and the samples are taken over one or several full periods, leakage will not occur. Otherwise, leakage is always a matter for consideration, and window functions generally need to be used.

---

[12] See J.A. Arnold, M. Frater and M. Pickering, *Digital Television; Technology and Standards*, John Wiley & Sons, N.Y., 2007.

[13] An excellent treatment of leakage and the use of window functions is given in E.O Brigham, *The Fast Fourier Transform and its Applications*, Prentice-Hall, N.J., 1988.

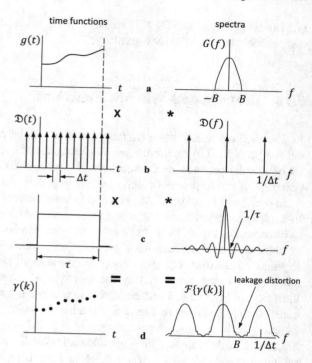

**Fig. 7.14** Demonstrating the effect of leakage distortion: **a** time signal and its spectrum, **b** infinite sequence of sampling impulses and its spectrum, **c** finite time sampling window and its spectrum, and **d** the result of finite time sampling as a product in the time domain and a convolution in the frequency domain

## 7.13 The Wavelet Transform

### 7.13.1 Background

In principle, the Fourier transform can be used to represent any signal by a collection, sometimes infinite, of sinusoidal functions. Likewise, the two-dimensional spatial Fourier transform can be used to model the distribution of brightness values in an image by using a collection of two dimensional sinusoidal basis functions.

Many of the image features of interest to us occur over a short distance, including edges and lines. Also, when dealing with functions of time, we are sometimes interested in representing short time signals rather than those that last for a long period. As an illustration, an organ playing a single, pure tone generates a signal that is well-modelled by simple sinusoids. In contrast, when a single note is played on a piano we have an approximately sinusoidal signal, at the frequency of the key played, which lasts for just a short time. We can still find the Fourier transform representation of the piano note—its spectrum—but there are other ways to represent such a short time signals, just as there are other ways of representing or modelling image features that change over a short distance. The wavelet transformation is generally more useful in such situations than the Fourier transform. It is based on the definition of a wavelet, which is a wavelike signal that is limited in time (or space, in the spatial domain). The theory of the wavelet transformation is quite detailed, especially when treated comprehensively.[14] Here we provide a simple introduction in which the mathematical detail is kept to a minimum and some concepts are simplified, so that its common usage in image processing can be understood. It finds most application in image compression and coding, and in the detection of localised features such as edges and lines.

### 7.13.2   *Orthogonal Functions and Inner Products*

The Fourier series and transform expansions of (7.10a) and (7.11b) are special cases of the more general representation of a function by a sum of other functions, expressible as

$$g(t) = \sum_n a_n \psi_n(t) \tag{7.34}$$

The $\psi_n(t)$ are called *basis functions* and the $a_n$ are expansion coefficients. We saw how to find the expansion coefficients for complex exponential basis functions in (7.10b). To do so depends on a property of the basis functions called *orthogonality*, which means:

$$\int \psi_m(t)\psi_n(t)dt = 1 \quad \text{for } m = n, \text{ and zero otherwise} \tag{7.35}$$

The range of the integral depends on the actual basis functions themselves. In (7.10b) the range extends over one period of the sinusoidal basis functions. If (7.35) holds, then using (7.34) we can see

$$\int g(t)\psi_m dt = \int \sum_n a_n \psi_n(t)\psi_m dt$$

$$= \sum_n a_n \int \psi_n(t)\psi_m dt = a_m$$

which gives us the procedure for calculating values for the expansion coefficients. That is seen in explicitly (7.10b), and in the Fourier transform formula of (7.11a).

It is fundamental to many functional representations of the form of (7.34) that an orthogonal basis set is chosen so that the expansion coefficients are easily established. In the general theory of the wavelet transform, in which we will seek to represent practical functions and images by basis functions that exist over only a limited interval, the same is essentially true.

---

[14] For a detailed treatment of wavelets see G. Strang and T. Nguyen, *Wavelets and Filter Banks*, Wellesley-Cambridge, Mass., 1996, and K.R. Castleman, *Digital Image Processing*, 2nd ed., Prentice Hall, N.J., 1996.

Operations like that in (7.35) are called inner products and are written symbolically as

$$\langle \psi_m(t), \psi_n(t) \rangle = \int \psi_m(t)\psi_n(t)dt$$

### 7.13.3 Wavelets as Basis Functions

What sorts of wavelet basis functions should we use in practice? Whatever functions we choose they have to be able to model events that occur at different positions in time, or space when we look at images, and to accommodate events that, while being localised, can occur over different ranges of time or position. To achieve that the wavelet basis functions generally have to have two indices (one for location and one for spread) so that a function can be represented, or expanded, as

$$g(t) = \sum_{j,k} c_{j,k}\psi_{j,k}(t) \tag{7.36}$$

Figure 7.15 shows such a set of functions. We see that a fundamental function is translated and scaled so that it can cover instances at different times and over different durations.

Rather than define translations and scalings arbitrarily we restrict attention to binary scalings, in which the range is shrunk by progressive factors of two, and to dyadic translations, in which the shift amount is an integral multiple of the binary scaling factor. That means that the set of wavelet functions we are dealing with are built up from a basis function $\psi(t)$, sometimes called the *mother wavelet* or generating wavelet, such that all other wavelets are defined by

$$\psi_{j,k}(t) = 2^{j/2}\psi(2^j t - k) \tag{7.37}$$

in which $j$ is the scaling factor and $k$ is the integral multiple of the scaling by which the shift occurs. The factor $2^{j/2}$ is included so that the integral of the squared amplitude of the wavelet is unity, one of the requirements of a wavelet basis function. The relationship in (7.37) is used in Fig. 7.15, although the amplitude scaling is omitted for clarity. Note that, apart from being integers, $j$ and $k$ are arbitrary at this stage. Putting (7.37) in (7.36) we have

$$g(t) = \sum_{j,k} c_{j,k} 2^{j/2}\psi(2^j t - k) \tag{7.38}$$

The set of coefficients $c_{j,k}$ is sometimes called the wavelet transform of $g(t)$, or its wavelet spectrum.

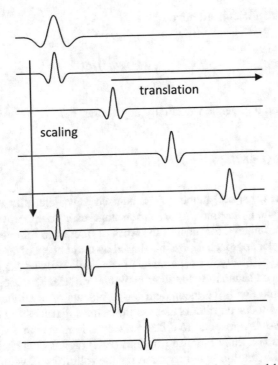

**Fig. 7.15** Some members of a family of wavelets (Mexican hat) created by time scaling and translating a basic form (without any amplitude scaling)

## 7.13.4 Dyadic Wavelets with Compact Support

It is possible to restrict further the set of wavelets of interest by establishing a relationship between $j$ and $k$. If we constrain our attention to so-called *compact* functions $g(t)$ that are zero outside the interval $[0, 1]$ we can use a single index $n$ to describe the set of basis functions, where

$$n = 2^j + k$$

The basis functions, while still having the general form in (7.37), can then be indexed by the single integer $n$, in which case the wavelet expansion, or representation, of the time restricted signal is

$$g(t) = \sum_{n=0}^{\infty} c_n \psi_n(t) \tag{7.39}$$

The expansion coefficients are given by

$$c_n = \int\limits_{-\infty}^{\infty} g(t)\psi_n(t)dt \qquad (7.40)$$

We do not pursue this version explicitly any further here.

### 7.13.5  Choosing the Wavelets

Not all finite time (or space) functions can be used as wavelets; it is only those that satisfy the so-called *admissibility criterion* that can be employed in wavelet expansions.[15] Fortunately, for most of the work of interest to remote sensing image processing a simpler approach is possible, based on the concept of *filter banks*, which avoids the need specifically to treat a range of candidate wavelet families. Although originally developed separately, the filter bank and wavelet approaches are related. Rather than continue with the theoretical developments of continuous wavelets as such, we will now focus on filter banks. As the name suggests a filter bank is made up of a set of filters that respond to different frequency ranges in a signal or image. Each of the filters is a finite impulse response (FIR) digital filter. A background in that material, while helpful, is not required for the following development.

### 7.13.6  Filter Banks

#### 7.13.6.1  Sub Band Filtering, and Downsampling

Suppose we want to represent the signal $g(t)$ by a wavelet model. The first step in the filter bank approach is to separate the signal into its low and high frequency components by passing it through two filters as illustrated in Fig. 7.16; one extracts the low frequencies from the signal and the other the high frequencies. In this case the filters are chosen so that they divide the spectrum of the signal into two halves, as indicated, which makes the following development simple. In practice other arrangements are also used.

We can represent the filtering operation diagrammatically as in Fig. 7.17, which shows the filters as blocks. The output of each filter block is given by the convolution between the input signal and the function of time—the filter function—that describes the operation of the filter.[16] The filter function is also called its *impulse*

---

[15] See Castleman, loc. cit., Sect. 14.2.1 and Problem 7.14.
[16] See also Sect. 5.8 and Fig. 5.13.

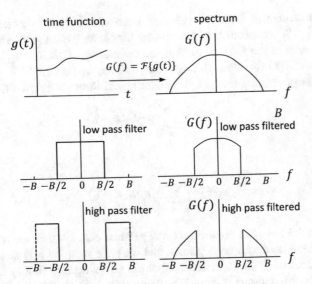

**Fig. 7.16** Filtering out the low and high frequency components of a signal

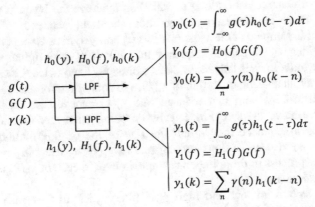

**Fig. 7.17** Signal transmission through low pass and high pass filters, showing three ways of describing signals and system functions: as continuous in time, by their frequency spectra or as discrete time (sampled) versions

*response*. That name comes about because if an impulse is fed into the filter then the output (the response) is the filter function, as seen from Sect. 7.6.2.

If we use the convolution theorem of (7.18a) we can represent the output of the filter in the frequency domain as the product of the Fourier transform of the input signal and the Fourier transform of the filter impulse response, also called the filter's *transfer function*.

We now assume that the input signal has been sampled and is represented by the sequence $\gamma(k)$. When dealing with sampled functions we must also represent the impulse response of the filter by its sampled counterpart. Call this $h_0(k)$ for the low pass filter and $h_1(k)$ for the high pass filter, as seen in Fig. 7.17.

The results of sending the sampled input signal through the two filters are

$$y_0(k) = \sum_n \gamma(n)h_0(k-n) \tag{7.41a}$$

and

$$y_1(k) = \sum_n \gamma(n)h_1(k-n) \tag{7.41b}$$

For simplicity we have left out limits on the summations, but they are understood to cover the full range of samples. Assume that we have sampled $g(t)$ at or just above the Nyquist limit of (7.23), which says that the sampling rate should exceed twice the bandwidth of the signal if there is to be no (aliasing) distortion, and so that the signal can be fully recovered from the samples. Such an assumption of near Nyquist rate sampling helps simplify the following development. It can be shown that the results hold in general just so long as sampling is above the Nyquist rate.

We now make an interesting observation about the sequence of samples in (7.41a); $y_0(k)$ is sampled at the same rate as $g(t)$ and yet, because it is the output of a low pass filter with a cut-off frequency that is half of the original signal bandwidth, the Nyquist limit is halved. In other words, we have twice the sampling rate necessary to avoid aliasing distortion in $y_0(k)$. We could therefore halve the rate and still avoid distortion; with that reduced rate $y_0(k)$ can still be reconstructed if necessary. We halve the sampling rate by dropping out every second sample in $y_0(k)$, leaving a sequence that is half the size of the original. Although not as obvious, we can also halve the sampling rate of $y_1(k)$ by dropping every second sample.[17] The process of dropping one sample in two, in each of $y_0(k)$ and $y_1(k)$, is called *downsampling*.

The next step is to low and high pass filter $y_0(k)$. In this case the cut-off frequency of the filter $h_0(k)$ is set to $B/4$ because the bandwidth of $y_0(k)$ is $B/2$. Similarly, the band pass of $h_1(k)$ is from $B/4$ to $B/2$. Thus, both filters now have a bandwidth that is half that used in the first stage; that affects the impulse response as we show below in Fig. 7.19.

Again, the result is downsampled by dropping every second sample in the resultant signals. We continue in that fashion, by filtering the low frequency signals

---

[17] See Castleman, loc. cit., Sect. 14.4.3.2. As an alternative explanation for those with a signal processing background, the ability to drop every second sample in $y_1(k)$ can be appreciated if we recognise that because the spectrum below $B/2$ is unfilled, the high pass signal can be frequency downshifted (translated) to the origin without distortion, whereupon its highest frequency component is then at $B/2$ and not $B$. The downshifted version can then also be represented without distortion at the downsampled rate.

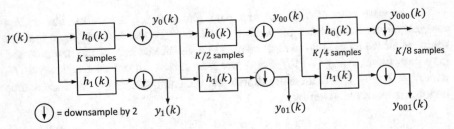

**Fig. 7.18** Successive low and high pass filtering, followed by downsampling; the impulse responses at each stage are successively dilated (scaled)

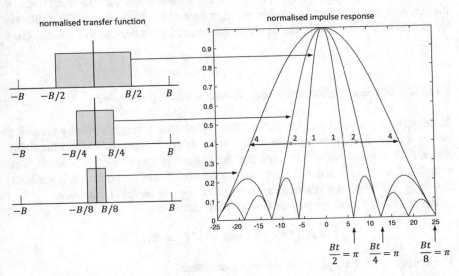

**Fig. 7.19** Relationship between the transfer function and the normalised magnitude of the impulse response for an ideal low pass filter

at each step, using filters with bandwidths half of their previous values, and downsampling, until the point is reached where the resulting low frequency signal is a single number. That process is shown in Fig. 7.18 for a three-stage *filter bank*. It has decomposed the sequence $\gamma(k)$ into the set $\{y_1(k), y_{01}(k), y_{001}(k), y_{000}(k)\}$ as a result of the chosen $h_0(k)$ and $h_1(k)$. The latter play the part of basis functions while the $\{y_1(k), y_{01}(k), y_{001}(k), y_{000}(k)\}$ act as expansion coefficients. That is the wavelet spectrum of $\gamma(k)$ on the $h_0(k), h_1(k)$ basis. At first it seems odd having two apparently different basis functions. Indeed, there aren't, as we will see soon in Sect. 7.13.6.3.

The high pass filters $h_1(k)$ constitute the wavelets, with the first—that with a bandwidth of $B/2$—being the mother wavelet. As we move from left to right in Fig. 7.18 the impulse responses broaden (dilate) as expected of a set of wavelet basis functions, even though that is opposite to the scaling from coarse to fine in the family shown in Fig. 7.15.

The impulse responses of the low pass filters, $h_0(k)$, are called the *scaling vectors* and their outputs are the *scaling functions*.[18] To see the changes in impulse response all we have to do is find the time equivalent versions of the low pass and high pass filters, using (7.11b). For a low pass filter with cut off frequency $B/N$, with $N = 2^j$, $j = 1, 2 \ldots$ it can be shown that the amplitude of the continuous time impulse response is given by

$$g(t) = \frac{B}{N\pi} |\text{sinc}(Bt/N)|$$

which is shown plotted in Fig. 7.19. It can be demonstrated that the magnitude of the impulse response for the equivalent high pass filters is the same. Time dilation by factors of 2 is evident in the figure as the bandwidth is decreased by those same factors.

### 7.13.6.2  Reconstruction from the Wavelets, and Upsampling

In principle we can reconstruct the filtered signal after a single stage by choosing two new filters in the manner seen in Fig. 7.20, shown in terms of transfer functions rather than impulse responses since that leads quickly to an important result. The signals are also expressed in their frequency domain (Fourier transformed) versions. By using the frequency domain representation, we can avoid convolution.

As in Fig. 7.17 the outputs of the left-hand filters are

$$Y_0(f) = H_0(f)G(f) \quad \text{and} \quad Y_1(f) = H_1(f)G(f)$$

The output from the right-hand summing device is

$$\mathbb{Y}(f) = R_0(f)Y_0(f) + R_1(f)Y_1(f)$$

which, when substituting for $Y_0(f)$ and $Y_1(f)$, gives

$$\begin{aligned} \mathbb{Y}(f) &= R_0(f)H_0(f)G(f) + R_1(f)H_1(f)G(f) \\ &= [R_0(f)H_0(f) + R_1(f)H_1(f)]G(f) \end{aligned}$$

Now if

$$R_0(f)H_0(f) + R_1(f)H_1(f) = 1 \tag{7.42a}$$

----

[18] They are also referred to as the *approximations* to the original signal; they are a successive set of reduced resolutions of the original and are sometimes said to form an image pyramid. They can be employed in multi-resolution analysis.

**Fig. 7.20** Decomposition
followed by reconstruction

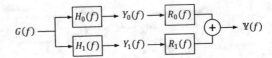

then

$$\mathbb{Y}(f) = G(f)$$

Equation (7.42a) shows a general relationship that must hold for perfect reconstruction. However, since the filtered samples were downsampled, they must be upsampled before the reconstruction filters can generate the correct result.[19] That is done by inserting zeros between each successive sample in $y_0(k)$ and $y_1(k)$. An outcome of doing that is that (7.42a) generalises to

$$R_0(f)H_0(f) + R_1(f)H_1(f) = 2 \tag{7.42b}$$

The reconstruction segment of Fig. 7.20 can be cascaded as was the analysis block in Fig. 7.18 to create a synthesis or reconstruction filter bank.

It would be desirable if the same filter bank could be used in reverse order to reconstruct the original signal from the transformed version. That turns out to be possible when the impulse response samples are symmetric, which is the simple case we treat in this section. In general, the samples have to be reversed when reconstructing.[20] The special case is shown in Fig. 7.21. As noted above, since, in Fig. 7.18, the sequences have been downsampled after each stage of filtering we have to upsample during reconstruction to restore each successive set of sequences to the right number of samples.

### 7.13.6.3 Relationship Between the Low and High Pass Filters

We now consider the interdependence of $h_0(k)$ and $h_1(k)$. Clearly, they are intimately related because one is the complement of the other in the frequency domain, as observed in Fig. 7.16. What does that mean for the samples themselves? If we view the high pass filter in the frequency domain as a frequency shifted version of the low pass filter, we can employ the frequency shift property of the Fourier transform to show the $h_0(k)$, $h_1(k)$ relationship. To use that theorem, we note first that the spectrum of the high pass filter is essentially the same as that of the low pass filter but shifted along the frequency axis by an amount $B$. That is expressed

---

[19] G. Strang and T. Nguyen, *Wavelets and Filter Banks*, Wellesley-Cambridge, Mass., 1996, Sect. 1.4 gives a nice example of reconstruction involving down- and upsampling.
[20] See P.S. Addison, *The Illustrated Wavelet Handbook*, IOP Publishing, Bristol, 2002, Fig. 3.18, and Strang and Nguyen, loc. cit.

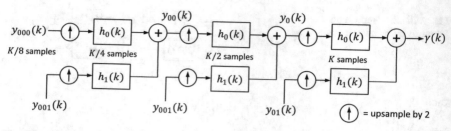

**Fig. 7.21** Reconstruction of the original time sequence by upsampling and recombining the low and high pass filtered versions at each stage; in this case the same impulse responses are shown for reconstruction as were used for analysis in Fig. 7.18

$$H_1(f) = H_0(f - B)$$

From the frequency shift theorem.[21]

$$h_1(t) = e^{j2\pi Bt} h_0(t)$$

We now have to express this in discrete form. The filter bandwidth B in a digital filter is set by the time between samples. A higher sampling rate leads to a greater bandwidth, and vice versa. If we write the continuous time variable as an integral multiple of the sampling interval $t = k\Delta t$ then $B = 1/2\Delta t$ so that, in discrete form, the last expression becomes

$$h_1(k) = e^{j\pi k} h_0(k)$$

The complex exponential with an exponent that is an integral multiple of $\pi$ is $\pm 1$. That tells us that the impulse response of the high pass filter is that same as that of the low pass filter, except that every odd numbered sample has its sign reversed. That applies for the case of the ideal low and high pass filters. More generally, it also requires a reversal of the order of the samples.[22]

### 7.13.7   *Choice of Wavelets*

In the filter bank development of Sect. 7.13.6 we have presented a decomposition and synthesis methodology that emulates wavelet analysis. It is based on the specification of the impulse responses of the low and high pass filters, which we

---

[21] See Brigham, loc. cit.

[22] See Strang and Nguyen, loc. cit.

now need to be more specific about because they describe the wavelets that are the basis functions for a given situation.

Although (7.38) is acceptable as a general expression of a wavelet expansion, it is more convenient, when comparing it to the filter bank approach, if we re-express the decomposition as the combination of an expansion in wavelets $\psi(t)$ plus a companion *scaling function* $\varphi(t)$ in the form

$$g(t) = A\varphi(t) + \sum_{j,k} c_{j,k} 2^{j/2} \psi(2^j t - k) \tag{7.43}$$

The scaling function satisfies the *scaling* or *dilation equation*

$$\varphi(t) = \sum_k h_0(k)\varphi(2t - k) \tag{7.44}$$

while the mother wavelet satisfies the *wavelet equation*

$$\psi(t) = \sum_k h_1(k)\varphi(2t - k) \tag{7.45}$$

As an illustration of these suppose the scaling function is the constant between 0 and 1 shown in Fig. 7.22 and the filter impulse response is a simple two sample average for the low pass filter $[h_0(0) = 1, h_0(1) = 1]$ and a simple two sample difference $[h_1(0) = 1, h_1(1) = -1]$ for the high pass filter. Then (7.44) and (7.45) become, respectively

$$\varphi(t) = \varphi(2t) + \varphi(2t - 1) \tag{7.46a}$$

$$\psi(t) = \varphi(2t) - \varphi(2t - 1) \tag{7.46b}$$

which are also shown plotted in Fig. 7.22. The recurrence relationship in (7.37) then allows others in the set to be generated. These are the Haar wavelets. When used as the basis for a filter bank implementation of the wavelet transform the expressions in (7.46) need to be scaled by $1/\sqrt{2}$ to give the correct square integral for the wavelets.

Haar wavelets are generated by the simple sum and difference filters above. A variety of other types is available, many of which are examples of the Daubechies wavelets[23] that can also be implemented readily as filter banks. The family of Daubechies wavelets includes the Haar wavelet as its simplest case.

---

[23] See Strang and Nguyen, loc. cit., and Addison, loc. cit., Fig. 3.15.

**Fig. 7.22** The Haar scaling function and wavelets generated with the scaling, dilation and recurrence relations—this is the first subset of 8 Haar wavelets; also shown at the top is the scaling function on a half time scale; note that the $2^{j/2}$ amplitude scaling in (7.37) has not been included

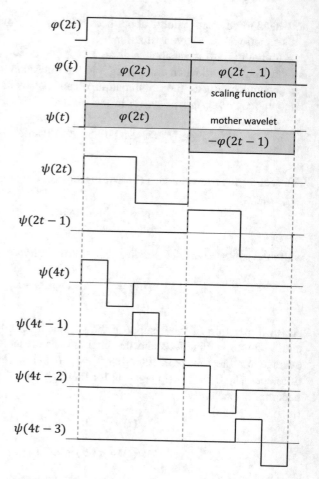

## 7.14 The Wavelet Transform of an Image

The application of the discrete wavelet transformation to imagery is similar to the manner in which the discrete Fourier transform is applied. First, the rows of the image are transformed using the first stage process in Fig. 7.18. Every second column is then discarded, which corresponds to downsampling the row transformed data. Next the columns are transformed and every second row discarded.

As illustrated in Fig. 7.23 that leads to four new images:

- a version that has been low pass filtered in both row and column, and which is referred to as the *approximation image*
- a version that has been high pass filtered to emphasise the horizontal edge information
- a version that has been high pass filtered to emphasise the vertical edge information

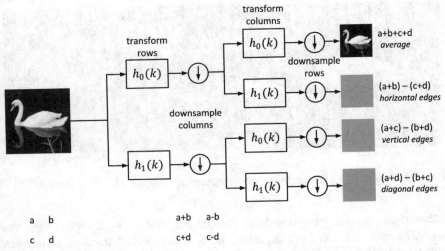

**Fig. 7.23** One stage discrete wavelet transform using filter banks: including a simple $2 \times 2$ example (a, b, c, d without amplitude scaling) to allow the operations to be tracked; and a complete image example (swan)

- a version that has been high pass filtered in both the vertical and horizontal directions, thereby emphasising diagonal edge information.

Because of the downsampling operations by column and row, the dimensions of those transformed images are half those of the original, giving us the same number of pixels overall. This is demonstrated with the small $2 \times 2$ image embedded in Fig. 7.23, which does not require downsampling, using the operations in (7.46). The single stage of Fig. 7.18 can be extended by transforming the low resolution image to produce yet a new low frequency approximation and high pass detail images, such as seen in Fig. 7.24 for two stages. As with the one dimensional wavelet transform, image reconstruction can be carried out by using the filter bank in reverse if the impulse response is symmetric.

## 7.15   Applications of the Wavelet Transform in Remote Sensing Image Analysis

Since the original image, such as the swan in the example of Fig. 7.24, can be completely reconstructed from any stage of the discrete wavelet transform using a reconstructing filter bank, it is clear that any level of decomposition contains all the original information content. Specifically, the low pass approximation image and the three high pass detail images in that figure, among them hold all the

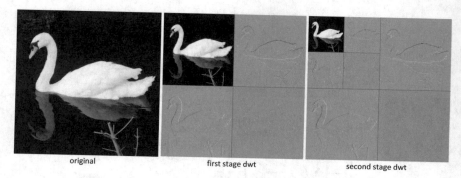

original                      first stage dwt                    second stage dwt

**Fig. 7.24** Two stages of the discrete wavelet transform of an image using Haar wavelets[24]

information. Interestingly, the detail images are much less complex than the approximation and can be compressed using a number of standard techniques without significant information loss. That allows the image data to be stored or transmitted using far fewer bits of digital data than the original. Compression based on the discrete wavelet transform is one of its principal applications in image processing.[25]

The wavelet transform can also be applied to decompose a single pixel spectrum into a more compact form as a feature reduction tool prior to classification.[26]

[24] MATLAB® contains a large number of wavelet bases that can be used in signal and image processing.

[25] See B. Li, R. Yang and H. Jiang, Remote-sensing image compression using two-dimensional oriented wavelet transform, *IEEE Transactions on Geoscience and Remote Sensing*, vol. 49, no. 1, January 2011, pp. 236–250. Sometimes the principal components transform has been found better for compression than the DWT: see, for example, B. Penna, T. Tillo, E. Magli and G. Olmo, Transform coding techniques for lossy hyperspectral data compression, *IEEE Transactions on Geoscience and Remote Sensing*, vol. 45, no. 5, May 2007, pp. 1408–1421. For a treatment of three dimensional wavelets and other transform techniques for compression of hyperspectral imagery in the two spatial and one spectral dimension see B. Penna, T. Tillo, E. Magli and G. Olmo, Transform coding techniques for lossy hyperspectral data compression, *IEEE Transactions on Geoscience and Remote Sensing*, vol. 45, no. 5, May 2007, pp. 1408–1421. An overview and comparison of compression techniques is E. Christophe, Hyperspectral data compression tradeoff, in S. Prasad, L.M. Bruce and J. Chanussot, eds., *Optical Remote Sensing: Advances in Signal Processing and Exploitation Techniques*, Springer, Berlin, 2011.

[26] See J. Zheng and E. Regentova, Wavelet based feature reduction method for effective classification of hyperspectral data, *Proc. Int. Conference on Information Technology, Computers and Communications* (ITCC 2003), April 2003, pp. 483–487, and L. M. Bruce, C.H. Kroger and J. Li, Dimensionality reduction of hyperspectral data using discrete wavelet transform feature extraction, *IEEE Transactions on Geoscience and Remote Sensing*, vol. 40, no. 10, October 2002, pp. 2331–2338.

## 7.16  Bibliography on Spatial Domain Image Transforms

Many texts in image processing for engineering and science cover much of the material presented in this chapter. The difficulty, however, is that the level of mathematical detail is sometimes quite high. Books that are comprehensive but not overly complex are

K. R. Castleman, *Digital Image Processing*, Prentice Hall, NJ, 1996, and

R.C. Gonzalez and R.E. Woods, *Digital Image Processing*, 4th ed. Pearson Prentice-Hall, Upper Saddle River, N.J., 2018.

Introductory treatments can also be found in books on remote sensing image processing including

R.A. Schowengerdt, *Remote Sensing: Models and Methods for Image Processing*, 3rd ed., Academic, Burlington, Mass., 2006.

Easily read accounts of Fourier series and the Fourier transform will be found in

L.A. Pipes, *Applied Mathematics for Engineers and Physicists*, 2nd ed., McGraw-Hill, N. Y., 1958, and the later edition

L.A. Pipes and L.R. Harvill, *Applied Mathematics for Engineers and Physicists*, 3rd ed., Dover, N.Y., 2014.

Material on fast Fourier transform is given in

E.O. Brigham, *The Fast Fourier Transform and its Applications*, Prentice-Hall, N. J., 1988.

which remains one of the best treatments available, with a good emphasis on sampling.

A. Papoulis, *Circuits and Systems: A Modern Approach*, Holt-Saunders, Tokyo, 1980

provides a detailed treatment of the Fourier transform and convolution for those with a higher level of mathematics expertise, as does the standard text in the field

R.N. Bracewell, *The Fourier Transform and its Applications*, 3rd ed., McGraw-Hill, N.Y., 1999.

Wavelets are covered in books that range from detailed mathematical treatments through to those which focus on applications. Castleman (above) has a good but difficult to read section. An idiosyncratic, but very comprehensive and readable account, is given in

G. Strang and T. Nguyen, *Wavelets and Filter Banks*, Wellesley-Cambridge, Mass., 1996

A good overview of the application of wavelets in a wide range of physical situations including medicine, finance, fluid flow, geophysics and mechanical engineering will be found in

P.S. Addison, *The Illustrated Wavelet Handbook*, IOP Publishing, Bristol, 2002

while a very good discussion on wavelet applications in astronomical image processing is given in

R. Berry and J. Burnell, *The Handbook of Astronomical Image Processing*, Willman-Bell, Richmond, VA, 2006.

This book also includes a helpful section on the use of the Fourier transform for image filtering. It shows the use of wavelets for multiresolution image analysis and how, by filtering specific wavelet components, features at particular levels of detail can be enhanced. Other texts that could be consulted are

C. Sidney Burrus, R. A. Gopinath and H. Guo, *Introduction to Wavelets and Wavelet Transforms*, Prentice Hall, Upper Saddle River, NJ, 1998, and

A.K. Chan and C. Peng, *Wavelets for Sensing Technologies*, Artech House, Norwood, MA, 2003

the last of which has a focus on SAR imagery and medical imaging.

## 7.17   Problems

7.1   Using (7.35) demonstrate that the complex exponentials $e^{jm\omega t}$, where $m$ is an integer, are an orthogonal set.

7.2   Verify the results of Sect. 7.6.2 using a simple sketch.

7.3   Using the Fourier transform of an impulse and the convolution theorem, verify the result of Sect. 7.6.2 mathematically.

7.4   Using (7.15a) compute the discrete Fourier transform of a square wave using 2, 4 and 8 samples per period respectively.

7.5   Compute the discrete Fourier transform of a unit pulse of width $2a$. Use 2, 4 and 8 samples over a time interval equal to $8a$. Compare the results to those obtained in Problem 7.4.

7.6   Image smoothing can be undertaken by computing averages over a square or rectangular window or by filtering in the spatial frequency domain. Consider just a single line of image data. Determine the corresponding spatial frequency domain filter function for a simple 3 pixel averaging filter to be used on that line of data. That requires the calculation of the discrete Fourier transform of a unit pulse.

7.7 As in Problem 7.6 consider a single line of image data. One way of applying a low pass filter to that data is to choose an ideal filter function in the spatial frequency domain that has a sharp cut-off, such as that shown in Fig. 7.16. Determine the corresponding function in the image domain by calculating the inverse Fourier transform of the ideal filter. Taking into account the discrete pixel nature of the image, approximate the inverse transform by an appropriate one dimensional template.

7.8 Are window functions required if a periodic signal is sampled over an integral number of periods?

7.9 In Fig. 7.9 suppose the function $g(t)$ is a sinusoid of frequency $B$ Hz. Its spectrum will consist of two impulses, one at $+B$ Hz and the other at $-B$ Hz. Produce the spectrum of the sinusoid obtained by taking three samples every two periods. Suppose the sinusoid is to be reconstructed from the samples by feeding them through a low pass filter that will pass all frequency components up to $1/2\Delta t$, where $\Delta t$ is the sampling interval, and will exclude all other frequencies. Describe the nature of the reconstructed signal; this will give an appreciation of aliasing distortion.

7.10 The periodic sequence of impulses in Fig. 7.9 is an idealised sampling function. In practice it is not possible to generate infinitesimally short samples of a function; rather, the samples will have a finite, although short, duration. That can be modelled mathematically by replacing the infinite sequence of impulses by a periodic sequence of finite-width pulses. One way of representing that periodic sequence of pulses is as the convolution of a single pulse and a periodic sequence of impulses. Using that model, describe what modifications are required in Fig. 7.9 to account for samples of finite duration.

7.11 Explain the appearance of the spectrum shown in Fig. 7.12 for the sequence of uniformly spaced vertical lines.

7.12 By examining Fig. 7.24 how many pixels are there in the collected set of images after the 1st and 2nd stages of the discrete wavelet transformation? What would be the case without downsampling?

7.13 The histogram of an image before wavelet transformation generally will have many filled bins. The same would be expected of the approximation images at each stage of the transformation. Qualitatively, what might the general shape of the histograms of the detail images look like in, say, Fig. 7.24? The simple example illustrated in Fig. 7.23 may help in answering this question.

7.14   The time function $g(t)$, with Fourier transform $G(f)$, is acceptable as a basic wavelet if it satisfies the admissibility criterion

$$\int_{-\infty}^{\infty} \frac{|G(f)|^2}{|f|} df < \infty$$

Show that the wavelets associated with the high pass filter in Fig. 7.16 satisfy the criterion.

7.15   Find the Fourier spectrum of a short, time-limited pure sinusoid, such as might occur when hitting a single note on a piano. In time that can be modelled by a sinusoidal signal that commences at a given time and stops a short time later. You may find this easier to handle using the convolution theorem, rather than calculating the Fourier transform from scratch.

7.16   The square waveform of Fig. 7.4 can be generated by convolving a unit pulse, such as that shown in Fig. 7.6a, with a periodic sequence of impulses. Using that model, and the convolution theorem, verify the spectrum of Fig. 7.5 from the spectrum of Fig. 7.6b.

# Chapter 8
# Supervised Classification Techniques

**Abstract** The variety of supervised classification techniques used with remotely sensed image data are presented in detail, commencing with the maximum likelihood decision rule and minimum distance classification, and progressing to the support vector classifier and neural networks, including the convolutional neural networks and recurrent neural networks used in deep learning. Emphasis is given to the development and properties of each of the algorithms, and examples are used to demonstrate how the various procedures are applied. The importance of the Hughes phenomenon (also called overfitting or the curse of dimensionality), in which insufficient training samples can lead to poor classification results, is highlighted as a key consideration in using any approach. Context classification is also covered in which the label attached to a pixel by a classifier can be made sensitive to the labels attached to neighbouring pixels.

## 8.1 Introduction

Supervised classification is the technique most often used for the quantitative analysis of remote sensing image data. At its core is the concept of segmenting the spectral domain into regions that can be associated with the ground cover classes of interest to a particular application. In practice those regions may sometimes overlap. A variety of algorithms is available for the task, and it is the purpose of this chapter to cover those most commonly encountered.

The development of different methods for supervised classification has perhaps been the area of greatest change in remote sensing image analysis over the past several decades. In this chapter we capture that development by commencing with the earliest benchmark procedure—the maximum likelihood classifier—and then progress through more recent methods to the currently popular deep learning procedures.

Essentially, the different methods vary in the way they identify and describe the regions in spectral space. Some seek a simple geometric segmentation while others adopt statistical models with which to associate spectral measurements and the

© The Author(s), under exclusive license to Springer Nature Switzerland AG 2022
J. A. Richards, *Remote Sensing Digital Image Analysis*,
https://doi.org/10.1007/978-3-030-82327-6_8

classes of interest. Some can handle user-defined classes that overlap each other spatially and are referred to as *soft classification* methods; others generate firm boundaries between classes and are called *hard classification* methods, in the sense of establishing boundaries rather than having anything to do with difficulty in their use. Hard classifiers produce only a single candidate class label for a pixel whereas soft classifiers produce several, often ordered in likelihood of being correct.

Sometimes the data from a *set* of sensors is available to help in the analysis task. Classification methods suited to multi-sensor or multi-source analysis are the subject of Chap. 12.

The techniques we are going to develop in this chapter come from a field that has had many names over the years, often changing as the techniques themselves develop. Most generically it is probably called pattern recognition or pattern classification but, as the field has evolved, the names

learning machines
pattern recognition
classification, and
machine learning

have been used. Learning machine theory commenced in the late 1950s in an endeavour to understand brain functioning and to endow machines with a degree of decision-making intelligence[1]; so, the principle of what we are going to develop here is far from new, although many of the procedures are.

## 8.2  The Essential Steps in Supervised Classification

Recall from Fig. 3.3 that supervised classification is essentially a mapping from the measurement space of the sensor to a field of labels that represent the ground cover types of interest to the user. It depends on having enough pixels available, whose class labels are known, with which to train the classifier. In this context "training" refers to the estimation of the parameters that the classifier needs in order to be able to recognise and label unseen pixels. The labels represent the classes on the map that the user requires. The map is called a *thematic map*, meaning a map of themes.

An important concept must be emphasised here, which is often overlooked in practice and about which we will have much to say in Chap. 11. That concerns whether the classes of interest to the user occupy unique regions in spectral space, and whether there is a one-to-one mapping between the measurement vectors and class labels set by the user. A simple illustration is shown in Fig. 8.1, which is a near infrared versus visible red spectral space of a geographical region used for growing crops. The data is shown to be in three moderately distinct clusters,

---

[1] See N.J. Nilsson, *Learning Machines*, McGraw-Hill, N.Y., 1965 for a good coverage of the earlier developments in the field.

corresponding to crops (on loam), a loamy soil and a sandy soil. One user might be interested in a map of crops and soil—just two classes, but the soil class has two sub-classes—loam and sand. Another user might be interested in which parts of the region are most suitable for cropping, and those which are not. The former will have two sub-classes—crops and loam—but they are different from those of the first user. We call the classes into which the data naturally clusters the *spectral classes* (in this simple example the crops, loam and sand) whereas those which match user requirements are called *information classes.*[2]

Spectral classes are a property of the recorded image data, whereas information classes are labels defined by the analyst to suit a particular application. The challenge in operational remote sensing is to gather information about the information classes in which we are interested through a discovery of the spectral class structure of the recorded data. Essentially, we undertake a two stage mapping as illustrated in Fig. 8.1.

The simple illustration just presented has been used to introduce the distinction between spectral and information classes. In practice the difference may be more subtle. Spectral classes are groupings of pixel vectors in spectral space that are matched to the particular classifier algorithm being used. With some supervised classification algorithms, it will be beneficial to segment even simple classes like vegetation into sub-groups of data and let the classifier work on the sub-groups. After classification is complete the analyst then maps the sub-groups (i.e., the spectral classes) to the information classes of interest. We will see that in action in examples in Chap. 11. For most of this chapter, though, we will not make the distinction between the two class types very often. For clarity and simplicity, we will assume the information and spectral classes are the same.

Many different algorithms are available for supervised classification. Irrespective of the one chosen, the essential practical steps in applying a classifier usually include:

1. Deciding on the set of ground cover type classes into which to segment the image.
2. Choosing representative pixels for each of the classes, for which the class labels are known. Those pixels are said to form *training data*. Training data sets for each class can be established using site visits, maps, air photographs or even photointerpretation of image products formed from the data. Sometimes the training pixels for a given class will lie in a common region enclosed by a border. That region is called a *training field*.
3. Using the training data to estimate the parameters of the particular classifier algorithm to be employed; the set of parameters is sometimes called the *signature* of that class.
4. Using the trained classifier to label *every* pixel in the image as belonging to one of the classes specified in step 1. Here the whole image is classified. Whereas in

---

[2] The important distinction between information and spectral classes was first made in P.H. Swain and S.M. Davis, eds., *Remote Sensing: The Quantitative Approach*, McGraw-Hill, N.Y., 1978.

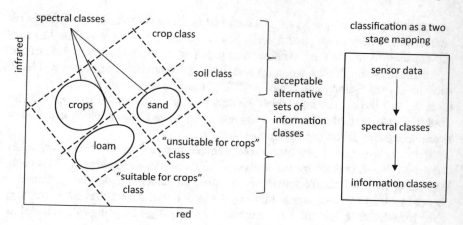

**Fig. 8.1** Simple illustration of the difference between spectral and information classes: in this case the spectral classes have identifiable names; in practice they are more likely to be groupings of data that match the characteristics of the classifier to be used

step 2 the user may need to know the labels for about 1% or so of the pixels, the remainder are now labelled by the classifier. This step is called classification, labelling or allocation. In the terminology of machine learning it is called *generalisation*. It is in this step that we see the major benefit of supervised classification in remote sensing. Gathering labelled training data for each class can be expensive and time-consuming. However, that investment is rewarded significantly by being able to label all the other pixels in the image, with just the cost of running the classifier algorithm.

5. In connection with step 4, producing thematic (class) maps and tables which summarise class memberships of all pixels in the image, from which the areas of the classes can be measured.

6. Assessing the accuracy of the final product using a labelled *testing data* set. This is a crucial step if the results of classification are to be relied upon in practice. We address this particular requirement in Chap. 11.

In practice it might be necessary to decide, on the basis of the results obtained at step 6, to refine the training process in order to improve classification accuracy. Sometimes it might even be desirable to classify just the training samples themselves to ensure that the parameters generated at step 3 are acceptable.

It is our objective now to consider the range of algorithms that could be used in steps 3 and 4. Recall that we are assuming that each information class consists of just one spectral class, unless specified otherwise, and we will use the names interchangeably. That allows the development of the basic algorithms to proceed without the added complexity of considering spectral classes, which are also called sub-classes. Sub-classes are treated in Sect. 8.4 and in Chaps. 9 and 11.

## 8.3 Maximum Likelihood Classification

Maximum likelihood classification is one of the most common supervised classification techniques used with remote sensing image data and was the first rigorous algorithm to be employed widely. In many ways it has become the benchmark against which later algorithms have been compared. It is developed in the following in a statistically acceptable manner. A more general derivation is given in Appendix E, although the present approach is sufficient for most remote sensing purposes.

### 8.3.1  Bayes' Classification

Let the classes be represented by $\omega_i, i = 1 \ldots M$ where $M$ is the number of classes. In determining the class or category to which a pixel with measurement vector $\mathbf{x}$ belongs, the conditional probabilities

$$p(\omega_i|\mathbf{x}), i = 1 \ldots M$$

play a central role. The vector $\mathbf{x}$ is a column vector of the brightness values for the pixel in each measurement band. It describes the pixel as a point in spectral space. The probability $p(\omega_i|\mathbf{x})$ tells us the likelihood that $\omega_i$ is the correct class for the pixel at position $\mathbf{x}$ in the spectral space. If we knew the complete set of $p(\omega_i|\mathbf{x})$ for the pixel, one for each class, then we could label the pixel—classify it—according to the *decision rule*

$$\mathbf{x} \in \omega_i \text{ if } p(\omega_i|\mathbf{x}) > p(\omega_j|\mathbf{x}) \text{ for all } j \neq i \tag{8.1}$$

This says that the pixel with measurement vector $\mathbf{x}$ is a member of class $\omega_i$ if $p(\omega_i|\mathbf{x})$ is the largest of the set of probabilities. This intuitive statement is a special case of a more general rule in which the decisions can be biased according to different degrees of significance being attached to different incorrect classifications. That general approach is called Bayes' classification which is the subject of Appendix E.

The rule of (8.1) is sometimes called a *maximum posterior* or MAP *decision rule*, a name which will become clearer in the next section.

### 8.3.2  The Maximum Likelihood Decision Rule

Despite the simplicity of (8.1) the $p(\omega_i|\mathbf{x})$ are unknown. What we can find relatively easily, though, are the set of class conditional probabilities $p(\mathbf{x}|\omega_i)$, which describe the chances of finding a pixel at position $\mathbf{x}$ in spectral space from each of the classes $\omega_i$. Those conditional probability functions are estimated from labelled training data for each class and are the class distribution functions we saw in Fig. 3.6. Later we will adopt a specific form for the distribution function, but for the moment we will retain it in general form.

The desired $p(\omega_i|\mathbf{x})$ in (8.1) and the available $p(\mathbf{x}|\omega_i)$, estimated from training data, are related by Bayes' theorem[3]

$$p(\omega_i|\mathbf{x}) = p(\mathbf{x}|\omega_i)p(\omega_i)/p(\mathbf{x}) \qquad (8.2)$$

in which $p(\omega_i)$ is the probability that pixels from class $\omega_i$ appear anywhere in the image. If 15% of the pixels in the scene belong to class $\omega_i$ then we could use $p(\omega_i) = 0.15$. The $p(\omega_i)$ are referred to as *prior probabilities*—or sometimes just *priors*. If we knew the complete set, then they are the probabilities with which we could guess the class label for a pixel before (*prior to*) doing any analysis. By contrast the $p(\omega_i|\mathbf{x})$ are the *posterior probabilities* since, in principle, they are the probabilities we find for the class labels of a pixel at position $\mathbf{x}$, *after* analysis. What we strive to do in statistical classification is to estimate, via (8.2), the set of class posterior probabilities for each pixel so that the pixels can be labelled according to the largest of the posteriors using the decision rule of (8.1).

The term $p(\mathbf{x})$ is the probability of finding a pixel with measurement vector $\mathbf{x}$ in the image, from any class. Although $p(\mathbf{x})$ is not important in the following development it should be noted that

$$p(\mathbf{x}) = \sum_{i=1}^{M} p(\mathbf{x}|\omega_i)p(\omega_i)$$

On substituting from (8.2) the decision rule of (8.1) reduces to

$$\mathbf{x} \in \omega_i \ \text{ if } \ p(\mathbf{x}|\omega_i)p(\omega_i) > p(\mathbf{x}|\omega_j)p(\omega_j) \ \text{ for all } \ j \neq i \qquad (8.3)$$

in which $p(\mathbf{x})$ has been removed as a common factor, since it is not class dependent and thus doesn't contribute to decision making. The decision rule of (8.3) is more acceptable than that of (8.1) since the $p(\mathbf{x}|\omega_i)$ are known from training data, and it is conceivable that the priors $p(\omega_i)$ are also known or can be estimated.

It is mathematically convenient now to define the *discriminant function*

$$g_i(\mathbf{x}) = \ln\{p(\mathbf{x}|\omega_i)p(\omega_i)\} = \ln p(\mathbf{x}|\omega_i) + \ln p(\omega_i) \qquad (8.4)$$

---

[3] J.E. Freund, *Mathematical Statistics*, 5th ed., Prentice Hall, N.J., 1992, or C.M. Bishop, *Pattern Recognition and Machine Learning*, Springer Science + Business Media, LLC, N.Y., 2006.

Because the natural logarithm is a monotonic function we can use (8.4) in (8.3) to give as the decision rule:

$$\mathbf{x} \in \omega_i \text{ if } g_i(\mathbf{x}) > g_j(\mathbf{x}) \text{ for all } j \neq i \tag{8.5}$$

### 8.3.3 Multivariate Normal Class Models

To develop the maximum likelihood classifier further we now choose a particular probability model for the class conditional density function $p(\mathbf{x}|\omega_i)$. The most common choice is to assume $p(\mathbf{x}|\omega_i)$ is a multivariate normal distribution, also called a Gaussian distribution. This tacitly assumes that the classes of pixel of interest in spectral space are normally distributed. That is not necessarily a real property of natural spectral or information classes, but the Gaussian is a simple distribution to handle mathematically, and its multivariate properties are well known. For an $N$ dimensional space the specific form of the multivariate Gaussian distribution function is[4]

$$p(\mathbf{x}|\omega_i) = (2\pi)^{-N/2}|\mathbf{C}_i|^{-1/2}\exp\left\{-1/2(\mathbf{x} - \mathbf{m}_i)^{\mathsf{T}}\mathbf{C}_i^{-1}(\mathbf{x} - \mathbf{m}_i)\right\} \tag{8.6}$$

where $\mathbf{m}_i$ and $\mathbf{C}_i$ are the mean vector and covariance matrix of the data in class $\omega_i$. We will sometimes write the normal distribution in the shorthand form $\mathcal{N}(\mathbf{x}|\mathbf{m}_i, \mathbf{C}_i)$.

Substituting (8.6) into (8.4) gives the discriminant function

$$g_i(\mathbf{x}) = -1/2 N \ln 2\pi - 1/2 \ln|\mathbf{C}_i| - 1/2(\mathbf{x} - \mathbf{m}_i)^{\mathsf{T}}\mathbf{C}_i^{-1}(\mathbf{x} - \mathbf{m}_i) + \ln p(\omega_i)$$

Since the first term is not class dependent is doesn't aid discrimination and can be removed, leaving as the discriminant function

$$g_i(\mathbf{x}) = \ln p(\omega_i) - 1/2 \ln|\mathbf{C}_i| - 1/2(\mathbf{x} - \mathbf{m}_i)^{\mathsf{T}}\mathbf{C}_i^{-1}(\mathbf{x} - \mathbf{m}_i) \tag{8.7}$$

If the analyst has no useful information about the values of the prior probabilities, they are assumed all to be equal. The first term in (8.7) is then ignored, allowing the $1/2$ to be removed as well, leaving

$$g_i(\mathbf{x}) = -\ln|\mathbf{C}_i| - (\mathbf{x} - \mathbf{m}_i)^{\mathsf{T}}\mathbf{C}_i^{-1}(\mathbf{x} - \mathbf{m}_i) \tag{8.8}$$

---

[4] See Appendix D.

which is the discriminant function for the *Gaussian maximum likelihood classifier*, so-called because it determines class membership of a pixel based on the highest of the class conditional probabilities, or likelihoods. Its implementation requires use of either (8.7) or (8.8) in (8.5). There is an important practical consideration concerning whether all the classes have been properly represented in the training data available. Because probability distributions like the Gaussian exist over the full domain of their argument (in this case the spectral measurements) use of (8.5) will yield a class label even in the remote tails of the distribution functions. In Sect. 8.3.5 the use of thresholds allows the user to avoid such inappropriate labelling and thereby also to identify potentially missing classes.

### 8.3.4   Decision Surfaces

The probability distributions in the previous section have been used to discriminate among the candidate class labels for a pixel. We can also use that material to see how the spectral space is segmented into regions corresponding to the classes. That requires finding the shapes of the separating boundaries in vector space. As will become evident, an understanding of the boundary shapes will allow us to assess the relative classification abilities, or strengths, of different classifiers.

Spectral classes are defined by those regions in spectral space where their discriminant functions are the largest. Those regions are separated by surfaces where the discriminant functions for adjoining spectral classes are equal. The $i$th and $j$th spectral classes are separated by the surface

$$g_i(\mathbf{x}) - g_j(\mathbf{x}) = 0$$

which is the actual equation of the surface. If all such surfaces were known, then the class membership of a pixel can be made on the basis of its position in spectral space relative to the set of surfaces.

The construction $(\mathbf{x} - \mathbf{m}_i)^{\mathrm{T}} \mathbf{C}_i^{-1} (\mathbf{x} - \mathbf{m}_i)$ in (8.7) and (8.8) is a quadratic function of $\mathbf{x}$. Consequently, the decision surfaces generated by Gaussian maximum likelihood classification are quadratic and thus take the form of parabolas, ellipses and circles in two dimensions, and hyperparaboloids, hyperellipsoids and hyperspheres in higher dimensions.

If an information class were represented by a set of spectral classes then, even though the surfaces between spectral classes were hyperquadratic, those between the information classes can be quite complex.[5]

---

[5] See J.A. Richards and N.G. Kingsbury, Is there a preferred classifier for operational thematic mapping? *IEEE Transactions on Geoscience and Remote Sensing*, 52, no 2, May 2014, pp. 2715–2725.

## 8.3.5  *Thresholds*

It is implicit in the rules of (8.1) and (8.5) that pixels at every point in spectral space will be allocated to one of the available classes $\omega_i$, irrespective of how small the actual probabilities of class membership are. This is illustrated for one dimensional data in Fig. 8.2a. Poor classification can result as indicated. Such situations can arise if spectral classes have been overlooked or, if knowing that other classes existed, enough training data was not available to estimate the parameters of their distributions with any degree of accuracy (see Sect. 8.3.6 following).

In situations such as those it is sensible to apply thresholds to the decision process in the manner depicted in Fig. 8.2b. Pixels which have probabilities for all classes below the threshold are not classified. In practice, thresholds are applied to the discriminant functions and not the probability distributions, since the latter are never actually computed. With the incorporation of a threshold the decision rule of (8.5) becomes

$$\mathbf{x} \in \omega_i \quad \text{if} \quad g_i(\mathbf{x}) > g_j(\mathbf{x}) \text{ for all } j \neq i \tag{8.9a}$$

$$\text{and } g_i(\mathbf{x}) > T_i \tag{8.9b}$$

where $T_i$ is the threshold to be used on class $\omega_i$.

We now need to understand how a reasonable value for $T_i$ can be chosen. Substituting (8.7) into (8.9b) gives

$$\ln p(\omega_i) - \tfrac{1}{2}\ln|\mathbf{C}_i| - \tfrac{1}{2}(\mathbf{x} - \mathbf{m}_i)^{\mathrm{T}}\mathbf{C}_i^{-1}(\mathbf{x} - \mathbf{m}_i) > T_i$$

or

$$(\mathbf{x} - \mathbf{m}_i)^{\mathrm{T}}\mathbf{C}_i^{-1}(\mathbf{x} - \mathbf{m}_i) < -2T_i + 2\ln p(\omega_i) - \ln|\mathbf{C}_i| \tag{8.10}$$

If $\mathbf{x}$ is normally distributed, the expression on the left-hand side of (8.10) has a $\chi^2$ distribution with $N$ degrees of freedom,[6] where $N$ is the number of bands in the data. We can, therefore, use the properties of the $\chi^2$ distribution to determine that value of $(\mathbf{x} - \mathbf{m}_i)^{\mathrm{T}}\mathbf{C}_i^{-1}(\mathbf{x} - \mathbf{m}_i)$ below which a desired percentage of pixels will exist. Larger values of $(\mathbf{x} - \mathbf{m}_i)^{\mathrm{T}}\mathbf{C}_i^{-1}(\mathbf{x} - \mathbf{m}_i)$ correspond to pixels lying further out in the tails of the normal distribution. That is shown in Fig. 8.3.

As an example of how this is used consider the need to choose a threshold such that 95% of all pixels in a class will be classified, or such that the 5% least likely pixels for the class will be rejected. From the $\chi^2$ distribution we find that 95% of all pixels have $\chi^2$ values below 9.488. Thus, from (8.10)

---

[6] See Swain and Davis, loc. cit.

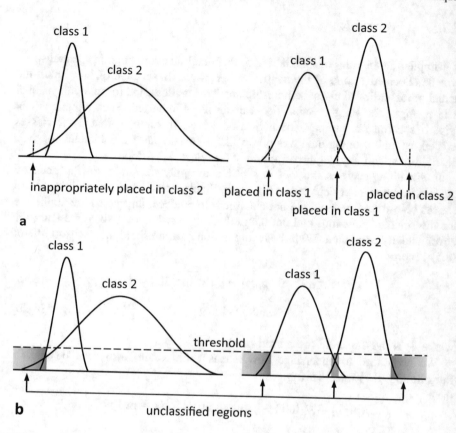

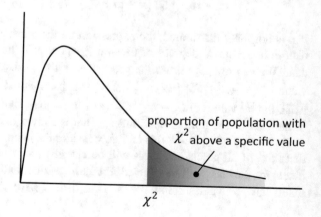

**Fig. 8.2** Illustration of the use of thresholds in **b** to avoid questionable classification decisions in **a**

**Fig. 8.3** Use of the $\chi^2$ distribution for obtaining classifier thresholds

$$T_i = -4.744 + \ln p(\omega_i) - \frac{1}{2} \ln |\mathbf{C}_i|$$

which can be calculated from a knowledge of the prior probability and covariance matrix for the $i$th spectral class.

### 8.3.6 Number of Training Pixels Required

Enough training pixels for each spectral class must be available to allow reasonable estimates to be obtained of the elements of the class conditional mean vector and covariance matrix. For an $N$ dimensional spectral space the mean vector has $N$ elements. The covariance matrix is symmetric of size $N \times N$; it has $\frac{1}{2}N(N+1)$ distinct elements that need to be estimated from the training data. To avoid the matrix being singular, at least $N(N+1)$ independent samples are needed. Fortunately, each $N$ dimensional pixel vector contains $N$ samples, one in each waveband; thus, the minimum number of independent training *pixels* required is $(N+1)$. Because of the difficulty in assuring independence of the pixels, usually many more than this minimum number are selected. A practical minimum of $10N$ training pixels per spectral class is recommended, with as many as $100N$ per class if possible.[7] If the covariance matrix is well estimated then the mean vector will be also, because of its many fewer elements.

For data with low dimensionality, say up to 10 bands, $10N$ to $100N$ can usually be achieved, but for higher order data sets, such as those generated by hyperspectral sensors, finding enough training pixels per class is often not practical, making reliable training of the traditional maximum likelihood classifier difficult. In such cases, dimensionality reduction procedures can be used, including methods for feature selection. They are covered in Chap. 10.

Another approach that can be adopted when not enough training samples are available is to use a classification algorithm with fewer parameters, several of which are treated in later sections. It is the need to estimate all the elements of the covariance matrix that causes problems for the maximum likelihood algorithm. The minimum distance classifier, covered in Sect. 8.5, depends only on knowing the mean vector for each training class—consequently there are only $N$ unknowns to estimate.

---

[7] See Swain and Davis, loc. cit., although some authors regard this as a conservatively high number of samples.

### 8.3.7 The Hughes Phenomenon and the Curse of Dimensionality

Although recognised as a potential problem since the earliest application of computer-based interpretation to remotely sensed imagery, the Hughes phenomenon had not been a significant consideration until the availability of hyperspectral data. It concerns the number of training samples required per class to train a supervised classifier reliably and is thus related to the material of the previous section. Although that was focussed on Gaussian maximum likelihood classification, in principle any classifier that requires parameters to be estimated is subject to the problem.[8] It manifests itself in the following manner.

Clearly, if too few samples are available good estimates of the class parameters cannot be found; the classifier will not be properly trained and will not predict well on data from the same classes that it has not already seen. The extent to which a classifier predicts on previously unseen data, referred to as *testing data*, is called *generalisation*. Increasing the set of random training samples would be expected to improve parameter estimates, which indeed is generally the case, and thus lead to better generalisation.

Suppose now we have enough samples for a given number of dimensions to provide reliable training and good generalisation. Instead of changing the number of training samples, imagine now we can add more features, or dimensions, to the problem. On the face of it that suggests that the results should be improved because more features should bring more information into the training phase. However, if we don't increase the number of training samples commensurately, we may not be able to get good estimates of the larger set of parameters created by the increased dimensionality, and both training and generalisation will suffer. That is the essence of the Hughes phenomenon.

The effect of poorly estimated parameters resulting from too few training samples with increased dimensionality was noticed very early in the history of remote sensing image classification. Figure 8.4 shows the results of a five-class agricultural classification using 12 channel aircraft scanner data.[9] Classification accuracy is plotted as a function of the number of features (channels) used. As noted, when more features are added there is an initial improvement in classification accuracy until such time that the estimated maximum likelihood statistics apparently become less reliable.

As another illustration consider the determination of a reliable linear separating surface; here we approach the problem by increasing the number of training samples for a given dimensionality, but the principle is the same. Figure 8.5 shows

---

[8] See M. Pal and G.F. Foody, Feature selection for classification of hyperspectral data for SVM, *IEEE Transactions on Geoscience and Remote Sensing*, vol. 48, no. 5, May 2010, pp. 2297–2307 for a demonstration of the Hughes phenomenon with the support vector machine of Sect. 8.14.

[9] Based on results presented in K.S. Fu, D.A. Landgrebe and T.L. Phillips, Information processing of remotely sensed agricultural data, *Proc. IEEE*, vol. 57, no. 4, April 1969, pp. 639–653.

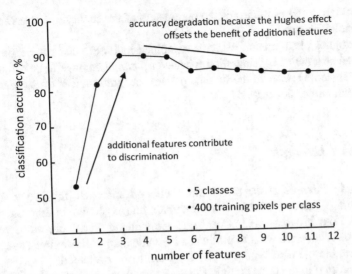

**Fig. 8.4** Actual experimental classification results showing the Hughes phenomenon—the loss of classification accuracy as the dimensionality of the spectral domain increases

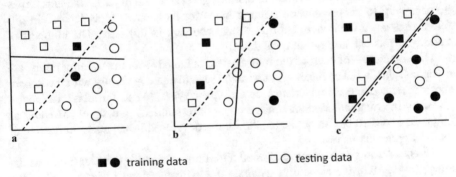

■● training data          □○ testing data

**Fig. 8.5** Estimating a linear separating surface using an increasing number of training samples, showing poor generalisation if too few per number of features are used; the dashed line represents the optimal surface while the full lines are the surfaces generated by the training data in each case

three different training sets of data for the same two-dimensional data set. The first has only one training pixel per class, and thus the same number of training pixels as dimensions. As seen, while a separating surface can easily be found it may not be accurate. The classifier performs at the 100% level on the training data but is very poor on testing data.

Having two training pixels per class, as in the case of the middle diagram, provides a better estimate of the separating surface, but it is not until we have many pixels per class compared to the number of channels, as in the right-hand figure, that we obtain good estimates of the parameters of the supervised classifier, so that generalisation is good.

The Hughes phenomenon was first examined in 1968[10] and has become widely recognised as a major consideration whenever parameters have to be estimated in a high dimensional space. In the field of pattern recognition, it is often referred to as the curse of dimensionality.[11]

## 8.3.8   An Example

As a simple example of the maximum likelihood approach, the $256 \times 276$ pixel segment of a Landsat Multispectral Scanner image shown in Fig. 8.6 is to be classified. Four broad ground cover types are evident: water, fire burn, vegetation and urban. Assume we want to produce a thematic map of those four cover types in order to enable the distribution of the fire burn to be evaluated.

The first step is to choose training data. For such a broad classification, suitable sets of training pixels for each of the four classes are easily identified visually in the image data. The locations of four training fields for this purpose are seen in solid colours on the image. Sometimes, to obtain a good estimate of class statistics it may be necessary to choose several training fields for each cover type, located in different regions of the image, but that is not necessary in this case. The numbers of training pixels are in the figure caption.

The signatures for each of the four classes, obtained from the training fields, are given in Table 8.1. The mean vectors can be seen to agree generally with the known spectral reflectance characteristics of the cover types. Also, the class variances, given by the diagonal elements in the covariance matrices, are small for water as might be expected but on the large side for the developed/urban class, indicative of its heterogeneous nature.

When used in a maximum likelihood algorithm to classify the four bands of the image in Fig. 8.6, the signatures generate the thematic map of Fig. 8.7. The four classes, by area, are given in Table 8.2. Note that there are no unclassified pixels, since a threshold was not used in the labelling process. The area estimates are obtained by multiplying the number of pixels per class by the effective area of a pixel. In the case of the Landsat Multispectral Scanner the pixel size is 0.4424 hectare.

---

[10] G. F Hughes, On the mean accuracy of statistical pattern recognizers, *IEEE Transactions on Information Theory*, vol. IT-14, no.1, 1968, pp. 55–63.

[11] C.M. Bishop, *Pattern Recognition and Machine Learning,* Springer Science + Business Media, LLC, N.Y., 2006.

**Fig. 8.6** Image segment to be classified, consisting of a mixture of natural vegetation, waterways, urban regions, and vegetation damaged by fire. Four training areas are identified:

| | | |
|---|---|---|
| water | violet | 847 pixels |
| vegetation | green | 1,293 pixels |
| fire burn | red | 2,347 pixels |
| urban | dark blue | 781 pixels |

Those pixels were used to generate the signatures in Table 8.1

water   vegetation

fire burn   urban

**Table 8.1** Class signatures generated from the training areas in Fig. 8.6

| Class | Mean vector | Covariance matrix | | | |
|---|---|---|---|---|---|
| Water | 44.27 | 14.36 | 9.55 | 4.49 | 1.19 |
| | 28.82 | 9.55 | 10.51 | 3.71 | 1.11 |
| | 22.77 | 4.49 | 3.71 | 6.95 | 4.05 |
| | 13.89 | 1.19 | 1.11 | 4.05 | 7.65 |
| Fire Burn | 42.85 | 9.38 | 10.51 | 12.30 | 11.00 |
| | 35.02 | 10.51 | 20.29 | 22.10 | 20.62 |
| | 35.96 | 12.30 | 22.10 | 32.68 | 27.78 |
| | 29.04 | 11.00 | 20.62 | 27.78 | 30.23 |
| Vegetation | 40.46 | 5.56 | 3.91 | 2.04 | 1.43 |
| | 30.92 | 3.91 | 7.46 | 1.96 | 0.56 |
| | 57.50 | 2.04 | 1.96 | 19.75 | 19.71 |
| | 57.68 | 1.43 | 0.56 | 19.71 | 29.27 |
| Urban | 63.14 | 43.58 | 46.42 | 7.99 | −14.86 |
| | 60.44 | 46.42 | 60.57 | 17.38 | −9.09 |
| | 81.84 | 7.99 | 17.38 | 67.41 | 67.57 |
| | 72.25 | −14.86 | −9.09 | 67.57 | 94.27 |

One of the benefits of classifiers such as the maximum likelihood rule is that the mean vector elements represent the average spectral reflectance curves of the cover types, as seen in Fig. 8.7.

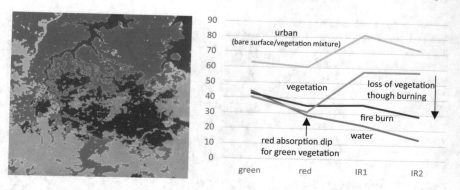

**Fig. 8.7** Thematic map produced using the maximum likelihood classifier in which blue represents water, red is fire damaged vegetation, green is natural vegetation and yellow is urban development; also shown are the spectral reflectance curves for the four cover types produced from the four components of each of the class mean vectors

**Table 8.2** Tabular summary of the thematic map of Fig. 8.7

| Class | Number of pixels | Area (ha) |
|---|---|---|
| Water | 4830 | 2137 |
| Fire burn | 14,182 | 6,274 |
| Vegetation | 28,853 | 12,765 |
| Urban | 22,791 | 10,083 |

## 8.4 Gaussian Mixture Models

In order that the maximum likelihood classifier algorithm work optimally it is important that the classes be as close to Gaussian as possible. In Sect. 8.2 that was stated in terms of knowing the number of spectral classes, sometimes also referred to as modes or sub-classes, of which each information class is composed. In Chap. 11 we will employ the unsupervised technique of clustering, to be developed in the next chapter, to help us do that. Here we present a different approach, based on the assumption that the image data is composed of a set of Gaussian distributions, some subsets of which may constitute information classes. This material is included at this point because it fits logically into the flow of the chapter, but it is a little more complex and could be passed over on a first reading.[12]

We commence by assuming that the pixels in a given image belong to a probability distribution that is a linear mixture of $K$ Gaussian distributions of the form

$$p(\mathbf{x}) = \sum_{k=1}^{K} \alpha_k \mathcal{N}(\mathbf{x}|\mathbf{m}_k, \mathbf{C}_k) \tag{8.11}$$

---

[12] A very good treatment of Gaussian mixture models will be found in C.M. Bishop, loc. cit.

The $\alpha_k$ are a set of mixture proportions, which are also considered to be the prior probabilities of each of the components that form the mixture. They satisfy

$$0 \leq \alpha_k \leq 1 \tag{8.12a}$$

$$\text{and} \quad \sum_{k=1}^{K} \alpha_k = 1 \tag{8.12b}$$

Equation (8.11) is the probability of finding a pixel at position $\mathbf{x}$ in spectral space from the mixture. It is the mixture equivalent of the probability $p(\mathbf{x}|\omega_i)$ of finding a pixel at position $\mathbf{x}$ from the single class $\omega_i$. Another way of writing the expression for the class conditional probability for a single distribution is $p(\mathbf{x}|\mathbf{m}_i, \mathbf{C}_i)$, in which the class has been represented by its parameters—the mean vector and covariance matrix—rather than by the class label itself. Similarly, $p(\mathbf{x})$ in the mixture equation above could be shown as conditional on the parameters of the mixture explicitly, viz.

$$p(\mathbf{x}|\boldsymbol{\alpha}_\mathbf{K}, \mathbf{m}_\mathbf{K}, \mathbf{C}_\mathbf{K}) = \sum_{k=1}^{K} \alpha_k \mathcal{N}(\mathbf{x}|\mathbf{m}_k, \mathbf{C}_k) \tag{8.13}$$

in which $\mathbf{m}_\mathbf{K}$ is the set of all $K$ mean vectors, $\mathbf{C}_\mathbf{K}$ is the set of all $K$ covariance matrices and $\boldsymbol{\alpha}_\mathbf{K}$ is the set of mixture parameters.

What we would like to do now is find the parameters in (8.13) that give the best match of the model to the data available. In what follows we assume we know, or have estimated, the number of components $K$; what we then have to find are the sets of mean vectors, covariance matrices and mixture proportions.

Suppose we have available a data set of $J$ pixels $\mathbf{X} = \{\mathbf{x}_1 \ldots \mathbf{x}_J\}$, assumed to be independent samples from the mixture. Their joint likelihood can be expressed

$$p(\mathbf{X}|\boldsymbol{\alpha}_\mathbf{K}, \mathbf{m}_\mathbf{K}, \mathbf{C}_\mathbf{K}) = \prod_{j=1}^{J} p(\mathbf{x}_j|\boldsymbol{\alpha}_\mathbf{K}, \mathbf{m}_\mathbf{K}, \mathbf{C}_\mathbf{K})$$

the logarithm[13] of which is

$$\ln p(\mathbf{X}|\boldsymbol{\alpha}_\mathbf{K}, \mathbf{m}_\mathbf{K}, \mathbf{C}_\mathbf{K}) = \sum_{j=1}^{J} \ln p(\mathbf{x}_j|\boldsymbol{\alpha}_\mathbf{K}, \mathbf{m}_\mathbf{K}, \mathbf{C}_\mathbf{K}) \tag{8.14}$$

We now want to find the parameters that will maximise the log likelihood that $\mathbf{X}$ comes from the mixture. In other words, we wish to maximise (8.14) with respect to the members of the sets $\mathbf{m}_\mathbf{K} = \{\mathbf{m}_k, k = 1 \ldots K\}$, $\mathbf{C}_\mathbf{K} = \{\mathbf{C}_k, k = 1 \ldots K\}$, $\boldsymbol{\alpha}_\mathbf{K} = \{\alpha_k, k = 1 \ldots K\}$. Substituting (8.13) into (8.14) we have

---

[13] As in Sect. 8.3, using the logarithmic expression simplifies the analysis to follow.

$$\ln p(\mathbf{X}|\boldsymbol{\alpha_K}, \mathbf{m_K}, \mathbf{C_K}) = \sum_{j=1}^{J} \ln \sum_{k=1}^{K} \alpha_k \mathcal{N}(\mathbf{x}_j|\mathbf{m}_k, \mathbf{C}_k) \qquad (8.15)$$

Differentiating this with respect to the mean vector of, say, the $k$th component, and equating the result to zero gives

$$\sum_{j=1}^{J} \frac{\partial}{\partial \mathbf{m}_k} \ln \sum_{k=1}^{K} \alpha_k \mathcal{N}(\mathbf{x}_j|\mathbf{m}_k, \mathbf{C}_k) = 0$$

$$\sum_{j=1}^{J} \frac{1}{\sum_{k=1}^{K} \alpha_k \mathcal{N}(\mathbf{x}_j|\mathbf{m}_k, \mathbf{C}_k)} \frac{\partial}{\partial \mathbf{m}_k} \sum_{k=1}^{K} \alpha_k \mathcal{N}(\mathbf{x}_j|\mathbf{m}_k, \mathbf{C}_k) = 0 \qquad (8.16)$$

$$\sum_{j=1}^{J} \frac{\alpha_k \mathcal{N}(\mathbf{x}_j|\mathbf{m}_k, \mathbf{C}_k)}{\sum_{k=1}^{K} \alpha_k \mathcal{N}(\mathbf{x}_j|\mathbf{m}_k, \mathbf{C}_k)} \frac{\partial}{\partial \mathbf{m}_k} \left\{ -\tfrac{1}{2}(\mathbf{x}_j - \mathbf{m}_k)^{\mathsf{T}} \mathbf{C}_k^{-1} (\mathbf{x}_j - \mathbf{m}_k) \right\} = 0$$

Using the matrix property[14] $\frac{\partial}{\partial \mathbf{x}} \{\mathbf{x}^{\mathsf{T}} \mathbf{A} \mathbf{x}\} = 2\mathbf{A}\mathbf{x}$ this can be written

$$\sum_{j=1}^{J} \frac{\alpha_k p(\mathbf{x}_j|\mathbf{m}_k, \mathbf{C}_k)}{p(\mathbf{x}_j|\boldsymbol{\alpha_K}, \mathbf{m_K}, \mathbf{C_K})} \mathbf{C}_k^{-1}(\mathbf{x}_j - \mathbf{m}_k) \equiv \sum_{j=1}^{J} \frac{\alpha_k p(\mathbf{x}_j|\mathbf{m}_k, \mathbf{C}_k)}{p(\mathbf{x}_j)} \mathbf{C}_k^{-1}(\mathbf{x}_j - \mathbf{m}_k) = 0$$

If we regard $\alpha_k$ as the prior probability of the $k$th component then the fraction just inside the sum will be recognised, from Bayes' Theorem, as $p(\mathbf{m}_k, \mathbf{C}_k|\mathbf{x}_j)$, the posterior probability for that component given the $\mathbf{x}_j$th sample,[15] so that the last equation becomes

$$\sum_{j=1}^{J} p(\mathbf{m}_k, \mathbf{C}_k|\mathbf{x}_j) \mathbf{C}_k^{-1}(\mathbf{x}_j - \mathbf{m}_k) = 0$$

The inverse covariance matrix cancels out; rearranging the remaining terms gives

$$\mathbf{m}_k = \frac{\sum_{j=1}^{J} p(\mathbf{m}_k, \mathbf{C}_k|\mathbf{x}_j) \mathbf{x}_j}{\sum_{j=1}^{J} p(\mathbf{m}_k, \mathbf{C}_k|\mathbf{x}_j)}$$

The posterior probabilities in the denominator, $p(\mathbf{m}_k, \mathbf{C}_k|\mathbf{x}_j)$, express the likelihood that the correct component is $k$ for the sample $\mathbf{x}_j$. It would be high if the sample belongs to that component, and low otherwise. The sum over all samples of the

---

[14] See K.B. Petersen and M.S. Pedersen, *The Matrix Cookbook*, 15 Nov 2012, https://www.math. uwaterloo.ca/~hwolkowi/matrixcookbook.pdf accessed 2021.

[15] In C.M. Bishop, loc. cit., it is called the *responsibility*.

posterior probability for component $k$ in the denominator is a measure of the effective number of samples likely to belong to that component,[16] and we define it as $N_k$:

$$N_k = \sum_{j=1}^{J} p(\mathbf{m}_k, \mathbf{C}_k|\mathbf{x}_j) \qquad (8.17)$$

Note that the numerator in the expression for $\mathbf{m}_k$ above is a sum over all the samples weighted by how likely it is that the $k$th component is the correct one in each case. The required values of the $\mathbf{m}_k$ that maximise the log likelihood in (8.15) can now be written

$$\mathbf{m}_k = \frac{1}{N_k} \sum_{j=1}^{J} p(\mathbf{m}_k, \mathbf{C}_k|\mathbf{x}_j)\mathbf{x}_j \qquad (8.18)$$

To find the covariance matrices of the components that maximise (8.15) we proceed in a similar manner by taking the first derivative with respect to $\mathbf{C}_k$ and equating the result to zero. Proceeding as before we have, similar to (8.16),

$$\sum_{j=1}^{J} \frac{\alpha_k}{\sum_{k=1}^{K} \alpha_k \mathcal{N}(\mathbf{x}_j|\mathbf{m}_k, \mathbf{C}_k)} \frac{\partial}{\partial \mathbf{C}_k} \mathcal{N}(\mathbf{x}_j|\mathbf{m}_k, \mathbf{C}_k) = 0 \qquad (8.19)$$

The derivative of the normal distribution with respect to its covariance matrix is a little tedious but relatively straightforward in view of two more useful matrix calculus properties[17]:

$$\frac{\partial}{\partial \mathbf{M}} |\mathbf{M}| = |\mathbf{M}| (\mathbf{M}^{-1})^{\mathrm{T}} \qquad (8.20a)$$

$$\frac{\partial}{\partial \mathbf{M}} \mathbf{a}^{\mathrm{T}} \mathbf{M}^{-1} \mathbf{a} = -(\mathbf{M}^{-1})^{\mathrm{T}} \mathbf{a}\mathbf{a}^{\mathrm{T}} (\mathbf{M}^{-1})^{\mathrm{T}} \qquad (8.20b)$$

Using these we find

$$\frac{\partial}{\partial \mathbf{C}_k} \mathcal{N}(\mathbf{x}_j|\mathbf{m}_k, \mathbf{C}_k) = \mathcal{N}(\mathbf{x}_j|\mathbf{m}_k, \mathbf{C}_k)\left\{(\mathbf{x}_j - \mathbf{m}_k)(\mathbf{x}_j - \mathbf{m}_k)^{\mathrm{T}}(\mathbf{C}_k^{-1})^{\mathrm{T}} - 1\right\}$$

---

[16] See Bishop, loc. cit.
[17] See K.B. Petersen and M.S. Pedersen, loc. cit.

When substituted in (8.19) this gives

$$\sum_{j=1}^{J} \frac{\alpha_k \, \mathcal{N}\left(\mathbf{x}_j | \mathbf{m}_k, \mathbf{C}_k\right)}{\sum_{k=1}^{K} \alpha_k \mathcal{N}\left(\mathbf{x}_j | \mathbf{m}_k, \mathbf{C}_k\right)} \left\{ \left(\mathbf{x}_j - \mathbf{m}_k\right)\left(\mathbf{x}_j - \mathbf{m}_k\right)^{\mathrm{T}}\left(\mathbf{C}_k^{-1}\right)^{\mathrm{T}} - 1 \right\} = 0$$

Recognising as before that the fraction just inside the summation is the posterior probability, multiplying throughout by $\mathbf{C}_k^{-1}$, and recalling that the covariance matrix is symmetric, we have

$$\sum_{j=1}^{J} p\left(\mathbf{m}_k, \mathbf{C}_k | \mathbf{x}_j\right)\left\{ \left(\mathbf{x}_j - \mathbf{m}_k\right)\left(\mathbf{x}_j - \mathbf{m}_k\right)^{\mathrm{T}} - \mathbf{C}_k \right\} = 0$$

so that, with (8.17), this gives

$$\mathbf{C}_k = \frac{1}{N_k} \sum_{j=1}^{J} p\left(\mathbf{m}_k, \mathbf{C}_k | \mathbf{x}_j\right)\left\{ \left(\mathbf{x}_j - \mathbf{m}_k\right)\left(\mathbf{x}_j - \mathbf{m}_k\right)^{\mathrm{T}} \right\} \qquad (8.21)$$

which is the expression for the covariance matrix of the $k$th component that will maximise (8.15). Note how similar this is in structure to the covariance matrix definition for a single Gaussian component in (6.3). In (8.21) the terms are summed with weights that are the likelihoods that the samples come from the $k$th component. If there were only one component then the weight would be unity, leading to (6.3).

The last task is to find the mixing parameter of the $k$th component that contributes to maximising the log likelihood of (8.15). The $\alpha_k$ are also constrained by (8.12b). Therefore, we maximise the following expression with respect to $\alpha_k$ using Lagrange multipliers $\lambda$ to constrain the maximisation[18]:

$$\sum_{j=1}^{J} \ln \sum_{k=1}^{K} \alpha_k \mathcal{N}\left(\mathbf{x}_j | \mathbf{m}_k, \mathbf{C}_k\right) + \lambda\left\{ \sum_{k=1}^{K} \alpha_k - 1 \right\}$$

Putting the first derivative with respect to $\alpha_k$ to zero gives

$$\sum_{j=1}^{J} \frac{\mathcal{N}\left(\mathbf{x}_j | \mathbf{m}_k, \mathbf{C}_k\right)}{\sum_{k=1}^{K} \alpha_k \mathcal{N}\left(\mathbf{x}_j | \mathbf{m}_k, \mathbf{C}_k\right)} + \lambda = 0$$

---

[18] For a good treatment of Lagrange multipliers see C.M. Bishop, loc. cit., Appendix E.

which, using (8.13), is

$$\sum_{j=1}^{J} \frac{\mathcal{N}(\mathbf{x}_j | \mathbf{m}_k, \mathbf{C}_k)}{p(\mathbf{x}_j | \alpha_k, \mathbf{m}_K, \mathbf{C}_K)} + \lambda = 0$$

and, using Bayes' Theorem

$$\sum_{j=1}^{J} \frac{p(\mathbf{m}_k, \mathbf{C}_k | \mathbf{x}_j)}{\alpha_k} + \lambda = 0$$

Multiplying throughout by $\alpha_k$ and using (8.17) this gives

$$N_k + \alpha_k \lambda = 0 \tag{8.22}$$

so that if we take the sum

$$\sum_{k=1}^{K} \{N_k + \alpha_k \lambda\} = 0$$

we have, in view of (8.12b)

$$\lambda = -N$$

Thus, from (8.22),

$$\alpha_k = \frac{N_k}{N} \tag{8.23}$$

With (8.18), (8.21) and (8.23) we now have expressions for the parameters of the Gaussian mixture model that maximise the log likelihood function of (8.14). Unfortunately, though, they each depend on the posterior probabilities $p(\mathbf{m}_k, \mathbf{C}_k | \mathbf{x}_j)$ which themselves are dependent on the parameters. The results of (8.18), (8.21) and (8.23) do not therefore represent solutions to the Gaussian mixing problem. However, an iterative procedure is available that allows the parameters and the posteriors to be determined progressively. It is called *Expectation–Maximisation* (EM) and is outlined in the algorithm of Table 8.3. It consists of an initial guess for the parameters, following which values for the posteriors are computed—that is called the expectation step. Once the posteriors have been computed, new estimates for the parameters are computed from (8.18), (8.21) and (8.23). That is the maximisation step. The loop is repeated until the parameters don't change any further. It is also helpful at each iteration to estimate the log likelihood of (8.14) and use it to check convergence.

**Table 8.3** The expectation–maximisation algorithm for mixtures of Gaussians

| | |
|---|---|
| **Step 1: Initialisation** | Choose initial values for $\alpha_k, \mathbf{m}_k, \mathbf{C}_k$. Call these $\alpha_k^{old}, \mathbf{m}_k^{old}, \mathbf{C}_k^{old}$ |
| **Step 2: Expectation** | Evaluate the posterior probabilities using the current estimates for $\alpha_k, \mathbf{m}_k, \mathbf{C}_k$, according to: $$p(\mathbf{m}_k, \mathbf{C}_k | \mathbf{x}_j) = \frac{\alpha_k^{old} p(\mathbf{x}_j | \mathbf{m}_k^{old}, \mathbf{C}_k^{old})}{p(\mathbf{x}_j | \alpha_{\mathbf{K}}^{old}, \mathbf{m}_{\mathbf{K}}^{old}, \mathbf{C}_{\mathbf{K}}^{old})}$$ i.e. $$p(\mathbf{m}_k, \mathbf{C}_k | \mathbf{x}_j) = \frac{\alpha_k^{old} p(\mathbf{x}_j | \mathbf{m}_k^{old}, \mathbf{C}_k^{old})}{\sum_{k=1}^{K} \alpha_k^{old} \mathcal{N}(\mathbf{x}_j | \mathbf{m}_k^{old}, \mathbf{C}_k^{old})}$$ |
| **Step 3: Maximisation** | Compute new values for $\alpha_k, \mathbf{m}_k, \mathbf{C}_k$ from (8.18), (8.21) and (8.23) in the following way: First compute $$N_k = \sum_{j=1}^{J} p(\mathbf{m}_k^{old}, \mathbf{C}_k^{old} | \mathbf{x}_j)$$ then $$\mathbf{m}_k^{new} = \frac{1}{N_k} \sum_{j=1}^{J} p(\mathbf{m}_k^{old}, \mathbf{C}_k^{old} | \mathbf{x}_j) \mathbf{x}_j$$ $$\mathbf{C}_k^{new} = \frac{1}{N_k} \sum_{j=1}^{J} p(\mathbf{m}_k^{old}, \mathbf{C}_k^{old} | \mathbf{x}_j) \{ (\mathbf{x}_j - \mathbf{m}_k^{new})(\mathbf{x}_j - \mathbf{m}_k^{new})^{\mathrm{T}} \}$$ $$\alpha_k^{new} = \frac{N_k}{N}$$ Also, evaluate the log likelihood $$\ln p(\mathbf{X} | \alpha_{\mathbf{k}}, \mathbf{m}_{\mathbf{K}}, \mathbf{C}_{\mathbf{K}}) = \sum_{j=1}^{J} \ln \sum_{k=1}^{K} \alpha_k^{old} \mathcal{N}(\mathbf{x}_j | \mathbf{m}_k^{old}, \mathbf{C}_k^{old})$$ |
| **Step 4: Evaluation** | If $\mathbf{m}_k^{new} \approx \mathbf{m}_k^{old}$, $\mathbf{C}_k^{new} \approx \mathbf{C}_k^{old}$, $\alpha_k^{new} \approx \alpha_k^{old}$ then terminate the process. Otherwise put $\mathbf{m}_k^{old} = \mathbf{m}_k^{new}$, $\mathbf{C}_k^{old} = \mathbf{C}_k^{new}$, $\alpha_k^{old} = \alpha_k^{new}$ and return to Step 2 This check can also be carried out on the basis of little or no change in the log likelihood calculation |

## 8.5 Minimum Distance Classification

### 8.5.1 The Case of Limited Training Data

The effectiveness of maximum likelihood classification depends on reasonably accurate estimation of the mean vector $\mathbf{m}$ and the covariance matrix $\mathbf{C}$ for each spectral class. This in turn depends on having a sufficient number of training pixels for each of those classes. In cases where that is not possible, inaccurate estimates of the elements of $\mathbf{C}$ result, leading to poor classification.

When the number of training samples per class is limited, it may sometimes be more effective to resort to an algorithm that does not make use of covariance information but instead depends only on the mean positions of the spectral classes, noting that for a given number of samples they can be more accurately estimated than covariances. The minimum distance classifier, or more precisely, minimum distance to class means classifier, is such an approach. With this algorithm, training data is used only to determine class means; classification is then performed by placing a pixel in the class of the nearest mean.

The minimum distance algorithm is also attractive because it is faster than maximum likelihood classification, as will be seen in Sect. 8.5.6. However, because it does not use covariance data it is not as flexible. In maximum likelihood classification each class is modelled by a multivariate normal class model that can account for spreads of data in particular spectral directions. Since covariance data is not used in the minimum distance technique class models are symmetric in the spectral domain. Elongated classes will, therefore, not be well modelled. Instead, several spectral classes may need to be used with this algorithm in cases where one might be suitable for maximum likelihood classification.

## 8.5.2   *The Discriminant Function*

Suppose $\mathbf{m}_i, i = 1 \ldots M$ are the means of the $M$ classes determined from training data, and $\mathbf{x}$ is the position of the pixel in spectral space to be classified. Compute the set of squared Euclidean distances of the unknown pixel to each of the class means:

$$d(\mathbf{x}, \mathbf{m}_i)^2 = (\mathbf{x} - \mathbf{m}_i)^{\mathrm{T}}(\mathbf{x} - \mathbf{m}_i) = (\mathbf{x} - \mathbf{m}_i) \cdot (\mathbf{x} - \mathbf{m}_i) \quad i = 1 \ldots M$$

Expanding the dot product form gives

$$d(\mathbf{x}, \mathbf{m}_i)^2 = \mathbf{x} \cdot \mathbf{x} - 2\mathbf{m}_i \cdot \mathbf{x} + \mathbf{m}_i \cdot \mathbf{m}_i$$

Classification is performed using the decision rule

$$\mathbf{x} \in \omega_i \text{ if } d(\mathbf{x}, \mathbf{m}_i)^2 < d(\mathbf{x}, \mathbf{m}_j)^2 \text{ for all } j \neq i$$

Since $\mathbf{x} \cdot \mathbf{x}$ is common to all squared distance calculations it can be removed from both sides in the decision rule. The sign of the remainder can then be reversed so that the decision rule can be written in the same way as (8.5) to give

$$\mathbf{x} \in \omega_i \text{ if } \quad g_i(\mathbf{x}) > g_j(\mathbf{x}) \text{ for all } j \neq i$$

in which

$$g_i(\mathbf{x}) = 2\mathbf{m}_i \cdot \mathbf{x} - \mathbf{m}_i \cdot \mathbf{m}_i \tag{8.24}$$

Equation (8.24) is the discriminant function for the minimum distance classifier.[19]

## 8.5.3  Decision Surfaces for the Minimum Distance Classifier

The implicit surfaces in spectral space separating adjacent classes are defined by the respective discriminant functions being equal. The surface between the $i$th and $j$th spectral classes is given by

$$g_i(\mathbf{x}) - g_j(\mathbf{x}) = 0$$

which, when substituting from (8.24), gives

$$2(\mathbf{m}_i - \mathbf{m}_j) \cdot \mathbf{x} - (\mathbf{m}_i \cdot \mathbf{m}_i - \mathbf{m}_j \cdot \mathbf{m}_j) = 0$$

This defines a linear surface—a *hyperplane* in more than three dimensions. The surfaces between each pair of classes define a set of first-degree separating hyperplanes that partition the spectral space linearly. The quadratic decision surface generated by the maximum likelihood rule in Sect. 8.3.4 renders that algorithm potentially more powerful than the minimum distance rule if properly trained; the minimum distance classifier nevertheless is effective when the number of training samples is limited or if linear separation of the classes is suspected.

## 8.5.4  Thresholds

Thresholds can be applied to minimum distance classification by ensuring not only that a pixel is closest to a candidate class but also that it is within a prescribed distance of that class in spectral space. Such an assessment is often used with the minimum distance rule. The distance threshold is usually specified in terms of a number of standard deviations from the class mean.

## 8.5.5  Degeneration of Maximum Likelihood to Minimum Distance Classification

The major difference between the minimum distance and maximum likelihood classifiers lies in the use, by the latter, of the sample covariance information. Whereas the

---

[19] It is possible to implement a minimum distance classifier using distance measures other than Euclidean: see A.G. Wacker and D.A. Landgrebe, Minimum distance classification in remote sensing, *First Canadian Symposium on Remote Sensing*, Ottawa, 1972.

minimum distance classifier labels a pixel as belonging to a particular class on the basis only of its distance from the relevant mean in spectral space, irrespective of its direction from that mean, the maximum likelihood classifier modulates its decision with direction, based on the information contained in the covariance matrix. Furthermore, the entry $-\frac{1}{2}\ln|C_i|$ in its discriminant function shows explicitly that patterns have to be closer to some means than others to have equivalent likelihoods of class membership. As a result, superior performance is expected of the maximum likelihood classifier. The following situation however warrants consideration since there is then no advantage in maximum likelihood procedures. It could occur in practice when class covariance is dominated by systematic noise rather than by the natural spectral spreads of the individual spectral classes.

Consider the covariance matrices of all classes to be diagonal and equal, and the variances in each component to be identical, so that

$$C_i = \sigma^2 I \text{ for all } i$$

in which $I$ is the identity matrix. The discriminant function for the maximum likelihood classifier in (8.7) then becomes

$$g_i(\mathbf{x}) = -\frac{1}{2}\ln \sigma^{2N} - \frac{1}{2\sigma^2}(\mathbf{x} - \mathbf{m}_i)^T(\mathbf{x} - \mathbf{m}_i) + \ln p(\omega_i)$$

The first term on the right-hand side is common to all discriminant functions and can be ignored. The second term can be expanded in dot product form in which the resulting $\mathbf{x} \cdot \mathbf{x}$ terms can also be ignored, leaving

$$g_i(\mathbf{x}) = \frac{1}{2\sigma^2}(2\mathbf{m}_i \cdot \mathbf{x} - \mathbf{m}_i \cdot \mathbf{m}_i) + \ln p(\omega_i)$$

If all the prior probabilities are assumed to be equal then the last term can be ignored, allowing $\frac{1}{2\sigma^2}$ to be removed as a common factor, leaving

$$g_i(\mathbf{x}) = 2\mathbf{m}_i \cdot \mathbf{x} - \mathbf{m}_i \cdot \mathbf{m}_i$$

which is the same as (8.24), the discriminant function for the minimum distance classifier. Therefore, minimum distance and maximum likelihood classification are the same for identical and symmetric spectral class distributions.

### 8.5.6  Classification Time Comparison of the Maximum Likelihood and Minimum Distance Rules

For the minimum distance classifier, the discriminant function in (8.24) must be evaluated for each pixel and each class. In practice $2\mathbf{m}_i$ and $\mathbf{m}_i \cdot \mathbf{m}_i$ would be calculated beforehand, leaving the computation of $N$ multiplications and $N$

additions to check the potential membership of a pixel to one class, where $N$ is the number of components in $\mathbf{x}$. By comparison, evaluation of the discriminant function for maximum likelihood classification in (8.7) requires $N^2 + N$ multiplications and $N^2 + 2N + 1$ additions to check a pixel against one class, given that $\ln p(\omega_i) - \frac{1}{2}\ln|\mathbf{C}_i|$ would have been calculated beforehand. Ignoring additions by comparison to multiplications, the maximum likelihood classifier takes $N + 1$ times as long as the minimum distance classifier to perform a classification. It is also significant to note that classification time, and thus cost, increases quadratically with the number of spectral components for the maximum likelihood classifier, but only linearly for minimum distance classification.

## 8.6   Parallelepiped Classification

The parallelepiped classifier is a very simple supervised classifier that is trained by finding the upper and lower brightness values in each spectral dimension. Often that is done by inspecting histograms of the individual spectral components in the available training data, as shown in Fig. 8.8. Together the upper and lower bounds in each dimension define a multidimensional box or *parallelepiped*; Fig. 8.9 shows a set of two-dimensional parallelepipeds. Unknown pixels are labelled as coming from the class of the parallelepiped within which they lie.

While it is very simple and fast, it has several limitations. First, there can be considerable gaps between the parallelepipeds in spectral space; pixels in those regions cannot be classified. By contrast, the maximum likelihood and minimum distance rules will always label unknown pixels unless thresholds are applied. Secondly, for correlated data some parallelepipeds can overlap, as illustrated in Fig. 8.10, because their sides are always parallel to the spectral axes. As a result, there are some parts of the spectral domain that can't be separated. Finally, as with the minimum distance classifier, there is no provision for prior probability of class membership with the parallelepiped rule.

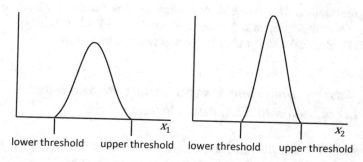

**Fig. 8.8** Setting the parallelepiped boundaries by inspecting class histograms in each band

**Fig. 8.9** A set of two-dimensional parallelepipeds

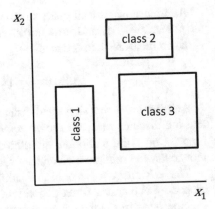

**Fig. 8.10** Classification of correlated data showing regions of inseparability

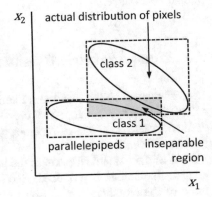

## 8.7 Mahalanobis Classification

Consider the discriminant function for the maximum likelihood classifier for the special case of equal prior probabilities in (8.8). If the sign is reversed the function can be considered as a distance squared measure because the quadratic entry has those dimensions and the other term is a constant. Thus, we can define

$$d(\mathbf{x}, \mathbf{m}_i)^2 = \ln|\mathbf{C}_i| + (\mathbf{x} - \mathbf{m}_i)^T \mathbf{C}_i^{-1}(\mathbf{x} - \mathbf{m}_i) \qquad (8.25)$$

and classify an unknown pixel on the basis of the smallest $d(\mathbf{x}, \mathbf{m}_i)$, as for the Euclidean minimum distance classifier. Thus, the maximum likelihood classifier (with equal priors) can be considered as a minimum distance algorithm but with a distance measure that is sensitive to direction in spectral space.

Assume now that all class covariances are the same and given by $\mathbf{C}_i = \mathbf{C}$ for all $i$. The $\ln|\mathbf{C}|$ term is now not discriminating and can be ignored. The squared distance measure then reduces to

$$d(\mathbf{x}, \mathbf{m}_i)^2 = (\mathbf{x} - \mathbf{m}_i)^{\mathrm{T}} \mathbf{C}^{-1}(\mathbf{x} - \mathbf{m}_i) \tag{8.26}$$

A classifier using this simplified distance measure is called a *Mahalanobis classifier* and the distance measure shown squared in (8.26) is called the *Mahalanobis distance*. Under the additional simplification of $\mathbf{C} = \sigma^2 \mathbf{I}$ the Mahalanobis classifier reduces to the minimum Euclidean distance classifier of Sect. 8.5.

The advantage of the Mahalanobis classifier over the maximum likelihood procedure is that it is faster and yet retains a degree of directional sensitivity in the spectral domain via the covariance matrix $\mathbf{C}$, which could be a class average or a pooled covariance.

## 8.8  Non-parametric Classification

Classifiers such as the maximum likelihood and minimum distance rules are often called *parametric* because the class definitions and discriminant functions are defined in terms of sets of parameters, such as the mean vectors and covariance matrices for each class.

One of the valuable aspects of the parametric, statistical approach is that a set of relative likelihoods is produced. Even though, in the majority of cases, the maximum of the likelihoods is chosen to indicate the most probable label for a pixel, there exists nevertheless information in the remaining likelihoods that could be made use of in some circumstances, either to initiate processes such as relaxation labelling (Sect. 8.23.4) and Markov random fields (Sect. 8.23.5) or simply to provide the user with some feeling for the other likely classes. Those situations are not common however and, in most remote sensing applications, the maximum selection is made. That being so, the material in Sects. 8.3.4 and 8.5.3 shows that the decision process has a geometric counterpart in that a comparison of statistically derived discriminant functions leads equivalently to a decision rule that allows a pixel to be classified on the basis of its position in spectral space compared with the location of a decision surface.

There are also classifiers that, in principle, don't depend on parameter sets and are thus called *non-parametric*. Two simple non-parametric methods are given in Sects. 8.9 and 8.10, although the table look up approach of Sect. 8.9 is now rarely used.

Non-parametric methods based on finding geometric decision surfaces were popular in the very early days of pattern recognition[20] but fell away over the 1980s

---

[20] See N.J. Nilsson, *Learning Machines*, McGraw-Hill, N.Y., 1965.

and 1990s because flexible training algorithms could not be found. They have, however, been revived in the past several decades with the popularity of the support vector classifier and the neural network, which we treat in Sects. 8.14 and 8.20, 8.21 after we explore some non-parametric approaches and their limitations.

## 8.9 Table Look Up Classification

Because the set of discrete brightness values that can be taken by a pixel in each spectral band is limited by the radiometric resolution of the data, there is a finite, although often very large, number of distinct pixel vectors in any particular image. For a given class there may not be very many different pixel vectors if the radiometric resolution is not high, as was the case with the earliest remote sensing instruments. In such situations a viable classification scheme is to note the set of pixel vectors corresponding to a given class based on representative training data, and then use those vectors to classify the image by comparing unknown pixels with each pixel in the training data until a match is found. No arithmetic operations are required and, notwithstanding the number of comparisons that might be necessary to determine a match, it is a fast classifier. It is referred to as a *look up table* approach since the training pixel brightness values are stored in tables that point to the corresponding classes.

An obvious drawback with this technique is that the chosen training data must contain an example of every possible pixel vector for each class. Should some be missed then the corresponding pixels in the image will be left unclassified. With modern data sets, in which there can be billions of individual data vectors, this approach is impractical.

## 8.10 *k*NN (Nearest Neighbour) Classification

A classifier that is particularly simple in concept, but can be time consuming to apply, is the $k$ Nearest Neighbour ($k$NN) classifier. It assumes that pixels close to each other in spectral space are likely to belong to the same class. In its simplest form an unknown pixel is labelled by examining the available training pixels in the spectral domain and choosing the class most represented among a pre-specified number of nearest neighbours in the training set. The comparison essentially requires the distances from the unknown pixel to all training pixels to be computed.

Suppose there are $k_i$ neighbours labelled as class $\omega_i$ among the $k$ nearest neighbours of a pixel vector $\mathbf{x}$, noting that $\sum_{i=1}^{M} k_i = k$ where $M$ is the total number of classes. In the basic $k$NN rule we define the discriminant function for the $i$th class as

$$g_i(\mathbf{x}) = k_i$$

and the decision rule as $\qquad \mathbf{x} \in \omega_i$ if $g_i(\mathbf{x}) > g_j(\mathbf{x})$ for all $j \neq i$.

The basic rule does not take distance into account apart from identifying the neighbours. An improvement is to distance-weight and normalise the discriminant function:

$$g_i(\mathbf{x}) = \frac{\sum_{l=1}^{k_i} 1/d\left(\mathbf{x}, \mathbf{x}_i^l\right)}{\sum_{i=1}^{M} \sum_{l=1}^{k_i} 1/d\left(\mathbf{x}, \mathbf{x}_i^l\right)} \qquad (8.27)$$

in which $d\left(\mathbf{x}, \mathbf{x}_i^l\right)$ is the distance (usually Euclidean) from the unknown pixel vector $\mathbf{x}$ to its neighbour $\mathbf{x}_i^l$, the $i$th of the $k_i$ pixels in class $\omega_i$.

If the training data for each class is not in proportion to its respective population $p(\omega_i)$ in the image, a Bayesian version of the simple nearest neighbour discriminant function is

$$g_i(\mathbf{x}) = \frac{p(\mathbf{x}|\omega_i)p(\omega_i)}{\sum_{i=1}^{M} p(\mathbf{x}|\omega_i)p(\omega_i)} = \frac{k_i p(\omega_i)}{\sum_{i=1}^{M} k_i p(\omega_i)} \qquad (8.28)$$

For each unknown pixel to be labelled in the $k$NN algorithm, as many spectral distances as there are training pixels must be evaluated. That requires an impractically high computational load, particularly when the number of spectral bands and/or the number of training samples is large. The method is not well-suited therefore to hyperspectral datasets, although it is possible to improve the efficiency of the distance search process.[21]

## 8.11  The Spectral Angle Mapper

A classifier sometimes used with data of high spectral dimensionality, such as that recorded by imaging spectrometers, is the spectral angle mapper[22] (SAM) which segments the spectral domain on the basis of the angles of vectors measured from

---

[21] See B.V. Dasarathy, Nearest neighbour (NN) norms, *NN Pattern Classification Techniques*, IEEE Computer Society Press, Los Alamitos, California, 1991.

[22] F.A. Kruse, A.B. Letkoff, J.W. Boardman, K.B. Heidebrecht, A.T.Shapiro, P.J. Barloon and A. F.H. Goetz, The spectral image processing system (SIPS)—interactive visualization and analysis of imaging spectrometer data, *Remote Sensing of Environment*, vol. 44, 1993, pp. 145–163.

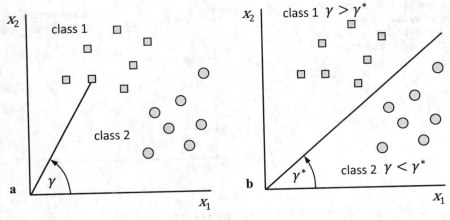

**Fig. 8.11** **a** Representing pixel vectors by their angles; **b** segmenting the spectral space by angle

the origin, as illustrated in Fig. 8.11 for two dimensions. Every pixel point in spectral space has both a magnitude and angular direction when expressed in polar, as against the more usual Cartesian, form. The decision boundary shown in Fig. 8.11b is based on the best angular separation between the training pixels in different classes, usually in an average sense. Only first order parameters are estimated, putting the SAM in the same class as the minimum distance classifier in terms of its suitability for high dimensional imagery.

## 8.12 Non-parametric Classification from a Geometric Basis

### 8.12.1 Linear Classification and the Concept of a Weight Vector

Consider the simple two class spectral space shown in Fig. 8.12, which has been constructed intentionally so that a simple straight line can be drawn between the pixels as shown. This straight line will be a multidimensional linear surface, or hyperplane in general, and can function as a decision surface for classification. In the two dimensions shown, the equation of the line can be expressed

$$w_1 x_1 + w_2 x_2 + w_3 = 0$$

where the $x_i$ are the brightness value coordinates in the spectral space and the $w_i$ are a set of coefficients, usually called *weights*. There will be as many weights as the number of channels in the data, plus one. If the number of channels or bands is $N$, the equation of a general linear surface is

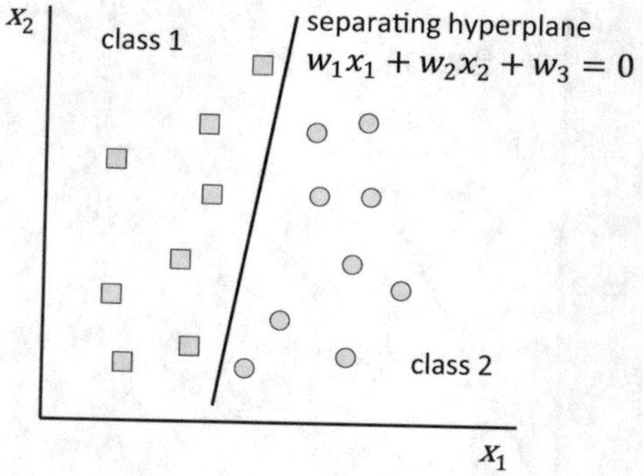

**Fig. 8.12** Two-dimensional spectral space, with two classes that can be separated by a simple linear surface

$$w_1 x_1 + w_2 x_2 + \cdots + w_N x_N + w_{N+1} = 0$$

which can be written

$$\mathbf{w}^{\mathsf{T}}\mathbf{x} + w_{N+1} \equiv \mathbf{w} \cdot \mathbf{x} + w_{N+1} = 0 \tag{8.29}$$

where $\mathbf{x}$ is the pixel measurement vector and $\mathbf{w}$ is called the *weight vector*. The transpose operation has the effect of turning the column vector into a row vector.

The position of the separating hyperplane would generally not be known initially; it would have to be found by training based on sets of reference pixels, just as the parameters of the maximum likelihood classifier are found by training. Note that there is not a unique solution; inspection of Fig. 8.12 suggests that any one of an infinite number of slightly different hyperplanes would be acceptable.

## 8.12.2 Testing Class Membership

The calculation in (8.29) will be zero only for values of $\mathbf{x}$ lying exactly on the hyperplane (the decision surface). If we substitute into that equation values of $\mathbf{x}$ corresponding to any of the pixel points shown in Fig. 8.12 the left-hand side of (8.29) will be non-zero. Pixels in one class will generate a positive inequality, while pixels in the other class will generate a negative inequality. Once the decision surface, or more specifically the weights $\mathbf{w}$ that define its equation, has been found through training then a decision rule for labelling unknown pixels is

$$\begin{aligned}
\mathbf{x} \in \text{class 1} \quad \text{if} \quad \mathbf{w}^{\mathsf{T}}\mathbf{x} + w_{N+1} > 0 \\
\mathbf{x} \in \text{class 2} \quad \text{if} \quad \mathbf{w}^{\mathsf{T}}\mathbf{x} + w_{N+1} < 0
\end{aligned} \tag{8.30}$$

## 8.13  Training a Linear Classifier

The simple linear classifier in Fig. 8.12 can be trained in several ways. A particularly simple approach is to choose the hyperplane as the perpendicular bisector of the line between the mean vectors of the two classes. That is the minimum distance to class means classifier of Sect. 8.5. Another is to guess an initial position for the separating hyperplane and then iterate it into position by reference, repetitively, to each of the training samples. Such a method has been known for over 50 years.[23] There is however a more elegant training method which forms the basis of the support vector classifier treated in the next section.

## 8.14  The Support Vector Machine: Linearly Separable Classes

Inspection of Fig. 8.12 suggests that the only training patterns that need to be considered in finding a suitable hyperplane are those from each class nearest the hyperplane. Effectively, they are the patterns closest to the border between the classes. If a hyperplane can be found that satisfies those pixels then the pixels further from the border must, by definition, also be satisfied. Moreover, again by inspecting Fig. 8.12, we can induce that the "best" hyperplane would be that which would be equidistant, on the average, between the nearest edge pixels for each of the two classes. This concept, along with the concentration just on the edge pixels, forms the basis of the support vector machine, which was introduced into remote sensing in 1998.[24]

If we expand the region in the vicinity of the hyperplane in Fig. 8.12, we can see that the optimal position and orientation of the separating hyperplane is when there is a maximum separation between the patterns of the two classes in the manner illustrated in Fig. 8.13. We can draw two more hyperplanes, parallel to the separating hyperplane, that pass through the nearest training (edge) pixels from the classes, as indicated. We call them marginal hyperplanes. Their equations are shown on the figure, suitably scaled so that the right-hand sides have a magnitude of unity. For pixels that lie on or beyond the marginal hyperplanes we have

$$\text{for class 1 pixels } \mathbf{w}^T\mathbf{x} + w_{N+1} \geq 1$$
$$\text{for class 2 pixels } \mathbf{w}^T\mathbf{x} + w_{N+1} \leq -1 \tag{8.31}$$

On the marginal hyperplanes

[23] See Nilsson, ibid.

[24] J.A. Gualtieri and R.F. Cromp, Support vector machines for hyperspectral remote sensing classification, *Proc. SPIE*, vol. 3584, 1998, pp. 221–232.

**Fig. 8.13** An optimal separating hyperplane can be determined by finding the maximum separation between classes; two marginal hyperplanes can be constructed using the pixel vectors closest to the separating hyperplane

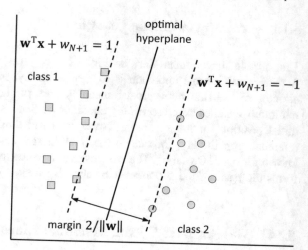

$$\text{for class 1 pixels } \mathbf{w}^T\mathbf{x} + w_{N+1} = 1$$
$$\text{for class 2 pixels } \mathbf{w}^T\mathbf{x} + w_{N+1} = -1$$

or

$$\text{for class 1 pixels } \mathbf{w}^T\mathbf{x} + w_{N+1} - 1 = 0$$
$$\text{for class 2 pixels } \mathbf{w}^T\mathbf{x} + w_{N+1} + 1 = 0$$

The perpendicular distances of these hyperplanes from the origin are, respectively, $|1 - w_{N+1}|/\|\mathbf{w}\|$ and $|-1 - w_{N+1}|/\|\mathbf{w}\|$ in which $\|\mathbf{w}\|$ is the Euclidean length, or norm, of the weight vector.[25] The separation of the hyperplanes is the difference in these perpendicular distances, which we call the *margin*:

$$\text{margin} = 2/\|\mathbf{w}\| \tag{8.32}$$

The best or optimal position for the separating hyperplane is that for which the margin of (8.32) is largest or, equivalently, when the weight vector norm $\|\mathbf{w}\|$ is smallest. This provides a goal for optimal training of the linear classifier. However, we must always ensure that every pixel vector stays on its correct side of the separating hyperplane, which gives us a set of constraints that must be observed when seeking the optimal separating hyperplane. We can capture those constraints mathematically in the following manner.

Describe the class label for the *i*th training pixel by the variable $y_i$, which takes the values of +1 for class 1 pixels and −1 for class 2 pixels. The two equations in (8.31) can then be written in single expression, valid for pixels from both classes:

---

[25] See Prob. 8.13.

for pixel $\mathbf{x}_i$ in its correct class   $y_i(\mathbf{w}^T\mathbf{x}_i + w_{N+1}) \geq 1$

$$y_i(\mathbf{w}^T\mathbf{x}_i + w_{N+1}) - 1 \geq 0 \tag{8.33}$$

or

In seeking to minimise $\|\mathbf{w}\|$ we must observe the constraints of (8.33), one for each training pixel. Constrained minimisation can be handled by the process of Lagrange multipliers.[26] This entails setting up a function, called the Lagrangian $\mathcal{L}$, which consists of the property to be minimised (the vector norm $\|\mathbf{w}\|$) but from which is subtracted a proportion of each constraint. For later convenience we will seek to minimise half the square of the vector norm, so that the Lagrangian has the form:

$$\mathcal{L} = \frac{1}{2}\|\mathbf{w}\|^2 - \sum_i \alpha_i \{y_i(\mathbf{w}^T\mathbf{x}_i + w_{N+1}) - 1\} \tag{8.34}$$

The $\alpha_i \geq 0$ for all $i$, are called the Lagrange multipliers and are positive by definition. How are they treated during the minimisation process? Suppose we chose a training pixel and find it is on the wrong side of the separating hyperplane, thus violating (8.30) and (8.33). Given that $\alpha_i$ is positive that would cause $\mathcal{L}$ to increase. What we need to do is find values for $\mathbf{w}$ and $w_{N+1}$ that minimise $\mathcal{L}$ while the $\alpha_i$ are trying to make it larger for incorrectly located training patterns. In other words, we are shifting the hyperplane via its weights to minimise $\mathcal{L}$ in the face of the $\alpha_i$ trying to make it bigger for wrongly classified training data.

Consider, first, the values of $\mathbf{w}$ and $w_{N+1}$ that minimise $\mathcal{L}$. That requires equating to zero the derivatives of $\mathcal{L}$ with respect to the weights. Noting $\|\mathbf{w}\|^2 \equiv \mathbf{w}^T\mathbf{w}$ we have

$$\frac{\partial \mathcal{L}}{\partial \mathbf{w}} = \mathbf{w} - \sum_i \alpha_i y_i \mathbf{x}_i = 0$$

which gives

$$\mathbf{w} = \sum_i \alpha_i y_i \mathbf{x}_i \tag{8.35}$$

Now

$$\frac{\partial \mathcal{L}}{\partial w_{N+1}} = -\sum_i \alpha_i y_i = 0 \tag{8.36}$$

We can use (8.35) and (8.36) to simplify (8.34). First, using (8.35), we can write

$$\|\mathbf{w}\|^2 = \mathbf{w}^T\mathbf{w} = \sum_j \alpha_j y_j \mathbf{x}_j^T \sum_i \alpha_i y_i \mathbf{x}_i$$

[26] See C.M. Bishop, loc. cit., Appendix E.

Substituting into (8.34) gives

$$\mathcal{L} = \frac{1}{2} \sum_j \alpha_j y_j \mathbf{x}_j^T \sum_i \alpha_i y_i \mathbf{x}_i - \sum_i \alpha_i \{ y_i \left( \sum_j \alpha_j y_j \mathbf{x}_j^T \mathbf{x}_i + w_{N+1} \right) - 1 \}$$

i.e., $\mathcal{L} = \frac{1}{2} \sum_{i,j} \alpha_i \alpha_j y_i y_j \mathbf{x}_j^T \mathbf{x}_i - \sum_{i,j} \alpha_i \alpha_j y_i y_j \mathbf{x}_j^T \mathbf{x}_i - w_{N+1} \sum_i \alpha_i y_i + \sum_i \alpha_i$

Using (8.36) this simplifies to

$$\mathcal{L} = \sum_i \alpha_i - \frac{1}{2} \sum_{i,j} \alpha_i \alpha_j y_i y_j \mathbf{x}_j^T \mathbf{x}_i \tag{8.37}$$

We are now in the position to find the $\alpha_i$. Remember they are trying to make the Lagrangian as large as possible, so we seek to maximise (8.37) with respect to the $\alpha_i$. This is referred to as a *dual representation* in the field of optimisation. Note that there are constraints on this optimisation too. They are that

$$\alpha_i \geq 0 \tag{8.38a}$$

and, from (8.36),
$$\sum_i \alpha_i y_i = 0 \tag{8.38b}$$

Equation (8.37), for any real problem, has to be solved numerically, following which the values of $\alpha_i$ have been found. The one remaining unknown is $w_{N+1}$ which we will come to in a moment. First, however, there is another constraint on (8.37) which, together with (8.38a, b), give what are called the *Karush–Kuhn–Tucker* conditions.[27] This further constraint is

$$\alpha_i \{ y_i \left( \mathbf{w}^T \mathbf{x}_i + w_{N+1} \right) - 1 \} = 0 \tag{8.38c}$$

This is very interesting because it says that either $\alpha_i = 0$ or $y_i(\mathbf{w}^T\mathbf{x}_i + w_{N+1}) = 1$. The latter expression is valid only for training vectors that lie on the marginal hyperplanes—what we call the *support vectors*. For any other training sample this condition is not valid so that (8.38c) can then only be satisfied if $\alpha_i = 0$. In other words, it seems that a whole lot of the training pixels are irrelevant. In a sense that is true but must be interpreted carefully. When maximising (8.37) we have no way of knowing beforehand which of the training samples will end up being support vectors because we don't yet know the value of $\mathbf{w}$. However once (8.37) has been optimised we know the optimal hyperplane and thus the support vectors. We are

---

[27] See Bishop, loc. cit.

then in the position to discard all the other training data. When is that important? It is in the classification phase. That is done via the test of (8.30) or (8.33). To test the class membership of a pixel at position $\mathbf{x}$ in multispectral space we evaluate the sign of $\mathbf{w}^T\mathbf{x} + w_{N+1}$ with $\mathbf{w}$ given by (8.35) but computed using only the support vectors. Thus, the test of class membership for pixel $\mathbf{x}$ is

$$\text{sgn}\{\mathbf{w}^T\mathbf{x} + w_{N+1}\} = \text{sgn}\left\{\sum_{i \in \mathcal{S}} \alpha_i y_i \mathbf{x}_i^T \mathbf{x} + w_{N+1}\right\} \tag{8.39}$$

where the symbol $\mathcal{S}$ in the sum refers only to the support vectors.

How do we now find the value of $w_{N+1}$? A simple approach is to choose two support vectors, one from each class; call these $\mathbf{x}(1)$ and $\mathbf{x}(-1)$ for which $y = 1$ and $y = -1$ respectively. From (8.33) we have for those vectors

$$\mathbf{w}^T\mathbf{x}(1) + w_{N+1} - 1 = 0$$
$$-\mathbf{w}^T\mathbf{x}(-1) - w_{N+1} - 1 = 0$$

so that

$$w_{N+1} = -\frac{1}{2}\mathbf{w}^T\{\mathbf{x}(1) + \mathbf{x}(-1)\} \tag{8.40}$$

We could alternately choose sets of $\mathbf{x}(1)$ and $\mathbf{x}(-1)$ and average the values of $w_{N+1}$ so generated. Following Bishop,[28] this can be generalised by noting from the argument of (8.39) that

$$y_\mathbf{x}\left(\sum_{i \in \mathcal{S}} \alpha_i y_i \mathbf{x}_i^T \mathbf{x} + w_{N+1}\right) = 1 \tag{8.41}$$

in which $y_\mathbf{x}$ is associated with the pixel $\mathbf{x}$. Multiplying throughout by $y_\mathbf{x}$, and noting $y_\mathbf{x}^2 = 1$, we have from (8.41)

$$w_{N+1} = y_\mathbf{x} - \sum_{i \in \mathcal{S}} \alpha_i y_i \mathbf{x}_i^T \mathbf{x}$$

This is for one support vector $\mathbf{x}$. We now average over all support vectors to give

$$w_{N+1} = \frac{1}{N_\mathcal{S}} \sum_{\mathbf{x} \in \mathcal{S}} \left\{y_\mathbf{x} - \sum_{i \in \mathcal{S}} \alpha_i y_i \mathbf{x}_i^T \mathbf{x}\right\} \tag{8.42}$$

in which $N_\mathcal{S}$ is the number of support vectors.

---

[28] ibid.

## 8.15    The Support Vector Machine: Overlapping Classes

It is unrealistic to expect that the pixel vectors from two ground cover classes will be completely separated, as implied in Fig. 8.13. Instead, there is likely to be class overlap more like the situation depicted in Fig. 8.2. Any classifier algorithm, to be effective, must be able to cope with such a situation by generating the best possible discrimination between the classes in the circumstances. As developed in the previous section the support vector classifier will not find a solution for overlapping classes and requires modification. That is done by relaxing the requirement on finding a maximum margin solution by agreeing that such a goal is not possible for *all* training pixels and that we will have to accept that some will not be correctly separated during training. Such a situation is accommodated by introducing a degree of "slackness" in the training step. To develop this variation, consider the set of training pixels in Fig. 8.14.

We introduce a set of positive "slack variables" $\xi_i$, one for each of the training patterns, which are used to modify the constraint of (8.33) such that it now becomes

$$(\mathbf{w}^T\mathbf{x}_i + w_{N+1})y_i \geq 1 - \xi_i \quad \forall i \tag{8.43a}$$

The slack variables are defined such that:

$\xi_i = 0,$        for training pixels that are on or on the correct side of the marginal hyperplane

$\xi_i = 1$        for a pixel on the separating hyperplane—the decision boundary—because $\mathbf{w}^T\mathbf{x}_i + w_{N+1} = 0$ and $|y_i| = 1$.

$\xi_i > 1$        for pixels that are on the wrong side of the separating hyperplane since $\mathbf{w}^T\mathbf{x}_i + w_{N+1}$ has the opposite sign to $y_i$ for misclassified pixels.

$\xi_i = |y_i - (\mathbf{w}^T\mathbf{x}_i + w_{N+1})| < 1$    for all other training pixels because they are between the marginal hyperplanes

When training the support vector machine with overlapping data we minimise the number of pixels in error while maximising the margin by minimising $\|\mathbf{w}\|$. A measure of the number of pixels in error is the sum of the slack variables over all the training pixels; they are all positive and thus their sum increases with misclassification. Minimising misclassification error and maximising the margin together can be achieved by seeking to minimise

$$\frac{1}{2}\|\mathbf{w}\|^2 + C\sum_i \xi_i \tag{8.43b}$$

in which the positive weight C, called the *regularisation parameter*, adjusts the relative importance of the margin versus misclassification error.

**Fig. 8.14** Slack variables and overlapping classes

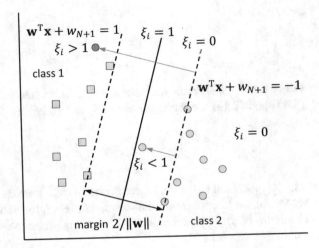

As before the minimisation is subject to constraints. One is (8.43a); the other is that the slack variables are positive. We again accomplish the minimisation by introducing Lagrange multipliers. However, now there is a different multiplier associated with each constraint, so that the Lagrangian is

$$\mathcal{L} = \frac{1}{2}\|\mathbf{w}\|^2 + C\sum_i \xi_i - \sum_i \alpha_i\{y_i(\mathbf{w}^{\mathrm{T}}\mathbf{x}_i + w_{N+1}) - 1 + \xi_i\} - \sum_i \mu_i\xi_i \quad (8.44)$$

in which the $\alpha_i$ and the $\mu_i$ are the Lagrange multipliers. We now equate to zero the first derivatives with respect to the weight vector and the slack variables in an attempt to find the values that minimise the Lagrangian.

First

$$\frac{\partial\mathcal{L}}{\partial\mathbf{w}} = \mathbf{w} - \sum_i \alpha_i y_i \mathbf{x}_i = 0$$

which gives

$$\mathbf{w} = \sum_i \alpha_i y_i \mathbf{x}_i \quad (8.45)$$

while

$$\frac{\partial\mathcal{L}}{\partial w_{N+1}} = -\sum_i \alpha_i y_i = 0 \quad (8.46)$$

Now

$$\frac{\partial \mathcal{L}}{\partial \xi_i} = C - \alpha_i - \mu_i = 0 \tag{8.47}$$

Remarkably, (8.47) removes the slack variables when substituted into (8.44). Since (8.45) and (8.46) are the same as (8.35) and (8.36), (8.44) reduces to

$$\mathcal{L} = \sum_i \alpha_i - \frac{1}{2} \sum_{i,j} \alpha_i \alpha_j y_i y_j \mathbf{x}_j^\mathrm{T} \mathbf{x}_i \tag{8.48}$$

which is identical to the dual formulation of (8.37) which has to be maximised with respect to the Langrange multipliers $\alpha_i$. However, the constraints on the $\alpha_i$ are now different. Since, by definition the Lagrange multipliers, both $\alpha_i$ and $\mu_i$, are non-negative we have

$$0 \le \alpha_i \le C \tag{8.49a}$$

and, from (8.46)

$$\sum_i \alpha_i y_i = 0 \tag{8.49b}$$

Again, (8.48) needs to be solved numerically subject to the constraints of (8.49). Once the $\alpha_i$ are found (8.39) is used to label unknown pixels. As with the linearly separable case some of the $\alpha_i$ will be zero, in which case the corresponding training pixels do not feature in (8.39). The training pixels for which $\alpha_i \ne 0$ are again the support vectors.

There are more *Karush–Kuhn–Tucker* conditions in the case of slack variables as seen in Bishop[29] and Burges[30]; one is $\mu_i \xi_i = 0$. We can use this to generate a means for finding $w_{N+1}$. We know that $\alpha_i > 0$ for support vectors. If, in addition, we have some vectors for which $\alpha_i < C$ then (8.47) shows that $\mu_i$ must be non-zero. Since $\mu_i \xi_i = 0$ this requires $\xi_i = 0$ so that (8.42) can again be used to find $w_{N+1}$ but with the sums over those support vectors for which also $\alpha_i < C$.

[29] ibid.

[30] C.J.C. Burges, A tutorial on support vector machines for pattern recognition, *Data Mining and Knowledge Discovery*, vol. 2, 1998, pp. 121–166.

## 8.16   The Support Vector Machine: Nonlinearly Separable Data and Kernels

When we examine the central equations for support vector classification, viz. (8.37), (8.39), (8.42) and (8.48) we see that the pixel vectors enter only via a scalar (or dot) product of the form $\mathbf{x}_i^T\mathbf{x}$. As is often the case in thematic mapping, it is possible to transform the original pixel vectors to a new set of features before we apply the support vector approach, in an endeavour to improve separability. For example, we could use a function $\phi$ to generate the transformed feature vector $\phi(\mathbf{x})$, so that equations such as (8.39) become

$$\mathrm{sgn}\left\{\phi(\mathbf{w})^T\phi(\mathbf{x}) + w_{N+1}\right\} = \mathrm{sgn}\left\{\sum_{i\in\mathcal{S}} \alpha_i y_i \phi(\mathbf{x}_i)^T\phi(\mathbf{x}) + w_{N+1}\right\} \qquad (8.50)$$

We refer to the scalar product of the transformed features as a *kernel function*, written as

$$k(\mathbf{x}_i, \mathbf{x}) = \phi(\mathbf{x}_i)^T\phi(\mathbf{x}) \qquad (8.51)$$

which has a scalar value. Since the pixel vectors enter the calculations only in this product form it is not necessary to know the actual transformation $\phi(\mathbf{x})$; all we have to do is specify a scalar kernel function $k(\mathbf{x}_i, \mathbf{x})$ of the two pixel vectors.

What functions are suitable as kernels? Effectively, any function that is expressible in the scalar (or dot) product form in (8.51) is suitable. Clearly, we could build up kernels by choosing the transformations first, but that defeats the purpose: the real benefit of the process known as *kernel substitution*, or sometimes the *kernel trick*, is that we don't need to know the transform but can just choose appropriate kernels. All we need do is satisfy ourselves that the kernel chosen is equivalent to a scalar product. The test comes in the form of satisfying the Mercer condition[31] which, for the kernels below, can always be assumed.

It is instructive to examine a classical example. Consider a kernel composed of the simple square of the scalar product:

$$k(\mathbf{x}, \mathbf{y}) = \left[\mathbf{x}^T\mathbf{y}\right]^2$$

To avoid a complication with subscripts in the following the kernel has been expressed in terms of the variables $\mathbf{x}$ and $\mathbf{y}$. We now restrict attention to the case of two-dimensional data, so that the column vectors can be written as $\mathbf{x} = [x_1, x_2]^T$ and $\mathbf{y} = [y_1, y_2]^T$. Expanding the kernel operation, we have

$$k(\mathbf{x}, \mathbf{y}) = \left[\mathbf{x}^T\mathbf{y}\right]^2 = [x_1 y_1 + x_2 y_2]^2$$

---

[31] ibid.

Expanding, we get $[x_1y_1 + x_2y_2]^2 = x_1^2y_1^2 + 2x_1y_1x_2y_2 + x_2^2y_2^2$
which can be expressed

$$[x_1y_1 + x_2y_2]^2 = [x_1^2, \sqrt{2}x_1x_2, x_2^2] \begin{bmatrix} y_1^2 \\ \sqrt{2}y_1y_2 \\ y_2^2 \end{bmatrix} = \begin{bmatrix} x_1^2 \\ \sqrt{2}x_1x_2 \\ x_2^2 \end{bmatrix}^{\mathrm{T}} \begin{bmatrix} y_1^2 \\ \sqrt{2}y_1y_2 \\ y_2^2 \end{bmatrix}$$

This shows that the quadratic kernel $k(\mathbf{x}, \mathbf{y}) = [\mathbf{x}^\mathrm{T}\mathbf{y}]^2$ can be written in the scalar product form of (8.51) and is thus valid. The transformation is now seen explicitly to be

$$\phi(\mathbf{x}) = \begin{bmatrix} x_1^2 \\ \sqrt{2}x_1x_2 \\ x_2^2 \end{bmatrix} = \begin{bmatrix} z_1 \\ z_2 \\ z_3 \end{bmatrix} \tag{8.52}$$

which transforms the original two-dimensional space into three dimensions defined by the squares and products of the original variables. Figure 8.15 shows how this transformation leads to linear separability of a two-class data set that is not linearly separable in the original coordinates. In this example the original data lies either side of the quadrant of a circle defined by $x_1^2 + x_2^2 = 2500$, shown dotted in Fig. 8.15a. Following transformation, the circle becomes the straight line $z_1 + z_3 = 2500$, which has a negative unity slope and intersects the axes at 2500, as shown in Fig. 8.15b. In this simple example the third dimension is not required for separation and can be thought of as coming out of the page; Fig. 8.15b is the two-dimensional sub-space projection in the $z_1, z_3$ plane.

The simple quadratic kernel is of limited value, but the above example serves to demonstrate the importance of using kernels as substitutions for the scalar product in the key equations. A more general polynomial kernel that satisfies the Mercer condition is of the form

$$k(\mathbf{x}_i, \mathbf{x}) = [\mathbf{x}_i^\mathrm{T}\mathbf{x} + b]^m \tag{8.53}$$

with $b > 0$. Values for the parameters $b$ and $m$ have to be found to maximise classifier performance.

A popular kernel in remote sensing applications is the Gaussian radial basis function kernel, which has one parameter $\gamma > 0$ and is based on the distance squared between the two vector arguments:

$$k(\mathbf{x}_i, \mathbf{x}) = \exp\left\{ -\gamma\|\mathbf{x} - \mathbf{x}_i\|^2 \right\} \tag{8.54}$$

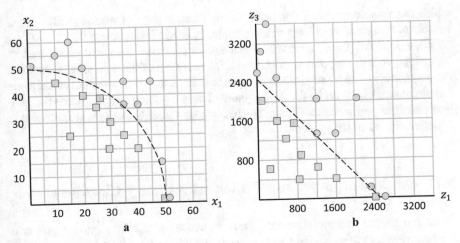

**Fig. 8.15** Use of (8.52) to transform the linearly inseparable data set shown in **a** into the separable case in **b**; the $z_2$ dimension is out of the page, but doesn't contribute to separation

The distance metric chosen is normally Euclidean, although others are also possible. It is interesting to note that the dimensionality of the transformed space with the radial basis function kernel is infinite, which explains its power and popularity. That can be seen by expanding

$$\|\mathbf{x} - \mathbf{x}_i\|^2 = (\mathbf{x} - \mathbf{x}_i)^\mathrm{T}(\mathbf{x} - \mathbf{x}_i) = \mathbf{x}^\mathrm{T}\mathbf{x} - 2\mathbf{x}^\mathrm{T}\mathbf{x}_i + \mathbf{x}_i^\mathrm{T}\mathbf{x}_i$$

so that

$$k(\mathbf{x}_i, \mathbf{x}) = \exp\{-\gamma\mathbf{x}^\mathrm{T}\mathbf{x}\}\ \exp\{2\gamma\mathbf{x}^\mathrm{T}\mathbf{x}_i\}\ \exp\{-\gamma\mathbf{x}_i^\mathrm{T}\mathbf{x}_i\}.$$

Taking just the first exponential and replacing it by its power series we have

$$\exp\{-\gamma\mathbf{x}^\mathrm{T}\mathbf{x}\} = 1 - \gamma\mathbf{x}^\mathrm{T}\mathbf{x} + \frac{\gamma^2}{2}\left(\mathbf{x}^\mathrm{T}\mathbf{x}\right)^2 - \frac{\gamma^3}{6}\left(\mathbf{x}^\mathrm{T}\mathbf{x}\right)^3 + \cdots$$

Note the quadratic transformation of the scalar product in (8.52) led to one more dimension than the power. It is straightforward to demonstrate for two-dimensional data that the cubic term leads to a four-dimensional transformed space. In this last case, the first exponential, consisting of an infinite set of terms, leads to an infinite transformed space, as do the other exponentials in $k(\mathbf{x}_i, \mathbf{x})$.

A third option, that has an association with neural network classifiers treated later, is the hyperbolic tangent kernel:

$$k(\mathbf{x}_i, \mathbf{x}) = \tanh\left(\kappa\mathbf{x}_i^\mathrm{T}\mathbf{x} + b\right) \tag{8.55}$$

which has two parameters to be determined. There is a set of rules that allow new kernels to be constructed from valid kernels, such as those above.[32]

When applying the support vector machine, we have a number of parameters to find, along with the support vectors. The latter come out of the optimisation step but the parameters—the regularisation parameter $C$ in (8.43b) and $\gamma$ for the radial basis function kernel, or $m$ and $b$ for the polynomial kernel—have to be estimated to give best classifier performance. Generally, that is done through a search procedure that is demonstrated in the example in Sect. 8.18.

## 8.17   Multi-category Classification with Binary Classifiers

Since the support vector classifier places unknown pixels into one of just two classes, a strategy is needed to allow its use in the multi-class situations encountered in remote sensing. Several approaches are possible. A simple method often used with binary classifiers in the past is to construct a decision tree, at each node of which one of $M$ classes is separated from the remainder as depicted in Fig. 8.16a. The disadvantage with this method is that the training classes are unbalanced, especially in the early decision nodes, and it is not known optimally which classes should be separated first.

More recently, parallel networks have been used to perform multiclass decisions from binary classifiers, as illustrated in Fig. 8.16b. They have been used principally in one of two ways. First, each classifier can be trained to separate one class from the rest, in which case there are as many classifiers as there are classes, in this case $M$. Again, this approach suffers from unbalanced training sets. If the classifiers are support vector machines there is evidence to suggest that such imbalances can affect classifier performance.[33] Also, some form of logic needs to be applied in the decision rule to choose the most favoured class recommendation over the others; that could involve selecting the class which has the largest argument in (8.39).[34] Despite these considerations, the method is often adopted with support vector classifiers and is known as the *one-against-the-rest* or *one-against-all* (OAA) technique.

A better approach, again using Fig. 8.16b, is to train $M(M-1)/2$ separate binary classifiers, each of which is designed to separate a pair of the classes of interest. This number covers all possible class pairs. Once trained, an unknown pixel is subjected to each classifier and is placed in the class with the highest overall number of classification recommendations in favour. Of course, each individual

---

[32] See Bishop, loc. cit. p. 296.

[33] K. Song, Tackling Uncertainties and Errors in the Satellite Monitoring of Forest Cover Change, *Ph.D. Dissertation*, The University of Maryland, 2010.

[34] See F. Melgani and L. Bruzzone, Classification of hyperspectral remote sensing images with support vector machines, *IEEE Transactions on Geoscience and Remote Sensing*, vol. 42, no. 8, August 2004, pp. 1778–1790.

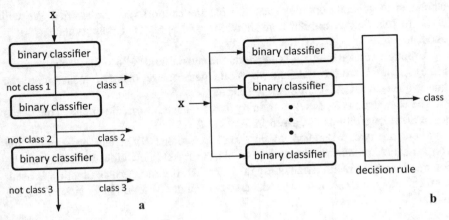

**Fig. 8.16** **a** A simple binary decision tree, and **b** multiclass decisions using binary classifiers in parallel, followed by decision logic

classifier will be presented with pixels from classes for which it was not trained. However, by training for every possible class pair, choosing the most recommended class always works, in principle.[35] The method is called *one-against-one* (OAO). The disadvantage of this method is the very large number of binary classifiers needed. For example, if there were 10 classes, the OAO multiclass strategy requires 45 separate binary classifiers, whereas the OAA only needs 10. Offsetting that problem, however, is the fact that the class-wise binary decisions are generally easier than those in the one against the rest strategy, usually involving simpler SVMs with fewer support vectors for each separate classifier.[36]

## 8.18 Applying the Support Vector Classifier

### 8.18.1 Initial Choices

When applying the support vector classifier in thematic mapping the user needs to make two initial decisions: which kernel to use to improve separability, and what multiclass strategy to adopt.

Generally, a polynomial or radial basis function kernel is chosen. Although many other possibilities exist, they are the two most often found in remote sensing studies and are known to work well. In this section we will focus on the radial basis function kernel of (8.54). It has one parameter $\gamma$, a value for which has to be

---

[35] See Prob. 8.7.

[36] See Melgani and Bruzzone, loc. cit.

estimated to generate optimum results from the support vector classifier. We will come to that below. Similar considerations will apply if a polynomial kernel is chosen.

Although several multiclass methods are available to allow the fundamentally binary support vector machine to work with many classes, most often the one-against-one (OAO) approach is selected.

The user also needs to select training data for each information class. That needs to be done carefully as discussed in Sect. 11.2.1.1.

There are two parameters to be found before the SVM can be applied: the regularisation parameter C which controls the misclassification error that can be tolerated, and the width parameter $\gamma$ in the kernel, if a radial basis function is used. Unfortunately, they are interdependent, so one cannot be found in isolation from the other.

## 8.18.2  Grid Searching for Parameter Determination

The best values of C and $\gamma$ will vary from data set to data set and thus have to be determined anew for each classification exercise. Unfortunately, they can vary over a wide range, particularly C, so an efficient search strategy is needed. A grid searching process is usually selected in which an initial large range of values for each parameter is chosen and the ranges discretised to give a matrix of C, $\gamma$ pairs. The SVM is trained on a representative set of data using each pair selected in turn, from which the best pair is selected. The grid spacing can then be narrowed in the vicinity of that pair and the process repeated, allowing more effective values to be found. This process should be used to find the best parameter pair for each binary classifier in the multiclass topology adopted but is sometimes done for the multiclass data set as a whole.

Once the user is satisfied with those values then the OAO multiclass network of SVMs can be fully trained. During training the final set of SVMs for each binary separation, the corresponding support vectors, and corresponding values for the coefficients $\alpha_i$ in (8.50) will be determined with the optimisation software used.[37] Once they are available the decision rule in (8.50) is then fully specified and can be used to label unseen pixels.

---

[37] Several packages are available for the support vector classifier, including LibSVM, described in Chih-Chung Chang and Chih-Jen Lin, LIBSVM: a library for support vector machines, *ACM Transactions on Intelligent Systems and Technology*, vol. 2, issue 3, 2011, pp. 27:1–27:27. The software is available at http://www.csie.ntu.edu.tw/cjlin/libsvm accessed 2021. Commercial image analysis packages also include SVM classifiers.

### 8.18.3   *Data Centering and Scaling*

Some, although not all, authors recommend that the data be shifted to a mean of zero and scaled to a common range in each band before training, as a means for improving performance and to speed up the optimisation step.[38]

### 8.18.4   *Examples*

We now look at a simple example of the use of the support vector machine for image classification.[39] Figure 8.17 shows a segment of a Quickbird 2 image taken over the city of Boumerdès in Algeria, located on the coast of the Mediterranean Sea. It was acquired on 22 April 2002 and consists of $500 \times 600$ pixels, with a spatial resolution of 0.6 m, achieved through pan-sharpening the sensor's 2.4 m multispectral bands. The Quickbird 2 sensor has the four bands indicated in the figure.

Seven principal information classes are evident in the scene: water, sand, tree, rock, bare soil, asphalt, roof. The last two were each separated into two spectral classes. In the case of asphalt, they are streets (asphalt 1) and pavements (asphalt 2), while for roof they are tiles (roof 1) and cement (roof 2). Table 8.4 shows the numbers of training and testing pixels chosen to support the exercise of mapping the image segment into the classes of interest.

A radial basis function kernel was used as was the OAO multi-class strategy. Since there were 9 (spectral) classes, the total number of classifiers to be trained with the OAO strategy was $9 \times 8/2 = 36$. The LibSVM package was used to perform the classification.

Although parameter determination is usually carried out with grid searching, in this case the experience of the researchers led to a slightly simpler procedure. First, $\gamma$ the kernel parameter was initially set to 0.25. Next, the regularisation parameter C was varied from 25 to 200 in steps of 25. Using the best value found for C, $\gamma$ was then varied over 0.25 to 2 in steps of 0.25. They found that the best parameter pair was $\gamma = 2$ and $C = 200$. These same parameter values were used for all 36 classifiers.

---

[38] See A.J. Gualtieri, S. R. Chettri, R.F. Cromp and L.F. Johnson, Support vector machine classifiers as applied to AVIRIS data, *Proc. 8th JPL Airborne Earth Sciences Workshop*, Pasadena, California, 1999, pp. 217–227, A.J. Gualtieri, The Support Vector Machine (SVM) algorithm for supervised classification of hyperspectral remote sensing data, in G. Camps-Valls and L. Bruzzone, eds., *Kernel methods for Remote Sensing Data Analysis*, John Wiley & Sons, Chichester, UK, 2009, and C-W Hsu, C–C Chang, and C-J Lin, A Practical Guide to Support Vector Classification, http://www.csie.ntu.edu.tw/cjlin/papers/guide/guide.pdf accessed 2021.

[39] This example was provided by Dr. Farid Melgani of the University of Trento, Italy. The analysis was carried out by Dr. Yakoub Bazi of King Saud University, Saudi Arabia. The ground truthing was done by Dr. Abdelhamid Daamouche of the University of Boumerdès, Algeria.

blue 450-520nm   green 520-600nm

red 630-690nm   NIR 760-890nm

**Fig. 8.17** Portion of a Quickbird 2 image, showing the four individual bands and a natural colour composite

**Table 8.4** Numbers of training and testing pixels used in the classification of the Quickbird 2 Boumerdès image segment with the support vector machine

| Class | Training pixels | Testing pixels |
|---|---|---|
| Water | 600 | 2400 |
| Sand | 600 | 2400 |
| Tree | 375 | 700 |
| Asphalt 1 | 105 | 200 |
| Asphalt 2 | 343 | 500 |
| Rock | 175 | 450 |
| Roof 1 | 75 | 200 |
| Roof 2 | 294 | 500 |
| Bare soil | 300 | 700 |
| Total | 2867 | 8050 |

Figure 8.18 shows the classification results obtained, both as a thematic map and as a table of class-wise accuracies. The overall average accuracy was 76.9%. Although the water class was handled perfectly the performance on the rock and bare soil classes was very poor and would not usually be acceptable. Several cases of these misclassifications have been identified in Fig. 8.18. If this were a real exercise the classification would be repeated, perhaps with a more careful choice of the values of $\gamma$ and C, and with a re-examination of the training data to ensure it is not in error.

This example has not demonstrated one of the key benefits of the SVM—the ability to classify hyperspectral imagery without being unduly affected by the

| | | |
|---|---|---|
| ■ | water | 100% |
| ■ | sand | 65.7% |
| ■ | tree | 95.6% |
| ■ | asphalt 1 | 63.5% |
| ■ | asphalt 2 | 85.4% |
| ■ | rock | 44.0% |
| ■ | roof 1 | 62.5% |
| ■ | roof 2 | 72.0% |
| ■ | bare soil | 44.1% |

**Fig. 8.18** SVM classification results, showing where major labelling errors have occurred

Hughes phenomenon. An example from Melgani and Bruzzone[40] is now considered to illustrate this point. This exercise applied a support vector classifier to the Indian Pines data set[41] recorded by AVIRIS (Airborne Visible and Infrared Imaging Spectrometer) in 1992. AVIRIS records 224 bands over the spectral range 0.4–2.5 μm. At the time of this work, it recorded 220 bands with a 10 bit radiometric resolution; for this exercise 20 bands were discarded because of atmospheric problems with the data. The remaining 200 bands were labelled into 9 of the available 16 Indian Pines classes. Table 8.5 shows the classes, along with the numbers of training and testing pixels available.

In this case the OAA multiclass strategy was used, and the radial basis function kernel was adopted. Through grid searching they chose $C = 40$ and $\gamma = 0.25$. The results are shown in Table 8.6, again on a class-wise basis. They are remarkably good on data of such high dimensionality.

The authors now did an interesting sensitivity analysis to check the importance of having precise values for the regularization and kernel parameters. One parameter was held constant while the other was varied over the range shown in Table 8.6. They then computed the average performance and the variance of performance over the tests, as indicated. The fact that the variance is small demonstrates that the parameters do not have to be determined with high precision in these ranges in order to get good results. Note though that the original (grid) searching operation was still needed to get the nominal values for $C$ and $\gamma$.

[40] F. Melgani and L. Bruzzone, Classification of hyperspectral remote sensing images with support vector machines, *IEEE Transactions on Geoscience and Remote Sensing*, vol. 42, no. 8, August 2004, pp. 1778–1790.

[41] Baumgardner, M. F., Biehl, L. L., Landgrebe, D. A. (2015). 220 Band AVIRIS Hyperspectral Image Data Set: June 12, 1992 Indian Pine Test Site 3. Purdue University Research Repository. https://doi.org/10.4231/R7RX991C.

**Table 8.5** The classes and the numbers of training and testing pixels used in the classification of the Indian Pines data set

| Class | Training pixels | Testing pixels |
|---|---|---|
| Corn-no till | 742 | 692 |
| Corn-min till | 442 | 392 |
| Grass/pasture | 260 | 237 |
| Grass/trees | 389 | 358 |
| Hay-windrowed | 236 | 253 |
| Soybean no-till | 487 | 481 |
| Soybean min-till | 1245 | 1223 |
| Soybean clean-till | 305 | 309 |
| Woods | 651 | 643 |
| Total | 4757 | 4588 |

**Table 8.6** Results of the classification and a sensitivity analysis on the regularisation and kernel parameters

| Class | Result |
|---|---|
| Corn-no till | 91.5% |
| Corn-min till | 87.8% |
| Grass/pasture | 94.9% |
| Grass/trees | 98.9% |
| Hay-windrowed | 100.0% |
| Soybean no-till | 88.6% |
| Soybean min-till | 91.3% |
| Soybean clean-till | 95.8% |
| Woods | 99.4% |
| Overall | 93.4% |

| Sensitivity tests | | |
|---|---|---|
| | Accuracy | |
| Range | mean | variance |
| $\gamma = 1$ $C = 1 - 100$ | 92.6% | 0.84% |
| $C = 40$ $\gamma = 0.1 - 3$ | 92.5% | 0.5% |

## 8.19 Committees of Classifiers

Classically, a committee classifier consists of a number of algorithms that all operate on the same data set to produce individual, sometimes competing, recommendations about the class membership of a pixel, as shown in Fig. 8.19. Those recommendations are fed to a decision maker, or chair, which resolves the final class label for the pixel.[42]

The decision maker resolves the conflicts by using one of several available logics. One is the *majority vote*, in which the decision maker decides that the class most recommended by the committee members is the appropriate one for the pixel. Another is *veto logic*, in which all classifiers have to agree about class membership before the decision maker will label the pixel. Yet another is *seniority logic*, in which the decision maker always consults one particular classifier first (the most

---

[42] See N.J. Nilsson, *Learning Machines*, McGraw-Hill, N.Y., 1965.

**Fig. 8.19** A committee of three classifiers

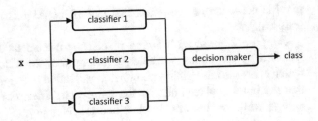

"senior") to determine the label for a pixel. If that classifier is unable to recommend a class label, then the decision maker consults the next most senior member of the committee, and so on until the pixel is labelled. Seniority logic has been used as a means for creating piecewise linear decision surfaces.[43]

## 8.19.1  *Bagging*

Apart from the simple approach of committee labelling based on logical decisions as in the previous section, several other procedures using a number of similar classifiers have been devised in an endeavour to improve thematic mapping accuracy. One is called *bagging* which entails training a set of classifiers on randomly chosen subsets of data, and then combining the results.

The different data sets are chosen by the process called *bootstrapping*, in which the available $K$ training pixels are used to generate $L$ different data sets each containing $N$ training pixels, in the following manner. For each data set $N$ pixels are chosen randomly from the $K$ available, with replacement; in other words, the same pixel could appear more than once among the $N$ chosen. Each of the $L$ data sets is used to train a classifier. The results of the individual classifiers are then combined by voting. The name **bagging** derives from **b**ootstrap **agg**regat**ing**.

## 8.19.2  *Boosting and AdaBoost*

**Ada**ptive **boost**ing, called **AdaBoost**, is a committee of binary classifiers in which the members are trained sequentially, in the following manner. The first classifier is trained. Training pixels that are found to be in error are then emphasised in the training set and the next classifier trained on that enhanced set. Training pixels unable to be separated correctly by the second classifier are given a further emphasis and the third classifier is trained, and so on. The final label allocated to a

---

[43] See T. Lee and J.A. Richards, A low cost classifier for multi-temporal applications, *Int. J. Remote Sensing*, vol. 6, 1985, pp. 1405–1417.

pixel is based on the outputs of all classifiers. Algorithmically, it proceeds in the following steps.

Suppose there are $K$ training pixels; the correct label for the $k$th pixel is represented by the binary variable $y_k$ which takes the values $\{+1, -1\}$ according to which class the $k$th training pixel belongs. Define $t_k \in \{+1, -1\}$ as the actual class that the pixel is placed into by the trained classifier. For correctly classified pixels $t_k = y_k$ while for incorrectly classified pixels $t_k \neq y_k$.

1. Initialise a set of weights for each of the training pixels in the first classifier step according to[44]

$$w_k^1 = 1/K$$

2. For $l = 1 \ldots L$, where $L$ is the number of classifiers in the committee, carry out the following steps in sequence

   a. Train a classifier using the available weighted training data. Initially each training pixel in the set is weighted equally, as in 1. above.
   b. Set up an error measure after the $l$th classification step according to

$$e^l = \frac{\sum_k w_k^l I(t_k, y_k)}{\sum_k w_k^l}$$

   in which $I(t_k, y_k) = 1$ for $t_k \neq y_k$
   $\qquad\qquad\qquad = 0$ otherwise

   c. Form the weights for the $(l+1)$th classifier according to either

$$w_k^{l+1} = w_k^l \exp\{\alpha^l I(t_k, y_k)\} \tag{8.56a}$$

   or

$$w_k^{l+1} = w_k^l \exp\{-\alpha^l t_k y_k\} \tag{8.56b}$$

   with

$$\alpha^l = \ln\{(1 - e^l)/e^l\} \tag{8.56c}$$

Both (8.56a) and (8.56b) enhance the weights of the incorrectly classified pixels; (8.56a) leaves the weights of the correct pixels unchanged, while (8.56b) de-emphasises them.

---

[44] All superscripts in this section are stage (iteration) indices and not powers.

d. Test the class membership of the $k$th pixel according to

$$T_L(\mathbf{x}_k) = \text{sgn} \sum_l \alpha^l t_k^l(\mathbf{x}_k) \tag{8.56d}$$

This weights the class memberships recommended by the individual classifiers to come up with the final label for the pixel. Note that $T_L \in \{1, -1\}$ for this binary classifier.

Bishop[45] gives a simple example of AdaBoost in which the improvement of accuracy is seen as more classifiers are added. Accuracy may not always improve initially and many, sometimes hundreds, of stages are needed to achieve good results.

## 8.20   Networks of Classifiers: The Artificial Neural Network

Decision trees, committee classifiers and processes such as boosting and bagging are examples of what more generally can be referred to as classifier networks, or sometimes layered classifiers.[46] An important example is the artificial neural network (ANN), which is available in several configurations. Here we will develop the most common—the multilayer Perceptron. The basic Perceptron is a simple, binary linear classifier that, once trained, places patterns into one of the two available classes by checking on which side of the linear separating surface they lie. The surface is defined in (8.29) as $\mathbf{w}^T\mathbf{x} + w_{N+1} = 0$, so that the class test is given by (8.30):

$$\mathbf{x} \in \text{class 1   if } \mathbf{w}^T\mathbf{x} + w_{N+1} > 0$$
$$\mathbf{x} \in \text{class 2   if } \mathbf{w}^T\mathbf{x} + w_{N+1} < 0$$

Diagrammatically this can be depicted as shown in Fig. 8.20, in which $z = \mathbf{w}^T\mathbf{x} + w_{N+1}$ the sign of which is checked via a thresholding operation. Together the weighting and thresholding operation are called a *threshold logic unit* (TLU), which is the essential building block of the Perceptron but which effectively limits its value to linearly separable, binary data unless layered procedures are used.

---

[45] See C.M. Bishop, loc. cit.
[46] Nilsson, loc. cit.

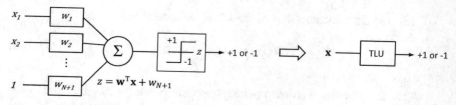

**Fig. 8.20** The threshold logic unit (TLU)

**Fig. 8.21** Neural network
processing element

## 8.20.1  The Processing Element

One of the breakthroughs that makes the ANN able to handle data that is not
linearly separable is that the hard thresholding of the TLU in the Perceptron is
replaced by the softer thresholding operation shown in Fig. 8.21. The building
block is then referred to as a *processing element* (PE) and is described by the
operation

$$g = f\left(\mathbf{w}^{\mathrm{T}}\mathbf{x} + \theta\right) \qquad (8.57)$$

where $\mathbf{w}$ is the vector of weighting coefficients, and $\mathbf{x}$ is the vector of inputs as
before; $\theta$ is a threshold, sometimes set to zero, which takes the place of the
weighting coefficient $w_{N+1}$. $f$ is called the *activation function* which can take many
forms, the most common of which is the sigmoid[47]

$$g = f(z) = \frac{1}{1 + \exp(-z/b)} \qquad (8.58)$$

where $z = \mathbf{w}^{\mathrm{T}}\mathbf{x} + \theta$ and $b$ is a constant. The activation function $g$ approaches 1 for $z$
large and positive, and 0 for $z$ large and negative, and is thus asymptotically
thresholding. For a very small value of $b$ it approaches a hard thresholding operation, and the PE behaves like a TLU. Usually, we choose $b = 1$. This is seen in
Fig. 8.22, along with the behaviour of the activation function for several other
values of $b$.

---

[47] See Fig. 3.2 of A. Graves, *Supervised Sequence Learning with Recurrent Neural Networks*,
Springer, Berlin, 2012, for a range of other possibilities.

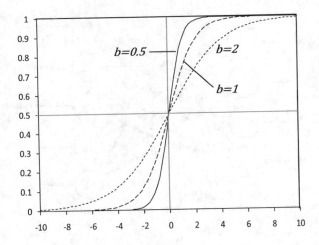

**Fig. 8.22** Plots of the activation function of (8.58)

The ReLU activation function of Fig. 8.35 is now more commonly being used because it leads to improved training performance, but we continue this development with the activation function of (8.58) since it is still often employed and demonstrates the original development of the training process for the MLP.

A typical neural network used in remote sensing applications will appear as shown in Fig. 8.23. It is a layered classifier composed of processing elements of the type in Fig. 8.21. It is conventionally drawn with an input layer of nodes, with one node per spectral measurement; it has the function only of distributing the inputs, possibly with scaling, to a set of processing elements that form the second layer. An output layer generates the class labels for the input provided. The central layer is referred to as a *hidden layer*. While we have shown only one here it is possible to use more than a single hidden layer. Even though one is often sufficient in most remote sensing applications of the ANN, choosing the number of processing elements to use in that layer is not simple and is often done by trial and error. We return to that in Sect. 8.20.3 below.

## 8.20.2  *Training the Neural Network—Backpropagation*

As with any supervised classifier the neural network must be trained before it can be employed for thematic mapping. Available training data is used to determine the elements of the weight vector **w** and the offset $\theta$ for each processing element in the network. The parameter $b$, which governs the slope of the activation function, as seen in Fig. 8.22, is generally pre-specified and does not need to be estimated from training data.

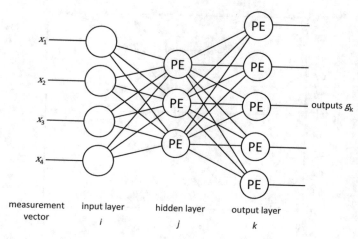

**Fig. 8.23**  A three-layer multilayer Perceptron neural network and the nomenclature used in the derivation of the backpropagation training algorithm

Part of the complexity in understanding the training process for a neural net is caused by the need to keep careful track of the parameters and variables over all layers and processing elements, how they vary with the presentation of training pixels and, as it turns out, with iteration count. This can be achieved with a detailed subscript convention, or by the use of a simpler generalised notation. We will adopt the latter approach, following essentially the development given by Pao.[48] The derivation will be focussed on a three-layer neural net, since this architecture has been found sufficient for many applications. Nevertheless, the results generalise to more layers, which is important when we come to convolutional neural networks later.

Figure 8.23 incorporates the nomenclature used. The three layers are lettered as $i, j, k$ with $k$ being the output. The set of weights (i.e., the components of the weight vectors) linking layer $i$ PEs with those in layer $j$ are represented by $w_{ji}$, while those linking layers $j$ and $k$ are represented by $w_{kj}$. There will be a very large number of those weights, but in deriving the training algorithm it is not necessary to refer to them all individually. Similarly, the activation function arguments $z_i$ and outputs $g_i$, can be used to represent all the arguments and outputs in the corresponding layer. For $j$ and $k$ layer PEs (8.58) is

$$g_j = f(z_j) \text{ with } z_j = \sum_j \left( w_{ji} g_i + \theta_j \right) \tag{8.59a}$$

[48] Y.H. Pao, *Adaptive Pattern Recognition and Neural Networks*, Addison-Wesley, Reading Mass., 1989.

$$g_k = f(z_k) \text{ with } z_k = \sum_k \left( w_{kj} g_j + \theta_k \right) \qquad (8.59b)$$

The sums in (8.59) are shown with respect to the indices $j$ and $k$. This should be read as meaning the *sums are taken over all inputs of the layer $j$ and layer $k$ PEs* respectively. Note also that the sums are expressed in terms of the outputs of the previous layer since those outputs form the inputs to the PEs in question.

An untrained or poorly trained network will give erroneous outputs. As a measure of how well a network is functioning during training, we assess the outputs at the last layer ($k$). A suitable performance measure is the sum of the squared output error. Using this, the error made by the network when presented with *a single training pixel* is expressed

$$E = \tfrac{1}{2} \sum_k (t_k - g_k)^2 \qquad (8.60)$$

where the $t_k$ represent the desired or target outputs[49] and $g_k$ represents the actual outputs from the output layer PEs in response to the training pixel. The factor of $\frac{1}{2}$ is included for arithmetic convenience in the following. The sum is taken over all output layer PEs.

A logical training strategy is to adjust the weights in the processing elements until the error has been minimised, at which stage the actual outputs are as close as possible to the desired outputs.

A common approach for adjusting weights to minimise the value of a function of which they are arguments, is to modify their values proportional to the negative of the partial derivative of the function. This is called a gradient descent technique.[50] Thus, for the weights linking the $j$ and $k$ layers let a revised estimate be

$$w'_{kj} = w_{kj} + \Delta w_{kj}$$

with

$$\Delta w_{kj} = -\eta \frac{\partial E}{\partial w_{kj}}$$

---

[49] These will be specified in the labelling of the training data pixels. The actual value taken by $t_k$ will depend on how the set of outputs is used to represent classes. Each individual output could be a specific class indicator, e.g., 1 for class 1 and 0 for class 2 as with the $y_i$ in (8.33); alternatively, some more complex coding of the outputs could be adopted. This is considered in Sect. 8.20.3.

[50] The conjugate gradient method can also be used: see J.A. Benediktsson, P.H. Swain and O.K. Esroy, Conjugate-gradient neural networks in classification of multisource and very high dimensional remote sensing data, *Int. J. Remote Sensing*, vol. 14, 1993, pp. 2883–2903.

in which $\eta$ is a positive constant that controls the degree of adjustment; it is called the *learning rate*. This requires an expression for the partial derivative, which can be found using the chain rule

$$\frac{\partial E}{\partial w_{kj}} = \frac{\partial E}{\partial g_k}\frac{\partial g_k}{\partial z_k}\frac{\partial z_k}{\partial w_{kj}} \qquad (8.61)$$

each term of which is now evaluated. From (8.58) and (8.59b) we see

$$\frac{\partial g_k}{\partial z_k} = \frac{1}{b}(1 - g_k)g_k \qquad (8.62a)$$

and

$$\frac{\partial z_k}{\partial w_{kj}} = g_j \qquad (8.62b)$$

From (8.60) we have

$$\frac{\partial E}{\partial g_k} = -(t_k - g_k) \qquad (8.62c)$$

so that the correction to be applied to the weights is

$$\Delta w_{kj} = \eta(t_k - g_k)(1 - g_k)g_k g_j \qquad (8.63)$$

in which we have chosen $b = 1$. For a given training pixel all the terms in (8.63) are known, so that a beneficial adjustment can be made to the weights that link the hidden layer to the output layer.

Now consider the weights that link layers $i$ and $j$. The weight adjustments are determined from

$$\Delta w_{ji} = -\eta\frac{\partial E}{\partial w_{ji}} = -\eta\frac{\partial E}{\partial g_j}\frac{\partial g_j}{\partial z_j}\frac{\partial z_j}{\partial w_{ji}}$$

In a manner similar to the above we can show, with $b = 1$, that

$$\Delta w_{ji} = -\eta\frac{\partial E}{\partial g_j}(1 - g_j)g_j g_i$$

Unlike the case with the output layer, we cannot generate an expression for $\frac{\partial E}{\partial g_j}$ directly because it requires an expression for the output error in terms of hidden layer responses. Instead, we proceed by the following chain rule

$$\frac{\partial E}{\partial g_j} = \sum_k \frac{\partial E}{\partial z_k} \frac{\partial z_k}{\partial g_j} = \sum_k \frac{\partial E}{\partial z_k} w_{kj}$$

The remaining partial derivative, from (8.62a) and (8.62c) with $b = 1$, is

$$\frac{\partial E}{\partial z_k} = -(t_k - g_k)(1 - g_k)g_k$$

so that

$$\Delta w_{ji} = \eta(1 - g_j)g_j g_i \sum_k (t_k - g_k)(1 - g_k)g_k w_{kj} \qquad (8.64)$$

Having determined the $w_{kj}$ via (8.63) it is now possible to find values for $w_{ji}$ using (8.64) since all the other entries in (8.64) are known or can be determined readily.

We now define

$$\delta_k = (t_k - g_k)(1 - g_k)g_k \qquad (8.65a)$$

and

$$\delta_j = (1 - g_j)g_j \sum_k \delta_k w_{kj} \qquad (8.65b)$$

so that

$$\Delta w_{kj} = \eta \delta_k g_j \qquad (8.66a)$$

and

$$\Delta w_{ji} = \eta \delta_j g_i \qquad (8.66b)$$

The thresholds $\theta_j$ and $\theta_k$ in (8.59) are found in exactly the same manner as for the weights, using (8.66), but with the corresponding inputs set to unity.

Sometimes "momentum" is included in the weight adjustment step to improve convergence during training. The gradient descent steps at (8.66) can be expressed generically as $\eta \delta g$ so that the weight adjustment step at either the hidden or output layers can be written $w' = w + \Delta w = w + \eta \delta g$. We now add another adjustment to the weights which assumes that the change at this iteration is likely to be not too different from the previous adjustment.

$$w' = w + \Delta w = w + \eta \delta g + \alpha \Delta w(-1)$$

$\alpha \Delta w(-1)$ is the momentum term in which $w(-1)$ is the weight adjustment from the previous iteration and $\alpha$ is a user-specified parameter that accounts for the degree to which momentum is used during training.

Now that we have the mathematics in place it is possible to describe how training is carried out. The network is initialised with an arbitrary set of weights so that it can generate a nominal output. The training pixels are then presented one at a time to the network. For a given pixel the output of the network is computed using the network equations. Almost certainly the output will be incorrect to start with— i.e., $g_k$ will not match the desired class $t_k$ for the training pixel. Correction to the output PE weights, described in (8.66a), is then carried out, using the definition of $\delta_k$ in (8.65a). With these new values of $\delta_k$, and thus $w_{kj}$, (8.65b) and (8.66b) can be applied to find the new weight values in the earlier layers. In this way the effect of the output being in error is propagated back through the network in order to correct the weights. The technique is thus referred to as *backpropagation*.

After the weights have been adjusted the training pixels are presented to the network again and the outputs re-calculated to see if they correspond better to the desired classes. Usually, they will still be in error and the process of weight adjustment is repeated. The process is iterated as many times as necessary in order that the network respond with the correct class for each of the training pixels, or until the number of errors in classifying the training pixels is reduced to an acceptable level.

Pao[51] recommends that the weights not be corrected on each presentation of a single training pixel; instead, he recommends that the corrections for all pixels in the training set should be aggregated into a single adjustment. For $p$ training patterns the bulk adjustments are

$$w'_{kj} = \sum_p w_{kj} \quad w'_{ji} = \sum_p w_{ji}$$

which is equivalent to deriving the algorithm with the error being calculated over all pixels $p$ in the training set, viz.

$$E = \sum_p E_p, \tag{8.67}$$

where $E_p$ is the error for a single pixel in (8.60). This is sometimes referred to as a *batch adjustment*, whereas updating with a single training pixel at a time using (8.60) is called *stochastic gradient descent* (SGD). In SGD the training pixels should be chosen randomly, but often in practice all available training samples are used. One of the advantages of SGD is that is can be used when the numbers of training samples is very large, whereas batch adjustment suffers with large data sets and is thus only used when numbers of training pixels is not too large. Also, batch processing generally requires all the training data to be available in machine memory simultaneously. However, convergence is smoother with batch processing but less so with SGD.

---

[51] Y.H. Pao, loc. cit.

The *mini-batch gradient descent technique* splits the training set into smaller batches of randomly chosen pixels for learning and is the recommended approach when the data sets are large. Essentially, it is mid-way between SGD and batch processing.

### 8.20.3 Choosing the Network Parameters

When considering the application of the artificial neural network to thematic mapping it is necessary to make several key decisions beforehand. First, the number of layers to use must be chosen. Generally, a three-layer network is selected, with the purpose of the first layer being simply to distribute, or fan out, the components of the input pixel vector to each of the processing elements in the second layer. The first layer does no processing as such, apart perhaps from scaling the input data if required. The second layer is the hidden layer and the third is the output layer.

Although one hidden layer is usually the case when applying an ANN to remote sensing problems, when we come to convolutional neural networks we will often use more than a single hidden layer.

The next choice relates to the number of elements in each layer. The input layer will generally be given as many nodes as there are spectral components (features) in the pixel vectors. The number to use in the output node will depend on how the outputs are used to represent the classes. The simplest method is to let each separate output signify a different class, in which case the number of output processing elements will be the same as the number of training classes. Alternatively, a single PE could be used to represent all classes, in which case a different value or level of the output variable will be attributed to a given class. A further possibility is to use the outputs as a binary code, so that two output PEs can represent four classes, three can represent 8 classes and so on.

As a general guide, the number of PEs to choose for the hidden or processing layers should be the same as, or larger than, the number of nodes in the input layer.[52]

### 8.20.4 Example

We now consider a simple example to see how a neural network is able to develop the solution to a classification problem. Figure 8.24 shows two classes of data, with three points in each, arranged so that they cannot be separated linearly. The network shown in Fig. 8.25 will be used to discriminate the data. The two PEs in the first

---

[52] R.P. Lippman, An introduction to computing with neural nets, *IEEE ASSP Magazine*, April 1987, pp. 4–22.

**Fig. 8.24** Non-linearly separable two class data set

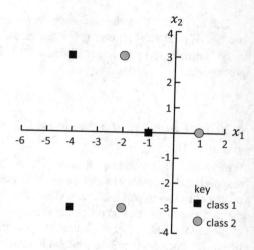

key
■ class 1
○ class 2

**Fig. 8.25** Neural network with two processing layers, to be applied to the data of Fig. 8.24

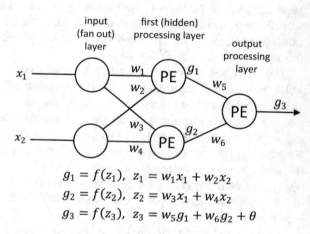

$$g_1 = f(z_1), \quad z_1 = w_1 x_1 + w_2 x_2$$
$$g_2 = f(z_2), \quad z_2 = w_3 x_1 + w_4 x_2$$
$$g_3 = f(z_3), \quad z_3 = w_5 g_1 + w_6 g_2 + \theta$$

processing layer are described by activation functions with no thresholds—i.e., $\theta = 0$ in (8.57), while the single output PE has a non-zero threshold in its activation function. The learning rate $\eta$ was set to 2, and $b = 1$.

Table 8.7 shows the results of training the network with the backpropagation method of the previous section, along with the error measure of (8.60) at each step. As seen, the network approaches a solution quickly (approximately 50 iterations) but takes more iterations (approximately 250) to converge to a final result.

Having trained the network, it is now possible to understand how it implements a solution to the nonlinear pattern recognition problem. The arguments of the activation functions of the PEs in the first processing layer each define a straight line (hyperplane in general) in the pattern space. Using the result at 250 iterations, these are:

**Table 8.7** Training the network of Fig. 8.25; iteration 0 shows an arbitrary set of initial weights

| Iteration | $w_1$ | $w_2$ | $w_3$ | $w_4$ | $w_5$ | $w_6$ | $\theta$ | Error |
|---|---|---|---|---|---|---|---|---|
| 0 | 0.050 | 0.100 | 0.300 | 0.150 | 1.000 | 0.500 | −0.500 | 0.461 |
| 1 | 0.375 | 0.051 | 0.418 | 0.121 | 0.951 | 0.520 | −0.621 | 0.424 |
| 2 | 0.450 | 0.038 | 0.455 | 0.118 | 1.053 | 0.625 | −0.518 | 0.408 |
| 3 | 0.528 | 0.025 | 0.504 | 0.113 | 1.119 | 0.690 | −0.522 | 0.410 |
| 4 | 0.575 | 0.016 | 0.541 | 0.113 | 1.182 | 0.752 | −0.528 | 0.395 |
| 5 | 0.606 | 0.007 | 0.570 | 0.117 | 1.240 | 0.909 | −0.541 | 0.391 |
| 10 | 0.642 | −0.072 | 0.641 | 0.196 | 1.464 | 1.034 | −0.632 | 0.378 |
| 20 | 0.940 | −0.811 | 0.950 | 0.882 | 1.841 | 1.500 | −0.965 | 0.279 |
| 30 | 1.603 | −1.572 | 1.571 | 1.576 | 2.413 | 2.235 | −1.339 | 0.135 |
| 50 | 2.224 | −2.215 | 2.213 | 2.216 | 3.302 | 3.259 | −1.771 | 0.040 |
| 100 | 2.670 | −2.676 | 2.670 | 2.677 | 4.198 | 4.192 | −2.192 | 0.010 |
| 150 | 2.810 | −2.834 | 2.810 | 2.835 | 4.529 | 4.527 | −2.352 | 0.007 |
| 200 | 2.872 | −2.919 | 2.872 | 2.920 | 4.693 | 4.692 | −2.438 | 0.006 |
| 250 | 2.901 | −2.976 | 2.902 | 2.977 | 4.785 | 4.784 | −2.493 | 0.005 |

**Fig. 8.26** Neural network solution for the data of Fig. 8.24

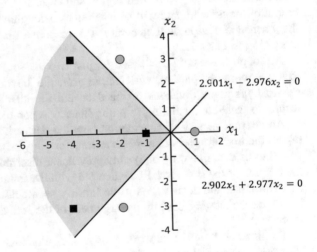

$$2.901x_1 - 2.976x_2 = 0$$

$$2.902x_1 + 2.977x_2 = 0$$

which are shown plotted in Fig. 8.26. Either one of these lines goes some way towards separating the data but cannot accomplish the task fully.

It is now important to consider how the output PE operates on the outputs of the first layer PEs to complete the discrimination of the two classes. For pattern points lying exactly on one of the above lines, the output of the respective PE will be 0.5, given that the activation function of (8.58) has been used. However, for patterns a little distance away from those lines the output of the first layer PEs will be close to 0 or 1 depending on which side of the hyperplane they lie.

**Table 8.8** Response of the output layer PE

| $g_1$ | $g_2$ | $g_3$ |
|---|---|---|
| 0 | 0 | $0.076 \approx 0$ |
| 0 | 1 | $0.908 \approx 1$ |
| 1 | 0 | $0.908 \approx 1$ |
| 1 | 1 | $0.999 \approx 1$ |

We can therefore regard the pattern space as being divided into two regions (0 and 1) by a particular hyperplane. Using these extreme values, Table 8.8 shows the possible responses of the output layer PE for patterns lying somewhere in the pattern space.

As seen, for this example the output PE functions in the nature of a logical OR operation; patterns that lie on the 1 side of *either* of the first processing layer PE hyperplanes are labelled as belonging to one class, while those that lie on the 0 side of *both* hyperplanes are labelled as belonging to the other class. Therefore, patterns which lie in the shaded region shown in Fig. 8.26 will generate a 0 at the output of the network and will be labelled as belonging to class 1, while patterns in the unshaded region will generate a 1 response and thus will be labelled as belonging to class 2.

Although this exercise is based on just two classes, similar functionality of the PEs in a more complex network can, in principle, be identified. The input PEs will set up hyperplane divisions of the data and the later PEs will operate on those results to generate a solution to a non-linearly separable problem.

An alternative way of considering how the network determines a solution is to regard the first processing layer PEs as transforming the data in such a way that later PEs (in this example only one) can apply linear discrimination. Figure 8.27 shows the outputs of the first layer PEs when fed with the training data of Fig. 8.24. After transformation the data is seen to be linearly separable. The hyperplane shown in the figure is that generated by the argument of the activation function of the output layer PE.

To illustrate how the network of Fig. 8.25 functions on unseen data Table 8.9 shows its response to the testing patterns indicated in Fig. 8.28. For this simple example all patterns are correctly classified.

## 8.21    The Convolutional Neural Network

For many decades, image analysts in remote sensing have been critically aware of the matter of spatial context. That is: when considering the label for a pixel there is a high likelihood that the surrounding pixels will be from the same class. That is especially the case for agricultural regions and many natural landscapes; and yet the classifiers we have treated with so far have ignored that property. In that sense they are called *point* (or pixel-specific) classifiers, because they just focus on a pixel, independently of its neighbours.

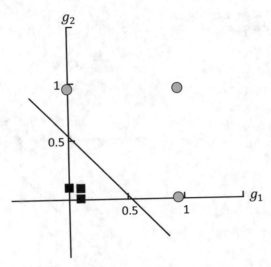

**Fig. 8.27** Illustration of how the first processing layer PEs transform the input data into a linearly separable set, which is then separated by the output layer hyperplane

**Table 8.9** Performance of the network of Fig. 8.25 on the test data of Fig. 8.28

| Pattern | $x_1$ | $x_2$ | $z_1$ | $g_1$ | $z_2$ | $g_2$ | $z_3$ | $g_3$ | Class |
|---------|-------|-------|-------|-------|-------|-------|-------|-------|-------|
| a | −3.0 | 2.8 | −17.036 | 0.000 | −0.370 | 0.408 | −0.539 | 0.368 | 1 |
| b | −3.0 | 2.0 | −14.655 | 0.000 | −2.752 | 0.056 | −2.206 | 0.099 | 1 |
| c | −2.0 | −1.0 | −2.826 | 0.056 | −8.781 | 0.000 | −2.224 | 0.098 | 1 |
| d | −2.0 | −1.94 | −0.029 | 0.493 | −11.579 | 0.000 | −0.135 | 0.466 | 1 |
| e | −1.0 | 1.1 | −6.175 | 0.002 | 0.373 | 0.592 | 0.350 | 0.587 | 2 |
| f | 1.0 | 2.0 | −3.051 | 0.045 | 8.856 | 1.000 | 2.506 | 0.925 | 2 |
| g | −1.0 | −2.0 | 3.051 | 0.955 | −8.856 | 0.000 | 2.077 | 0.889 | 2 |
| h | 2.0 | −2.0 | 11.754 | 1.000 | −0.150 | 0.463 | 4.505 | 0.989 | 2 |

In Sect. 8.23 we will consider a number of methods that have been developed over the years to bring the influence of neighbouring pixels—i.e., spatial context—into a classification. The convolutional neural network (CNN) that forms the subject of this section can also incorporate spatial context into a classification, by reason of its fundamental topology.

Historically, convolutional neural networks were developed after many of the procedures we will look at in Sect. 8.23. However, because the CNN is an extension of the original artificial neural network we have just treated, it is more logical to consider it first.[53]

---

[53] For a standard treatment of deep learning and CNNs see I. Goodfellow, Y. Bengio, and A Courville, *Deep Learning (Adaptive Computation and Machine Learning Series)*, The MIT Press, Cambridge Mass., 2016.

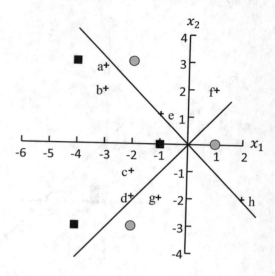

**Fig. 8.28** Location of the test data, indicated by the lettered crosses

## 8.21.1   The Basic Topology of the Convolutional Neural Network

In the neural networks of Sect. 8.20 the pixels of an image were fed one at a time into the network and a label was, thereby, found pixel by pixel.

Suppose now we make the seemingly bold move of inputting *all* the pixels of an image in one go as depicted in Fig. 8.29, so that we have to have enough input nodes to accommodate the *full set* of spectral measurements for the *full set* of image pixels. For a practical image that will be a very large number of inputs. We have allowed for a number of hidden layers and, for the moment, the network is fully connected in the manner of the ANN of the previous section. This means that the output of every node or processing element in one layer is fed to all the nodes or processing elements of the following layer. In this configuration there will be a huge number of unknown weight vectors and offsets to be learned through training.

One immediately obvious problem with feeding the network in this manner is that the spatial inter-relationships among the pixels appears to be lost. Even though this is really just a problem of how the pixels are addressed, it is more meaningful to arrange them as shown in Fig. 8.30. That doesn't change anything about the network, other than arranging the nodes (or processing elements) into an array rather than column format. For convenience we have shown the hidden layers to be the same size and shape as the input layer, but in general they could be any size. Note the output layer is still one dimensional, since it represents a set of classes.

For the moment we will consider an image with just a single band of data, so that each pixel in the input array is represented by a single scalar value. We will look at multiband data later once we have understood the operation of the CNN.

With the arrangement of Fig. 8.30 the number of potential connections is enormous. Consider the number of unknowns between just the input and the first

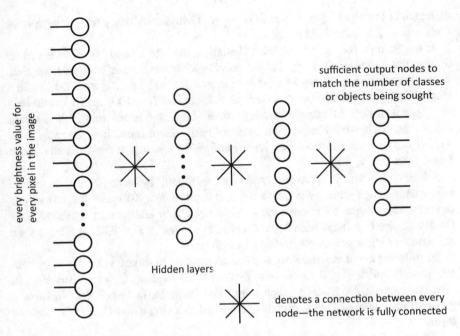

**Fig. 8.29** Feeding every pixel into the neural network in one step

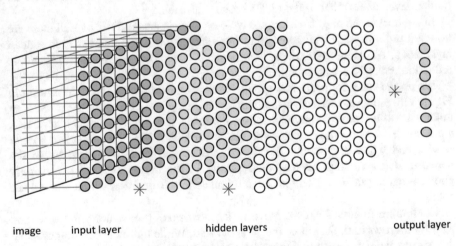

**Fig. 8.30** Arranging the nodes of the neural network into image format

hidden layer. The input to each processing element in the *hidden* layer is of the form $z = \mathbf{w}^T\mathbf{x} + \theta$, where $\mathbf{x}$ represents the array of input pixels but expressed in vector form and $\mathbf{w}$ is the corresponding vector of weights that connect each of the input pixels to a processing element. Its dimensionality will be equal to the number of elements in the input layer, which is $N \times N$. There are as many weight *vectors* as there are nodes in the hidden layer.

If we assume, for the sake of this calculation, that the hidden layer has the same dimensions as the input layer, that means altogether we have $N^4$ different weights, values for which have to be found during training to make the network usable. In a similar fashion there will be $N^2$ values of $\theta$. If we had $N = 100$, which would be a very small image in remote sensing, then there are more than 100 million unknowns. That would require an extraordinarily large amount of training data. Added to this is the fact that we have multiple bands, and images usually much larger than $100 \times 100$.

Clearly, a simpler approach needs to be found, but one in which spatial inter-relationships among the pixels are still represented. To make the rest of our development simpler we remove the explicit input layer and let it be represented by the image itself, perhaps with some scaling, as shown in Fig. 8.31. Again, we are still considering a simple single-band image.

In order to avoid the massive number of weights involved with the fully connected network of Fig. 8.31 we now explore what happens if we are selective in how we connect the layers to each other. For example, in Fig. 8.32a we show a group of only nine of the input pixels connected to a single node in the first hidden layer.

Because of the geometry, the $3 \times 3$ group of nine pixels is centred on the one which is in the second row and second column of the input. The PE element in the hidden layer is also that in the (2, 2) position as seen.

In contrast to the need to determine $N^4 + N^2$ weights and offsets overall there are now ten unknowns (9 weights $w_{ij}$ and one offset $\theta$) to determine for each hidden layer node. Altogether, therefore, there are $10N^2$ unknowns to find, a considerable reduction, but still a large number if $N$ is large.

We do the same for the $3 \times 3$ group which is one column to the right as seen in Fig. 8.32b. Now we take a decision *that significantly reduces again* the number of unknowns to be found in training: rather than use a new set of weights and offsets, we assume we can employ the same set as for Fig. 8.32a. This is called *weight re-use*, and while that sounds like it will reduce substantially the power of the network to learn complicated spatial patterns in the image, it gives surprisingly good results in practice. There is also a rationale to this decision which we will see soon.

Continuing in Fig. 8.32, we move to the next pixel group along the row, and then for all rows until the whole image is covered. While this example suggests that the actions happen sequentially, in fact all the operations are in parallel—they are just sets of connections. This is important to keep in mind.

Clearly, there is a problem with the edge pixels. Given the large numbers of pixels in an image we could ignore the edge problem. Sometimes an artificial border of zeros is created so that the edge PEs in the hidden layer can receive inputs and thus preserve dimensionality, if that is important. That is called *padding*.

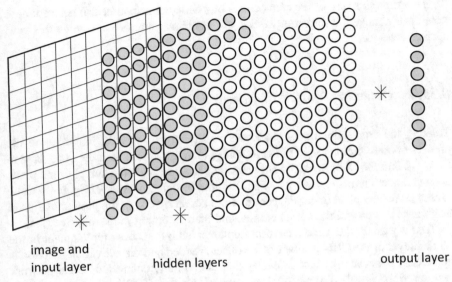

image and
input layer          hidden layers                    output layer

**Fig. 8.31** Combining the representation of the image and the input layer for simplicity

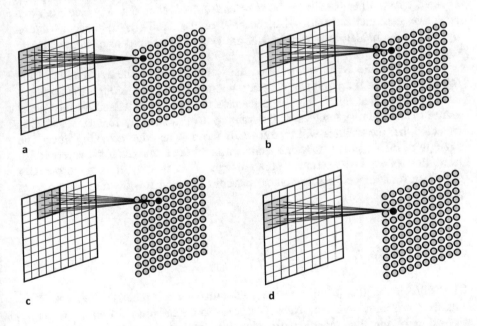

a

b

c

d

**Fig. 8.32** Restricting the connections between layers; groups of just nine pixels in one layer feed into a single node of the following layer

Even though many of the connections of a fully connected neural network have now been removed, it turns out we can still use back propagation to train this new, sparser network.

## 8.21.2   Detecting Spatial Structure

Treating the connections between layers as shown in Fig. 8.32 is similar to the process of convolution used to filter an image to detect spatial features as covered in Sect. 5.2 and Fig. 5.3. In spatial convolution a window, called a kernel or template, is moved over an image row by row and column by column. A new brightness value is created for the pixel under the centre of the kernel by taking the products of pixel brightness values and the kernel entries, and then summing the result. See (5.1).

That is exactly the same operation implemented by a processing element in the hidden layer of the CNN just before the offset is added and the activation function is applied. It is because of that similarity that the partially connected neural network just described is called a *convolutional neural network* (CNN).

In the CNN the kernel is usually called a *filter*, and the set of input pixels covered by the filter is called a *local receptive field*. Note that any size filter and receptive field can be used. Also, because of the similarity to the convolution operation, the hidden layers in a CNN are generally referred to as *convolutional layers*.

In the CNN the kernel entries (i.e., the weights prior to the application of the activation function) are initially chosen randomly. However, through training they take on values that match the image features that are characterized by the spatial nature of the training samples. If the training images strongly feature edges, it is expected that the weights will tend towards those of an edge detecting filter. The strength of the CNN is that with a sufficient number of convolutional layers it can learn the spatial characteristics of an image. That is why it is a particularly important tool for performing context classification and for picture processing in general.

## 8.21.3   Stride

The "shift" in the filter position can be greater than a single pixel, as in Fig. 8.32. In Fig. 8.33 it is shown as two pixels. That leads to the definition of *stride*, which is the offset of the filter position (or receptive field) that provides input to each successive node in the hidden layer. Note that a stride of 2 will approximately halve the size of the hidden layer.

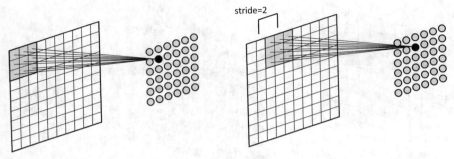

**Fig. 8.33** Defining stride by the pixel shift of the center of the filter in the receptive field

## 8.21.4   Pooling or Down-Sampling

Another topological element that is often used in a CNN is to add after each convolutional layer so-called *pooling layers* as seen in Fig. 8.34. This strengthens the dependence on neighbourhood spatial information and reduces further the number of parameters to be found though training, particularly when more than a single convolutional (hidden) layer is used. Pooling is sometimes called *down-sampling*. In pooling sets of (usually) four outputs from a convolutional layer are grouped, either by averaging or by choosing the maximum among them, to use in the pooled layer. The pooled layer values are fed to the next convolutional (hidden) layer or to the output layer, depending on the network design chosen.

## 8.21.5   The ReLU Activation Function

In order to speed up training,[54] it is of value to replace the sigmoid activation function of Fig. 8.22, by what is called a *Rectified Linear Unit*—ReLU—whose form is shown in Fig. 8.35. This improves the efficiency of the gradient descent operation used in back propagation because of the exceptional simplification of the derivative in (8.62a). Sometimes the ReLU is shown as a separate layer in a CNN. In our treatment we have assumed it is incorporated into each of the processing elements of the convolution layers.

---

[54] See A. Krizhevsky, I. Sutskever and G. Hinton, ImageNet classification with deep convolutional neural networks, *Proc. Advances in Neural Information Processing Systems*, vol. 25, 2012, pp. 1090–1098.

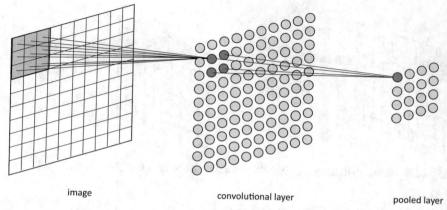

image                                    convolutional layer                         pooled layer

**Fig. 8.34** Using a pooled layer to improve spatial properties and to reduce the number of weights in the following layers

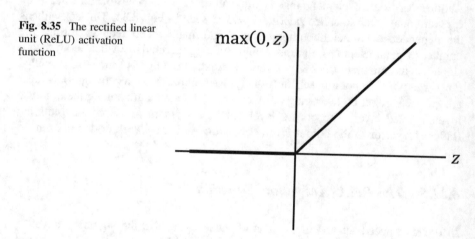

**Fig. 8.35** The rectified linear unit (ReLU) activation function

$$\max(0, z)$$

z

### 8.21.6 Handling the Outputs of a CNN

The convolutional/pooling layer pair is a fundamental building block of a more complex CNN. Often the output of the pooling layer is fed to the input of another convolutional layer (hidden layer in the terminology of the ANN). The network depth is defined by the number of convolutional layers—*deep learning* refers to the analytical benefits obtained by having a number of cascaded convolutional (and pooling) layers.

We now need to think about what to do with the outputs from the last convolutional or pooling layer. There are several possibilities. The pooled layer outputs can

1. feed into another convolutional layer, to provide a **deeper** network
2. feed into a set of output layer PEs, which give the class labels of interest as in Fig. 8.31
3. act as the inputs to a normal fully-connected neural network of the type of Sect. 8.20 and Fig. 8.23, in which case the CNN acts as a feature selector for the fully connected structure
4. generate a set of class probabilities.

The first two are straightforward and do not require further consideration, but consider now options 3 and 4. As an example of 3, Fig. 8.36 shows a common topology, in which the final output from the CNN is a set of processing elements that are arranged in vector form. Since the network has been handling the data in image form up to that point, rearranging the outputs into a vector is called *flattening*.

Consider now option 4 above. If the signals from the flattening layer in Fig. 8.36 are represented by $o_n, n = 1 \ldots N$ then we could compute a set of pseudo-probabilities, called the *softmax* probabilities, according to

$$p(o_n) = \frac{e^{o_n}}{\sum_{n=1}^{N} e^{o_n}}$$

These have the probability-like properties that $0 \leq p \leq 1$ and $\sum_n p = 1$.

### 8.21.7  *Multiple Filters in the Convolution Layer*

A very common extension of the CNN is to have several convolution pathways in parallel as seen in Fig. 8.37, so that as much spatial information as possible can be extracted, particularly at different spatial scales. The filters (and thus receptive

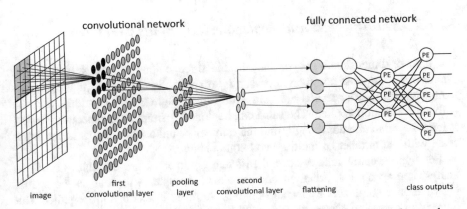

**Fig. 8.36** Using the CNN as a feature selector for input to a fully connected neural network

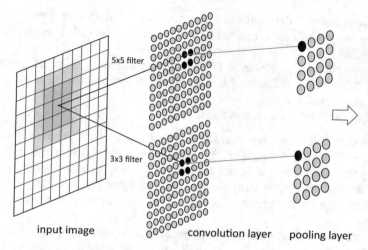

**Fig. 8.37** CNN with parallel convolution paths with different filters; the outputs can go on to further convolutional layers, an output layer or a fully connected neural network

fields) can be of the same or different sizes. At some stage the outputs of the different pathways will be combined, usually at the flattening step.

### 8.21.8 Simplified Representation of the CNN

We are now at the stage where we can introduce some simplification into the diagrammatic representation of the CNN, using the properties described in the previous sections. Although there is, as yet, no standard topological representation, that shown in Fig. 8.38 is similar to the form used by most authors.

### 8.21.9 Multispectral and Hyperspectral Inputs to a CNN

For a three-dimensional image, such as the three colour primaries in a colour photograph, the most common approach is to apply filters to each of the components separately. The results are then summed, a single bias $\theta$ is added, and the activation function applied. The same can be done for simple multispectral images; it must be remembered, though, that the number of unknown weights at the input scale with the number of multispectral components.

For hyperspectral images several different approaches are possible. One is to analyse the spectral information content alone. Another is to analyse the spatial information content alone (spatial context). Another is to do both together. But there is a processing challenge. We could, for example, treat hyperspectral imagery

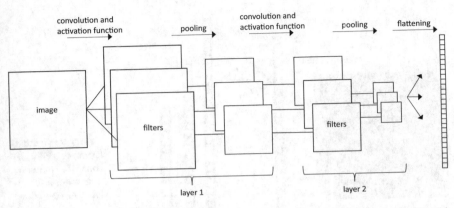

**Fig. 8.38** A simplified representation of a CNN with several convolution paths in parallel: this shows only two stages of convolution and pooling; there can be more, and some may or may not use pooling; the flattened output can be used as is, or fed to a fully connected neural network

by allocating one convolution filter (kernel) to each band, as for the three band colour picture. For a 200 channel image that requires 200 times as many weights as for a single band image. For an image with 200 bands, and $3 \times 3$ kernels, the total number of unknowns (weights plus offsets) connecting just the input image to the first convolutional layer is 2000, noting that the same weights are used in each filter right across a particular band. This, of course, gets multiplied upwards if sets of different filters are used in the convolutional layer.

If we are interested just in the spatial properties of a scene, we can reduce the dimensionality of the data, say, by performing a principal components transform and then keeping just the first or first few significant components. We would then proceed as for a simple colour photograph. With such an arrangement it is feasible to use all image pixels at the input to the CNN.

If we want to analyse hyperspectral data for spectral properties alone, we can use the CNN to find a label for each pixel based just upon its spectrum, and thus implicitly the correlations between bands. That is illustrated in Fig. 8.39, in which the prospect of several parallel paths is used to extract different correlations in the spectral domain.

Using the CNN to analyse both spatial and spectral content simultaneously for pixel labelling can be done with separate pathways, as shown in Fig. 8.40. In the spatial path a neighbourhood patch of pixels is defined around the pixel of interest. That patch can have reduced spectral dimensionality as in the spatial example referred to above or can have full spectral dimensionality. The outputs of the separate pathways are then combined at the flattening step.

An alternative to separating spectral and spatial processing in the manner just described is to apply three dimensional correlations to the image cube, again restricting the spatial sub-domain to a relevant neighbourhood of pixels about the pixel being labelled.

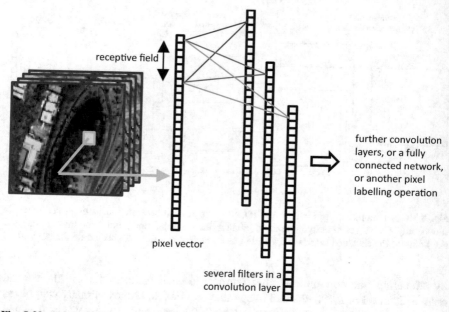

receptive field

further convolution
layers, or a fully
connected network,
or another pixel
labelling operation

pixel vector

several filters in a
convolution layer

**Fig. 8.39** Using a CNN to analyse a pixel based just on its spectral properties

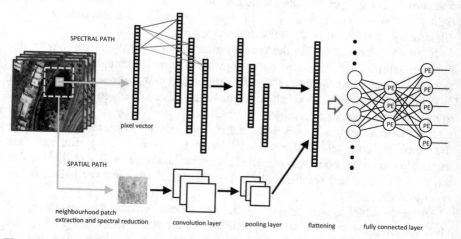

SPECTRAL PATH

pixel vector

SPATIAL PATH

neighbourhood patch
extraction and spectral reduction          convolution layer          pooling layer          flattening          fully connected layer

PE

**Fig. 8.40** Using a CNN to find the correct class label for a pixel using both spectral and spatial information; the neighbourhood patch about the pixel of interest is reduced in spectral dimensionality by taking the principal components of the original spectral data or even by using simple arithmetic averaging of the different spectral components for each pixel in the patch

## 8.21.10   A Spectral-Spatial Example of the Use of the CNN

As an example of the use of the topology shown in Fig. 8.40, we show the exercise of classifying the hyperspectral sample AVIRIS image of Salinas in California undertaken by Yang et al.[55] The image segment consists of $512 \times 217$ pixels, with 3.7 m spatial resolution. It has 224 recorded bands. The authors reduced those to 200 by removing channels with poor quality. There are 16 horticultural ground cover types present in the image, as seen in the ground truth map in Fig. 8.41.

The authors chose to train the network using different percentages of the ground truth pixels, but we show the results here for just for the training data being 25% of the total labelled pixels. They used all the available ground truth pixels to test the generalisation of the network—i.e., the classifier performance.

The CNN topology, or architecture, used by the authors consisted of a spectral path at the top and a spatial path at the bottom as in Fig. 8.40, although they used two convolution layers and one pooling layer per path. Each spatial path has 30 filters of size $3 \times 3$ for each convolution layer, with a $2 \times 2$ pooling filter. The spectral path has 20 filters of size $16 \times 1$ for each convolution layer, with a $5 \times 1$ pooling filter.

The spatial layer is required to capture the neighbourhood (or spatial) properties of a pixel. A patch of $21 \times 21$ pixels, centered on the pixel of interest, was used. The patch was created by averaging over all the spectral channels in that neighbourhood.

The outputs from the two paths are flattened, concatenated and then fed into a fully connected neural network with two hidden layers, each with 400 nodes. Thus, the two path CNN is acting as a feature selector to the neural network.

The output layer has 16 nodes, representing the 16 classes in the Salinas image. The outputs are in the form of class conditional probabilities computed with the softmax function.

The authors also used *transfer learning*. This is a technique based on the concept that networks previously trained on different images, but with the same sensor, will most likely perform acceptably on the image of interest. This is based on the assumption that the spatial properties are similar from image to image with about the same spatial resolution. The authors trained the CNN layers on a different AVIRIS image, and then used the weights so found to initialize the CNN weights for training on the Salinas scene. This is not necessary in general, but it is a useful approach, based on the concept that we, as humans, adapt our learning from past experience.

The results for the Salinas image are shown in Table 8.10 in which it is seen that the best performance is given when both spectral and neighbourhood properties of

---

[55] See J. Yang, Y-Q Zhao and J Cheung-Wai Chan, Learning and transferring deep joint spectral–spatial features for hyperspectral classification, *IEEE Transactions on Geoscience and Remote Sensing*, Volume 55, No 8, August 2017, pp. 4729–4742.

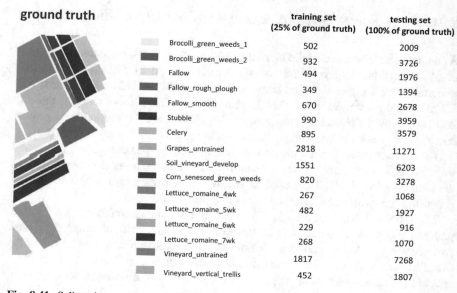

**ground truth**

| | | training set (25% of ground truth) | testing set (100% of ground truth) |
|---|---|---|---|
| | Brocolli_green_weeds_1 | 502 | 2009 |
| | Brocolli_green_weeds_2 | 932 | 3726 |
| | Fallow | 494 | 1976 |
| | Fallow_rough_plough | 349 | 1394 |
| | Fallow_smooth | 670 | 2678 |
| | Stubble | 990 | 3959 |
| | Celery | 895 | 3579 |
| | Grapes_untrained | 2818 | 11271 |
| | Soil_vineyard_develop | 1551 | 6203 |
| | Corn_senesced_green_weeds | 820 | 3278 |
| | Lettuce_romaine_4wk | 267 | 1068 |
| | Lettuce_romaine_5wk | 482 | 1927 |
| | Lettuce_romaine_6wk | 229 | 916 |
| | Lettuce_romaine_7wk | 268 | 1070 |
| | Vineyard_untrained | 1817 | 7268 |
| | Vineyard_vertical_trellis | 452 | 1807 |

**Fig. 8.41** Salinas image ground truth map and numbers of training and testing pixels per class used by Yang et al.[56]

**Table 8.10** Results of CNN classification of the Salinas image

| Configuration | Testing set accuracy (%) |
|---|---|
| Spectral only | 92.3 |
| Spatial only | 96.6 |
| Both spectral and spatial | 98.3 |

the pixel are used as features. The spectral and spatial tests were carried out separately by removing the other path.

The authors ran extensive trials to find the best topology for the network—the numbers of convolutions layers, the numbers of filters, the numbers of nodes in the hidden layers, and so on, which indicates that the preparatory stages in using a CNN can be quite extensive.

## 8.21.11 Avoiding Overfitting

We now come to an important practical consideration, similar to that we met with the maximum likelihood classifier when considering the Hughes phenomenon in Sect. 8.3.7. That is the problem of *overfitting*.

---

[56] ibid.

The concern arises because we have so many weights and offsets to be found through training; it is the availability of training data that determines how effectively those unknowns can be found. We must have sufficient training samples available to get reliable estimates of the unknown parameters, otherwise the network will not generalize well. In other words, it will not perform well on previously unseen pixels.

It is not sufficient to have a minimum of samples to estimate the unknowns, otherwise over-fitting will occur. This is illustrated in the example from curve fitting shown in Fig. 8.42. Fitting a high order curve through just three points, will guarantee good fits for those points, but the behaviour between the points can be *way out* in terms of being able to represent intervening points not used in generating the curve. If many "training" samples are used then the function found interpolates (generalizes) well.

Clearly, we need many more training pixels than the minimum to ensure we do not strike the same problem when training the neural network. If enough training data is not available so that overfitting may occur a number of procedures are available for minimizing the effect. The most common is to implement dropouts, which involves randomly dropping PEs during the training phase.[57]

Overfitting can also be controlled by regularization, as seen in the next section.

## *8.21.12   Variations*

There is a greater deal of variability in how CNNs are implemented in practice, more so than with most other classification techniques. Architectures and processes vary widely. To quote Goodfellow et al.[58]

> Research into convolutional network architectures proceeds so rapidly that a new best architecture for a given benchmark is announced every few weeks to months, rendering it impractical to describe in print the best architecture. Nonetheless, the best architectures have consistently been composed of the building blocks described here.

One common variation for hyperspectral image analysis, instead of the two-path approach of Fig. 8.40, is to use a three-dimensional convolution kernel (or filter) which has two spatial dimensions and one spectral dimension.[59] Again, the spatial dimensions are generally limited to a user-chosen patch size. The three-dimensional convolution filter is applied in each convolution layer, which could also contain several filters in parallel.

---

[57] See N. Srivastava, G. Hinton, A. Krizhevsky, I. Sutskever and R. Salakhutdinov, Dropout: A simple way to prevent neural networks from overfitting, *Journal of Machine Learning Research*, vol. 15, 2014, pp. 1929–1958.

[58] I. Goodfellow, et al., loc cit.

[59] See, for example, Y. Chen, H. Jiang, C. Li, X. Jia and P. Ghamisi, Deep feature extraction and classification of hyperspectral images based on convolutional neural networks, *IEEE Transactions on Geoscience and Remote Sensing*, vol. 54, no. 10, 2016, pp. 6232–6251.

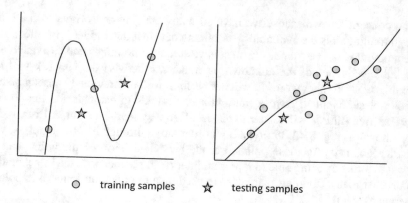

Fig. 8.42 Showing the problem with overfitting if not enough representative samples are used during training

Another variation is to use a different objective measure for training other than the sum of squared error measure of (8.60) and (8.67). Error measures are more generally called cost or objective functions. A common alternative to (8.60, 8.67) is, at the output layer, to minimise the cross-entropy

$$E = -\sum_{n=1}^{N}\sum_{m=1}^{M} t_{mn} ln g_{mn} \tag{8.68}$$

in which $N$ is the number of training pixels and $M$ is the number of classes; $t_{mn}$ is the actual class for the $n$th training pixel and $g_{mn}$ is the class currently indicated at the network output. Most often this measure, as with most error measures in CNN training, is based on the use of mini-batches, particularly when the training data sets are large.

As an alternative to using dropouts to control overfitting, *regularisation* can be employed. That involves adding a constraint to the error measures of (8.67) or (8.68) such that in the minimisation process the weights are reduced, if possible; the expectation is that some will become so small that they are not influential in classification. The most common constraint to add is the sum of weights squared, to form a new error measure

$$\mathcal{L} = E + \lambda \sum_{all\ weights} \|w\|^2 \tag{8.69}$$

in which $\lambda$ is the regularisation parameter. Because (8.69) reduces the weights it is known as *weight decay*.

## 8.22 Recurrent Neural Networks

### 8.22.1 Multi-temporal Remote Sensing

From the very start analysts recognised that better discrimination of some cover types was improved when image acquisitions at more than a single time were employed. Crop analysis, particularly, can be facilitated if several images over the growing season are used because their development with time can be an important discriminator.

Recognition of the importance of the time evolution of cover types in producing thematic maps is one thing but finding viable techniques for undertaking multi-temporal image classification is another. The tasseled cap model of Sect. 6.5 was one attempt, based on producing images of soil brightness, greenness and yellowness that could be tracked with time.

Another approach has been to take a so-called multi-temporal set of images and concatenate their pixel vectors which are then fed into standard supervised classification procures. That is sometimes called *sequence classification*, characterised by there being only one class for a pixel as the result of the analysis of the complete multi-temporal sequence for that pixel. In some ways it can be looked on as a brute force approach.

Possibly a more elegant technique is to use a recurrent neural network; we develop that algorithm in this section. In preparation for that we need to understand the importance of memory in making predictions. In this context remember that classification is a process of prediction—the trained algorithm is used to predict the correct class label for a previously unseen pixel.

### 8.22.2 Importance of Memory

To develop this concept, we will look at a simple example of text analysis, which is explained very well, with a number of examples in different languages, in a book on information theory now almost 60 years old.[60]

Suppose someone is spelling out words to you, one letter at a time, and you have to predict what letter comes next. For the first letter, the best you can do is to be guided by the frequency with which letters (and a space) occur in English. Suppose we ignore spaces and just concentrate on letters. The most frequent letters in English are, with their associated probabilities of occurrence: $e$ (0.103), $t$ (0.080),

---

[60] N. Abramson, *Information Theory and Coding*, McGraw-Hill, N.Y., 1963.

*a* (0.064), *o* (0.063), *i* (0.058), *n* (0.057) and *s* (0.051).[61] So, the best letter to guess is *e*. In the absence of any other information, you would also guess *e* for the second letter. However, in reality you remember what the first letter actually was so when it comes to predicting the second letter you use that "stored" knowledge.

If the first letter was *t*, then there are several possibilities for the second letter. In order of decreasing likelihood in English, they are *th*, *ti*, *te*. Knowing that, and remembering that the first letter was *t*, you would predict the second would be *h*. That is a much better guess than choosing the second letter based just on the simple frequency of occurrence of the letters. Your use of memory has allowed a much better prediction for the second letter.

You can see that your prediction of the third letter should be much better still, and so on, because you have memory of the previous letters.

We can transfer this idea to remote sensing image classification, and that is the basis of the recurrent neural network.

### 8.22.3  The Recurrent Neural Network (RNN) Architecture

The recurrent neural network (RNN) has been developed principally for applications such as speech recognition and prediction where the signal to be analysed consists of a time-series string of words. It can also be used to analyse sequences of data such a co-registered multi-temporal image data sets, which is the perspective we will follow in this treatment.

The RNN is effectively a simple ANN, such as that treated in Sect. 8.20.2, but with the addition of memory in the hidden layer. Memory is created by feeding the output of the hidden layer, computed from one member of the data set, into the input of the hidden layer when the next member of the data set (sequence) is presented to the hidden layer. Figure 8.43 depicts that diagrammatically. The loop over the top of the hidden layer with a unit time delay, and the summing element, is how memory is implemented. Essentially the network has access to the prior image data in the sequence as it carries out its computation—that is it is *recurrent* in the sense that as more images (actually individual pixels) are presented to the network during training it has benefitted from all the previous presentations.

In this analysis we use time to indicate the different images in the multi-temporal sequence, with *t* used to represent the current image in the set, *t* − 1 to represent the previous image, and so on.

---

[61] The well-known Morse code makes use of these probabilities to ensure an efficient and thus fast code in English: for example, the letter *e* is given the shortest code of a dot, the letter *t* is then coded as a dash, the letter *a* as a dot-dash, the letter *o* as a dash-dash-dash and the letter *i* as a dot-dot. By contrast, the lesser occurring letters have longer codes: *j* is coded as dot-dash-dash-dash, *q* as dash-dash-dot-dash and *z* as dash-dash-dot-dot.

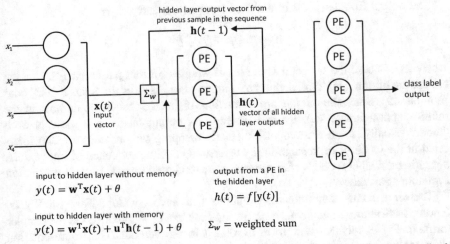

**Fig. 8.43** Showing the creation of a recurrent neural network through the use of delayed feedback at the hidden layer to provide memory in a standard ANN

Equation (8.57) describes the operation of a processing element in the hidden layer. The input to the processing element with an explicit time (i.e., sequence) dependence is

$$y(t) = \mathbf{w}^T\mathbf{x}(t) + \theta$$

Recall that $\mathbf{x}(t)$ is the pixel vector from the previous layer in the network and $\mathbf{w}$ is the set of weights in the hidden layer to be found through training, along with the offset $\theta$.

The output of the hidden layer, represented by $h(t)$, involves the operation of the activation function $f[..]$ on the PE input

$$h(t) = f[y(t)]$$

Collectively, the set of the outputs from all hidden layer PEs can be represented by the column vector $\mathbf{h}(t)$. If the previous outputs of the hidden layer PEs are fed back to their inputs in combination with the next set of inputs, then the input to an individual processing element in the hidden layer can be expressed

$$y(t) = \mathbf{w}^T\mathbf{x}(t) + \mathbf{u}^T\mathbf{h}(t-1) + \theta$$

in which the vector $\mathbf{u}$ is a set of weights on each of the delayed PE outputs when they are fed back. As with $\mathbf{w}$, the elements in $\mathbf{u}$ have to be learnt during training.

The output from each PE in the *final* layer of the network is

$$g(t) = \mathbf{w}^{\mathrm{T}}\mathbf{h}(t) + \theta$$

where in this case the $\mathbf{w}$ vector is the set of weights for that particular PE. It is the collection of the $g(t)$ that define the output class, just as in Sect. 8.20.3 (and footnote 49). Since the outputs are a function of time (or sequence number in the multi-temporal image set), we can generate an output label for a pixel at each "time." Usually, for multi-temporal thematic mapping we are not so much interested in the labels at each stage in the presentation of the sequence of image data, but rather the label found by the network at the end, making use of the information available at all dates.

There are some diagrammatic variations that have developed around RNNs. Figure 8.44a shows how the network of Fig. 8.43 can be summarised in the simpler form. In Fig. 8.44b that new form is rotated to the vertical. That representation allows the adjacent so-called unfolded structure to be drawn, in which each repetition of the fundamental network represents subsequent time steps. In that form the delayed hidden layer outputs are seen explicitly to feed into the next time step.

## 8.22.4  Training the RNN

The recurrent neural network can be trained using backpropagation, but the process has to propagate back through all the time steps. As a result, it is called backpropagation through time (BPTT). However, a problem arises because of the recurrent nature of the network and the fact that the weights are the same at each time step. The problem is characterised as either a vanishing gradient or exploding gradient problem, the details of which are beyond this introductory treatment, as are their remedies, but can be found in a number of detailed treatments.[62]

## 8.23  Context Classification

## 8.23.1  The Concept of Spatial Context

Apart from the CNN of the previous section, the classifiers treated in the earlier sections are often categorised as point, or pixel-specific, in that they label a pixel on

---

[62] See I. Goodfellow, Y. Bengio, and A Courville, *Deep Learning (Adaptive Computation and Machine Learning Series)* The MIT Press, Cambridge Mass., 2016, A. Graves, *Supervised Sequence Learning with Recurrent Neural Networks,* Springer, Berlin, 2012 and R. Pascanu, T. Mikolov and Y. Benigo, On the difficulty of training recurrent neural networks, *Proc 30th Int. Cong. Machine Learning,* Atlanta, Georgia, 2013.

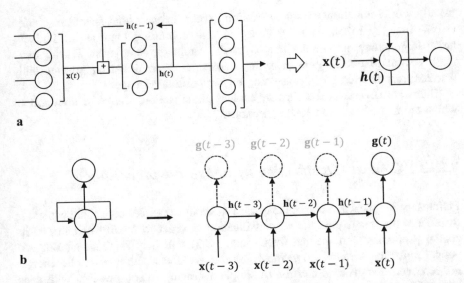

**Fig. 8.44** **a** Developing a simpler representation of the basic RNN and **b** unfolding that simpler representation to make the signal flow through time clearer; outputs at each time step are shown greyed out since they are not normally used in remote sensing

the basis of its spectral properties alone. No account is taken of how their neighbours might be labelled.

In any real image adjacent pixels are related or correlated because imaging sensors acquire significant portions of energy from adjacent pixels[63] and because ground cover types generally occur over a region that is large compared with the size of a pixel. Knowledge of neighbourhood relationships is a rich source of information that is not exploited in simple, traditional classifiers.

In this section we consider further the importance of spatial relationships—spatial context—and see the benefit of taking context into account when making classification decisions. Not only is the inclusion of context important because it exploits spatial properties, but sensitivity to the correct context for a pixel can improve a thematic map by removing individual pixel labelling errors that might result from noisy data, or from unusual classifier performance (see Problem 8.6).

Classification methods that take into account the labelling of neighbours when seeking to determine the most appropriate class for a pixel are said to be context sensitive and are called context classifiers. They attempt to develop a thematic map that is consistent both spectrally and spatially.

The degree to which adjacent pixels are strongly correlated will depend on the spatial resolution of the sensor and the scale of natural and cultural features on the earth's surface. Adjacent pixels over an agricultural region will be strongly correlated, whereas for the same sensor, adjacent pixels over a busier, urban region

---

[63] This is known as the point spread function effect.

usually would not show strong correlation except for very high spatial resolution (VHR) imagery Likewise, for a given area, neighbouring Landsat MSS pixels, being larger, may not demonstrate as much correlation as adjacent Worldview-2 multispectral pixels. In general terms, context classification techniques usually warrant consideration when processing higher resolution imagery.

There are several approaches to context classification, the most common of which are treated in the following sections.[64]

## 8.23.2  Context Classification by Image Pre-processing

Perhaps the simplest and one of the earliest methods for exploiting spatial context is to process the image data before classification in order to modify or enhance its spatial properties.[65] A median filter (Sect. 5.3.2) will help in reducing salt and pepper noise that would lead to inconsistent class labels with a point classifier, if not removed first. The application of simple averaging filters, possibly with edge preserving thresholds, can be used to impose a degree of homogeneity among the brightness values of adjacent pixels, thereby increasing the chance that neighbouring pixels may be given the same label.

An alternative is to generate a separate channel of data that associates spatial properties with pixels. For example, a texture channel could be added and classification carried out on the combined spectral and texture channels. See Sect. 5.10.2.

One of the more interesting historical spatial pre-processing techniques is the ECHO (Extraction and Classification of Homogeneous Objects) methodology[66] in which regions of similar spectral properties are "grown" before classification is performed. Several region growing techniques are available, possibility the simplest of which is to aggregate pixels into small regions by comparing their brightness values in each channel; smaller regions are then combined into bigger regions in a similar manner. When that is done ECHO classifies the regions as single objects. It only resorts to point classification when it has to treat individual pixels that could not be put into regions. ECHO is included in the *Multispec* image analysis software package.[67]

---

[64] Although now not often used, for statistical context methods see P.H. Swain, S.B. Varderman and J.C. Tilton, Contextual classification of multispectral image data, *Pattern Recognition*, vol. 13, 1981, pp. 429–441, and N. Khazenie and M.M. Crawford, A spatial–temporal autocorrelation model for contextual classification, *IEEE Transactions on Geoscience and Remote Sensing*, vol. 28, no. 4, July 1990, pp. 529–539.

[65] See P. Atkinson, J.L. Cushine, J.R. Townshend and A. Wilson, Improving thematic map land cover classification using filtered data, *Int. J. Remote Sensing*, vol. 6, 1985, pp. 955–961.

[66] R.L. Kettig and D.A. Landgrebe, Classification of multispectral image data by extraction and classification of homogeneous objects, *IEEE Transactions on Geoscience Electronics*, vol. GE-14, no. 1, 1976, pp. 19–26.

[67] https://engineering.purdue.edu/~biehl/MultiSpec/ accessed 2021.

## 8.23.3  Post Classification Filtering

If a thematic map has been generated using a simple point classifier, a degree of spatial context can be developed by logically filtering the map.[68] If the map is examined in, say, $3 \times 3$ windows, the label at the centre can be changed to that most represented in the window. Clearly this must be done carefully, with the user controlling the minimum size region of a given cover type that is acceptable in the filtered image product.

## 8.23.4  Probabilistic Relaxation Labelling

Spatial consistency in a classified image product can also be improved using the process of label relaxation. While it has little theoretical foundation and is more complex than the methods outlined in the previous sections, it does allow the spatial properties of a region to be carried into the classification process in a logical manner.

### 8.23.4.1  The Algorithm

The process commences by assuming that a point classification has already been carried out, based on spectral data alone. We then assume that, for each pixel, we have a set of posterior probabilities available that describe the likelihoods that the pixel belongs to each of the possible ground cover classes under consideration. That set could be computed from (8.6) or (8.7) if maximum likelihood classification had been used. If another classification method had been employed, some other assignment process will be required, which could be as simple as allocating a high probability to the most favoured class label and lower probabilities to the others. It could also use the softmax approach covered in Sect. 8.21.6.

Although we should perhaps use the posterior probability notation $p(\omega_i|\mathbf{x})$ in what is to follow, for notational simplicity we instead adopt the following expression for the set of label probabilities on a pixel $m$:

$$p_m(\omega_i) \quad i = 1 \ldots M \tag{8.70}$$

where $M$ is the total number of classes; $p_m(\omega_i)$ should be read as the probability that $\omega_i$ is the correct class for pixel $m$. As posteriors, the full set of $p_m(\omega_i)$ for the pixel sum to unity:

---

[68] F.E. Townsend, The enhancement of computer classifications by logical smoothing, *Photogrammetric Engineering and Remote Sensing*, vol. 52, 1986, pp. 213–221.

**Fig. 8.45** A neighbourhood
about pixel $m$

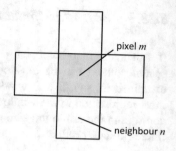

$$\sum_i p_m(\omega_i) = 1$$

Suppose now that a neighbourhood is defined surrounding pixel $m$. This can be of any size and, in principle, should be large enough to ensure that all the pixels considered to have any spatial correlation with $m$ are included. For high resolution imagery this is not practical and a simple neighbourhood such as that shown in Fig. 8.45 is often adopted.

Now assume that a *neighbourhood function* $Q_m(\omega_i)$ can be found, by means to be described below, through which the pixels in the prescribed neighbourhood can influence the classification of pixel $m$. This influence is exerted by multiplying the label probabilities in (8.70) by the $Q_m(\omega_i)$. The results can be turned into a new set of label probabilities for the pixel by dividing by their sum:

$$p'_m(\omega_i) = \frac{p_m(\omega_i)Q_m(\omega_i)}{\sum_i p_m(\omega_i)Q_m(\omega_i)} \tag{8.71}$$

The modification is made to the set of label probabilities for all pixels. In the following it will be seen that $Q_m(\omega_i)$ depends on the label probabilities of the neighbouring pixels, so that if all pixel probabilities are modified in the manner just described then the neighbours of any given pixel have also been altered. Consequently (8.71) should be applied again to give newer estimates still of the label probabilities. Indeed, (8.71) is applied as many times as necessary to ensure that the $p'_m(\omega_i)$ have stabilised—i.e., that they do not change with further iteration. It is assumed that the $p'_m(\omega_i)$ then represent the correct set of label probabilities for the pixel, having taken account both of spectral data, in the initial determination of label probabilities, and spatial context via the neighbourhood functions. Since the process is iterative, (8.71) is usually written as an explicit iteration formula:

$$p_m^{k+1}(\omega_i) = \frac{p_m^k(\omega_i)Q_m^k(\omega_i)}{\sum_i p_m^k(\omega_i)Q_m^k(\omega_i)} \tag{8.72}$$

where $k$ is the iteration count. Depending on the size of the image and its spatial complexity, the number of iterations required to stabilise the label probabilities may be quite large. However, most change in the probabilities occurs in the first few iterations and there is good reason to believe that proceeding beyond say 5 to 10 iterations may not be necessary in most cases (see Sect. 8.23.4.4).

### 8.23.4.2 The Neighbourhood Function

Consider just one of the neighbours of pixel $m$ in Fig. 8.45—call it pixel $n$. Suppose there is available some measure of compatibility of the current labelling on pixel $m$ and its neighbouring pixel $n$. Let $r_{mn}(\omega_i, \omega_j)$ describe numerically how compatible it is to have pixel $m$ classified as $\omega_i$ and neighbouring pixel $n$ classified as $\omega_j$. It would be expected, for example, that this measure would be high if the adjoining pixels are both wheat in an agricultural region, but low if one of the neighbours was snow. There are several ways these *compatibility coefficients*, as they are called, can be defined. An intuitively appealing definition is based on conditional probabilities. The compatibility measure $p_{mn}(\omega_i|\omega_j)$ is the probability that $\omega_i$ is the correct label for pixel $m$ if $\omega_j$ is the correct label for pixel $n$. A small piece of evidence in favour of $\omega_i$ being correct for pixel $m$ is $p_{mn}(\omega_i|\omega_j)p_n(\omega_j)$—i.e., the probability that $\omega_i$ is correct for pixel $m$ if $\omega_j$ is correct for pixel $n$ multiplied by the probability that $\omega_j$ is correct for pixel $n$. This is the joint probability of pixel $m$ being labelled $\omega_i$ **and** pixel $n$ being labelled $\omega_j$.

Since probabilities for all possible labels on pixel $n$ are available (even though some might be very small) the total evidence from pixel $n$ in favour of $\omega_i$ being the correct class for pixel $m$ will be the sum of the contributions from all pixel $n$'s labelling possibilities. viz.

$$\sum_j p_{mn}(\omega_i|\omega_j)p_n(\omega_j)$$

Consider now the full neighbourhood of the pixel $m$. All the neighbours contribute evidence in favour of labelling pixel $m$ as belonging to class $\omega_i$. These contributions are added[69] via the use of *neighbour weights* $d_n$ that recognise that some neighbours may be more influential than others. Thus, at the $k$th iteration, the total neighbourhood support for pixel $m$ being classified as $\omega_i$ is:

$$Q_m^k(\omega_i) = \sum_n d_n \sum_j p_{mn}(\omega_i|\omega_j)p_n^k(\omega_j) \tag{8.73}$$

This is the definition of the neighbourhood function. In (8.72) and (8.73) it is common to include pixel $m$ in its own neighbourhood so that the modification process is not entirely dominated by the neighbours, particularly if the number of

---

[69] An alternative way of handling the full neighbourhood is to take the geometric mean of the neighbourhood contributions.

iterations is so large as to take the process quite a long way from its starting point. Unless there is good reason to do otherwise the neighbour weights are generally chosen all to be the same.

### 8.23.4.3   Determining the Compatibility Coefficients

Several methods are possible for determining values for the compatibility coefficients $p_{mn}(\omega_i|\omega_j)$. One is to have available a spatial model for the region under consideration, derived from some other data source. In an agricultural area, for example, some general idea of field sizes, along with a knowledge of the pixel size of the sensor, should make it possible to estimate how often one particular class occurs simultaneously with a given class on an adjacent pixel. Another approach is to compute values for the compatibility coefficients from ground truth pixels, although the ground truth needs to be in the form of training regions that contain heterogeneous and spatially representative cover types.

### 8.23.4.4   Stopping the Process

While the relaxation process operates on label probabilities, the user is interested in the actual labels themselves. At the completion of relaxation, or at any intervening stage, each of the pixels can be classified according to the highest label probability. Thought has to be given as to how and when the iterations should be terminated. As suggested earlier, the process can be allowed to go to a natural completion at which continued iteration leads to no further changes in the label probabilities for all pixels. That however presents two difficulties. First, up to several hundred iterations may be involved. Secondly, it is observed in practice that classification accuracy improves in the first few iterations but often deteriorates later in the process.[70] If the procedure is not terminated, the thematic map, after a large number of iterations, can be worse than before the technique was applied. To avoid those difficulties, a stopping rule or some other controlling mechanism is needed. As seen in the example following, stopping after just a few iterations may allow most of the benefit to be drawn from the process. Alternatively, the labelling errors remaining at each iteration can be checked against ground truth, if available, and the iterations terminated when the labelling error is seen to be minimised.[71]

---

[70] J.A. Richards, D.A. Landgrebe and P.H. Swain, On the accuracy of pixel relaxation labelling, *IEEE Transactions on Systems, Man and Cybernetics*, vol. SMC-11, 1981, pp. 303–309.

[71] P. Gong and P.J. Howarth, Performance analyses of probabilistic relaxation methods for land-cover classification in remote sensing, *Remote Sensing of Environment*, vol. 30, 1989, pp. 33–42.

Another approach is to control the propagation of contextual information as iteration proceeds.[72] In the first iteration, only the immediate neighbours of a pixel have an influence on its labelling. In the second iteration the neighbours two away will now have an influence via the intermediary of the intervening pixels. As iterations proceed, information from neighbours further out is propagated into the pixel of interest to modify its label probabilities. If the user has a view of the separation between neighbours at which the spatial correlation has dropped to negligible levels, then the appropriate number of iterations should be able to be estimated at which to terminate the process without unduly sacrificing any further improvement in labelling accuracy. Noting also that the nearest neighbours should be most influential, with those further out being less important, a useful variation is to reduce the values of the neighbour weights $d_n$ as iteration proceeds so that after, say, 5–10 iterations they have been brought to zero.[73] Further iterations will have no effect, and degradation in labelling accuracy cannot occur.

### 8.23.4.5   Examples

Figure 8.46 shows a simple application of relaxation labelling,[74] in which a hypothetical image of 100 pixels has been classified into just two classes—grey and white. The ground truth for the region is shown, along with the thematic map assumed to have been generated from a point classifier, such as the maximum likelihood rule. That map functions as the "initial labelling." The compatibility coefficients are shown as conditional probabilities, computed from the ground truth map. Label probabilities were assumed to be 0.9 for the favoured label in the initial labelling and 0.1 for the less likely label. The initial labelling, by comparison with the ground truth, can be seen to have an accuracy of 88%—there are 12 pixels in error. The labelling at significant stages during iteration, selected on the basis of the largest current label probability, is shown, illustrating the reduction in classification error owing to the incorporation of spatial information into the process. After 15 iterations all initial labelling errors have been removed, leading to a thematic map 100% in agreement with the ground truth. In this case the relaxation process was allowed to proceed to completion and there have been no ill effects. This is an exception and stopping rules have to be applied in most cases.[75]

---

[72] T. Lee, Multisource context classification methods in remote sensing, *PhD Thesis*, The University of New South Wales, Kensington, Australia, 1984.

[73] T. Lee and J.A. Richards, Pixel relaxation labelling using a diminishing neighbourhood effect, *Proc. Int. Geoscience and Remote Sensing Symposium, IGARSS89*, Vancouver, 1989, pp. 634–637.

[74] Other simple examples will be found in Richards, Landgrebe and Swain, loc. cit.

[75] ibid.

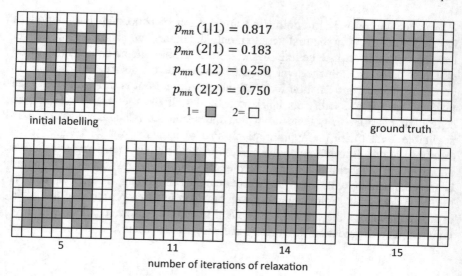

$$p_{mn}(1|1) = 0.817$$
$$p_{mn}(2|1) = 0.183$$
$$p_{mn}(1|2) = 0.250$$
$$p_{mn}(2|2) = 0.750$$

1= ■  2= □

initial labelling

ground truth

5

11

14

15

number of iterations of relaxation

**Fig. 8.46** Simple example of pixel relaxation labelling

As a second example, the leftmost $82 \times 100$ pixels of the agricultural image shown in Fig. 5.11 have been chosen[76] to demonstrate the benefit of diminishing the neighbourhood contribution with iteration. A maximum likelihood classification of the four Landsat multispectral scanner bands was carried out to label the image into 7 classes and to initialise the relaxation process. The accuracy achieved was 65.6%.

Instead of using conditional probabilities as compatibility coefficients the slightly different mechanism proposed by Peleg and Rosenfeld was adopted.[77] To control the propagation of context information, and thereby avoid the deleterious effect of allowing the relaxation process to proceed unconstrained, the neighbourhood weights were diminished with iteration count according to

$$d_n(k) = d_n(1)e^{-\alpha(k-1)}$$

in which $\alpha$ controls how the neighbour weights change with iteration. If $\alpha = 0$ there is no reduction and normal relaxation applies. For $\alpha$ large the weights drop quickly with iteration The central pixel was not included in the neighbourhood definition in this example.

Table 8.11 shows how the relaxation performance depends on $\alpha$. Irrespective of the value chosen the optimal result is achieved after about 4 iterations, giving an accuracy of 72.2%. The table also shows the result achieved if relaxation is left to run for more iterations (final result). As seen, without diminishing the neighbour

---

[76] Full details of this example will be found in T. Lee and J.A. Richards, loc. cit.

[77] S. Peleg and A. Rosenfeld, A new probabilistic relaxation procedure, *IEEE Transactions on Pattern Analysis and Machine Intelligence*, vol. PAMI-2, 1980, pp. 362–369.

**Table 8.11** Performance of relaxation labelling with a diminishing neighbourhood influence

| $\alpha$ | Optimal result | | Final result | |
|---|---|---|---|---|
| | Accuracy | At iteration | Accuracy | At iteration |
| 0.0 | 72.2 | 4 | 70.6 | 32 |
| 1.0 | 72.2 | 4 | 71.4 | 17 |
| 1.8 | 72.2 | 4 | 72.1 | 10 |
| 2.0 | 72.2 | 4 | 72.2 | 9 |
| 2.2 | 72.2 | 4 | 72.2 | 8 |
| 2.5 | 72.2 | 5 | 72.2 | 7 |
| 3.0 | 72.2 | 4 | 72.2 | 6 |

weights or without diminishing them sufficiently, the final result deteriorates. For values of $\alpha$, in the vicinity of 2 the result is fixed at 72.2% from iteration 4 onwards.

## 8.23.5 Handling Spatial Context by Markov Random Fields

Another method for incorporating the effect of spatial context is to use the construct of the Markov Random Field (MRF). When developing this approach, it is useful to commence by considering the whole image, rather than just a local neighbourhood. We will restrict our attention to a neighbourhood once we have established some fundamental concepts.

Suppose there is a total of $M$ pixels in the image, with measurement vectors $\mathbf{x}_m, m = 1 \ldots M$ in which $m = (i,j)$. We describe the full set of vectors by $\mathbf{X} = \{\mathbf{x}_1 \ldots \mathbf{x}_M\}$.

Let the labels on each of the $M$ pixels, derived from a classification, be represented by the set $\Omega = \{\omega_{c1} \ldots \omega_{cM}\}$ in which $\omega_{cm}$ is the label on pixel $m$, drawn from a set of $c = 1 \ldots C$ available classes. We refer to $\Omega$ as the *scene labelling*, because it looks at the classification of every pixel in the scene.

We want to find the scene labelling that best matches the ground truth—the actual classes of the pixels on the earth's surface, which we represent by $\Omega^*$. There will be a probability distribution $p\{\Omega\}$ associated with a labelling $\Omega$ of the scene, which describes the likelihood of finding that distribution of labels over the image. $\Omega$ is sometimes referred to as a *random field*. In principle, what we want to do is find the scene labelling $\hat{\Omega}$ that maximises the global posterior probability $p(\Omega|\mathbf{X})$, and thus that best matches $\Omega^*$. $p(\Omega|\mathbf{X})$ is the probability that $\Omega$ is the correct overall scene labelling given that the full set of measurement vectors for the scene is $\mathbf{X}$. By using Bayes' theorem, we can express it as

$$\hat{\Omega} = \underbrace{\text{argmax}}_{\Omega} \{p(\mathbf{X}|\Omega)p(\Omega)\} \qquad (8.74)$$

in which the argmax function says that we choose the value of $\Omega$ that maximises its argument. The distribution $p(\Omega)$ is the prior probability of the scene labelling.

What we need to do now is to perform the maximisation in (8.74) recognising, however, that the pixels are contextually dependent; in other words, noting that there is spatial correlation among them. To render the problem tractable, we now confine our attention to the neighbourhood immediately surrounding a pixel of interest. Our task is to find the class $c$ that maximises the conditional posterior probability $p(\omega_{cm}|\mathbf{x}_m, \omega_{\partial m})$, where $\omega_{\partial m}$ is the labelling on the pixels in the neighbourhood of pixel $m$. A possible neighbourhood is that shown in Fig. 8.45, although often the immediately diagonal neighbours about $m$ are also included. Now we note

$$
\begin{aligned}
p(\omega_{cm}|\mathbf{x}_m, \omega_{\partial m}) &= p(\mathbf{x}_m, \omega_{\partial m}, \omega_{cm})/p(\mathbf{x}_m, \omega_{\partial m}) \\
&= p(\mathbf{x}_m|\omega_{\partial m}, \omega_{cm})p(\omega_{\partial m}, \omega_{cm})/p(\mathbf{x}_m, \omega_{\partial m}) \\
&= p(\mathbf{x}_m|\omega_{\partial m}, \omega_{cm})p(\omega_{cm}|\omega_{\partial m})p(\omega_{\partial m})/p(\mathbf{x}_m, \omega_{\partial m})
\end{aligned}
$$

The first term on the right-hand side is a class conditional distribution function, conditional also on the labelling of the neighbouring pixels. While we do not know that distribution, it is reasonable to assume that the probability of finding a pixel with measurement vector $\mathbf{x}_m$ from class $\omega_{cm}$ is not dependent on the class membership of spatially adjacent pixels so that $p(\mathbf{x}_m|\omega_{\partial m}, \omega_{cm}) = p(\mathbf{x}_m|\omega_{cm})$, the simple class conditional distribution function compiled for class $\omega_{cm}$ pixels from the remote sensing measurements. We also assume that the measurement vector $\mathbf{x}_m$ and the neighbourhood labelling are independent, so that $p(\mathbf{x}_m, \omega_{\partial m}) = p(\mathbf{x}_m)p(\omega_{\partial m})$. Substituting these simplifications into the last expression above gives

$$
\begin{aligned}
p(\omega_{cm}|\mathbf{x}_m, \omega_{\partial m}) &= p(\mathbf{x}_m|\omega_{cm})p(\omega_{cm}|\omega_{\partial m})p(\omega_{\partial m})/p(\mathbf{x}_m)p(\omega_{\partial m}) \\
&= p(\mathbf{x}_m|\omega_{cm})p(\omega_{cm}|\omega_{\partial m})/p(\mathbf{x}_m)
\end{aligned}
$$

Since $p(\mathbf{x}_m)$ is not class dependent it does not contribute any discriminating information so that the maximum posterior probability rule of (8.74) at the local neighbourhood level becomes

$$
\begin{aligned}
\hat{\omega}_{cm} &= \underset{\omega_{cm}}{\operatorname{argmax}}\, p(\omega_{cm}|\mathbf{x}_m, \omega_{\partial m}) \\
&= \underset{\omega_{cm}}{\operatorname{argmax}}\, p(\mathbf{x}_m|\omega_{cm})p(\omega_{cm}|\omega_{\partial m})
\end{aligned}
$$

or, expressed in the form of a discriminant function for class $m$,

$$
g_{cm}(\mathbf{x}_m) = \ln p(\mathbf{x}_m|\omega_{cm}) + \ln p(\omega_{cm}|\omega_{\partial m}) \tag{8.75}
$$

We now need to consider the meaning of the conditional probability $p(\omega_{cm}|\omega_{\partial m})$. It is the probability that the correct class is $c$ for pixel $m$ given the labelling of the neighbouring pixels. It is analogous to the neighbourhood function in probabilistic relaxation given in (8.73). Because it describes the labelling on pixel $m$, conditional on the neighbourhood labels, the random field of labels we are looking for is called a Markov Random Field (MRF).

How do we determine a value for $p(\omega_{cm}|\omega_{\partial m})$? It can be expressed in the form of a Gibbs distribution[78]

$$p(\omega_{cm}|\omega_{\partial m}) = \frac{1}{Z}\exp\{-U(\omega_{cm})\} \qquad (8.76)$$

in which $U(\omega_{cm})$ is called an *energy function* and $Z$ is a normalising constant, referred to as the *partition function,* which ensures that the sum of (8.76) over all classes $c$ on pixel $m$ is unity. To this point we have said nothing special about the style of the neighbourhood or how the individual neighbours enter into the computation in (8.76). In relaxation labelling they entered through the compatibility coefficients. In the MRF approach the neighbour influence is encapsulated in the definition of sets of *cliques* of neighbours. The cliques of a given neighbourhood are sets of adjacent pixels that are linked (or related in some sense of being correlated), as shown in Fig. 8.47 for the two neighbourhoods given.[79]

Once a neighbourhood definition has been chosen the energy function is evaluated as a sum of clique *potentials* $V_C(\omega_{cm})$, each one of which refers to a particular clique $C$ in the neighbourhood:

$$U(\omega_{cm}) = \sum_{C} V_C(\omega_{cm}) \qquad (8.77)$$

The neighbourhood most often chosen is that of Fig. 8.45 for which the cliques are just the neighbour pairs vertically and horizontally, and the pixel itself, as seen in the left-hand side of Fig. 8.47. That leads to the so-called Ising model for which

$$V_C(\omega_{cm}) = \beta[1 - \delta(\omega_{cm}, \omega_{Cm})]$$

where $\delta(\omega_{cm}, \omega_{Cm})$ is the Kronecker delta; it is unity when the arguments are equal and zero otherwise. $\omega_{Cm}$ is the labelling on the member of the binary clique other than the pixel $m$ itself. $\beta$ is a parameter to be chosen by the user, as below. Thus $U(\omega_{cm})$ is found by summing over the neighbourhood:

$$U(\omega_{cm}) = \sum_{C} \beta[1 - \delta(\omega_{cm}, \omega_{Cm})] \qquad (8.78)$$

---

[78] Although a little complex in view of the level of treatment here, see S. German and D. German, Stochastic relaxation, Gibbs distributions, and the Bayesian restoration of images, *IEEE Transactions on Pattern Analysis and Machine Intelligence,* vol. PAMI-6, no. 6, 1984, pp. 721–741.

[79] For the simple first order (four neighbour) neighbourhood the concept of cliques is not important, since there are only the four neighbourhood relationships.

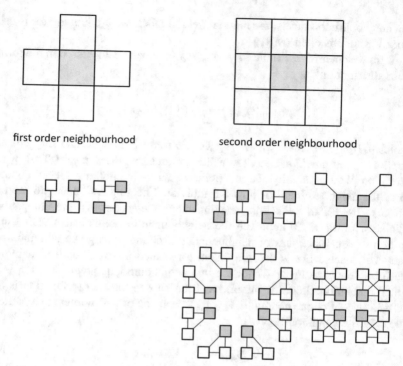

first order neighbourhood                    second order neighbourhood

**Fig. 8.47** First and second order neighbourhoods and their cliques

which is now substituted into (8.76) and (8.75) to give, assuming a multivariate normal class model,

$$
\begin{aligned}
g_{cm}(\mathbf{x}_m) \\
= -\tfrac{1}{2}\ln|\mathbf{C}_c| - \tfrac{1}{2}(\mathbf{x}_m - \mathbf{m}_c)^{\mathrm{T}}\mathbf{C}_c^{-1}(\mathbf{x}_m - \mathbf{m}_c) - \sum_{\mathcal{C}}\beta[1 - \delta(\omega_{cm}, \omega_{\mathcal{C}m})]
\end{aligned}
$$

$$(8.79)$$

where the constant $Z$ has been removed as non-discriminating. It is interesting to understand the structure of this last equation. Whenever a neighbour has the same label as the central pixel the spatial term is zero and thus the spectral evidence in favour of the labelling on the central pixel, via the class conditional distribution function, is unaffected. If a neighbour has a different label the discriminant function for the currently favoured class on the central pixel is reduced, thus making that class slightly less likely.

To use (8.79) there needs to be an allocation of classes over the scene before the last term can be computed. Accordingly, an initial classification would be performed, say with the maximum likelihood classifier of Sect. 8.3. Equation (8.79) is then used to modify the labels on the individual pixels to incorporate the effect of context. However, in so doing some (initially many) of the labels on the pixels will be modified. The process is then run again, and indeed as many times presumably until there are no further changes. This process is referred to as *iterated conditional modes*.[80]

## 8.24   Bibliography on Supervised Classification Techniques

The machine learning techniques that underpin thematic mapping in remote sensing are covered in many texts at different levels of mathematical complexity. Although it does not have a remote sensing focus, the standard first treatment for many researchers, which covers the basis of much of the material treated in this chapter, is

R.O. Duda, P.E. Hart and D.G. Stork, *Pattern Classification*, 2nd ed., John Wiley & Sons, N.Y., 2001.

A comprehensive, and slightly more mathematical, coverage containing good examples, but again without a remote sensing focus, is

C.M. Bishop, *Pattern Recognition and Machine Learning*, Springer Science + Business Media, N.Y., 2006.

Although its treatment of the more standard procedures is not as easily assimilated by those with a limited mathematical background, it is nevertheless an important work, especially for some of the more advanced topics covered in this chapter. A book at a comparable level of mathematical detail and which is a very good complement to those above is

T. Hastie, R. Tibshirani and J. Friedman, *The Elements of Statistical Learning*, Springer Science + Business Media, N.Y., 2009.

For a focus on the classification challenges of high dimensional (hyperspectral) image data see

D.A. Landgrebe, *Signal Theory Methods in Multispectral Remote Sensing*, John Wiley & Sons, Hoboken, N.J., 2003.

This book is written for the remote sensing practitioner, student and researcher and keeps its mathematical detail at the level needed for understanding algorithms and techniques. It is a very practical guide on how high dimensional image analysis should be performed.

---

[80] See J. Besag, On the statistical analysis of dirty pictures, *J. Royal Statistical Society B*, vol. 48, no. 3, 1986, pp. 259–302.

Although now largely of historical interest

N.J. Nilsson, *Learning Machines*, McGraw-Hill, N.Y., 1965

and its successor

N.J. Nilsson, *The Mathematical Foundations of Learning Machines*, Morgan Kaufmann, San Francisco, 1990.

are helpful to consult on the foundations of linear classifiers. Nilsson also has good sections on the maximum likelihood rule and the committee (layered) classifier concept.

One cannot escape the Hughes phenomenon when considering the classification of remotely sensed data, also called the curse of dimensionality in machine learning treatments. The original paper is

G. F. Hughes, On the mean accuracy of statistical pattern recognizers, *IEEE Transactions on Information Theory*, vol. IT-14, no.1, 1968, pp. 55–63,

while Landgrebe above also contains a good general discussion of the topic. Its influence on the support vector machine is examined in

M. Pal and G.F. Foody, Feature selection for classification of hyperspectral data for SVM, *IEEE Transactions on Geoscience and Remote Sensing*, vol. 48, no. 5, May 2010, pp. 2297–2307.

The development of kernel transformation methods, and particularly their application to remote sensing, has been rapid over the past two decades. The standard, non-remote sensing treatment, which includes coverage of support vector machines, is

B. Schölkopf and A. Smola, *Learning with Kernels: Support Vector Machines, Regularisation, Optimisation and Beyond*, MIT Press, Cambridge, 2002.

For a concentration on remote sensing topics see

G. Camps-Valls and L. Bruzonne, *Kernel Methods for Remote Sensing Data Analysis*, John Wiley & Sons, Chichester, U.K., 2009

A short review will be found in

L. Gómez-Chove, J Muñoz-Marí, V. Laparra, J. Malo-López and G. Camps-Valls, A review of kernel methods in remote sensing data analysis, in S. Prasad, L.M. Bruce and J. Chanussot, eds., *Optical Remote Sensing: Advances in Signal Processing and Exploitation Techniques*, Springer, Berlin, 2011.

The formative paper for support vector machines in remote sensing is

J.A. Gualtieri and R.F. Cromp, Support vector machines for hyperspectral remote sensing classification, *Proc. SPIE*, vol. 3584, 1998, pp. 221–232 while a good tutorial treatment is

C.T.C. Burges, A tutorial on support vector machines for pattern recognition, *Data Mining and Knowledge Discovery*, vol. 2, 1998, pp. 121–167.

Papers which demonstrate the application of kernel methods and support vector classifiers to remote sensing problems are

F. Melgani and L. Bruzzone, Classification of hyperspectral remote sensing images with support vector machines, *IEEE Transactions on Geoscience and Remote Sensing*, vol. 42, no. 8., August 2004, pp. 1778–1790,

G. Camps-Valls and L. Bruzzone, Kernel-based methods for hyperspectral image classification, *IEEE Transactions on Geoscience and Remote Sensing*, vol. 43, no. 6, June 2005, pp. 1351–1362.

A review of the suitability of support vector classifiers in land cover classification is

C. Huang, L.S. Davis and J.R.G. Townshend, An assessment of support vector machines for land cover classification, *Int. J. Remote Sensing*, vol. 23, no. 4, 2002, pp. 725–749,

while

G. Mountrakis, J. Im and C. Ogole, Support vector machines in remote sensing: a review, *ISPRS Journal of Photogrammetry and Remote Sensing*, vol. 66, 2011, pp. 247–259

is an extensive critical literature review, although with some omissions.

Neural networks are covered in many machine learning texts, including Duda et al., Bishop and Hastie et al., above. Other texts on neural networks which are very readable, are

Y.H. Pao, *Adaptive Pattern Recognition and Neural Networks*, Addison-Wesley, Reading, Mass., 1989

R.D. Reed and R.J. Marks, *Neural Smithing: Supervised Learning in Feedforward Artificial Neural Networks*, MIT Press, Cambridge Mass., 1999.

A good introduction will be found in

R.P. Lippman, An introduction to computing with neural nets, *IEEE ASSP Magazine*, April 1987, pp. 4–22.

Readable accounts of the performance of artificial neural networks in remote sensing image classification are found in

J.A. Benediktsson, P.H. Swain and O.K. Ersoy, Neural network approaches versus statistical methods in classification of multisource remote sensing data, *IEEE Transactions on Geoscience and Remote Sensing*, vol. 28, no. 4, July 1990, pp. 540–552,

J.D. Paola and R.A. Schowengerdt, A review and analysis of backpropagation neural networks for classification of remotely sensed multispectral imagery, *Int. J. Remote Sensing*, vol. 16, 1995, pp. 3033–3058, and

J.D. Paola and R.A. Schowengerdt, A detailed comparison of backpropagation neural network and maximum likelihood classifiers for urban land use classification, *IEEE Transactions on Geoscience and Remote Sensing*, vol. 33, no. 4, July 1995, pp. 981–996.

A good theoretical and background coverage of convolutional neural networks is

I. Goodfellow, Y. Bengio, and A Courville, *Deep Learning (Adaptive Computation and Machine Learning series)* The MIT Press, Cambridge Mass., 2016.

although the material relevant to the treatment here is distributed over several chapters. Bishop (above) has sections on regularisation, cost functions and other elemental aspects of CNN that are very readable, and a short section on CNNs overall.

The formative paper for CNNs is

Y. LeCun, Y. Bengio and G. Hinton, Deep learning, *Nature*, vol. 521, 28 May 2015, pp. 436–444

which is largely non-mathematical but very detailed, and with commentary on a range of non remote sensing applications, and

A.Krizhevsky, I. Sutskever and G. Hinton, ImageNet classification with deep convolutional neural networks, *Proc. Advances in Neural Information Processing Systems*, vol. 25, 2012, pp. 1090–1098.

Reviews of the application of deep learning techniques to remotely sensed data will be found in

L. Zhang, L. Zhang and B. Du, Deep learning for remote sensing data, *IEEE Geoscience and Remote Sensing Magazine*, June 2016, pp. 22–39, and

X. Zhu, D. Tuia, L. Mou, G-S Xia, L. Zhang, F. Xu and F. Fraundorfer, Deep learning in remote sensing, *IEEE Geoscience and Remote Sensing Magazine*, December 2017, pp. 8–35.

The use of CNNs for hyperspectral image classification will be found in many papers, including

H. Petersson, D. Gustafsson and D. Bergstrom, Hyperspetral image analysis using deep learning—a review, *Sixth Int. Conf. on Image Processing Theory, Tools and Applications*, Oulu Finland, 12–15 Dec 2016, https://doi.org/10.1109/IPTA.2016.7820963

S. Li, W. Song, L. Fang, Y. Chen, P. Ghamisi and J. Benediktsson, Deep learning for hyperspectral imaeg classification: an overview, *IEEE Transactions on Geoscience and Remote Sensing*, Vol. 57, No 9, September 2019, pp. 6690–6709

Y. Chen, Z. Lin, X. Zhao, G. Wang and Y. Gu, Deep learning-based classification of hyperspectral data, *IEEE Journal of Selected Topics in Applied Earth Observation and Remote Sensing*, vol. 7, June 2014, pp. 2094–2107,

Y. Li, H. Zhang and Q. Shen, Spectral-spatial classification of hyperspectral imagery with 3D convolutional neural network, *Remote Sensing*, vol. 9, 2017, 21 pages, https://doi.org/10.3390/rs9010067,

W. Hu, Y. Huang, L. Wei, F. Zhang, and H. Li, Deep convolutional neural networks for hyperspectral image classification, *Journal of Sensors*, Volume 2015, Article ID 258619,

J. Yang, Y-Q Zhao and J Chan, Learning and transferring deep joint spectral–spatial features for hyperspectral classification, *IEEE Transactions on Geoscience and Remote Sensing*, Vol. 55, No 8, August 2017, pp. 4729–4742.

The standard treatment on Recurrent Neural Networks is

A. Graves, *Supervised Sequence Learning with Recurrent Neural Networks*, Springer, Berlin, 2012.

Material on Markov Random Fields (MRF) can be found in Bishop above, but the fundamental papers, both of which are quite detailed, are

S. German and D. German, Stochastic relaxation, Gibbs distributions, and the Bayesian restoration of images, *IEEE Transactions on Pattern Analysis and Machine Intelligence*, vol. PAMI-6, no. 6, 1984, pp. 721–741, and

J. Besag, On the statistical analysis of dirty pictures, *J. Royal Statistical Society B*, vol. 48, no. 3, 1986, pp. 259–302.

One of the earliest applications of the MRF in remote sensing is in

B. Jeon and D.A. Landgrebe, Classification with spatio-temporal interpixel class dependency contexts, *IEEE Transactions on Geoscience and Remote Sensing*, vol. 30, 1992, pp. 663–672.

Other remote sensing applications that could be consulted are

A.H.S. Solberg, T. Taxt and A.K. Jain, A Markov Random Field model for classification of multisource satellite imagery, *IEEE Transactions on Geoscience and Remote Sensing*, vol. 34, no. 1, January 1996, pp. 100–113, and

Y. Jung and P.H. Swain, Bayesian contextual classification based on modelling M-estimates and Markov Random Fields, *IEEE Transactions on Geoscience and Remote Sensing*, vol. 34, no. 1, January 1996, pp. 67–75.

Development of the fundamental probabilistic label relaxation algorithm is given in

A. Rosenfeld, R. Hummel and S. Zuker, Scene labeling by relaxation algorithms, *IEEE Transactions on Systems, Man and Cybernetics*, vol. SMC-6, 1976, pp. 420–433.

## 8.25 Problems

8.1 Suppose you have been given the training data in Table 8.12 for three spectral classes, in which each pixel is characterised by only two spectral components $\lambda_1$ and $\lambda_2$. Develop the discriminant functions for a maximum likelihood classifier and use them to classify the patterns

**Table 8.12** Training data for three classes, each with two measurement dimensions (wavebands)

| Class 1 | | Class 2 | | Class 3 | |
|---|---|---|---|---|---|
| $\lambda_1$ | $\lambda_2$ | $\lambda_1$ | $\lambda_2$ | $\lambda_1$ | $\lambda_2$ |
| 16 | 13 | 8 | 8 | 19 | 6 |
| 18 | 13 | 9 | 7 | 19 | 3 |
| 20 | 13 | 6 | 7 | 17 | 8 |
| 11 | 12 | 8 | 6 | 17 | 1 |
| 17 | 12 | 5 | 5 | 16 | 4 |
| 8 | 11 | 7 | 5 | 14 | 5 |
| 14 | 11 | 4 | 4 | 13 | 8 |
| 10 | 10 | 6 | 3 | 13 | 1 |
| 4 | 9 | 4 | 2 | 11 | 6 |
| 7 | 9 | 3 | 2 | 11 | 3 |

$$\mathbf{x}_1 = \begin{bmatrix} 5 \\ 9 \end{bmatrix} \quad \mathbf{x}_2 = \begin{bmatrix} 9 \\ 8 \end{bmatrix} \quad \mathbf{x}_3 = \begin{bmatrix} 15 \\ 9 \end{bmatrix}$$

under the assumption of equal prior probabilities.

8.2 Repeat Problem 8.1 but with the prior probabilities

$$p(1) = 0.048$$
$$p(2) = 0.042$$
$$P(3) = 0.910$$

8.3 Using the data of Problem 8.1 develop the discriminant functions for a minimum distance classifier and use them to classify the patterns $\mathbf{x}_1$, $\mathbf{x}_2$ and $\mathbf{x}_3$.

8.4 Develop a parallelepiped classifier from the training data given in Problem 8.1 and compare its classifications with those of the maximum likelihood classifier for the patterns $\mathbf{x}_1$, $\mathbf{x}_2$ and $\mathbf{x}_3$ and the new pattern

$$\mathbf{x}_4 = \begin{bmatrix} 3 \\ 7 \end{bmatrix}$$

At the conclusion of the tests in Problems 8.1, 8.3 and 8.4, it would be worthwhile sketching a spectral space and locating in it the positions of the training data. Use this to form a subjective impression of the performance of each classifier in Problems 8.1, 8.3 and 8.4.

8.5 The training data in Table 8.13 represents a subset of that in problem 8.1 for just two of the classes. Develop discriminant functions for both the maximum likelihood and minimum distance classifiers and use them to classify the patterns

**Table 8.13** Training data for two classes, each with two measurement dimensions (wavebands)

| Class 1 | | Class 3 | |
|---|---|---|---|
| $\lambda_1$ | $\lambda_2$ | $\lambda_1$ | $\lambda_2$ |
| 11 | 12 | 17 | 8 |
| 10 | 10 | 16 | 4 |
| 14 | 11 | 14 | 5 |
| | | 13 | 1 |

$$\mathbf{x}_5 = \begin{bmatrix} 14 \\ 7 \end{bmatrix} \quad \mathbf{x}_6 = \begin{bmatrix} 20 \\ 13 \end{bmatrix}$$

Also classify these patterns using the minimum distance and maximum likelihood classifiers developed on the full training sets of problem 8.1 and compare the results.

8.6 Suppose a particular scene consists of just water and soil, and that a classification into these cover types is to be carried out on the basis of near infrared data using the maximum likelihood rule. When the thematic map is produced it is noticed that some water pixels are erroneously labelled as soil. How can that happen, and what steps could be taken to avoid it? It may help to sketch some typical one-dimensional normal distributions to represent the soil and water in infrared data, noting that soil would have a very large variance while that for water would be small. Remember the mathematical distribution functions extend to infinity.

8.7 Figure 8.48 shows 3 classes of data, that are linearly separable. Draw three linear separating surfaces on the diagram, each of which separates just two of the classes, in a maximum margin sense. Then demonstrate that the one-against-one multiclass process in Sect. 8.17 will generate the correct results. Extend the example to 4 and 5 classes.

8.8 Compare the properties of probabilistic relaxation and Markov Random Fields based on iterated conditional modes, as methods for context classification in remote sensing.

8.9 If you had to use a simple method for embedding spatial context would you choose pre-classification filtering of the data or post-classification filtering of the thematic map?

8.10 Compare the steps you would need to take to perform thematic mapping based on the maximum likelihood rule, an artificial neural network (ANN), a convolutional neural network (CNN) and a support vector machine with kernel.

8.11 Would kernels be of value in minimum distance classification?

8.12 (This is a bit complex) With probabilistic relaxation treated in Sect. 8.23.4 the results can improve with early iterations and then deteriorate. That is because processes such as simple averaging take over when relaxation nears what is called a fixed point, at which individual labels are strongly favoured for the pixels, but the iterations are then allowed to continue. See J.A. Richards, P.H. Swain and D.A. Landgrebe, On the accuracy of pixel

**Fig. 8.48** Three linearly separable classes

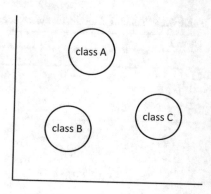

relaxation labelling, *IEEE Transactions on Systems, Man and Cybernetics*, vol. SMC-11, 1981, pp. 303–309. Would you expect the same to happen with the iterated conditional modes of the MRF approach?

8.13 In two dimensions the linear surface of (8.29) is written $w_1x_1 + w_2x_2 + w_3 = 0$. Show that its perpendicular distance to the origin is $w_3/\sqrt{w_1^2 + w_2^2}$. Can you generalise that expression to any number of dimensions?

8.14 Explain from the point of view of spectral measurements why the covariance matrix used with the maximum likelihood classifier is a symmetric matrix.

8.15 Without using mathematics describe how the back propagation method is used to train a multi-layer Perceptron neural network.

8.16 What mechanisms are used in the convolutional neural network (CNN) to reduce the numbers of connections between layers and thus the numbers of unknowns that have to be found during training?

8.17 Discuss the role of the CNN as a feature selector when its output is fed into a fully connected neural network (such as the multilayer Perceptron).

8.18 Why is the ReLU activation function in a convolutional neural network used instead of the sigmoid and other functions used in the multi-layer Perceptron? Is it because:

    (a) It generates more accurate results?
    (b) It allows faster convergence during training?
    (c) It improves the multi-class capabilities of the algorithm?

8.19 What is the reason for including a pooling layer in a convolutional neural network?

    (a) Does it further reduce the number of connections and imbeds spatial context?
    (b) Do convolutional layers require a pooling layer as their inter-connection?
    (c) Does it remove the need for an activation function?

8.25   Problems

8.20   Why is the softmax function sometimes used with convolutional neural networks?

8.21   Does the stride operation in a convolutional neural network

   (a) Have any effect of the spatial properties to be learnt?
   (b) Allow the number of connections and thus unknowns to be reduced?

8.22   Compared with a fully connected neural network, can convolutional neural nets

   (a) Have more hidden layers because of the fewer connections between nodes?
   (b) Always perform better when spatial context is a consideration?
   (c) Have fewer output nodes?

8.23   What is the purpose of flattening in a convolutional neural network?

8.24   In its fundamental operation (based on its underlying equations) is the neural network most like

   (a) The maximum likelihood classifier?
   (b) The support vector machine?
   (c) The minimum distance classifier?

8.25   Is the role of the kernel in the support vector machine to

   (a) Allow the SVM to handle binary data sets that are not linearly separable?
   (b) Minimise errors caused by classes which overlap the linear decision surface?
   (c) Make it into a multi-class machine?

8.26   Is the fundamental support vector machine algorithm

   (a) A multi-class classifier?
   (b) A binary linear classifier?
   (c) A binary non-linear classifier?

# Chapter 9
# Clustering and Unsupervised Classification

**Abstract** The analytical process of clustering is introduced as a means for discovering the structure of remote sensing image data in the spectral domain. This is done by placing pixel vectors into distinct spectral classes or clusters. Most of the techniques covered are developed by example and include iterative and streaming methods. Unsupervised classification, which is employed when training data is not available, is based on clustering. Application of unsupervised methods is discussed, including how clusters are associated with ground cover classes after clustering is complete. Special attention is paid to procedures that can be applied to very large data sets.

## 9.1 How Clustering is Used

The classification techniques treated in Chap. 8 all require the availability of labelled training data with which the parameters of the respective class models are estimated. As a result, they are called supervised techniques because, in a sense, the analyst supervises an algorithm's learning about those parameters. Sometimes labelled training data is not available and yet it would still be of interest to convert remote sensing image data into a thematic map of labels. Such an approach is called *unsupervised classification* since the analyst, in principle, takes no part in an algorithm's learning process. Several methods are available for unsupervised learning. Perhaps the most common in remote sensing are based on the use of clustering algorithms, which seek to identify pixels in an image that are spectrally similar. That is one of the applications of clustering treated in this chapter.

Clustering techniques find other applications in remote sensing, particularly for resolving sets of Gaussian modes (single multivariate normal distributions) in image data before the Gaussian maximum likelihood classifier can be used successfully. This is necessary since each class has to be modelled by a single normal probability distribution, as discussed in Chap. 8. If a class happens to be multimodal, and that is not resolved, then the modelling will not be very effective. Users of remotely sensed data can only specify the information classes. Occasionally it

© The Author(s), under exclusive license to Springer Nature Switzerland AG 2022
J. A. Richards, *Remote Sensing Digital Image Analysis*,
https://doi.org/10.1007/978-3-030-82327-6_9

might be possible to guess the number of spectral classes in a given information class but, in general, the user would have little idea of the number of distinct unimodal groups into which the data falls in spectral space. Gaussian mixture modelling can be used for that purpose (Sect. 8.4) but the complexity of estimating simultaneously the number of Gaussian components, and their parameters, can make that approach difficult to use. Clustering procedures are practical alternatives and have been used in many fields to enable inherent data structures to be determined.

There are many clustering methods with many variations within each method.[1] In this chapter only those commonly employed with remote sensing data are treated, including those which can be used with the increasing availability of very large data sets.

## 9.2   Similarity Metrics and Clustering Criteria

In clustering we try to identify groups of pixels that are somehow similar to each other. The only real attributes that we can use to check similarity are the spectral measurements recorded by the sensor used to acquire the data.[2] Here, therefore, clustering will imply a grouping of pixels in the spectral domain. Pixels belonging to a particular cluster will be spectrally similar. In order to quantify their spectral proximity, it is necessary to devise a measure of similarity. Many similarity measures, or metrics, have been proposed but those used commonly in clustering procedures are usually simple distance measures in spectral space. The most frequently encountered are Euclidean distance($L_2$) and the *city block* or *Manhattan* ($L_1$) distance. If $\mathbf{x}_1$ and $\mathbf{x}_2$ are the measurement vectors of two pixels whose similarity is to be checked then the Euclidean distance between them is

$$
\begin{aligned}
d(\mathbf{x}_1, \mathbf{x}_2) &\triangleq \|\mathbf{x}_1 - \mathbf{x}_2\| \\
&= \left\{ (\mathbf{x}_1 - \mathbf{x}_2) \cdot (\mathbf{x}_1 - \mathbf{x}_2) \right\}^{1/2} \\
&= \left\{ (\mathbf{x}_1 - \mathbf{x}_2)^{\mathrm{T}} (\mathbf{x}_1 - \mathbf{x}_2) \right\}^{1/2} \\
&= \left\{ \sum_{n=1}^{N} (x_{1n} - x_{2n})^2 \right\}^{1/2}
\end{aligned}
\tag{9.1}
$$

where $N$ is the number of spectral components.

---

[1] See C.C. Aggarwal and C. K. Reddy, *Data Clustering Algorithms and Applications*, CRC Press, Roca Baton 2014.

[2] We might also decide that spatially neighbouring pixels are likely to belong to the same group; some clustering algorithms use combined spectral and spatial similarity.

The city block ($L_1$) distance between the pixels is just the accumulated difference along each spectral dimension, similar to walking between two locations in a city laid out on a rectangular street grid. It is given by

$$d_{L_1}(\mathbf{x}_1, \mathbf{x}_2) = \sum_{n=1}^{N} |x_{1n} - x_{2n}| \qquad (9.2)$$

Clearly the latter is faster to compute but questions must be raised as to how spectrally similar all the pixels within a given $L_1$ distance of each other will be.

The Euclidean and city block distance measures are two special cases of the Minkowski $L_p$ distance metric[3]

$$d_{L_p}(\mathbf{x}_1, \mathbf{x}_2) = \left\{ \sum_{n=1}^{N} |x_{1n} - x_{2n}|^p \right\}^{1/p}$$

When p = 1 we have the city block distance, while when p = 2 we have Euclidean distance.

By using a distance measure, it should be possible to determine clusters in data. Often there can be several acceptable clusters assignments, as illustrated in Fig. 9.1, so that once a candidate clustering has been identified it is desirable to have a means by which the "quality" of that clustering can be measured. The availability of such a measure should allow one cluster assignment of the data to be chosen over all others.

A common clustering criterion, or quality indicator, is the sum of squared error (SSE) measure; when based on Euclidean distance it is defined as

$$\text{SSE} = \sum_{C_i} \sum_{\mathbf{x} \in C_i} \|\mathbf{x} - \mathbf{m}_i\|^2 = \sum_{C_i} \sum_{\mathbf{x} \in C_i} (\mathbf{x} - \mathbf{m}_i)^{\mathrm{T}} (\mathbf{x} - \mathbf{m}_i) \qquad (9.3)$$

in which $\mathbf{m}_i$ is the mean vector of the $i$th cluster and $\mathbf{x} \in C_i$ is a pixel assigned to that cluster. The inner sum is over all pixels in a cluster and the outer sum is taken over all clusters. SSE will be small for tightly grouped clusters and large otherwise, thereby allowing an assessment of the quality of clustering.

Note that SSE has a theoretical minimum of zero, which corresponds to all clusters containing a single data point. As a result, if an iterative method is used to seek the natural clusters or spectral classes in a set of data then it has a guaranteed termination point, at least in principle. In practice it may be too expensive to allow natural termination. Instead, iterative procedures are often stopped when an acceptable degree of clustering has been achieved.

[3] See J. Lattin, J. Douglas Carroll and P. E. Green, *Analyzing Multivariate Data*, Thompson, Brooks/Cole, Canada, 2003.

**Fig. 9.1** Two apparently
acceptable cluster
assignments of a set of two
dimensional data

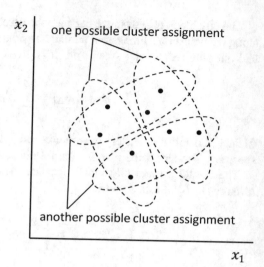

It is possible now to consider the implementation of an actual clustering algorithm. While it should depend on finding the clustering that minimises SSE that is impracticable since it requires the calculation of an enormous number of values of SSE to evaluate all candidate clusterings. For example, there are approximately $C^K/K!$ ways of placing $K$ pixel vectors into $C$ clusters[4]; that will be an enormous number for any practical image size. Rather than embark on such a rigorous and computationally expensive approach the heuristic procedure of the following section is usually adopted in remote sensing practice unless the number of data points is huge.

## 9.3   *k* Means Clustering

The $k$ means clustering method, also called *migrating means* and *iterative optimisation*, is one of the most common approaches used in image analysis applications. With certain refinements it becomes the *Isodata* technique treated in the next section. It also underpins the K Trees clustering technique for large data sets in Sect. 9.13.1.

The $k$ means approach requires an initial assignment of the available measurement vectors into a user-specified number of clusters. That is done using an arbitrarily specified set of initial cluster centres or means; a very crude set of clusters is generated by assigning the pixels to those arbitrary means. After that first assignment the means are recomputed. The pixel vectors are then reassigned to the cluster

---

[4] See R.O. Duda, P.E. Hart and D.G. Stork, *Pattern Classification*, 2nd ed., John Wiley & Sons, N.Y., 2001.

with the closest mean, the means are again recomputed, and the pixels again reassigned. The process is repeated as many times as necessary such that there is no further movement of pixels between clusters. In practice, with large data sets, the process is not run to completion and some other stopping rule is used, as discussed in the following. However, the SSE measure progressively reduces with iteration as we will see by example.

## 9.3.1  The k *Means Algorithm*

The *k* means or iterative optimisation algorithm is implemented in the following steps.

1. Select a value for $C$, the number of clusters into which the pixels are to be grouped.[5] This requires some feel beforehand as to the number of clusters that might naturally represent the image data set. Depending on the reason for using clustering some guidelines are available (see Sect. 11.4.2).
2. Initialise cluster generation by selecting $C$ points in spectral space to serve as candidate cluster centres. Call these

$$\hat{\mathbf{m}}_c \quad c = 1 \ldots C$$

In principle the choice of the $\hat{\mathbf{m}}_c$ at this stage is arbitrary with the exception that no two can be the same. To avoid anomalous cluster generation with unusual data sets it is generally best to space the initial cluster centres uniformly over the data (see Sect. 9.5). That can also aid convergence.
3. Assign each pixel vector $\mathbf{x}$ to the candidate cluster of the nearest mean using an appropriate distance metric in the spectral domain between the pixel and the cluster means. Euclidean distance is commonly used. That generates a cluster of pixel vectors about each candidate cluster mean.
4. Compute a new set of cluster means from the groups formed in Step 3; call these

$$\mathbf{m}_c \quad c = 1 \ldots C$$

5. If $\mathbf{m}_c = \hat{\mathbf{m}}_c$ for all $c$ then the procedure is complete. Otherwise, the $\hat{\mathbf{m}}_c$ are set to the current values of $\mathbf{m}_c$ and the procedure returns to step 3.

This process is illustrated with a simple two-dimensional data set in Fig. 9.2.

---

[5] By its name the *k* means algorithm actually searches for *k* clusters; here however we use $C$ for the total number of clusters but retain the common name by which the method is known.

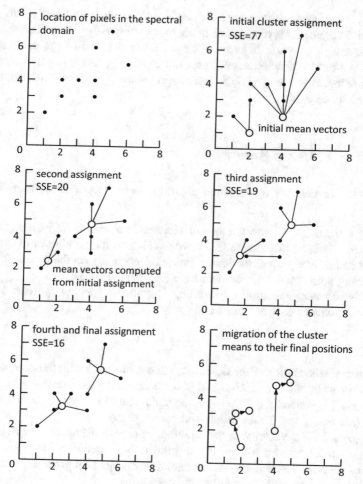

**Fig. 9.2** Illustration of clustering with the *k* means, or iterative optimisation, algorithm, showing a progressive reduction in SSE; also shown is how the cluster means migrate during the process

## 9.4   Isodata Clustering

The Isodata clustering algorithm[6] builds on the *k* means approach by introducing a number of checks on the clusters formed, either during or at the end of the iterative assignment process. Those checks relate to the number of pixels assigned to clusters and their shapes in the spectral domain.

---

[6] G.H. Ball and D.J. Hall, A novel method of data analysis and pattern recognition, *Stanford Research Institute*, Menlo Park, California, 1965.

## 9.4.1   Merging and Deleting Clusters

At any suitable stage clusters can be examined to see whether:

(a) any contain so few points as to be meaningless; for example, if the statistical distributions of pixels within clusters are important, as they might be when clustering is used as a pre-processing operation for maximum likelihood classification (see Sect. 11.4.1), sufficient pixels per cluster must be available to generate reliable estimates of the class means and covariance matrices;
(b) any are so close together that they represent an unnecessary or inappropriate division of the data, in which case they should be merged.

In view of the material of Sect. 8.3.6 a guideline exists for (a). A cluster cannot reliably be modelled by a multivariate normal distribution unless it contains about $10N$ members, where $N$ is the number of spectral components.

Decisions in (b) about when to merge adjacent clusters can be made by assessing how similar they are spectrally. Similarity can be assessed simply by the distance between them in the spectral domain, although more sophisticated similarity measures are available (see Chap. 10).

## 9.4.2   Splitting Elongated Clusters

Another test sometimes incorporated in the Isodata algorithm concerns the shapes of clusters in spectral space. Clusters that are elongated can be split in two, if required. Such a decision can be made on the basis of pre-specifying a standard deviation in each spectral band beyond which a cluster should be halved.

## 9.5   Choosing the Initial Cluster Centres

Initialising the $k$ means and Isodata procedures requires specification of the number of clusters and their initial mean positions. In practice the actual or optimum number of clusters to choose will not be known. Therefore, it is often chosen conservatively high, having in mind that any spectrally similar clusters that result can be consolidated after the process is completed, or at intervening iterations, if a merging option is available.

The choice of the initial locations of the cluster centres is not critical, although it can influence the time it takes to reach a final, acceptable clustering. In some extreme cases it might influence the final set of clusters found. Several procedures are in common practice. In one, the initial cluster centres are chosen uniformly spaced along the multidimensional diagonal of the spectral space. That is a line from the origin to the point corresponding to the maximum brightness value in each spectral component. The choice can be refined if the user has some idea of the actual range of brightness values in each spectral component, say by having previously computed histograms. The cluster centres would then be initialised along a diagonal through the actual multidimensional extremities of the data. An alternative, implemented as an option in the MultiSpec package,[7] is to distribute the initial centres uniformly along the first eigenvector (principal component) of the data. Since most data exhibits a high degree of correlation, the eigenvector approach is essentially a refined version of the first method.

The choice of the initial cluster locations using these methods is reasonable and effective since they are then spread over a region of the spectral space in which many classes occur, particularly for correlated data such as that for soils, rocks, concretes, etc. A range of other approaches is available.[8]

## 9.6  Cost of $k$ Means and Isodata Clustering

The need to check every pixel against all cluster centres at each iteration means that the basic $k$ means algorithm can be time consuming to operate, particularly for large data sets. For $C$ clusters and $P$ pixels, $P \times C$ distances have to be computed at each iteration, and the smallest found. For $N$ band data, each Euclidean distance calculation will require $N$ multiplications and $N$ additions, ignoring the square root operation, since that need not be carried out. Thus for 20 classes and 10,000 pixels, 100 iterations of $k$ means clustering requires 20 million multiplications per band of data. When the number of bands is high, as in hyperspectral imagery, $k$ means clustering is very computationally intensive.

## 9.7  Unsupervised Classification

At the completion of clustering, pixels belonging to each cluster are usually given a symbol or colour to indicate that they belong to the same group or spectral class. Based on those symbols, a cluster map can be produced; that is a map corresponding to the image which has been clustered, but in which the pixels are represented by their cluster symbol rather than by the original measurement vector. Sometimes only part of an image is used to generate the clusters, but all pixels in the image can be allocated to one of the clusters through, say, a minimum distance assignment of pixels to clusters.

The availability of a cluster map allows a classification to be made. If some pixels with a given label can be identified with a particular ground cover type

---

[7] See www.engineering.purdue.edu/~biehl/MultiSpec/ accessed 2021.
[8] See Aggarwal and Reddy, loc. cit., Sect. 4.3.2.1.

(by means of maps, site visits or other forms of reference data) then all pixels with the same label can be assumed to be from that same class. Cluster identification is often aided by the spatial patterns evident; elongated features, such as roads and rivers, are usually easily recognisable. This method of image classification, depending as it does on *a posteriori*[9] recognition of the classes, is as noted earlier, called unsupervised classification since the analyst plays no part in class definition until the computational aspects are complete.

Unsupervised classification can be used as a stand-alone technique, particularly when reliable training data for supervised classification cannot be obtained or is too expensive to acquire. However, it is also of value, as noted earlier, to determine the spectral classes that should be considered in a subsequent supervised approach. This is pursued in detail in Chap. 11. It also forms the basis of the cluster space method for handling high dimensional data sets, treated in Sect. 9.14.

## 9.8   An Example of Clustering with the *k* Means Algorithm

To illustrate the nature of the results produced by the *k* means algorithm consider the segment of HyMap imagery in Fig. 9.3a, which shows a highway interchange near the city of Perth in Western Australia. It was recorded in January 2010 and consists of vegetation, roadway pavements, water and bare and semi-bare areas. Figure 9.3b shows a scatter diagram for the image in which a near infrared channel (29) is plotted against a visible red channel (15). This is a subspace of the five channels used for the clustering, as summarised in Table 9.1.

The data was clustered using the *k* means (Isodata) procedure available in MultiSpec. The algorithm was asked to determine six clusters, since a visual inspection of the image showed that to be reasonable. No merging and splitting options were employed, but any clusters with fewer than 125 pixels at the end of the process were eliminated. The results shown in Fig. 9.3c were generated after 8 iterations. The cluster means are plotted in just two dimensions in Fig. 9.3d, while Table 9.2 shows the full five-dimensional means which, as a pattern, exhibit the spectral reflectance characteristics of the class names assigned to the clusters. The class labels were able to be found in this exercise both because of the spatial distributions of the clusters and the spectral dependences seen in Table 9.2.

It is important to realise that the results generated in this example are not unique but depend on the clustering parameters chosen, and the starting number of clusters.

In practice the user may need to run the algorithm a number of times to generate a segmentation that matches the needs of a particular analysis. Also, in this simple case, each cluster is associated with a single information class; that is usually not the case in more complex situations.

---

[9] That is, *after the event*.

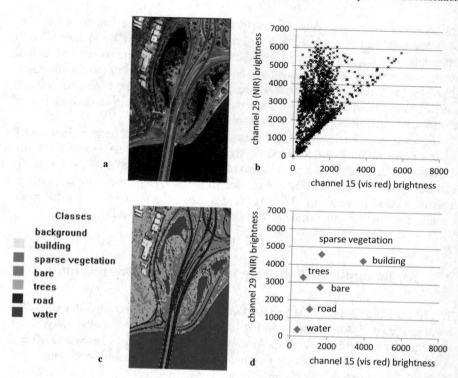

**Fig. 9.3 a** Segment of a HyMap image of Perth, Western Australia; **b** scatterplot of the image in a near infrared–visible red subspace; **c** $k$ means clustering result, searching for 6 clusters using the channels specified in Table 9.1; **d** cluster means in the near infrared–visible red subspace

**Table 9.1** HyMap channels used in the $k$ means clustering example

| Channel | Band centre (nm) | Band width (nm) |
|---|---|---|
| 7 (visible green) | 511.3 | 17.6 |
| 15 (visible red) | 634.0 | 16.4 |
| 29 (near infrared) | 846.7 | 16.3 |
| 80 (middle infrared) | 1616.9 | 14.8 |
| 108 (middle infrared) | 2152.7 | 30.2 |

## 9.9  A Single Pass Clustering Technique

Alternatives to the $k$ means and Isodata algorithms have been proposed and are widely implemented in software packages for remote sensing image analysis. One, which requires only a single pass through the data, is described in the following.

**Table 9.2**  Cluster centres for the $k$ means (Isodata) exercise in Fig. 9.3

| Cluster | Label | Cluster mean vectors (on 16 bit scale) | | | | |
|---|---|---|---|---|---|---|
| | | Channel 7 | Channel 15 | Channel 29 | Channel 80 | Channel 108 |
| 1 | Building | 3511.9 | 3855.7 | 4243.7 | 4944.2 | 4931.6 |
| 2 | Sparse veg | 1509.6 | 1609.3 | 4579.5 | 3641.7 | 2267.0 |
| 3 | Bare | 1333.9 | 1570.7 | 2734.3 | 2715.1 | 2058.7 |
| 4 | Trees | 725.6 | 650.6 | 3282.4 | 1676.2 | 866.6 |
| 5 | Road | 952.3 | 1037.1 | 1503.7 | 1438 5 | 1202.3 |
| 6 | Water | 479.2 | 391.1 | 354.8 | 231.0 | 171.6 |

## 9.9.1   The Single Pass Algorithm

The single pass process was designed originally to be a faster alternative to iterative procedures when the image data was only available in sequential format, such as on a magnetic tape. Nevertheless, the method is still used in some remote sensing image applications, particularly when the data sets are too large to accommodate iteration. Single pass techniques are also called *streaming methods* and have the added attraction that clustering can be refined as further samples become available.

A randomly selected set of samples is often chosen to generate the clusters, rather than using the full image segment of interest. The samples are arranged into a two-dimensional array. The first row of samples is employed to obtain a starting set of cluster centres. This is initiated by adopting the first sample as the centre of the first cluster. If the second sample in the first row is further away from the first sample than a user-specified critical spectral distance, then it is used to form another cluster centre. Otherwise, the two samples are said to belong to the same cluster and their mean is computed as the new cluster centre. This procedure, which is illustrated in Fig. 9.4, is applied to all samples in the first row. Once that row has been exhausted the multidimensional standard deviations of the clusters are computed.

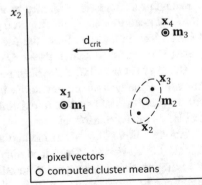

**Fig. 9.4**  Generation of cluster centres using the first row of samples

**Fig. 9.5** If later samples lie within a set number of standard deviations (dotted circles) they are included in existing clusters, otherwise they start new clusters

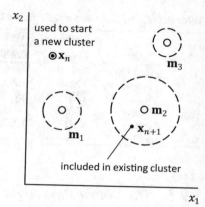

Each sample in the second and subsequent rows is checked to see to which cluster it is closest. It is assigned to that cluster if it lies within a user-prescribed number of standard deviations; the cluster statistics are then recomputed. Otherwise, that sample is used to form a new cluster centre (which is assigned a nominal standard deviation), as shown in Fig. 9.5. In that manner all the samples are clustered, and clusters with fewer than a prescribed number of pixels are deleted. Should a cluster map be required then the original segment of image data is scanned pixel by pixel and each pixel labelled according to the cluster to which it is closest, on the basis usually of Euclidean distance. Should it be an outlying pixel, in terms of the available cluster centres, it is not labelled.

### 9.9.2  Advantages and Limitations of the Single Pass Algorithm

Apart from speed, a major advantage of this approach over the iterative Isodata and $k$ means procedures is its ability to create cluster centres as it proceeds. The user does not need to specify the required number of clusters beforehand. However, the method has limitations. First, the user has to have a feel for the necessary parameters. The critical distance parameter needs to be specified carefully to enable a satisfactory set of initial cluster centres to be established. In addition, the user has to know how many standard deviations to use when assigning pixels in the second and subsequent lines of samples to existing clusters. With experience, those parameters can be estimated reasonably.

Another limitation is the method's dependence on the first line of samples to initiate the clustering. Since it is only a one pass algorithm, and has no feedback checking mechanism by way of iteration, the final set of cluster centres can depend significantly on the character of the first line of samples.

### 9.9.3   *Strip Generation Parameter*

Adjacent pixels along a line of image data frequently belong to the same cluster, particularly for images of cultivated regions. A method for enhancing the speed of clustering is to compare a pixel with its predecessor and immediately assign it to the same cluster if it is similar. The similarity measure used can be straightforward, consisting of a check of the brightness difference in each spectral band. The difference allowable for two pixels to be part of the same cluster is called the strip generation parameter.

### 9.9.4   *Variations on the Single Pass Algorithm*

The single pass technique has a number of variations. For example, the initial cluster centres can be specified by the analyst as an alternative to using the critical distance parameter in Fig. 9.4. Also, rather than use a multiplier of standard deviation for assigning pixels from the second and subsequent rows of samples, some algorithms proceed exactly as for the first row, without employing standard deviation. Some algorithms use the $L_1$ metric of (9.2), rather than Euclidean distance, and some check inter-cluster distances and merge if desired; periodically small clusters can also be eliminated. MultiSpec uses critical distance parameters over the full range, although the user can specify a different critical distance for the second and later rows of samples.

### 9.9.5   *An Example of Clustering with the Single Pass Algorithm*

The single pass option available in MultiSpec was applied to the data set of Fig. 9.3. The critical distances for the first and subsequent rows were chosen as 2500 and 2800 respectively. Those numbers are so large because the data we are dealing with is 16 bit (on a scale of 0 to 65,535) and there are five bands involved. The image was not sampled prior to clustering; all pixels were used. The results are shown in Fig. 9.6 and Table 9.3.

Several points are important to note. First, the image and map as displayed were rotated 90 degrees clockwise after clustering to bring them to a north–south orientation from the east–west flight line recorded by the HyMap instrument for this mission. (The same was the case for the data of Fig. 9.3). Therefore, the line of pixels used to generate the original set of cluster centres is that down the right-hand side of the image. Secondly, the clusters are different from those in Fig. 9.3, and a

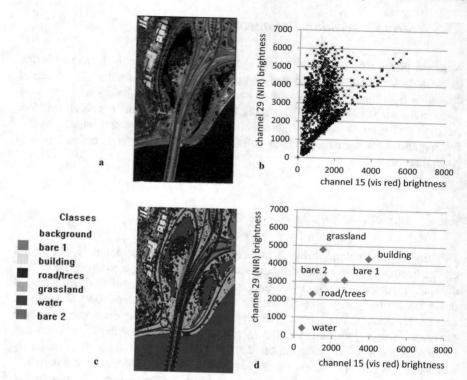

**Fig. 9.6** **a** Segment of a HyMap image of Perth, Western Australia; **b** scatterplot of the image in a near infrared–visible red subspace; **c** Single pass clustering result using the channels specified in Table 9.1; **d** cluster means in the near infrared–visible red subspace

**Table 9.3** Cluster centres for the single pass exercise in Fig. 9.6

| Cluster | Label | Cluster mean vectors (on 16 bit scale) | | | | |
|---|---|---|---|---|---|---|
| | | Channel 7 | Channel 15 | Channel 29 | Channel 80 | Channel 108 |
| 1 | Bare 1 | 2309.7 | 2632.9 | 3106.1 | 3713.2 | 3663.4 |
| 2 | Building | 3585.8 | 3901.5 | 4300.5 | 4880.7 | 4870.2 |
| 3 | Road/trees | 900.4 | 940.0 | 2307.6 | 1640.2 | 1143.4 |
| 4 | Grassland | 1441.3 | 1447.2 | 4798.6 | 3455.6 | 2028.6 |
| 5 | Water | 490.4 | 408.2 | 409.0 | 274.9 | 207.5 |
| 6 | Bare 2 | 1372.7 | 1630.5 | 3105.7 | 3033.3 | 2214.8 |

slightly different class naming has been adopted. Again, it was possible to assign information class labels to the clusters because of the mean vector behaviour seen in Table 9.3, and the spatial distribution of the clusters. In this case, compared with Fig. 9.3, there are two spectral classes called "bare."

## 9.10 Hierarchical Clustering

Another approach that does not require the user to specify the number of classes beforehand is hierarchical clustering. This method produces an output that allows the user to decide on the set of natural groupings into which the data falls. There are two types of hierarchical process. The first commences by assuming that all the pixels individually are distinct clusters; it then systematically merges neighbouring clusters by checking distances between means. That is continued until all pixels have been grouped into one single, large cluster. The approach is known as *agglomerative hierarchical clustering*. The second method, called *divisive hierarchical clustering*, starts by assuming all the pixels belong to one large, single cluster, which is progressively subdivided until all pixels form individual clusters. This is a computationally more expensive approach than the agglomerative method and is not considered further here.

### 9.10.1 Agglomerative Hierarchical Clustering

In the agglomerative approach a history of the mergings, or fusions, is displayed on a *dendrogram*. That is a diagram that shows at what distances between centres particular clusters are merged. An example of hierarchical clustering, along with its fusion dendrogram, is shown in Fig. 9.7. It uses the same two-dimensional data set as in Fig. 9.2 but note that the ultimate cluster compositions are slightly different. This demonstrates again that different algorithms can and do produce different cluster results.

The fusion dendrogram of a particular hierarchical clustering exercise can be inspected to determine the intrinsic number of clusters or spectral classes in the data. Long vertical sections between fusions in the dendrogram indicate regions of "stability" which reflect natural data groupings. In Fig. 9.7 the longest stretch on the distance scale between fusions corresponds to two clusters in the data. One could conclude therefore that this data falls most naturally into two groups. In the example presented, similarity between clusters was judged on the basis of Euclidean distance. Other similarity measures are sometimes used, as noted below.

## 9.11 Other Clustering Metrics

Clustering metrics other than simple distance measures exist. One derives a *within cluster scatter measure* by computing the average covariance matrix over all the clusters, and a *between cluster scatter measure* by looking at how the means of the clusters scatter about the global mean of the data. Those two measures are

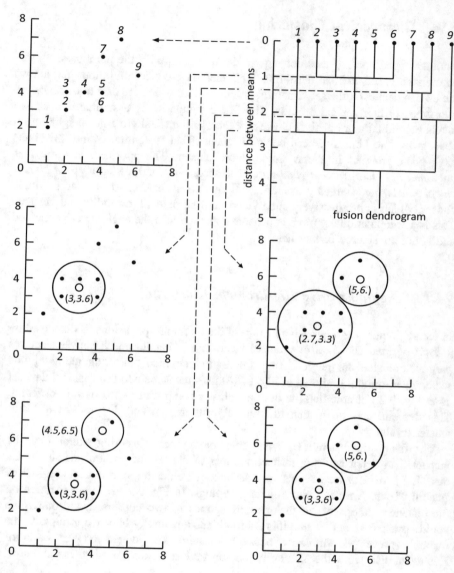

**Fig. 9.7** Agglomerative hierarchical clustering of the data in Fig. 9.2

combined into a single figure of merit[10] based on minimising the within cluster scatter while attempting to maximise the among cluster measure. It can be shown that figures of merit such as those are similar to the sum of squared error criterion.

---

[10] See Duda, Hart and Stork, loc. cit., or G.B. Coleman and H.C. Andrews, Image segmentation by clustering, *Proc. IEEE*, vol. 67, no. 5, 1979, pp. 773–785.

Similarity metrics can incorporate measures other than spectral likeness. Spatial proximity might be important in some applications as might properties that account for categorical information. For example, clustering crop pixels might be guided by all of spectral measurements, soil type and spatial contiguity.

## 9.12   Some Alternative Clustering Techniques

### 9.12.1   Histogram Peak Selection

From time-to-time some other clustering algorithms have been applied to remote sensing image data, although with the increasing spectral dimensionality of imagery some have fallen into disuse. If the dimensionality is small—say 3 or 4 data channels—and the radiometric resolution is limited, clustering by histogram peak selection is viable.[11] That is the multidimensional form of histogram thresholding often used to segment scenes in picture processing.[12]

### 9.12.2   Mountain Clustering

Not unlike histogram peak selection is the technique of mountain clustering. It seeks to define cluster centres as local density maxima in the spectral domain. In its original form the spectral space was overlaid with a grid; the grid intersections were then chosen as sites for evaluating density.[13] More recently, the density maxima have been evaluated at each pixel site rather than at overlaid grid positions.[14] The method, which could be used as a clustering technique in its own right, or as a process to initialise cluster centres for algorithms such as Isodata, sets up a function, called a *mountain function*, that measures the local density about each pixel. A typical mountain function for indicating density in the vicinity of pixel $\mathbf{x}_i$ could be

$$m(\mathbf{x}_i) = \sum_j \exp\left[-\beta d\left(\mathbf{x}_j, \mathbf{x}_i\right)^2\right]$$

---

[11] See P.A. Letts, Unsupervised classification in the Aries image analysis system, *Proc. 5th Canadian Symposium on Remote Sensing*, 1978, pp. 61–71, or J.A. Richards and X. Jia, *Remote Sensing Digital Image Analysis*, 4th ed., Springer, Berlin, 2006, Sect. 9.8.

[12] See K.R. Castleman, *Digital Image Processing*, 2nd ed., Prentice-Hall, N.J., 1996.

[13] R.R. Yager and D.P. Filev, Approximate clustering via the mountain method, *IEEE Transactions on Systems, Man Cybernetics*, vol. 24, 1994, pp. 1279–1284.

[14] S.L. Chiu, Fuzzy model identification based on cluster estimation, *J Intelligent Fuzzy Systems*, vol. 2, 1994, pp. 267–278, and M-S. Yang and K-L Wu, A modified mountain clustering algorithm, *Pattern Analysis Applications*, vol. 8, 2005, pp. 125–138.

in which $d(\mathbf{x}_j, \mathbf{x}_i)$ is the distance from that pixel to another pixel $\mathbf{x}_j$, and $\beta$ is a constant that effectively controls the region of neighbours. Once the largest $m(\mathbf{x}_i)$ has been found, that density maximum is removed or de-emphasised and the next highest density is found, and so on.

### 9.12.3  k Medians Clustering

In Sect. 5.3.2 median filtering of an image was seen as an alternative to mean value smoothing in cases where there were outlying or atypical pixel brightness values. If a data set to be clustered is suspected of having outlying samples then some authors, for the same reason, adopt the use of the cluster medians in the $k$ means algorithm of Sect. 9.3 instead of the cluster means. Then called $k$ medians clustering it has the same algorithm of Sect. 9.3.1 but based on medians.

This approach requires the definition of a multidimensional median. As noted in Sect. 5.3.2 the (one dimensional) median of a set of numbers is that member of the set which sits in the middle. In effect, it is more similar to all the other members of the set than any other member.

When we come to more than one dimension, there is no single definition of a median in common use. One approach is to find the point in the multiple dimensional space which is least different from every point in the set. That requires using a measure, such as the sum of squared error criterion of (9.3), or even the sum of Manhattan distances from a point to all members of the set, and then finding the point at which that measure is a minimum. This can lead to a median position (point) which is not one of the set members, which is acceptable under this definition. If we wish to restrict attention just to set members then that leads to the concept of the *medoid*, in the next section.

Another method for computing a multidimensional median is to find the actual single dimensional medians in each of the spectral axes, from which the multidimensional median is set by using those *marginal medians* as its spectral components.

### 9.12.4  k Medoids Clustering

More common than $k$ medians clustering is $k$ medoids. The *medoid* is that member of a multidimensional set, which is most typical of the set. It more closely resembles the concept of a median than the definitions given in the previous section. Again, $k$ medoids clustering can be advantageous when the data set has outlying or atypical members.

There are several variations of the $k$ medoids approach, but most operate along the lines of the following algorithm (Fig. 9.8).

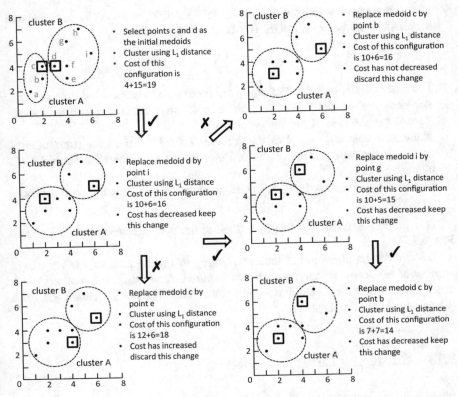

**Fig. 9.8** An example of the application of $k$ medoids clustering using the data set of Fig. 9.2. Interestingly, the same clusters are generated. The arrows shown as ✓ and ✗ indicate acceptable and unacceptable reallocation of the medoids. The two separate numbers shown in the cost sum are for cluster A and cluster B respectively; there is no further improvement beyond the bottom right hand cluster assignment

1. Select a value for $C$, the number of clusters into which the pixels are to be grouped.
2. Initialise cluster generation by selecting $C$ points in spectral space to serve as candidate medoids. Call these

$$\hat{\mathbf{m}}_c \quad c = 1 \ldots C$$

In principle the choice of the $\hat{\mathbf{m}}_c$ at this stage is arbitrary with the exception that no two can be the same.

3. Assign each pixel vector $\mathbf{x}$ to the candidate cluster of the nearest medoid using an appropriate distance metric in the spectral space. Manhattan distance is often used with medoids, because of the ease of computation. That generates a cluster of pixel vectors about each candidate medoid.

4. Compute a cost function for the current configuration, based on the accumulated Manhattan distance of all pixel points from their medoid:

$$\text{cost} = \sum_{C_c} \sum_{\mathbf{x} \in C_c} |\mathbf{x} - \mathbf{m}_c|$$

The outer sum is over clusters while the inner sum is taken within the clusters.
5. Replace one of the medoids with a randomly selected pixel vector $\mathbf{x}^*$. Recompute the cost of the configuration.
6. If the new cost is lower than the previous cost, keep the new configuration, including the replacement of the old medoid by $\mathbf{x}^*$. Otherwise keep the old configuration.
7. Repeat steps 5 and 6 until the configuration with the lowest cost has been achieved.

This process is illustrated in Fig. 9.8 using the two dimensional data set from Fig. 9.2.

The $k$ medoids algorithm is time consuming to run because of the very large number of replacements and comparisons required. As a result, it is not used with large data sets unless modified—one such modification is called CLARA (Clustering Large Applications).[15]

## 9.13  Clustering Large Data Sets

We now want to explore how to cluster image data sets that are very large, such that iterative procedures like $k$ means in their standard form present challenges.

Consider the time demand of the $k$ means algorithm. For $P$ pixels, $C$ clusters and $I$ iterations, the $k$ means algorithm requires $PCI$ distance calculations. For $N$ bands, the distance calculations involve $N$ multiplications each, giving a total of $PCIN$ multiplications to complete a $k$ means clustering exercise. For a $1000 \times 1000$ image segment, involving 200 bands and searching for 15 clusters, if 100 iterations were required then $30 \times 10^{10}$ multiplications are needed! Further, all the data needs to be held in memory to make the iterative process acceptable. Nevertheless, because of its simplicity, the $k$ means algorithm is still used with big data sets if:

- A more powerful processor can be employed.

  From an operational point of view this can be a difficult choice since most remote sensing practitioners would require access to readily available, and not specialized, computer hardware.

- A careful method for initiating the cluster centres could be used to speed up convergence by reducing the number of iterations needed.

---

[15] C.C. Aggarwal and C. K. Reddy, *Data Clustering Algorithms and Applications*, CRC Press, Roca Baton 2014. p. 94.

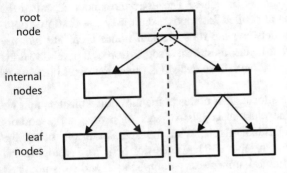

K trees operation

pixels are fed in at the root node, in tree building and use

algorithm determines tree structure and node properties

the leaf nodes hold the clusters generated

root node

internal nodes

leaf nodes

a decision to follow a particular branch is made in the in the node from which the branches emanate

**Fig. 9.9** Definitions for a decision tree and the structure of the K trees process

- A multi-processor (multi-core) machine could be used to speed up the computation.

    This is a good approach when multi-core machines are available; however, steps need to be taken to parallelise the $k$ means algorithm which, because of its iterative nature, requires some innovative approaches.[16]

- A more efficient version of the $k$ means algorithm might be possible.

    We will examine one such technique in the next section, which speeds up significantly the time required to undertake clustering and to allocate a pixel to a cluster class.

Some algorithms by their nature can handle larger data sets, such as those which don't iterate but stream the input data to the algorithm. The single pass method of Sect. 9.9 is one such approach. Another is DBSCAN, which is outlined in Sect. 9.13.2.

### 9.13.1   The K Trees Algorithm

As the name implies K trees is a tree-based approach to clustering. We met decision trees in the context of the support vector machine, but now we want to look at them again in a different setting. We start with some nomenclature, defined in Fig. 9.9, which also sets up the basis of the K trees process.

---

[16] See for example X. Cui, P. Zhu, K. Li and C. Ji, Optimised big data K-means clustering using MapReduce, *J. Supercomputing*, vol. 70, 2014, pp. 1249–1259, and A. Mohebi, S. Aghabozorgi, T.Y. Wah, T. Herewan and R. Yahyapour, Iterative big data algorithms: a review, *Software Practice and Experience*, vol. 46, 2016, pp. 107–129.

Trees consist of *nodes*, linked by branches. The uppermost node is called the *root*, and the lowermost nodes are called *leaf* nodes, curiously upside-down compared with a physical tree. In between the root and leaf nodes there are *internal* nodes. The nodes are arranged in layers, as shown. Progression of a pixel down the tree is based on decisions at the nodes; those decisions direct the pixel into one of the available branches.

In the K trees algorithm, we allocate leaf nodes to the individual clusters that we are trying to find. Because the clusters consist of sets of pixels, some authors allocate leaf nodes to the individual pixels, with the actual clusters being the internal nodes in the layer directly above. That does not help in developing the algorithm and just adds an additional, unnecessary complication and so is not used here.

Most clustering algorithms require some user-specified parameters. For the $k$ means technique it is the number of clusters to be generated; for the single pass algorithm they are the critical distance and standard deviation multiplier. For the K trees method, it is the maximum population of the nodes. Called the *tree order*, $m$, it specifies that any node cannot have more than that number of members. Full details of the algorithm will be found in Geva.[17] That treatment is a little hard to understand in the remote sensing context since it is written in the language of computer science; so, we will develop the algorithm by example, using a simple two-dimensional set of data, *and* using remote sensing terminology.

Our example uses the set of eight vector samples shown in Fig. 9.10. A tree order of 3 will be used. Specification of the order controls the structure of the tree, as we will see.

As shown at the left of Fig. 9.11a the tree starts with a single root node and a single leaf. We then feed in the first sample, say $c$. This is called *insertion*. Since we have no other information, it simply flows down to the leaf node, as does the second sample $a$. We now insert a third sample, say $g$; it can be accommodated, but it fills the leaf node, since we have specified a tree order of 3. A fourth sample, say $d$, *cannot* be accommodated in the current tree because the leaf node cannot contain more than 3 samples by design. That leaf node has to be split. The K trees algorithm does the split into two by doing a $k$ means clustering of the four samples, as shown in Fig. 9.12. The split is shown in Fig. 9.11b and the root node contains the mean vectors of the (leaf) nodes directly below it.

The tree now has capacity to absorb more pixels, so a 5th sample $f$ can be inserted as shown in Fig. 9.11c. That pixel vector is checked against the two mean vectors in the root node. It is closest to $\mathbf{m}_{cd}$ so it is allocated to the left-hand leaf node and the corresponding cluster mean is updated and used in the root node as $\mathbf{m}_{cdf}$.

---

[17] S. Geva, K-tree: a height balanced tree structured vector quantizer. *Neural Networks for Signal Processing X, 2000. Proceedings of the 2000 IEEE Signal Processing Society Workshop*, IEEE, Sydney, 2000, pp. 271–280.

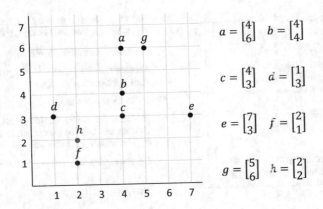

$$a = \begin{bmatrix} 4 \\ 6 \end{bmatrix} \quad b = \begin{bmatrix} 4 \\ 4 \end{bmatrix}$$

$$c = \begin{bmatrix} 4 \\ 3 \end{bmatrix} \quad d = \begin{bmatrix} 1 \\ 3 \end{bmatrix}$$

$$e = \begin{bmatrix} 7 \\ 3 \end{bmatrix} \quad f = \begin{bmatrix} 2 \\ 1 \end{bmatrix}$$

$$g = \begin{bmatrix} 5 \\ 6 \end{bmatrix} \quad h = \begin{bmatrix} 2 \\ 2 \end{bmatrix}$$

**Fig. 9.10** Samples to be used to demonstrate K trees clustering

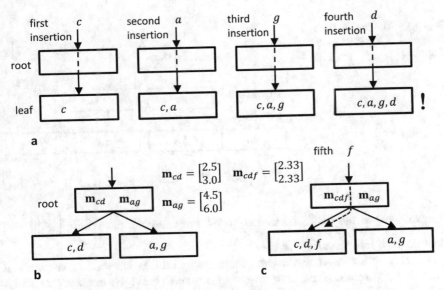

**Fig. 9.11 a** Initialisation and insertion of the first set of pixels, the fourth of which cannot be accommodated in the leaf node; **b** splitting the leaf node by $k$ means clustering as in Fig. 9.12; the root is now populated with the mean vectors of the nodes below; **c** accommodating the fifth pixel, following which the relevant mean vector in the root is updated

Continuing in this manner of splitting nodes by $k$ means clustering when they become full, and using mean vectors in place of all the pixels which go down a particular path, leads to the final K tree shown in Fig. 9.13. It has three layers, with two internal nodes and four leaf nodes. Any pixel vector fed into the top of the tree will make its way down to one of the clusters via the decisions (Euclidean distance

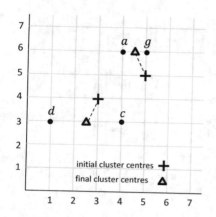

We need to find two clusters using the *k* means algorithm, which for this simple case is trivial.

- Choose the two initial cluster centres

- Allocate the pixels to the clusters; one cluster is (c,d) and the other is (a,g)

- Calculate the new means

- Reallocate the pixels—in this simple case the cluster memberships don't change, so the means as calculated are the final values

- Call the means $\mathbf{m}_{cd}$ and $\mathbf{m}_{ag}$, where

$$\mathbf{m}_{cd} = \begin{bmatrix} 2.5 \\ 3.0 \end{bmatrix} \quad \mathbf{m}_{ag} = \begin{bmatrix} 4.5 \\ 6.0 \end{bmatrix}$$

**Fig. 9.12** Using *k* means clustering to separate the four sample pixels in the leaf node of Fig. 9.11a, allowing the node to be split as seen in Fig. 9.11b

**Fig. 9.13** The final K trees clustering of the samples in Fig. 9.10

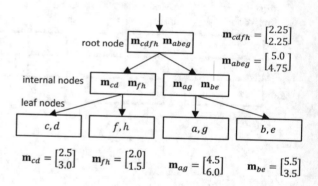

comparisons) at the root and internal nodes. For example, the vector $\begin{bmatrix} 3.0 \\ 3.0 \end{bmatrix}$ will flow through into the cluster $c, d$.

Apart from *how well they cluster*, there are two things we want to know about clustering algorithms. First, how long does it take to build the tree and, secondly, especially with unsupervised classification in mind, how quick is it at allocating unseen data to a cluster?

If we look at the speed of allocation first, we can do so by counting the number of distance comparisons. In the simple example here, both the K trees and the equivalent *k* means approach require the same number of comparisons. But what about with bigger data sets?

If we take the simplest case of each node in the K tree requiring two distance comparisons, the number of comparisons increases by 2 for each new layer added, which in this case also doubles the number of clusters. By contrast, the number of distance comparisons for the *k* means algorithm goes up as powers of 2. So, for larger numbers of clusters the K trees algorithm is much faster when allocating an unseen sample to an existing cluster.

Getting a meaningful comparison of the times to build the K tree and the $k$ means clusters is not straightforward. We can make comments on the number of nodes to be built and the checks within them, but the complexity introduced by the effect of different tree orders makes meaningful theoretical comparisons difficult. However, a number of trials by Geva[18] shows that the K trees method is significantly faster than $k$ means in developing the clusters, although he comments that the $k$ means approach performed slightly better.

A significant factor in favour of the K trees approach is that it can be adapted to run on multi-core processors and does not require all samples to be held in core memory during clustering.[19]

## 9.13.2   DBSCAN

The DBSCAN algorithm[20] (Density Based Spatial Clustering of Applications with Noise) is another approach that can be applied to large data sets. In some ways it is similar to the single pass algorithm of Sect. 9.9 but, whereas the single pass approach favours the generation of hyperspherical clusters because of its full reliance on distance metrics, DBSCAN can grow elongated clusters if the data is so distributed. It can also reject data points from any available cluster if they are not strongly attached to a cluster. Those rejected points are called noise in the terminology of the algorithm. Often they are outliers and should not be put into a cluster in any case. The other algorithms we treated earlier incorporate them and they can, therefore, distort the clusters.

As the name implies, DBSCAN relies on the assumption that clusters are dense regions of spectral space.[21] Characterisation of density is central to cluster development; all acceptable clusters have to have a minimum density of data points. This is controlled through the use of two parameters: the minimum number of points per primitive cluster, usually called MinPts, and a radius defining a primitive cluster size called Eps, or sometimes $\varepsilon$. The term "primitive" here is taken to mean the smallest possible, or starting cluster, size from which larger clusters can be grown. DBSCAN encourages clusters to merge into elongated groups if the density of the data points is more smeared out, as we will see in the following description of its

---

[18] loc cit., Fig. 4.

[19] A. Woodley, L-X Trang, S. Geva, R Nayak and T. Chappell, Parallel K Tree: A multicore, multimode solution to extreme clustering, *Future Generation Computing Systems*, Vol 99, October 2019, pp. 333–345.

[20] M. Ester, H-P. Kriegel, J. Sander and X. Xu A density-based algorithm for discovering clusters in large spatial databases with noise, *Proc Second Int Conf on Knowledge Discovery and Data Mining*, Aug. 1996, pp. 226–231.

[21] The histogram peak selection and mountain clustering algorithms in Sect. 9.12 are also effectively density-seeking techniques.

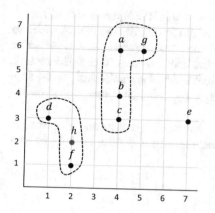

1. Randomly choose *b*.   Neighbours are *a* and *c*. Therefore, *b* is core, and cluster 1 is initiated.

2. Check *a*. It is a core sample with neighbours *b* and *g*. Since *a* is directly density reachable from *b* its neighbours are added to cluster 1.

3. Check neighbour *c* of *b*. The density test fails (Eps OK but MinPts not). Thus, it is a border point.

4. Check neighbour *g* of *a*. The density test fails (Eps OK but MinPts not). Thus, it is a border point.

5. We can go no further with the first cluster. Choose randomly another point from the data set that has not yet been considered, say *d*.  It fails the density test because there are not at least two samples within distance 2 but it could be a border point.

6. Randomly choose another point from the data set that has not yet been considered, say *h*. The density test is satisfied, and it is thus a core point, used to start cluster 2. Point *d* now becomes confirmed as a border point.

7. Randomly choose another point from the data set that has not yet been considered, say *e*. It fails the density test and is not close enough to a cluster to be a border point, so it is noise (an outlier)

8. Finally choose the last point from the data set *f*. It fails the density test but is close enough to *h* to be another border point of cluster 2

**Fig. 9.14** Illustration of the operation of the DBSCAN algorithm. Here the parameter values Eps = 2 and MinPts = 2 were used, along with Manhattan distance

operation, and in the example in Fig. 9.14 which uses the same data set we adopted to demonstrate K Trees.

The method proceeds as follows:

1. One of the points in the data set to be clustered is chosen randomly. Its spectral neighbourhood, set by the density radius Eps, is searched to find other data points in that neighbourhood. If the number of points found equals or exceeds MinPts, then the selected point starts a cluster and is said to be a *core point*. Members of the cluster are all the neighbours. This is what is meant by a primitive cluster.

2. If the point selected in 1 does not have any neighbour within Eps it is called a *noise point*. It is not assigned to any cluster, effectively removing it from the process.

3. If the point selected in 1 is a core point, then each of its neighbours is checked to see whether they are also core points: that is, they have more than MinPts neighbours within Eps. If they are core, then their neighbours are added to those found in step 1 to create a larger cluster still. As can be envisaged, while the initial (primitive) cluster would be hyperspherical, growing the cluster using the neighbours of the other core points makes possible the generation of very elongated clusters.

4. If at step three the neighbours do not satisfy the requirements to be core points, they are called *border points* and while regarded as members of the cluster they cannot be used to find more neighbours because they are not part of a dense

enough neighbourhood; instead, they seem to be on the edge or border of a region of density.

There is some nomenclature that will be found in treatments of DBSCAN. Apart from *core point*, *border point* and *noise point* defined above, the following terms are also often used. Note that by referring to density they are just reminding us that the fundamental approach is density-seeking.

A point $q$ is *directly density-reachable* from another point $p$ if $p$ is in the neighbourhood of $q$ and $q$ is a core point of its neighbourhood.

A point $q$ is *density-reachable* from another point $p$ if they can be linked via a chain of directly density-reachable points.

Two points are *density-connected* if both are density reachable from another common point.

## 9.14   Cluster Space Classification

Whereas the high dimensionality of hyperspectral data sets presents a processing challenge to statistical supervised classifiers such as the maximum likelihood rule (see Sect. 8.3.7), clustering and thus unsupervised classification with hyperspectral imagery is less of a problem because there are no parameters to be estimated. As a result, clustering can be used as a convenient bridge to assist in thematic mapping with high dimensional data.

The cluster space technique now to be developed is based first on clustering image data and then using reference data to link clusters with information classes.[22] Importantly, the power of the method rests on the fact that there does not need to be a one-to-one association of clusters (spectral classes) and information classes. That has the benefit of allowing the analyst the flexibility of generating as many clusters as needed to segment the spectral domain appropriately without worrying too much about the precise class meanings of the clusters produced. The significance of that lies in the fact that the spectral domain is rarely naturally composed of discrete groups of pixels; rather it is generally more of the nature of a multidimensional continuum, with a few density maxima that might be associated with spectrally well-defined classes such as water.[23]

The method starts by assuming that we have clustered the spectral domain as shown in the two-dimensional illustration of Fig. 9.15. Suppose, by the use of

---

[22] See X. Jia and J.A. Richards, Cluster space representation for hyperspectral classification, *IEEE Transactions on Geoscience and Remote Sensing*, vol. 40, no.3, March 2002, pp. 593–598. This approach is a generalisation of that given by M.D. Fleming, J.S. Berkebile and R.M. Hofer, Computer aided analysis of Landsat-1 MSS data: a comparison of three approaches, including a modified clustering approach, *Information Note 072,475*, Laboratory for Applications of Remote Sensing, Purdue University, West Lafayette, Indiana, 1979.

[23] See J.A. Richards & D.J. Kelly, On the concept of spectral class. *Remote Sensing Letters*, vol 5, 1984, pp. 987–991.

**Fig. 9.15** Relationship
between data, clusters and
information classes

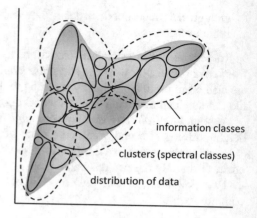

information classes

clusters (spectral classes)

distribution of data

reference data, we are able to associate clusters with information classes, as shown
by the overlaid boundaries in the figure. An information class can include more than
one cluster and some clusters appear in more than one information class.

By counting the pixels in each cluster, we can estimate the set of cluster conditional probabilities

$$p(\mathbf{x}|c) \quad c = 1\ldots C$$

in which $C$ is the total number of clusters. For convenience we might assume the
clusters are normally distributed so this cluster conditional density function is
represented by its mean and covariance, which can be estimated from the relevant
pixels if the dimensionality is acceptable. Clustering algorithms such as $k$ means
and Isodata tend to generate hyperspherical clusters so we generally assume a
diagonal covariance matrix with identical elements, in which case there are many
fewer parameters to estimate.

From Bayes' theorem we can find the posterior probability of a given cluster
being correct for a particular pixel measurement vector

$$p(c|\mathbf{x}) = \frac{p(\mathbf{x}|c)p(c)}{p(\mathbf{x})} \quad c = 1\ldots C \qquad (9.4)$$

in which $p(c)$ is the "prior" probability of the existence of cluster $c$. That can be
estimated from the relative populations of the clusters.

By examining the distribution of the information classes over the clusters we can
generate the class conditional probabilities

$$p(c|\omega_i) \quad i = 1\ldots M$$

where $M$ is the total number of information classes. Again, from Bayes' theorem we
have

$$p(\omega_i|c) = \frac{p(c|\omega_i)p(\omega_i)}{p(c)} \quad i = 1 \ldots M \tag{9.5}$$

We are interested in the likely information class for a given pixel, expressed in the set of posterior probabilities $p(\omega_i|\mathbf{x}) \; i = 1 \ldots M$. These can be written

$$p(\omega_i|\mathbf{x}) = \sum_{c=1}^{C} p(\omega_i|c)p(c|\mathbf{x}) \quad i = 1 \ldots M$$

Substituting from (9.4) and (9.5) gives

$$p(\omega_i|\mathbf{x}) = \frac{1}{p(\mathbf{x})} \sum_{c=1}^{C} p(c|\omega_i)p(\mathbf{x}|c)p(\omega_i) \quad i = 1 \ldots M \tag{9.6}$$

Since $p(\mathbf{x})$ does not aid discrimination we can use the decision rule

$$\mathbf{x} \in \omega_i \text{ if } p'(\omega_i|\mathbf{x}) > p'(\omega_j|\mathbf{x}) \text{ for all } j \neq i \tag{9.7}$$

to determine the correct class for the pixel at $\mathbf{x}$, where $p'(\omega_i|\mathbf{x}) = p(\mathbf{x})p(\omega_i|\mathbf{x})$.

It is instructive now to consider a simple example to see how this method works.[24] For this we use the two-dimensional data in Fig. 9.16 which contains two information classes $A$ and $B$. For the clustering phase of the exercise the class labels are not significant. Instead, the full set of training pixels, in this case 500 from each class, is clustered using the $k$ means algorithms. Here we have searched for just 4 clusters. The results for more clusters are given in Table 9.5. The resulting class mean positions are seen in the figure. Although two appear among the pixels of information class $A$ and two among those for information class $B$ that is simply the result of the distribution of the training pixels and has nothing to do with the class labels as such.

Using reference data (a knowledge of which pixels are class A and which are class B) it is possible to determine the mapping between information classes and clusters. Table 9.4 demonstrates that, both in terms of the number of pixels and the resulting posterior probabilities $p(c|\omega_i)$ derived from a normalisation of the counts. Also shown are the prior probabilities of the clusters.

From (9.6) and (9.7), and using the data in Table 9.4, we have (using $c1$ etc. to represent the clusters)

$$p'(A|\mathbf{x}) = p(c1|A)p(\mathbf{x}|c1)p(c1) + p(c2|A)p(\mathbf{x}|c2)p(c2)$$
$$+ p(c3|A)p(\mathbf{x}|c3)p(c3) + p(c4|A)p(\mathbf{x}|c4)p(c4)$$

and

---

[24] The computing for this example was carried out by Associate Professor Xiuping Jia.

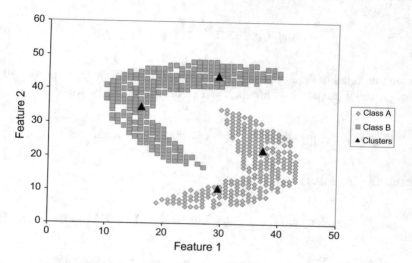

**Fig. 9.16** Two-dimensional data set used to illustrate the cluster space technique, and the four cluster centres generated by applying the $k$ means algorithm to the data

**Table 9.4** Association of clusters and information classes

| Number of pixels | Cluster 1 | Cluster 2 | Cluster 3 | Cluster 4 |
|---|---|---|---|---|
| Class A | 7 | 0 | 300 | 194 |
| Class B | 185 | 283 | 0 | 32 |
| $p(cluster|class)$ | Cluster 1 | Cluster 2 | Cluster 3 | Cluster 4 |
| Class A | 0.014 | 0.000 | 0.600 | 0.386 |
| Class B | 0.370 | 0.566 | 0.000 | 0.064 |
| $p(cluster)$ | 0.192 | 0.283 | 0.300 | 0.226 |

**Table 9.5** Performance of the cluster space method as a function of the number of clusters

| No. of clusters | 2 | 3 | 4 | 5 | 6 | 10 | 14 | 18 |
|---|---|---|---|---|---|---|---|---|
| On training set (%) | 93.4 | 90.3 | 95.7 | 96.0 | 99.0 | 99.3 | 99.6 | 100 |
| On testing set (%) | 93.4 | 90.7 | 95.9 | 96.0 | 99.2 | 99.6 | 99.5 | 100 |

$$p'(B|\mathbf{x}) = p(c1|B)p(\mathbf{x}|c1)p(c1) + p(c2|B)p(\mathbf{x}|c2)p(c2)$$
$$+ p(c3|B)p(\mathbf{x}|c3)p(c3) + p(c4|B)p(\mathbf{x}|c4)p(c4)$$

The cluster conditional distribution functions $p(\mathbf{x}|c), c \in \{c1, c2, c3, c4\}$ are obtained from the pixels in each cluster and, in this example, have been modelled by spherical Gaussian distributions in which the diagonal covariances for each cluster are assumed to be the same and equal to the average covariance over the four clusters.

Equation (9.7) can now be used to label the pixels. Based on the 1000 pixels of training data used to generate the cluster space model, an overall accuracy of 95.7% is obtained. Using a different testing set of 500 pixels from each class an accuracy of 95.9% is obtained. Clearly, the performance of the method depends on how effectively the data space is segmented during the clustering step. Table 9.5 shows how the results depend on the numbers of clusters used.

## 9.15   Bibliography on Clustering and Unsupervised Classification

Cluster analysis is a common tool in many fields that involve large amounts of data. As a result, material on clustering algorithms will be found in the social and physical sciences, and particularly fields such as numerical taxonomy. Because of the enormous amounts of data used in remote sensing, the range of viable techniques is limited so that some treatments contain methods not generally encountered in remote sensing. Standard texts on image processing and remote sensing could be consulted. Perhaps the most comprehensive of these treatments is

R.O. Duda, P.E. Hart and D.G. Stork, *Pattern Classification*, 2nd ed., John Wiley & Sons, N.Y., 2001.

Other, more recent, coverages of clustering and unsupervised learning are in

C.C. Aggarwal and C. K. Reddy, *Data Clustering Algorithms and Applications*, CRC Press, Roca Baton 2014.

T. Hastie, R. Tibshirani and J. Friedman, *The Elements of Statistical Learning: Data Mining, Inference and Prediction*, Springer Science + Business Media, N.Y., 2009.

H. Bangui, M. Ge and B. Buhnova, Exploring Big Data Clustering Algorithms for Internet of Things Applications, *Proc. 3rd Int. Conf. on Internet of Things, Big Data and Security (IoTBDS 2018)*, 2019, pp. 269–276.

The seminal work on the Isodata algorithm is

G.H. Ball and D.J. Hall, A novel method of data analysis and pattern recognition, *Stanford Research Institute*, Menlo Park, California, 1965.

Some more general treatments are

M.R. Andberg, Cluster Analysis for Applications, Academic, N.Y., 1973

B.S. Everitt, S. Landau, M. Leese and D. Stahl, *Cluster Analysis*, 5th ed., John Wiley and Sons, N.Y., 2011

G. Gan, C. Ma and J. Wu, *Data Clustering: Theory, Algorithms and Applications*, ASA-SIAM Series on Statistics and Applied Probability, SIAM, Philadelphia, ASA, Alexandria, Virginia, 2007

J.A. Hartigan, *Clustering Algorithms*, John Wiley & Sons, N.Y., 1975

J. van Ryzin, *Classification and Clustering*, Academic, N.Y., 1977.

## 9.16   Problems

9.1   Find the Euclidean and city block distances between the following three
pixel vectors. which two are similar?

$$\mathbf{x}_1 = \begin{bmatrix} 4 \\ 4 \\ 4 \end{bmatrix} \quad \mathbf{x}_2 = \begin{bmatrix} 5 \\ 5 \\ 5 \end{bmatrix} \quad \mathbf{x}_3 = \begin{bmatrix} 4 \\ 4 \\ 7 \end{bmatrix}$$

9.2   Repeat the Exercise of Fig. 9.2 but with

 (a)  two initial cluster centres at (2, 3) and (5, 6),
 (b)  three initial cluster centres at (1, 1), (3, 3) and (5, 5), and
 (c)  three initial cluster centres at (2, 1), (4, 2) and (15, 15).

9.3   From a knowledge of how a particular clustering algorithm works it is
sometimes possible to infer the multidimensional spectral shapes of the
clusters generated. For example, methods that depend entirely on Euclidean
distance as a similarity metric would tend to produce hyperspheroidal clusters. Comment on the cluster shapes you would expect to be generated by the
migrating means technique based on Euclidean distance, the single pass
procedure, also based on Euclidean distance, DBSCAN based on Manhattan
distance and DBSCAN based on Euclidean distance.

9.4   Suppose two different techniques have given two different cluster assignments of a particular set of data and you wish to assess which of the two
segmentations is the better. One approach might be to evaluate the sum of
square errors measure treated in Sect. 9.2. Another could be based on
covariance matrices. For example, it is possible to define an "among clusters"
covariance matrix that describes how the clusters themselves are scattered
about the data space, and an average "within class" covariance matrix that
describes the average shape and size of the clusters. Let these matrices be
called $\mathbf{C}_A$ and $\mathbf{C}_B$ respectively. How could they be used together to assess the
quality of the two clustering results? (See G.R. Coleman and H.C. Andrews,
Image segmentation by clustering, *Proc IEEE*, vol. 67, 1979, pp. 773–785.)
Here you may wish to use measures of the "size" of a matrix, such as its trace
or determinant (see Appendix C).

9.5   Different clustering methods often produce quite different segmentations of
the same set of data, as illustrated in the examples of Figs. 9.3 and 9.6. Yet
the results generated for remote sensing applications are generally usable.
Why do you think that is the case? (Is it related to the number of clusters
generated?).

9.6   The Mahalanobis distance of (8.26) can be used as the similarity metric for a
clustering algorithm. Invent a possible clustering technique based on (8.26)
and comment on the nature of the clusters generated.

9.7   Do you see value in having a two-stage clustering process in which a single pass procedure is used to generate initial clusters and then an iterative technique is used to refine them?

9.8   Recompute the agglomerative hierarchical clustering example of Fig. 9.7 but use the $L_1$ distance measure in (9.2) as a similarity metric.

9.9   Consider the two dimensional data shown in Fig. 9.2 and suppose the three pixels at the upper right form one cluster and the remainder another cluster. Such an assignment might have been generated by some clustering algorithm other than the $k$ means method. Calculate the sum of squared error for this new assignment and compare with the value of 16 found in Fig. 9.2.

9.10   In the cluster space technique, how is (9.6) modified if there is uniquely only one cluster per information class?

9.11   Is the K trees method for clustering preferable to the $k$ means approach because

   (a)   It finds a better set of clusters
   (b)   It is more accurate, or
   (c)   It is faster to train (find the clusters)?

9.12   Does the tree order parameter used with K trees clustering

   (a)   Specify the number of clusters to be found
   (b)   Specify the number of layers in the tree, or
   (c)   Set an upper limit on the population of each node?

9.13   How is a medoid different from a median? When are they always the same?

9.14   Clustering procedures are often used as the basis of unsupervised classification

   (a)   Describe how
   (b)   Are the spectral classes (clusters) unique?
   (c)   How can the information classes be determined from the clusters generated?

9.15   If you had undertaken clustering with the K trees method, based on a given set of pixel vectors, and then another representative pixel became available for training, how would insertion of that new pixel be handled by the tree; in other words, how would the tree be modified? What would be the case if you had used the $k$ means approach instead?

# Chapter 10
# Feature Reduction

**Abstract** Feature reduction is seen to be important for performing efficient and effective classification, in that no more features than necessary should be used to describe the pixel vectors. The range of commonly employed feature reduction techniques are presented including those based on transforming the data beforehand, those that exploit the correlations between near-neighbouring bands (or features) and those based on measures that allow the least significant bands to be identified. The principal components transform is shown to be suitable as the basis of feature reduction in some circumstances, as is the transform based on canonical analysis. The application of many of the procedures given is illustrated by hand-worked examples.

## 10.1 The Need for Feature Reduction

Many remote sensing instruments record more channels or bands of data than are usually needed for most applications. As a simple example, even though the Hyperion sensor on EO-1 produces 220 channels of image data over the wavelength range 0.4–2.4 μm, it is unlikely that channels beyond about 1.0 μm would be relevant for water studies, unless the water were especially turbid. Furthermore, unless the actual reflectance spectrum of the water was essential for the task at hand, it may not even be necessary to use all the contiguous bands recorded in the range 0.4–1.0 μm; instead, a representative subset may be sufficient in most cases.

One of the first tasks facing an analyst is to determine whether all the available bands need to be used in any particular study and, if not, how should a subset be chosen that minimises any loss of information essential to the analysis. It is the purpose of this chapter to consider methods for selecting acceptable subsets of bands in the context of classification. We will also explore other means by which the dimensionality of the data can be reduced. In this context the recorded bands are called *features*, a term which has a more general meaning, as we will see soon.

© The Author(s), under exclusive license to Springer Nature Switzerland AG 2022
J. A. Richards, *Remote Sensing Digital Image Analysis*,
https://doi.org/10.1007/978-3-030-82327-6_10

This raises an intriguing question. If we have to go to all the trouble of reducing the feature subset for analysis, why not just record a smaller number of bands in the first place? To answer that question, we need to examine the rationale behind hyperspectral imaging.

Originally, hyperspectral imaging was referred to as *imaging spectroscopy*. Spectroscopy means identifying a substance through the use of its spectral response —or spectrum. Mass spectroscopy, visible spectroscopy and electron spectroscopy are areas we are familiar with, in which an absorption or emission spectrum is used to help us identify a substance.

Imaging spectroscopy is an application of the familiar field of (visible) absorption spectroscopy at the pixel level. For each pixel we record the full reflectance spectrum of the substance using reflected sunlight, usually over the wavelength range of about 0.4–2.5 μm. What we are measuring is the sunlight reflected to the sensor on a remote sensing platform after a component of the sunlight has been absorbed by the surface material.

The reason the field is called imaging spectroscopy is because such a measurement is done for every pixel in an image, as depicted in Fig. 10.1. For a particular pixel, the large number of recorded bands represents samples of the reflectance spectrum of the corresponding region on the ground. From those samples we can reconstruct the reflectance spectrum of the pixel. Notice that the reconstructed spectrum, if there are sufficient fine samples, shows various important diagnostic features such as the dips corresponding to chlorophyll absorption in the blue and red regions, and the water absorption bands. There are other finer absorption features too, especially for soils and minerals. They do not show up on this vegetation example.

One of the benefits of recording the full reflectance spectrum for a pixel is that we can use scientific knowledge and spectral libraries to identify the pixel rather than depend on supervised learning and machine classification. That is discussed in Sect. 11.8.

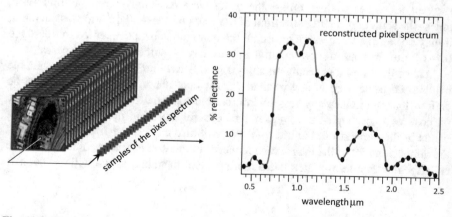

**Fig. 10.1** Reconstructing a pixel spectrum from hyperspectral image data

Although spectroscopic analysis is regularly applied in the earth sciences using recorded hyperspectral image data, in many applications we still wish to use machine learning methods for labelling a pixel. This approach is convenient and does not rely on expert knowledge or recorded spectral libraries. But we then have to face the problem that there are too many bands recorded to allow all of them to be used as features in a supervised classification exercise. We need to find methods, therefore, that will allow us to identify a subset of the recorded bands that is still sufficient for the development of an accurate classifier.

In the early years of remote sensing image analysis, it was considered sensible economically to ensure that no more features than necessary were utilised when performing a classification, in order to contain computation costs.

That is now less important, and other factors drive the need for feature reduction. Chief among these is the number of training pixels required to ensure that reliable estimates of class signatures or classifier parameters can be generated. We have met that problem under several names—the Hughes phenomenon, the curse of dimensionality and over-fitting. See Sects. 8.3.7 and 8.21.11. The number of training samples needed with most classification algorithms increases significantly with the number of bands or channels. For high dimensional data, such as hyperspectral imagery, that requirement presents a particular challenge. Keeping the number of informative features to as few as possible is especially important if reliable results are to be expected when using affordable numbers of training pixels.

## 10.2  Approaches to Feature Reduction

We need to be clear on nomenclature. *Features* is the name given to the input data to our various classification algorithms—most usually they are the recorded bands, or some transformed version of the bands after, say, the application of the principal components transform. In object detection they may be spatial descriptors. In some applications the features sets may include both spectral and spatial descriptors.

In this chapter we are embarking on a search for methods to reduce the number of features; this is called *feature reduction* in general, although some authors use the term *feature extraction*. One approach to feature reduction is to select subsets of bands, which still function effectively. That is called *feature selection*.

An enormous number of feature reduction techniques has been proposed over the past few decades, particularly since the availability of hyperspectral image data.[1] Some approaches are combined with the classification method being used by the

[1] See D. Singh, S. Appavu and E. Leavline, Literature review on feature selection methods for high-dimensional data, *International Journal of Computer Applications* (0975-8887) Vol. 136, No. 1, February 2016, pp. 9–17 and B. Rasti, D. Hong, R. Hang, P. Ghasimi, X. Kang, J. Chanussot and J. Benediktsson, Feature extraction for hyperspectral imagery, *IEEE Geoscience and Remote Sensing Magazine*, December 2020, pp. 60–88.

analyst, whereas others are independent of any specific machine learning algorithm. Some of the methods transform the data sets so that feature selection is easier.

The methods we will look at can be put into three categories:

- Those that transform the data so that the least significant transformed bands can be discarded
- Those that exploit the correlations among the recorded bands allowing them to be treated in small groups
- Those that allow the least significant of the originally-recorded bands to be discarded.

Irrespective of the approach used it is important that we do not cause classes in the spectral domain to overlap by reducing features. Good classification results depend on maintaining separation between the classes—in feature reduction, that requirement is called *separability*. Later in this chapter we will introduce measures of separability, but for now we will use the term generally.

## 10.3   Feature Reduction by Spectral Transforms

A popular means for carrying out feature reduction is to transform the data to a new set of axes in which separability is higher in a subset of the transformed features than in any subset of the original data. That allows the remaining transformed features to be discarded. A number of different image transformations can be used this. The most commonly encountered in remote sensing are the principal components transform, the transformation associated with canonical analysis, and versions of discriminant analysis, each of which is treated in the following sections.

### 10.3.1   Feature Reduction Using the Principal Components Transform

The principal components transformation of Chap. 6 maps image data into a new, uncorrelated coordinate system or vector space. In doing so, it produces a data representation in which the most variance is along the first axis, the next largest variance is along a second mutually orthogonal axis, and so on. The later principal components would be expected, in general, to show little variance. They could be regarded as contributing little to separability and thus could be ignored, thereby reducing the essential dimensionality of the vector space, and leading to more effective classification. That is only of value, however, if the spectral class structure of the data is distributed substantially along the first few axes. Should that not be the case, it is possible that feature reduction using transformed data may be no better than with the original data. In such a situation the technique of canonical analysis may be better. Nevertheless, because of its simplicity the principal components transformation is frequently used in practice.

As an illustration of a case in which the principal components transformation does allow effective feature reduction, consider the two class, two dimensional data set illustrated in Fig. 10.2. Assume that the classes are not separable in either of the original data coordinates alone and that both dimensions are required for separability. Inspection of the data, however, suggests that the first component of a principal components transform will separate the two classes. That is now demonstrated mathematically by undertaking some manual calculations.

Notwithstanding the class structure of the data, the principal components transformation makes use of the global mean and covariance. Using (6.1) and (6.3) it can be shown that

$$\mathbf{m} = \begin{bmatrix} 4.50 \\ 4.25 \end{bmatrix} \text{ and } \mathbf{C} = \begin{bmatrix} 2.57 & 1.86 \\ 1.86 & 6.21 \end{bmatrix}$$

The eigenvalues of the covariance matrix are $\lambda_1 = 6.99$ and $\lambda_2 = 1.79$ so that the first principal component will contain 79.6% of the variance. The normalised eigenvectors corresponding to the eigenvalues are

$$\mathbf{g}_1 = \begin{bmatrix} 0.387 \\ 0.922 \end{bmatrix} \text{ and } \mathbf{g}_2 = \begin{bmatrix} -0.922 \\ 0.387 \end{bmatrix}$$

**Fig. 10.2** Two dimensional, two class data set in which feature reduction using the principal components transformation is possible

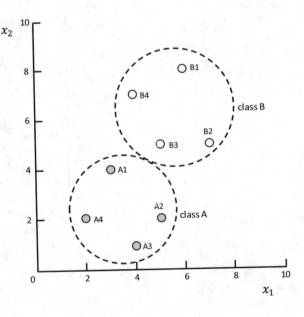

so that the principal components transformation matrix is

$$\mathbf{G} = \begin{bmatrix} 0.387 & 0.922 \\ -0.922 & 0.387 \end{bmatrix}$$

The first principal component for each pixel vector is then found from

$$y_1 = 0.387x_1 + 0.922x_2$$

The transformed pixel vectors are shown plotted in Fig. 10.3, in which it is seen that the data is separable into the two classes in this single coordinate. Figure 10.3 also shows both principal axes relative to the original image axes, again demonstrating separation in the first component but, for this example, significant class overlap in the second component.

**Fig. 10.3** Principal components transformation of the data in Fig. 10.2 showing class separation along the first principal component; that component alone is sufficient for discriminating between the classes

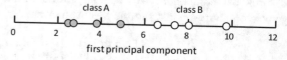

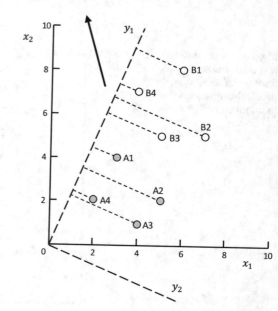

## 10.3.2  *Feature Reduction Using the Canonical Analysis Transform*

The principal components transformation is based on the global covariance matrix of the full set of image data and is thus not sensitive to class structure. The reason it often works well in remote sensing as a feature reduction tool is because classes are frequently distributed in the direction of maximum data scatter. However, should good separation not be given by the principal components transformation, derived from the global covariance matrix, then a subset of image data could be selected that contains the cover types of interest; that subset is then used to compute the covariance matrix. The resulting transformation will have its first principal axis oriented so that the cover types are well discriminated.

Another, more rigorous method for generating a transformed set of features, in which class separation is optimised, is based on the procedure called canonical analysis. To illustrate this approach, consider the two-dimensional, two class data distributions shown in Fig. 10.4a. The classes can be seen not to be separable in either of the original feature axes on their own. Nor will they be separable in only one of the two principal component axes because of the nature of the global data scatter compared with the scatter of data within the individual classes. It is clear, however, that the data can be separated with a single feature if an axis rotation (an image transformation) such as that shown in Fig. 10.4b is adopted. The primary axis in this new transformation will be oriented so that the classes have the largest possible separation between their means when projected onto that axis; at the same time they appear as small as possible in their individual spreads. We characterise the former by a measure $\sigma_A^2$ shown in the diagram, which we call the variance among the classes. The spreads of data within the classes in the new axis are characterised by $\sigma_{W1}^2$ and $\sigma_{W2}^2$. Our interest is in finding the new axis for which

$$\frac{\sigma_A^2}{\sigma_W^2} = \frac{\text{among class variance}}{\text{within class variance}} \tag{10.1}$$

is as large as possible; $\sigma_W^2$ is the average of $\sigma_{W1}^2$ and $\sigma_{W2}^2$ for this example.

### 10.3.2.1  Within-Class and Among-Class Covariance

To handle data with any number of dimensions it is necessary to define average data scatter within the classes, and the manner in which the classes themselves scatter about the multidimensional spectral domain. These properties are described by covariance matrices. The average within-class covariance matrix is defined as

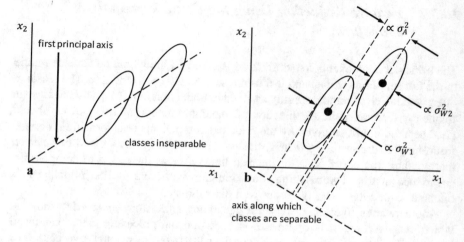

**Fig. 10.4** **a** Two-dimensional, two class data set showing lack of separability in either of the original axes or in either principal component, **b** demonstrating an axis along which the classes can be separated

$$\mathbf{C}_W = \frac{1}{M} \sum_{i=1}^{M} \mathbf{C}_i \tag{10.2a}$$

in which $\mathbf{C}_i$ is the covariance matrix of the data in class $i$ and where $M$ is the total number of classes. Equation (10.2a) applies only if the classes have equal populations. A better expression is

$$\mathbf{C}_W = \frac{1}{N} \sum_{i=1}^{M} (n_i - 1) \mathbf{C}_i \tag{10.2b}$$

where $n_i$ is the population of the $i$th class and $N = \sum_{i=1}^{M} n_i$.

The among-class covariance is given by

$$\mathbf{C}_A = \mathcal{E}\left\{ (\mathbf{m}_i - \mathbf{m}_o)(\mathbf{m}_i - \mathbf{m}_o)^T \right\} \tag{10.3}$$

in which $\mathbf{m}_i$ is the mean vector of the $i$th class and $\mathbf{m}_o$ is the global mean, given by

$$\mathbf{m}_o = \frac{1}{M} \sum_{i=1}^{M} \mathbf{m}_i \tag{10.4a}$$

when the classes have equal populations or, in general

$$\mathbf{m}_o = \frac{1}{N} \sum_{i=1}^{M} n_i \mathbf{m}_i \qquad (10.4b)$$

### 10.3.2.2 A Separability Measure

Let $\mathbf{y} = \mathbf{D}^T \mathbf{x}$ be the required transform that generates a new set of axes $\mathbf{y}$ in which the classes have optimal separation. By the same procedure that was used for the principal components transformation in Sect. 6.3.2 it is possible to show that the within-class and among-class covariance matrices in the new coordinate system are

$$\mathbf{C}_{W,y} = \mathbf{D}^T \mathbf{C}_{W,x} \mathbf{D} \qquad (10.5a)$$

$$\mathbf{C}_{A,y} = \mathbf{D}^T \mathbf{C}_{A,x} \mathbf{D} \qquad (10.5b)$$

where the subscripts $x$ and $y$ have been used to identify the matrices with their respective coordinate systems. It is significant to understand here, unlike with the case of principal components analysis, that the two new covariance matrices are not necessarily diagonal. However, as with principal components, the row vectors of $\mathbf{D}^T$ define the axis directions in $\mathbf{y}$ space. Let $\mathbf{d}^T$ be one particular vector, say that one which defines the first axis along which the classes will be optimally separated. Then the corresponding within class and among class variances will be

$$\sigma_{W,y}^2 = \mathbf{d}^T \mathbf{C}_{W,x} \mathbf{d}$$

$$\sigma_{A,y}^2 = \mathbf{d}^T \mathbf{C}_{A,x} \mathbf{d}$$

What we now want to do is to find the specific $\mathbf{d}$, and ultimately the full transformation matrix $\mathbf{D}^T$, that maximises

$$\lambda = \frac{\sigma_{A,y}^2}{\sigma_{W,y}^2} = \mathbf{d}^T \mathbf{C}_{A,x} \mathbf{d} / \mathbf{d}^T \mathbf{C}_{W,x} \mathbf{d} \qquad (10.6)$$

### 10.3.2.3 The Generalised Eigenvalue Equation

The ratio of variances in (10.6) is maximised by the value of $\mathbf{d}$ for which

$$\frac{\partial \lambda}{\partial \mathbf{d}} = 0$$

Noting the identity

$$\frac{\partial}{\partial \mathbf{x}} \{\mathbf{x}^T \mathbf{A} \mathbf{x}\} = 2\mathbf{A}\mathbf{x}$$

then

$$\frac{\partial \lambda}{\partial \mathbf{d}} = \frac{\partial}{\partial \mathbf{d}} \left\{ \left(\mathbf{d}^T \mathbf{C}_{A,x} \mathbf{d}\right) \left(\mathbf{d}^T \mathbf{C}_{W,x} \mathbf{d}\right)^{-1} \right\}$$

$$= 2\mathbf{C}_{A,x} \mathbf{d} \left(\mathbf{d}^T \mathbf{C}_{W,x} \mathbf{d}\right)^{-1} - 2\mathbf{C}_{W,x} \mathbf{d} \left(\mathbf{d}^T \mathbf{C}_{A,x} \mathbf{d}\right)$$

$$\left(\mathbf{d}^T \mathbf{C}_{W,x} \mathbf{d}\right)^{-2} = 0$$

This reduces to

$$\mathbf{C}_{A,x} \mathbf{d} - \mathbf{C}_{W,x} \mathbf{d} \left(\mathbf{d}^T \mathbf{C}_{A,x} \mathbf{d}\right) \left(\mathbf{d}^T \mathbf{C}_{W,x} \mathbf{d}\right)^{-1} = 0$$

which can be rewritten as

$$[\mathbf{C}_{A,x} - \lambda \mathbf{C}_{W,x}]\mathbf{d} = 0$$

$$(10.7)$$

Equation (10.7) is called a *generalised eigenvalue equation*, which has to be solved for the unknowns $\lambda$ and $\mathbf{d}$. The first canonical axis will be in the direction of $\mathbf{d}$, while $\lambda$ will give the ratio of among-class to within-class variance along that axis. Generalising (10.7) to incorporate all components we have

$$[\mathbf{C}_{A,x} - \Lambda \mathbf{C}_{W,x}]\mathbf{D} = 0 \qquad (10.8)$$

in which $\Lambda$ is a diagonal matrix of the full set of $\lambda s$ and $\mathbf{D}$ is a matrix of the vectors $\mathbf{d}$.

The development to this stage is usually referred to as discriminant analysis. One additional step is required in the case of canonical analysis. As with the equivalent step in the principal components transformation, solution of (10.7) amounts to finding the set of eigenvalues $\lambda$ and the corresponding eigenvectors $\mathbf{d}$. While unique values for the $\lambda$ can be determined, the components of $\mathbf{d}$ can only be found relative to each other. In the case of principal components, we introduced the additional requirement that the vectors have unit magnitude, thereby allowing them to be determined uniquely. For canonical analysis, the additional constraint is

$$\mathbf{D}^T \mathbf{C}_{W,x} \mathbf{D} = \mathbf{I} \qquad (10.9)$$

which says that the within-class covariance matrix after transformation must be the identity matrix—a unit diagonal matrix. In other words, after transformation the classes should appear spherical.

For $M$ classes and $N$ bands of data, if $N > M - 1$ there will only be $M - 1$ non-zero roots of (10.8) and thus $M - 1$ canonical axes.[2] For this example, in which

---

[2] See H. Seal, *Multivariate Statistical Analysis for Biologists*, Methuen, London, 1964, and N. A. Campbell and W. R. Atchley, The geometry of canonical variate analysis, *Systematic Zoology*, vol. 30, 1981, pp. 268–280.

$M$ and $N$ are both 2, one of the eigenvalues of (10.7) will be zero and the corresponding eigenvector will not exist. That implies that the dimensionality of the transformed space will be less than that of the original data. In general, the classes will have maximum separation along the first canonical axis, corresponding to the largest $\lambda$. The second axis, corresponding to the next largest $\lambda$, will give the next best degree of separation, and so on.

### 10.3.2.4 An Example

Consider the two dimensional, two category data shown in Fig. 10.5. Both of the original features are required to discriminate between the two categories. We will now perform a canonical analysis transformation on the data to show that the categories can be discriminated in the first canonical axis.

The individual class covariance matrices are

$$\mathbf{C}_C = \begin{bmatrix} 2.25 & 2.59 \\ 2.59 & 4.25 \end{bmatrix} \text{ and } \mathbf{C}_D = \begin{bmatrix} 4.25 & 3.00 \\ 3.00 & 6.67 \end{bmatrix}$$

so that the within-class covariance is

$$\mathbf{C}_{W,x} = \frac{1}{2}\{\mathbf{C}_C + \mathbf{C}_D\} = \begin{bmatrix} 3.25 & 2.80 \\ 2.80 & 5.46 \end{bmatrix}$$

The among-class covariance matrix is

$$\mathbf{C}_{A,x} = \begin{bmatrix} 8.00 & 5.50 \\ 5.50 & 3.78 \end{bmatrix}$$

The canonical transformation matrix $\mathbf{D}^T$ is given by the solution to (10.8) where $\mathbf{D}$ is a matrix of column vectors. Recall that those vectors are the axes in the transformed space, along the first of which the ratio of among-class variance to within-class variance is greatest. On this axis there is most chance of separating the classes. $\mathbf{\Lambda}$ is a diagonal matrix of scalar constants that are the eigenvalues of (10.8); numerically they are the ratios of variances along each of the canonical axes.

Each $\lambda$ and the accompanying $\mathbf{d}$ can be found by considering the individual component Eq. (10.7) rather than the more general form in (10.8). For (10.7) to have a non-trivial solution it is necessary that

$$\left| \mathbf{C}_{A,x} - \lambda \mathbf{C}_{W,x} \right| = 0$$

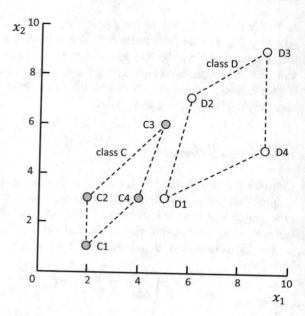

**Fig. 10.5** Two classes of two dimensional data, each containing four data points

Using the values for $\mathbf{C}_{A,x}$ and $\mathbf{C}_{W,x}$ above this is

$$\begin{vmatrix} 8.00 - 3.25\lambda & 5.50 - 2.80\lambda \\ 5.50 - 2.80\lambda & 3.78 - 5.46\lambda \end{vmatrix} = 0$$

which gives $\lambda = 2.54$ or 0. Thus there is only one canonical axis, described by the vector $\mathbf{d}$ which corresponds to $\lambda = 2.54$. This is found as the solution to

$$[\mathbf{C}_{A,x} - 2.54\mathbf{C}_{W,x}]\mathbf{d} = 0$$

i.e., $$\begin{bmatrix} -0.26 & -1.61 \\ -1.61 & -10.09 \end{bmatrix} \begin{bmatrix} d_1 \\ d_2 \end{bmatrix} = 0$$

from which we find $d_1 = -6.32d_2$. We now use (10.9), which for one vector component of $\mathbf{D}$, is

$$\begin{bmatrix} d_1 & d_2 \end{bmatrix} \begin{bmatrix} 3.25 & 2.80 \\ 2.80 & 5.46 \end{bmatrix} \begin{bmatrix} d_1 \\ d_2 \end{bmatrix} = \mathbf{I}$$

On expanding this gives $3.25d_1^2 + 5.60d_1d_2 + 5.46d_2^2 = 1$. Using $d_1 = -6.32d_2$ we see that $d_1 = 0.63$ and $d_2 = -0.10$ so that

$$\mathbf{d} = \begin{bmatrix} 0.63 \\ -0.10 \end{bmatrix}$$

which is shown plotted in Fig. 10.6. The projections of the pixel points onto the axis defined by that vector show the classes to be separable. The brightness values of the pixels along that axis are given by

$$\mathbf{y} = \mathbf{d}^T\mathbf{x} = 0.63x_1 - 0.10x_2$$

## 10.3.3 Discriminant Analysis Feature Extraction (DAFE)

A variation on the canonical analysis development of the previous section is to use the Fisher criterion

$$J = \mathrm{tr}\left\{ \mathbf{C}_{W,y}^{-1}\mathbf{C}_{A,y} \right\} \tag{10.10}$$

instead of the measure of (10.6). Again, we want to find an axis transformation that maximises $J$. Let that transformation be $\mathbf{y} = \mathbf{D}^T\mathbf{x}$. Then (10.10) can be written

$$J = \mathrm{tr}\left\{ \left( \mathbf{D}^T\mathbf{C}_{W,x}\mathbf{D} \right)^{-1} \left( \mathbf{D}^T\mathbf{C}_{A,x}\mathbf{D} \right) \right\}$$

It can be shown[3] that differentiating the last expression to find the transformation matrix $\mathbf{D}^T$ that maximises $J$ leads to

$$\mathbf{C}_{W,x}^{-1}\mathbf{C}_{A,x}\mathbf{D} = \mathbf{D}\mathbf{C}_{W,y}^{-1}\mathbf{C}_{A,y} \tag{10.11}$$

Consider now the transformation $\mathbf{z} = \mathbf{B}^T\mathbf{y}$ that diagonalises the transformed among-class covariance $\mathbf{C}_{A,y}$, viz.

$$\mathbf{B}^T\mathbf{C}_{A,y}\mathbf{B} = \mathbf{M}$$

in which $\mathbf{M}$ is diagonal. Thus $\mathbf{C}_{A,y} = \left( \mathbf{B}^T \right)^{-1}\mathbf{M}\mathbf{B}^{-1}$ and (10.11) becomes

$$\mathbf{C}_{W,x}^{-1}\mathbf{C}_{A,x}\mathbf{D} = \mathbf{D}\mathbf{C}_{W,y}^{-1}\left( \mathbf{B}^T \right)^{-1}\mathbf{M}\mathbf{B}^{-1} \tag{10.12}$$

---

[3] See K. Fukunaga, *Introduction to Statistical Pattern Recognition*, Academic, London, 1990.

**Fig. 10.6** The first canonical axis for the two class data of Fig. 10.9 showing that class discrimination is possible

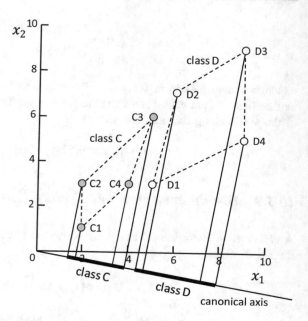

As with canonical analysis we now introduce the additional criterion that the transformed within-class covariance matrix be unity after transformation to the $\mathbf{z}$ space, so that the classes then appear hyperspherical. That requires[4]

$$\mathbf{B}^T \mathbf{C}_{W,y} \mathbf{B} \equiv \mathbf{B}^{-1} \mathbf{C}_{W,y}^{-1} \left( \mathbf{B}^T \right)^{-1} = \mathbf{I}$$

so that   $\mathbf{C}_{W,y}^{-1} \left( \mathbf{B}^T \right)^{-1} = \mathbf{B}$

which, when substituted into (10.12) gives

$$\mathbf{C}_{W,x}^{-1} \mathbf{C}_{A,x} \mathbf{D} = \mathbf{DBMB}^{-1}$$

and thus   $\mathbf{C}_{W,x}^{-1} \mathbf{C}_{A,x} \mathbf{DB} = \mathbf{DBM}$

which is an eigenfunction equation, in which $\mathbf{M}$ is a diagonal matrix of the eigenvalues of $\mathbf{C}_{W,x}^{-1} \mathbf{C}_{A,x}$ and $\mathbf{DB}$ is the matrix of eigenvectors of $\mathbf{C}_{W,x}^{-1} \mathbf{C}_{A,x}$. Eigenanalysis can be carried out by analysing $\mathbf{C}_{W,x}^{-1}$ and $\mathbf{C}_{A,x}$ separately.[5] The axis along which the classes have maximum separation corresponds to the largest eigenvalue of $\mathbf{C}_{W,x}^{-1} \mathbf{C}_{A,x}$, and so on.

---

[4] If a matrix is equivalent to the identity matrix then so is its inverse. Further $[\mathbf{ABC}]^{-1} = \mathbf{C}^{-1}\mathbf{B}^{-1}\mathbf{A}^{-1}$.

[5] See Fukunaga, loc. cit.

Deriving the transformed axes based on maximisation of (10.10) is called *discriminant analysis feature extraction* (DAFE). As with canonical analysis, it requires good estimates of the relevant covariance matrices. In the case of the within-class matrices, that could be difficult when the dimensionality of the data is high.

## 10.3.4    Non-parametric Discriminant Analysis (NDA)

As depicted in Fig. 10.4 we have assumed with the previously treated transformation-based methods that the spectral classes are clusters of similar pixel vectors, so that within-class covariance adequately describes how they spread about their mean positions, and among-class covariance, computed from the means, makes sense. If discrete spectral classes have been found beforehand, that is acceptable.

If, however, one or more of the classes were unusual in shape, such as an unresolved class that might be made up of a set of similar cover types or sub-classes, then feature reduction methods that depend on class means and covariance matrices may not work well. A class distribution such as that depicted in Fig. 10.7 is an example. Should such a case be suspected then it is better to avoid separability measures that depend on class statistics and, instead, try to find a method that is non-parametric. Non-parametric Discriminant Analysis (NDA) and its extension to Decision Boundary Feature Extraction (DBFE) are two such approaches. In this section we treat NDA while DBFE is the subject of Sect. 10.3.5.

In its simplest form NDA examines the relationship between the training pixels of one class and their nearest neighbour training pixels from another class. For example, let $\mathbf{x}_{j \in s, NN_i \in r}$ represent pixel $j$ from class $s$ that is the nearest neighbour of pixel $i$ from class $r$ as shown in Fig. 10.8. We can describe the distribution of class $r$ pixels with respect to their nearest neighbours in class $s$ by a covariance-like calculation. However, because we are now not describing the distribution function of pixels about a class mean (a parametric description), it is better to use a different term than covariance matrix. To talk about the scatter of pixels with respect to each other we use the term *scatter matrix*.

The scatter of all of the training pixels from class $r$ about their nearest neighbours from class $s$ is defined by the scattering matrix

$$S_{b1} = \mathcal{E}\left\{ \left(\mathbf{x}_{i \in r} - \mathbf{x}_{j \in s, NN_i \in r}\right)\left(\mathbf{x}_{i \in r} - \mathbf{x}_{j \in s, NN_i \in r}\right)^{\mathrm{T}} | \omega_r \right\}$$

where $\mathbf{x}_{i \in r}$ is the $i$th pixel from class $r$ and the $|\omega_r$ conditionality reminds us that the calculation is determined by pixels from class $r$.

We perform a similar calculation for the scatter of the training pixels from class $s$ about their class $r$ nearest neighbours, and then average the two measures. Usually, the average is weighted by the prior probabilities, or relative abundances, of the classes:

**Fig. 10.7** A situation in
which the DAFE technique
would not perform well since
the mean of class 2 may not
be too different from that of
class 1, and the within class
covariance matrix would not
reflect the actual scatter of the
data in class 2

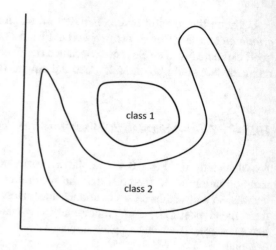

**Fig. 10.8** Identification of
the pixels used in the
development of the between
class scatter matrix in
non-parametric discriminant
analysis

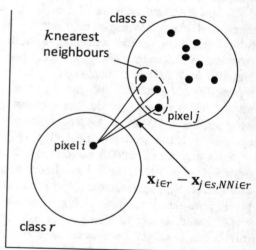

$$\mathbf{S}_b = \mathbf{S}_{b1} + \mathbf{S}_{b2} = p(\omega_r)\mathcal{E}\left\{ \left(\mathbf{x}_{i\in r} - \mathbf{x}_{j\in s, NNi\in r}\right)\left(\mathbf{x}_{i\in r} - \mathbf{x}_{j\in s, NNi\in r}\right)^{\mathrm{T}}|\omega_r\right\}$$
$$+ p(\omega_s)\mathcal{E}\left\{ \left(\mathbf{x}_{j\in s} - \mathbf{x}_{i\in r, NNj\in s}\right)\left(\mathbf{x}_{j\in s} - \mathbf{x}_{i\in r, NNj\in s}\right)^{\mathrm{T}}|\omega_s\right\}$$

Often NDA uses not just the nearest *neighbour* but instead a set of $k$ class $s$ training
pixels as a nearest *neighbourhood* for each class $r$ training pixel. The local mean
over that neighbourhood is then used in the calculation of the between class scattering matrix.

Let $\mathbf{x}_{l\in s, kNNi\in r}$ be the $l$th member of the $k$ nearest neighbours from class $s$ of pixel
$i$ in class $r$. Then the local class $s$ mean is defined as

$$\mathbf{m}_{s,kNNi\in r} = \frac{1}{k} \sum_{l=1}^{k} \mathbf{x}_{l\in s,kNNi\in r} \tag{10.13}$$

in which case the expression for the between-class scattering matrix becomes

$$\begin{aligned}
\mathbf{S}_b &= p(\omega_r)\mathcal{E}\left\{ \left(\mathbf{x}_{i\in r} - \mathbf{m}_{s,kNNi\in r}\right)\left(\mathbf{x}_{i\in r} - \mathbf{m}_{s,kNNi\in r}\right)^{\mathrm{T}}|\omega_r\right\} \\
&+ p(\omega_s)\mathcal{E}\left\{ \left(\mathbf{x}_{j\in s} - \mathbf{m}_{r,kNNj\in s}\right)\left(\mathbf{x}_{j\in s} - \mathbf{m}_{r,kNNj\in s}\right)^{\mathrm{T}}|\omega_s\right\}
\end{aligned} \tag{10.14}$$

Note from (10.13) that if $k$, the size of the neighbourhood, is the same as the total number of training pixels available in class $s$ then the local mean becomes the class mean, and the between class scatter matrices resemble covariance matrices, although taken around the mean of the opposite class rather than the mean of their own class.

Generalisation of (10.14) requires a little thought because there are as many weighted means of the pixels "from the other class" as they are "other classes." This is illustrated in Fig. 10.9 for the case of three classes: $r$, $s$ and $t$. It is easier to express the expectations in (10.14) in algebraic form, so that for $C$ total classes the among-class matrix is

$$\mathbf{S}_A = \sum_{r=1}^{C} p(\omega_r) \sum_{c=1,c\neq r}^{C} \frac{1}{N_r} \sum_{i=1}^{N_r} \left(\mathbf{x}_{i\in r} - \mathbf{m}_{c,kNNi\in r}\right)\left(\mathbf{x}_{i\in r} - \mathbf{m}_{c,kNNi\in r}\right)^{\mathrm{T}} \tag{10.15}$$

in which the inner sum computes the expected scatter between the $N_r$ training pixels from class $r$ and the mean of the nearest neighbours in class $c$ (different for each training pixel); the middle sum then changes the class ($c$), still relating to the

**Fig. 10.9** The k nearest neighbours of the $i$th pixel in class $r$ in each of the other two classes

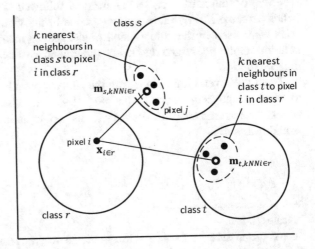

training pixels from class $r$; the outer sum changes the class ($r$) for which the training pixels are being considered. The latter computation is weighted by the prior probability for the class.

Having determined a non-parametric expression for the among-class scatter we now need to consider within-class scatter properties, so that we can use a criterion such as that in (10.10) to guide feature reduction.

The usual form of the within-class scatter matrix in (10.2a) can be used, although a transformation that also maps it to the identity matrix should be employed.[6] That leads to the following NDA transformation that ranks the transformed features by decreasing value of separability:

$$z = \mathbf{\Psi}^{\mathrm{T}} \mathbf{\Lambda}^{-1/2} \mathbf{\Phi}^{\mathrm{T}} \mathbf{x}$$

where $\mathbf{\Psi}$ is the matrix of eigenvectors of $\mathbf{S}_A$, $\mathbf{\Lambda}$ is the diagonal eigenvalue matrix and $\mathbf{\Phi}$ is the eigenvector matrix of the within-class scatter matrix.

Alternative expressions for the within-class scatter matrix can be used.[7] For example, the mean vector can be based on the $k$ nearest neighbours of the $i$th pixel in class $r$ from the same class

$$\mathbf{m}_{r,kNNi\in r} = \frac{1}{k} \sum_{l=1, l\neq i}^{k} \mathbf{x}_{l\in r,kNNi\in r}$$

so that the within-class scatter matrix in the two class case becomes

$$\mathbf{S}_W = p(\omega_r)\mathcal{E}\left\{ \left(\mathbf{x}_{i\in r} - \mathbf{m}_{r,kNNi\in r}\right)\left(\mathbf{x}_{i\in r} - \mathbf{m}_{r,kNNi\in r}\right)^{\mathrm{T}} | \omega_r \right\}$$
$$+ p(\omega_s)\mathcal{E}\left\{ \left(\mathbf{x}_{j\in s} - \mathbf{m}_{s,kNNj\in s}\right)\left(\mathbf{x}_{j\in s} - \mathbf{m}_{s,kNNj\in s}\right)^{\mathrm{T}} | \omega_s \right\}$$

The procedures of Sect. 10.3.3 are then followed. This process could lead to a better outcome if only those training pixels in the vicinity of the decision boundaries were used in the computation of the scatter matrices. Accordingly, the calculations could be weighted to lessen the influence of neighbours further away from the boundaries.

There are several limitations with the NDA approach, particularly given the need to identify the neighbours to be used. By comparison, even though canonical analysis is parametric in its basis, its computational demand is relatively straightforward.

---

[6] Fukunaga, ibid.

[7] See B-C Kuo and D. A. Landgrebe, Nonparametric weighted feature extraction for classification, *IEEE Transactions on Geoscience and Remote Sensing*, vol. 42, no. 5, May 2004, pp. 1096–1105.

**Fig. 10.10** Transformed axes
in which one of the new
features is of value in
separating the classes shown;
the other feature, being
parallel to the likely decision
boundary, does not assist
discrimination

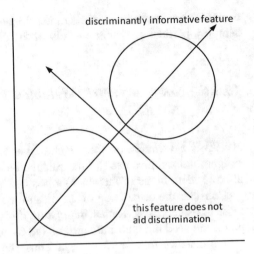

discriminantly informative feature

this feature does not
aid discrimination

## 10.3.5 Decision Boundary Feature Extraction (DBFE)

Another feature reduction procedure that uses training pixels only in the vicinity of
the decision boundary is Decision Boundary Feature Extraction (DBFE).[8] It is
based on the idea that transformed feature vectors normal to decision boundaries are
discriminantly informative, whereas feature vectors that are parallel to decision
boundaries don't help class separation. That is illustrated in Fig. 10.10 using a
two-dimensional, two class example. The problem is to find an effective representation of the normals to the segment of the decision boundary in the vicinity of
the training data.

DBFE is a parametric procedure. It commences by estimating the class conditional mean vectors and covariance matrices, which are then used to define the
actual decision surfaces and to classify the training pixels. Outlying pixels from
each class are removed using a Chi-squared test. A sample of each class in the
vicinity of the decision surface is selected by applying the Chi-squared test to the
pixels of the opposite class, using the statistics of the first class. From the sample
identified, the decision surface normals are estimated in the vicinity of the training
data, from which an effective decision boundary feature matrix is computed. While
the dimensionality of the matrix will be the same as the original feature space, its
rank may be smaller indicating the reduced number of discriminantly informative
features.

DBFE has a number of drawbacks including, again, the large number of calculations required, along with the need to obtain reliable estimates of the original
class signatures and the decision boundary feature matrix. Those parametric

---

[8] See Sect. 6.6 of D. A. Landgrebe, *Signal Theory Methods in Multispectral Remote Sensing*, John
Wiley & Sons, Hoboken N.J., 2003.

estimates are not reliable if the dimensionality is high and the number of training samples, particularly in the vicinity of the decision surface, is limited.

## 10.3.6  Non-parametric Weighted Feature Extraction (NWFE)

NWFE is a variation on the weighted version of NDA treated in Sect. 10.3.4. It weights the samples used in the calculations of the local means and uses slightly different definitions of the among-class and within-class scattering matrices.

Consider the calculation of the mean for class $r$ pixels in the vicinity of pixel $r$ from the same class. Rather than using a set of $k$ nearest neighbours, all training pixels are used but their influence on the computed value of the mean is diminished by the distance they are from $\mathbf{x}_{i\in r}$. Thus, the weighted $r$ class mean about the $i$th pixel in class $r$ is

$$\mathbf{m}_{r,i\in r} = \sum_{l=1}^{N_r} w_{l\in r, i\in r} \mathbf{x}_{l\in r}$$

where $N_r$ is the number of training pixels in class $r$, and the weight $w_{l\in r, i\in r}$ is defined by

$$w_{l\in r, i\in r} = \frac{d^{-1}(\mathbf{x}_{i\in r}, \mathbf{x}_{l\in r})}{\sum_{l=1}^{N_r} d^{-1}(\mathbf{x}_{i\in r}, \mathbf{x}_{l\in r})}$$

where $d^{-1}$ is the reciprocal of the distance between the pixel vectors in its argument. In a similar manner the local mean of class $s$ pixels, as far as the $i$th pixel from class $r$ is concerned, is

$$\mathbf{m}_{s,i\in r} = \sum_{l=1}^{N_s} w_{l\in s, i\in r} \mathbf{x}_{l\in s}$$

where the weight is now $\quad w_{l\in s, i\in r} = \dfrac{d^{-1}(\mathbf{x}_{i\in r}, \mathbf{x}_{l\in s})}{\sum_{l=1}^{N_s} d^{-1}(\mathbf{x}_{i\in r}, \mathbf{x}_{l\in s})}$

Using these new definitions of the means, the among-class and within-class scattering matrices, for the multiclass case, are now

$$\mathbf{S}_A = \sum_{r=1}^{C} p(\omega_r) \sum_{c=1, c\neq r}^{C} \frac{1}{N_r} \sum_{i=1}^{N_r} w_{i\in r, c} (\mathbf{x}_{i\in r} - \mathbf{m}_{c,i\in r})(\mathbf{x}_{i\in r} - \mathbf{m}_{c,i\in r})^{\mathrm{T}} \qquad (10.16a)$$

$$\mathbf{S}_W = \sum_{r=1}^{C} p(\omega_r) \frac{1}{N_r} \sum_{i=1}^{N_r} w_{i\in r,r}(\mathbf{x}_{i\in r} - \mathbf{m}_{r,i\in r})(\mathbf{x}_{i\in r} - \mathbf{m}_{r,i\in r})^{\mathrm{T}} \qquad (10.16b)$$

in which the weights are defined by $w_{l\in r,\xi} = \dfrac{d^{-1}(\mathbf{x}_{i\in r},\mathbf{m}_{\xi,i\in r})}{\sum_{l=1}^{N_r} d^{-1}(\mathbf{x}_{i\in r},\mathbf{m}_{\xi,i\in r})}$

with $\xi = c$ or $r$ in (10.16a) and (10.16b) respectively. To avoid problems with reliable estimation, or even singularity, the within-class scatter matrix of (10.16b) is sometimes replaced with the approximate form[9]

$$\mathbf{S}'_W = 0.5\mathbf{S}_W + 0.5\mathrm{diag}\mathbf{S}_W$$

Having established the form of the among class and within class scatter matrices, the required features can be found from the eigenvectors corresponding to the largest eigenvalues of

$$J = \mathbf{S}_W'^{-1}\mathbf{S}_A \qquad (10.17)$$

This is equivalent to using (10.8) but with the newly defined scatter matrices.

## 10.4  Feature Reduction by Block Diagonalising the Covariance Matrix

We have met the covariance matrix several times in the past. It is the starting point for the principal components transformation and, along with the mean vector, defines the class signature when we undertake maximum likelihood classification.

In the latter context we know that the accurate computation of the class covariance matrix can be a problem for hyperspectral data if we do not have enough training samples. Remember, we have to have enough independent training samples in each class in order to estimate reliably the elements of the covariance matrix. Generally, that is not a problem with the principal components transformation because, then, the covariance matrix is computed using all of the available training samples and not just those for an individual class.

So, it is feasible to examine the covariance matrix for the data as a whole; we will do that now and note that it has some interesting structural properties. Rather than the covariance matrix itself we will inspect the correlation matrix instead which, remember, is derived from the elements of the covariance matrix. See Eq. (6.4).

Figure 10.11 shows the correlation matrix for 192 bands of data recorded by the AVIRIS instrument over Jasper Ridge. Rather than display the data in numerical

---

[9] See Kuo and Landgrebe, loc. cit.

**Fig. 10.11** Correlation matrix for 196 bands of the Jasper Ridge AVIRIS image in which white represents high correlation and black represents zero correlation[10]

form, which would be usual if the matrix dimension were not too high, it is convenient and effective to show the matrix in the form of an image in which the colour or grey scale indicates the magnitude of the matrix elements (in this case, the degrees of correlation between corresponding band pairs). The coordinates of the image correspond to the band numbers, starting with the first band at the top left-hand corner, both horizontally and vertically.

In a grey scale representation, we usually choose black to indicate zero correlation and white to indicate total positive or negative correlation. Thus in Fig. 10.11, lighter entries indicate that the corresponding band pairs are highly correlated, while darker entries indicate low correlations. Not surprisingly, spectral regions of high correlation are scattered down the principal diagonal of the matrix, indicating the high degree of redundancy that exists between the adjacent and near neighbouring bands recorded with fine spectral resolution sensors.

The most striking feature of the correlation matrix is that high correlations exist in blocks, the strongest of which are usually along the diagonal. That allows us to make a useful assumption to simplify expressions which involve the *covariance* matrix, such as the discriminant function in maximum likelihood classification[11] and the eigen-analysis step in principal components transformation. If we assume that we can neglect the off-diagonal correlations, the correlation matrix, and more

---

[10] Overlapping bands, bands corresponding to significant water absorption, and bands with very small means have been deleted from the original 224 band set.

[11] See X. Jia, Classification Techniques for Hyperspectral Remote Sensing Image Data, *Ph.D. Thesis*, The University of New South Wales, University College, Australian Defence Force Academy, Canberra, 1996, and X. Jia and J. A. Richards Efficient maximum likelihood classification for imaging spectrometer data sets, *IEEE Transactions on Geoscience and Remote Sensing*, vol. 32, no. 2, March 1994.

**Fig. 10.12** Block diagonal approximation to the correlation matrix of Fig. 10.11

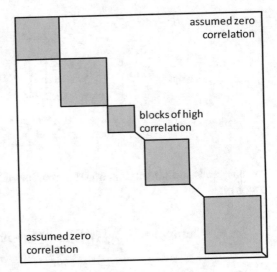

assumed zero correlation

blocks of high correlation

assumed zero correlation

importantly the associated covariance matrix, can be represented by a block diagonal approximation as illustrated in Fig. 10.12.

If we represent a block diagonal covariance matrix, with $H$ diagonal blocks, as

$$\mathbf{C} = \begin{bmatrix} \mathbf{C}_1 & 0 & \cdots & 0 \\ 0 & \mathbf{C}_2 & & \\ \vdots & & \ddots & \vdots \\ 0 & & \cdots & \mathbf{C}_H \end{bmatrix} \tag{10.18}$$

it is readily shown that $\quad |\mathbf{C}| = |\mathbf{C}_1||\mathbf{C}_2|...|\mathbf{C}_H| \tag{10.19a}$

so that $\quad \ln|\mathbf{C}| = \ln|\mathbf{C}_1| + \ln|\mathbf{C}_2|... + \ln|\mathbf{C}_H| \tag{10.19b}$

and $\quad \mathrm{tr}\mathbf{C} = \mathrm{tr}\mathbf{C}_1 + \mathrm{tr}\mathbf{C}_2... + \mathrm{tr}\mathbf{C}_H \tag{10.19c}$

also $\quad \mathbf{C}^{-1} = \begin{bmatrix} \mathbf{C}_1^{-1} & 0 & \cdots & 0 \\ 0 & \mathbf{C}_2^{-1} & & \\ \vdots & & \ddots & \vdots \\ 0 & & \cdots & \mathbf{C}_H^{-1} \end{bmatrix} \tag{10.19d}$

Consider now a column vector $\mathbf{z}$ that has the same dimensions as the covariance matrix $\mathbf{C}$ and which is partitioned into a set of sub-vectors, each of which corresponds in dimensionality with the sub-matrices in the block diagonal covariance matrix:

$$\mathbf{z} = \begin{bmatrix} \mathbf{z}_1 \\ \mathbf{z}_2 \\ \vdots \\ \mathbf{z}_H \end{bmatrix}$$

$$\text{then} \quad \mathbf{z}^{\mathrm{T}}\mathbf{C}\mathbf{z} = \sum_{h=1}^{H} \mathbf{z}_h^{\mathrm{T}}\mathbf{C}_h\mathbf{z}_h$$

so that the discriminant function for maximum likelihood classification in (8.7) becomes

$$g_i(\mathbf{x}) = \ln p(\omega_i) - {}^1\!/_2 \sum_{h=1}^{H} \left\{ \ln|\mathbf{C}_{ih}| + (\mathbf{x}_h - \mathbf{m}_{ih})^{\mathrm{T}} \mathbf{C}_{ih}^{-1}(\mathbf{x}_h - \mathbf{m}_{ih}) \right\} \qquad (10.20)$$

How does this affect the number of training pixels required for reliable training? Recall that the number of training pixels needed per class is related to the dimensionality of the covariance matrix and thus the need to estimate reliable values for each of the elements of the matrix. When the class conditional covariance is block diagonalised as above the component matrices are independent of each other and the number of training pixels required per class is now set by the dimensions of the largest of the component matrices down the diagonal. That results in a reduced requirement for the numbers of labelled training pixels, particularly for the image data generated by hyperspectral sensors.

Although strictly not a feature reduction technique, the process of block diagonalising the covariance matrix achieves the same purpose of reducing the computational demand when second order statistics are involved, and offsets the influence of the Hughes phenomenon of Sect. 8.3.7.

While the actual blocks to use in the decomposition of (10.20) is guided by the observed structure of the correlation and covariance matrices, such a data dependent selection is not strictly necessary. In fact, because adjacent bands are generally strongly correlated in high dimensional, high spectral resolution data sets it is feasible to block diagonalise the covariance matrix by selecting equal size blocks down the diagonal, of dimensions two or three for example. Although simple, such a process has been shown to lead to good classifier performance.[12]

Consider now how block diagonalising the covariance matrix impacts on the calculation of the principal components transformation, which entails finding the eigenvalues and eigenvectors of the covariance matrix. The eigenvalues are solutions to the characteristic equation

---

[12] Jia, loc. cit.

$$|\mathbf{C}_x - \lambda\mathbf{I}| = 0 \tag{10.21}$$

Since the original covariance matrix $\mathbf{C}_x$ is assumed to be block diagonal, then the matrix $\mathbf{C}_x - \lambda\mathbf{I}$ is also block diagonal. From (10.19a) Eq. (10.21) can be expressed

$$|\mathbf{C}_x - \lambda\mathbf{I}| = \prod_{h=1}^{H}|\mathbf{C}_{xh} - \lambda\mathbf{I}| = 0 \tag{10.22}$$

Thus, the roots and eigenvalues of (10.21) are the eigenvalues of the component matrices of the block diagonal form. Therefore, in order to compute a principal components transformation with data of high dimensionality we can introduce the approximation of the block diagonal form of the covariance matrix and then find the set of eigenvalues and eigenvectors, and thus the transformation matrix, by applying eigen-analysis to the blocks individually.

Having blocked the principal components transform in this manner we can select the most informative components from each transformed segment, group them and transform again. The selection of components to retain in each block can be made on the basis of variance, as is common with principal components, or by using some other form of separability analysis. The process can be repeated by again identifying the components that are highly correlated among the selected set, block diagonalising the respective covariance matrix and transforming. Ultimately a large data set, such as that generated by a hyperspectral sensor, can be reduced to a small number of components with better compression than if a single, and time-consuming principal components step were carried out. Such a segmented principal components process can be used for effective colour display of hyperspectral data and as a feature reduction tool.[13]

The block diagonalisation idea can be extended to canonical analysis. The complete set of bands is segmented into $H$ groups and canonical analysis is applied to each of the individual groups. Class statistics involving the complete set of bands are no longer needed so that the difficulties presented by limited numbers of training samples with high dimensionality data can be obviated.

Sometimes highly correlated blocks of bands will occur away from the principal diagonal of the covariance matrix, as seen in Figs. 10.11 and 10.13. If required, they can be moved onto the diagonal by reordering the bands before the matrix is computed.[14] Such an operation makes no difference to the matrix, or any subsequent analysis, but it does mean the bands are out of order.

---

[13] X. Jia and J. A. Richards, Segmented principal components transformation for efficient hyperspectral remote sensing image display and classification. *IEEE Transactions on Geoscience and Remote Sensing*, vol. 37, no. 1 pt. 2, 1999, pp. 538–542.

[14] See Jia, loc. cit.

[15] https://engineering.purdue.edu/~biehl/MultiSpec/hyperspectral.html accessed 2021.

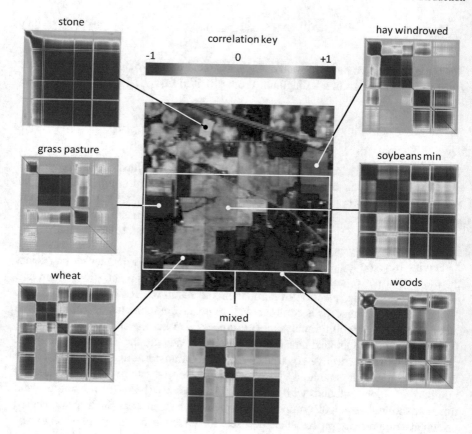

**Fig. 10.13** Demonstration of the class dependence of the correlation matrix using the Indian Pines data set available from Purdue University[15] and computed using MultiSpec; although the soybean and hay classes by name imply vegetation, they are only sparsely covered, as can be deduced from the colours in the composite image which has been formed by displaying channel 40 (745 nm) as red, channel 20 (577 nm) as green and channel 7 (449 nm) as blue

The very broad major water absorption bands can mask some of the features in the correlation matrices when displayed in image form. Those in Fig. 10.13 have been produced after channels 102–109 and 148–162 were removed, which correspond to the wavelengths of major water absorption. It is easier now to see the correlation structure. Moreover, by showing the matrices in colour, interpretability is enhanced, and both positive and negative correlations can be displayed as indicated by the colour key on the figure.

## 10.5  Feature Selection

### 10.5.1  Measures of Separability

If we can identify features that do not aid discrimination, by contributing little to the separability of spectral classes, we could discard them when training a classifier. Subsetting features by removal of those that are least effective is, as we noted earlier, referred to as feature selection, that being the simplest form of feature reduction.

In some situations, knowing which features to discard is relatively easy. In general, that is not the case and methods must be devised that allow the relative value of features to be assessed in a quantitative and rigorous way. A measure commonly used for this is to determine the *mathematical separability* of classes. Feature reduction is performed by checking how separable various spectral classes remain when reduced sets of features are used; provided separability is not lowered unduly by the removal of features then those features can be considered to be of little value. While this is a Gaussian-based measure which makes it more suited to the maximum likelihood approach, the lessons we learn by examining it flow on to a consideration of other techniques.

The material in this and the following sections has been based in part on Swain and Davis.[16] In order to assess whether a feature is important for discriminating among spectral classes we could examine the extent to which spectral classes overlap in that feature. Significant overlap needs to be avoided if we want the classes to be separable. It is useful to examine the effect that removing a feature has on the degree to which spectral classes overlap, and thus their separability, a situation we now consider in some detail.

Consider a two-dimensional spectral space with two spectral classes as shown in Fig. 10.14. Suppose we want to see whether the classes could be separated using only one feature—either $x_1$ or $x_2$. Of course, it is not known beforehand which feature is better. That is what has to be determined by a measure of separability. Consider an assessment of $x_1$. The spectral classes in the $x_1$ subset or subspace are shown in the figure in which some overlap of the single dimensional distributions is observed. If the distributions are well separated in the $x_1$ dimension, then the overlap will be small, and a classifier would be unlikely to make an error in discriminating between them on the basis of that feature alone. On the other hand, for a large degree of overlap substantial classifier error would be expected. The usefulness of the $x_1$ feature subset can be assessed therefore in terms of the overlap of the distributions in that dimension.

Consider now how to quantify the separation between a pair of probability distributions (spectral classes) as an indication of their degree of overlap. The distance between means is insufficient since overlap will also be influenced by the

---

[16] P. H. Swain and S. M. Davis, eds., *Remote Sensing: The Quantitative Approach*, McGraw-Hill, NY, 1978.

**Fig. 10.14** Two dimensional spectral space, with two spectral classes, and its one-dimensional marginal distribution; while the classes are apparently separable in two dimensions there is significant overlap in one dimension

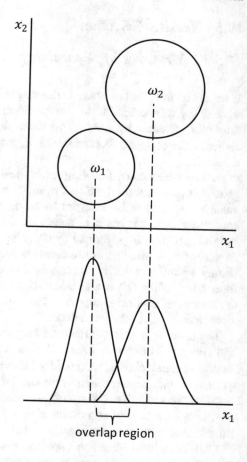

standard deviations of the distributions. Instead, a combination of both the distance between means and a measure of standard deviation is required. That must be a vector-based measure in order to be applicable to multidimensional subspaces. Several candidate measures of separability are available; only those commonly used with remote sensing data are treated in this chapter. Others may be found in books on statistics that treat similarities of probability distributions.

## 10.5.2   Divergence

### 10.5.2.1   Definition

Divergence is a measure of the separability of a pair of probability distributions that has its basis in their degree of overlap. It uses the definition of the likelihood ratio

**Fig. 10.15** The probabilities
used in the definition of the
likelihood ratio

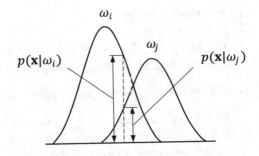

$$L_{ij}(\mathbf{x}) = p(\mathbf{x}|\omega_i)/p(\mathbf{x}|\omega_j) \qquad (10.23)$$

where $p(\mathbf{x}|\omega_i)$ and $p(\mathbf{x}|\omega_j)$ are the values of the $i$th and $j$th spectral class probability distributions at the position $\mathbf{x}$ in the spectral domain. The overlap region of Fig. 10.14 is shown exaggerated in Fig. 10.15 from which it is seen that $L_{ij}(\mathbf{x})$ is a measure of the overlap just at the point $\mathbf{x}$. For very separable classes $L_{ij}(\mathbf{x}) \rightarrow$ 0 or $\infty$ for all $\mathbf{x}$.

It is useful in what is to follow to take the logarithm of the likelihood ratio:

$$L'_{ij}(\mathbf{x}) = \ln p(\mathbf{x}|\omega_i) - \ln p(\mathbf{x}|\omega_j)$$

We can also define a log likelihood ratio with the numerator and denominators reversed in (10.23). We now define the *divergence* of the pair of class distributions as the sum of conditional expected values of the two definitions of the log likelihood ratio:

$$d_{ij} = \mathcal{E}\{L'_{ij}(\mathbf{x})|\omega_i\} + \mathcal{E}\{L'_{ji}(\mathbf{x})|\omega_j\} \qquad (10.24)$$

For continuous distributions the expectation operators are defined in terms of the integral of the relevant quantity weighted by the respective probability distribution, i.e.

$$\mathcal{E}\{L'_{ij}(\mathbf{x})|\omega_i\} = \int_{\mathbf{x}} L'_{ij}(\mathbf{x}) p(\mathbf{x}|\omega_i) d\mathbf{x}$$

This is the expected value of the likelihood ratio with respect to all pixels in the $i$th spectral class. Using this definition (10.24) becomes

$$d_{ij} = \int_{\mathbf{x}} \{p(\mathbf{x}|\omega_i) - p(\mathbf{x}|\omega_j)\} \ln \frac{p(\mathbf{x}|\omega_i)}{p(\mathbf{x}|\omega_j)} d\mathbf{x} \qquad (10.25)$$

from which the following properties of divergence can be established.

1. It is always positive
2. It is symmetric, $d_{ij} = d_{ji}$
3. If $p(\mathbf{x}|\omega_i) = p(\mathbf{x}|\omega_j)$, $d_{ii} = 0$; in other words, there is no divergence or difference between a probability distribution and itself
4. For statistically independent features $x_1, x_2 \ldots x_N$, $p(\mathbf{x}|\omega_i) = \prod_{n=1}^{N} p(x_n|\omega_i)$ so that divergence becomes $d_{ij} = \sum_{n=1}^{N} d_{ij}(x_n)$. Since divergence is never negative, it follows that $d_{ij}(x_1, ..x_n, x_{n+1}) \geq d_{ij}(x_1, ..x_n)$; that is, divergence never decreases as the number of features is increased.

### 10.5.2.2  Divergence of a Pair of Normal Distributions

Since spectral classes in remote sensing image data are often modelled by multi-dimensional normal distributions it is of interest to have available the specific form of (10.25) when $p(\mathbf{x}|\omega_i)$ and $p(\mathbf{x}|\omega_j)$ are normal distributions with means and covariances of $\mathbf{m}_i$ and $\mathbf{C}_i$, and $\mathbf{m}_j$ and $\mathbf{C}_j$, respectively. After substituting the full expressions for the normal distributions, we find

$$d_{ij} = \frac{1}{2}\mathrm{tr}\left\{ (\mathbf{C}_i - \mathbf{C}_j)(\mathbf{C}_i^{-1} - \mathbf{C}_j^{-1}) \right\}$$
$$+ \frac{1}{2}\mathrm{tr}\left\{ (\mathbf{C}_i^{-1} + \mathbf{C}_j^{-1})(\mathbf{m}_i - \mathbf{m}_j)(\mathbf{m}_i - \mathbf{m}_j)^{\mathrm{T}} \right\} = \text{term 1} + \text{term 2} \qquad (10.26)$$

Note that term 1 involves only the covariance matrices, while term 2 is a squared distance between the means of the distributions, normalised by the covariances. Equation (10.26) is the divergence between a *pair* of normally distributed spectral classes. Should there be more than two, which is generally the case in practice, all pairwise divergences need to be checked when evaluating which features might be discarded as relatively non-discriminative. An average indication of separability is given by the *average divergence*

$$d_{ave} = \sum_{i=1}^{M} \sum_{j=i+1}^{M} p(\omega_i)p(\omega_j)d_{ij}(\mathbf{x}) \qquad (10.27)$$

where M is the number of spectral classes and $p(\omega_i)$ and $p(\omega_j)$ are spectral class prior probabilities.

### 10.5.2.3  Using Divergence for Feature Selection

Consider the need to select the best three discriminating channels for an image recorded with four channels and in which three spectral classes exist. The pairwise divergence between each pair of spectral classes would be computed for all combinations of three out of the four channels. The feature subset chosen would be that which gives the highest average divergence.

In general, for $M$ spectral classes, $N$ total features, and a need to select the best $n$ feature subset, the following set of pairwise divergence calculations are necessary, leaving aside the need finally to compute the average divergence for each subset. First there are $^NC_n$, possible combinations of $n$ features from the total $N$, and for each combination there are $^MC_2$ pairwise divergence measures to be computed. For a complete evaluation $^NC_n \times {}^MC_2$ measures of pairwise divergence have to be calculated. Thus, to assess the best 4 of 7 Landsat ETM+ bands for an image involving 10 spectral classes $^7C_4 \times {}^{10}C_2 = 1575$ divergences have to be calculated. Inspection of (10.26) shows each divergence calculation to be considerable. That, together with the large number required in a typical problem, makes the use of divergence to check separability expensive computationally.

#### 10.5.2.4  A Problem with Divergence

As spectral classes become further separated in multispectral space, the probability of being able to classify a particular pattern moves asymptotically to 1.0 as depicted in Fig. 10.16a. If divergence is plotted it will be seen from its definition that it increases quadratically with separation between spectral class means, as illustrated in Fig. 10.16b. That behaviour is unfortunately misleading if divergence is to be used as an indication of how successfully pixels in the corresponding spectral classes could be mutually discriminated or classified. It implies, for example, that at large separations, further small increases will lead to vastly better classification accuracy whereas in practice that is not the case, as observed from the very slight increase in probability of correct classification implied by Fig. 10.16a. Also, outlying, easily separable classes will weight average divergence upwards in a misleading fashion to the extent that sub-optimal reduced feature subsets might be indicated as best.[17] This problem renders divergence, as it is presently defined, to be unsatisfactory. The Jeffries-Matusita distance in the next section does not suffer that drawback.

### 10.5.3  The Jeffries-Matusita (JM) Distance

#### 10.5.3.1  Definition

The JM distance between a pair of probability distributions (spectral classes) is defined as

---

[17] See Swain and Davis, loc. cit.

**Fig. 10.16  a** Probability of correct classification as a function of class separation, **b** divergence as a function of class separation

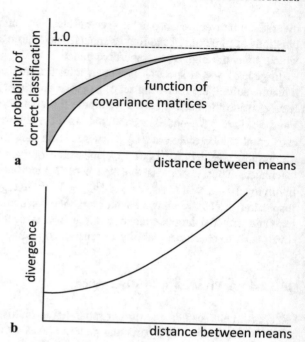

$$J_{ij} = \int\limits_{\mathbf{x}} \left\{ \sqrt{p(\mathbf{x}|\omega_i)} - \sqrt{p(\mathbf{x}|\omega_j)} \right\}^2 d\mathbf{x} \tag{10.28}$$

which is a measure of the average distance between the two distributions.[18] For normally distributed classes it becomes

$$J_{ij} = 2\left(1 - e^{-B_{ij}}\right) \tag{10.29}$$

in which

$$B_{ij} = \frac{1}{8}(\mathbf{m}_i - \mathbf{m}_j)^{\mathrm{T}} \left\{ \frac{\mathbf{C}_i + \mathbf{C}_j}{2} \right\}^{-1} (\mathbf{m}_i - \mathbf{m}_j) + \tfrac{1}{2}\ln\left\{ \frac{|(\mathbf{C}_i + \mathbf{C}_j)/2|}{|\mathbf{C}_i\mathbf{C}_j|^{1/2}} \right\} \tag{10.30}$$

which is referred to as the *Bhattacharyya distance*.[19] The first term in $B$ is like the square of a normalised distance between the class means. That is counteracted by

---

[18] See A. G. Wacker, The Minimum Distance Approach to Classification, *Ph.D. Thesis*, Purdue University, West Lafayette, Indiana, 1971.

[19] See T. Kailath, The divergence and Bhattacharyya distance measures in signal selection, *IEEE Transactions on Communications Theory*, vol. COM-15, 1967, pp. 52–60.

**Fig. 10.17** Jeffries-Matusita distance as a function of class separation

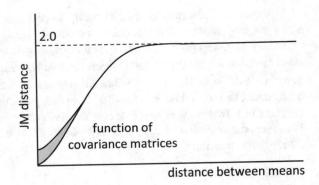

the exponential factor in (10.29) which gives an exponentially decreasing weight to increasing separations between spectral classes. As seen in Fig. 10.17, if plotted as a function of distance between class means, JM distance shows a saturating behaviour not unlike that expected for the probability of correct classification.

Equation (10.29) is asymptotic to 2.0, so that a JM distance of 2.0 between two spectral classes implies that a pixel could be placed into one of those classes with 100% accuracy. This saturating behaviour is highly desirable since it does not suffer the difficulty experienced with divergence.

As with divergence, an average pairwise JM distance can be defined according to

$$J_{ave} = \sum_{i=1}^{M} \sum_{j=i+1}^{M} p(\omega_i) p(\omega_j) J_{ij}(\mathbf{x}) \tag{10.31}$$

where $M$ is the number of classes and $p(\omega_i)$ and $p(\omega_j)$ are the prior probabilities.

The block diagonal approximation of Sect. 10.4 can also be used to compute separability measures such as divergence in (10.26) and the JM distance of (10.29) and (10.30). In particular, the Bhattacharyya distance can be expressed

$$B_{ij} = \sum_{h=1}^{H} \left[ \frac{1}{8} (\mathbf{m}_{ih} - \mathbf{m}_{jh})^{\mathrm{T}} \left\{ \frac{\mathbf{C}_{ih} + \mathbf{C}_{jh}}{2} \right\}^{-1} (\mathbf{m}_{ih} - \mathbf{m}_{jh}) + \frac{1}{2}\ln \left\{ \frac{|(\mathbf{C}_{ih} + \mathbf{C}_{jh})/2|}{|\mathbf{C}_{ih}\mathbf{C}_{jh}|^{1/2}} \right\} \right]$$

### 10.5.3.2  Comparison of Divergence and JM Distance

JM distance performs better as a feature selection criterion for multivariate normal classes than divergence for the reasons given above. However, it is computationally more expensive to use, as can be seen from a comparison of (10.26) and (10.30). Suppose a particular problem involves $M$ spectral classes. Consider the cost of computing all pairwise divergences and all pairwise JM distances. Costs can be assessed largely on the basis of having to compute matrix inverses.

In the case of divergence, it is necessary to compute only $M$ matrix inverses to allow all the pairwise divergences to be found. However, for JM distance it is necessary to compute $^MC_2 + M$ equivalent matrix inverses, since the individual class covariances appear as pairs which have to be added and then inverted. Note that $^MC_2 + M = \frac{1}{2}(M + 1)$ so that divergence is a factor of $\frac{1}{2}(M + 1)$ more economical to use. When it is recalled how many feature subsets may need to be checked in a feature selection exercise that is clearly an important consideration. However, the unbound nature of divergence as discussed previously calls its usefulness into question.

## 10.5.4   Transformed Divergence

### 10.5.4.1   Definition

A useful modification to divergence can be generated by noting the algebraic similarity of divergence and the parameter $B$ in the JM distance expression in (10.30). Since both involve terms which are functions of covariance alone, and terms which appear as normalised distances between class means, a heuristic *transformed divergence* measure can be defined[20]

$$d_{ij}^{T} = 2\left(1 - e^{-d_{ij}/8}\right) \qquad (10.32)$$

Because of its exponential character it will have a saturating behaviour with increasing class separation, as does JM distance, and yet it is computationally easier to generate. It been demonstrated to be almost as effective as JM distance in feature selection, and considerably better than simple divergence, or simple Bhattacharyya distance.[21]

### 10.5.4.2   Transformed Divergence and the Probability of Correct Classification

The probability of making a classification error when placing a pattern into one of two equal prior probability classes with a pairwise divergence $d_{ij}$ is bound by[22]

---

[20] See Swain and Davis, loc. cit.

[21] See P. H. Swain and R. C. King, Two effective feature selection criteria for multispectral remote sensing, *Proc. 1st Int. Joint Conference on Pattern Recognition*, November 1973, pp. 536–540, and P. W. Maunsell, W. J. Kramber and J. K. Lee, Optimum band selection for supervised classification of multispectral data, *Photogrammetric Engineering and Remote Sensing*, vol. 56, 1990, pp. 55–60.

[22] Kailath, loc. cit.

$$p_E > \frac{1}{8} e^{-d_{ij}/2}$$

so that the probability of correct classification is bound by

$$p_C < 1 - \frac{1}{8} e^{-d_{ij}/2}$$

$$\text{since from (10.32)} \quad d_{ij} = -8\ln\left[1 - \frac{1}{2}d_{ij}^{T}\right]$$

$$\text{then} \quad p_C < 1 - \frac{1}{8}[1 - \frac{1}{2}d_{ij}^{T}]^4 \tag{10.33}$$

This bound on classification accuracy is shown plotted in Fig. 10.18. Also shown is an empirical relationship between transformed divergence and probability of correct (pairwise) classification. This figure has considerable value in establishing *a priori* the upper bound on classification accuracy for an existing set of spectral classes.

### 10.5.4.3  Use of Transformed Divergence in Clustering

Clustering algorithms are the subject of Chap. 9. One of the last stages in clustering is to evaluate the size and relative locations of the clusters produced. If clusters are too close to each other in spectral space, they should be merged. The availability of the information in Fig. 10.18 allows merging decisions to be based on a pre-specified transformed divergence if cluster mean and covariance data is available. By establishing a desired accuracy level for the subsequent classification, from which the corresponding value of transformed divergence can be specified, classes with separabilities below that value should be merged.

## 10.5.5  *Separability Measures for Minimum Distance Classification*

The separability measures of the previous section relate to spectral classes modelled by multivariate normal distributions, which assumes that the classifier to be used is based on the Gaussian maximum likelihood rule. Should another classifier be chosen those measures are largely without meaning. If supervised classification is to be carried out using the minimum distance to class means technique there is no sense in using distribution-based separability measures, since distribution class models are not employed. Instead, it is better to use a simple measure of distance between the class means. Commonly, this is Euclidean distance. Consequently, when a set of spectral classes has been determined in preparation for classification, the complete set of pairwise Euclidean distances will provide an indication of class

**Fig. 10.18** Probability of correct classification as a function of pairwise transformed divergence; the empirical results were determined from 2790 sets of multivariate, normally distributed, two class data[23]

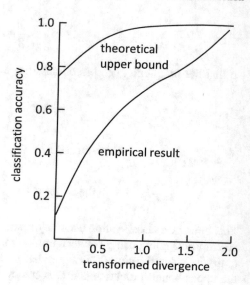

similarities. Unfortunately, this cannot be related to a classification error probability, but finds application as an indicator of what pairs of classes could be merged, if so desired.

## 10.6   Distribution Free Feature Selection—ReliefF

The methods for feature selection treated in Sect. 10.5 make the assumption that the individual classes in image data can be modelled by distribution functions, thereby requiring knowledge of class parameters such as the mean vector and covariance matrix. The application to high dimensional data is therefore limited.

We met some procedures that did not rely on distribution assumptions in Sect. 10.3, but they were transformation based. In this section we examine a method for selecting a reduced set of bands from those recorded, by making an assessment of the relative importance of those original features, without transformation or the need to know class statistics. The method is called ReliefF, which is a commonly used measure for feature selection.[24]

---

[23] Swain and King, *loc. cit.*

[24] See K. Kira and L. A. Rendell, The feature selection problem: traditional methods and a new algorithm, *Proc 10th Int. Conf. on Artificial Intelligence, AAAI-92*, 12–16 July 1992, San Jose, pp. 129–134, and Z. Wang, Y. Zhang, Z. Chen, H. Yang, X. Sun, J. Kang, Y. Kang and X. Liang, Application of ReliefF algorithm to selecting feature sets for classification of high resolution remote sensing image, *Proc IEEE Int. Geoscience and Remote Sensing Symposium*, 10–15 July 2016, Beijing, pp. 755–758.

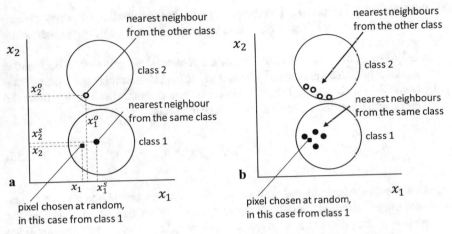

**Fig. 10.19  a** Defining the nearest neighbours, **b** defining the nearest neighbourhoods

ReliefF gives features a weight, which is adjusted as a measure of relevance by reference to the classes of data being used. Those features with weights below a user-specified threshold are discarded.

We will develop the process by reference to the two class, two dimensional data set shown in Fig. 10.19. The classes have been drawn intentionally so that one feature $x_1$ does not aid separation while the other $x_2$ does.

The process commences by selecting a pixel at random from one of the classes. We then find its nearest neighbours in the same and the other class, as shown. We now want to derive a measure that gives more weight to feature $x_2$ than feature $x_1$ in this example with respect to those chosen pixels.

We define a quantitative weight *for each feature* that tells us how important it is with regard to class separation. Call these weights $\omega_1$ and $\omega_2$ respectively. The weights are initially set to zero and then updated using the pixel we have chosen randomly. We will shortly choose further random pixels, to give us a better measure of the weights, but for the moment just concentrate on the one in Fig. 10.19.

For the weight corresponding to the $i$th feature we use the updating rule:

$$\omega_i^{new} = \omega_i^{old} - d\left(x_i - x_i^s\right) + d\left(x_i - x_i^o\right) \tag{10.34}$$

$x_i$ is the feature of the randomly chosen pixel, $x_i^s$ is the corresponding feature of the nearest neighbour from the same class and $x_i^o$ is the corresponding feature of the nearest neighbour from the other class; $d$ is a distance measure.

Applying the updating rule to Fig. 10.19a we see that the adjustment will be small and negative for feature $x_1$ but large and positive for feature $x_2$. That means the weight for the second feature increases, indicating its relative importance, while that for the first feature drops.

The same process is carried out several times, using a set of $m$ randomly chosen sample pixels, updating the weights each time. The distance values are normalised by the number of samples $m$. If there is a reasonable distribution of pixels in each class, a little thought will show that the weight for feature $x_2$ will go on increasing relative to that for $x_1$.

A helpful modification is to use a set of the $K$ nearest neighbours (in each class) to the randomly chosen pixels, as indicated in Fig. 10.19b. In this case the updating rule becomes as shown in (10.35), which indicates normalisation by both the number of trials and the number of nearest neighbours.

$$\omega_i^{new} = \omega_i^{old} - \frac{1}{mK} \sum_{k=1}^{K} \left[ d\left(x_i - x_i^{sk}\right) - d\left(x_i - x_i^{ok}\right) \right] \qquad (10.35)$$

Again, the process is initialised with all weights $\omega_i$ set to zero. Each weight is then updated by selecting $m$ random samples (pixels) using the above rule. At the completion of the process those features $x_i$ with weight values above a threshold are kept for subsequent quantitative analysis (classification).

The reliefF method, as originally formulated, applies to two classes. It is readily extended to the more usual multi-class situation by adding an "other class" term for each of the classes other than that from which the random sample is taken, leading to (10.36)

$$\omega_i^{new} = \omega_i^{old} - \frac{1}{mK} \sum_{k=1}^{K} d\left(x_i - x_i^{sk}\right) + \frac{1}{mK} \sum_{o \neq s} \frac{p(o)}{1 - p(s)} \sum_{k=1}^{K} d\left(x_i - x_i^{ok}\right) \qquad (10.36)$$

The $x_i^{ok}$ are now the $i$th features of the $k$th nearest neighbours of the current random sample in each of the other classes. The probability expression $\frac{p(o)}{1-p(s)}$ weights the contributions from the other classes in proportion to their (and the same class) prior probabilities.

In all these formulas the distance measure can be any convenient metric. Euclidean and city block distance are the most commonly used.

## 10.7   Improving Covariance Estimates Through Regularisation

Some of the techniques we have considered in this chapter require the use of covariance matrices, and yet there may not be enough data available to be comfortable that those estimates are reliable. In such cases it is sometimes helpful to use reliable approximations.

Regularisation is a technique used in mathematics and statistics to constrain estimates of parameters to within reasonable bounds. In the context of developing a reliable estimate of a class covariance matrix in the face of limited training data, regularisation refers to the mixing of several estimates of covariance in order to provide a more robust measure.

For example, let $\mathbf{C}_i$, $i = 1...M$ be estimates of the class covariance matrices obtained from available training data for the classes $\omega_i, i = 1...M$. If sufficient training samples are not available, the $\mathbf{C}_i$ will be poor estimates. Let $\mathbf{C}_M$ be the covariance matrix computed from the full set of training samples; in other words, it will be a global covariance matrix which reflects the scatter of the complete set of training data. Because it is based on a greater number of samples it is likely to be more accurate, for what it is, than any of the set of $\mathbf{C}_i$. A regularised approximation that can be used for the class conditional covariance matrices is

$$\mathbf{C}_i^{approx} = \alpha\mathbf{C}_i + (1 - \alpha)\mathbf{C}_M \qquad (10.37)$$

where $\alpha$ is a mixing parameter. Often diagonal versions of one of the constituent matrices are be used in (10.37), particularly for the original class covariance estimate. For example, the following approximations will be found in practice.

$$\mathbf{C}_i^{approx} = \alpha \, \text{diag}\mathbf{C}_i + (1 - \alpha)\mathbf{C}_M \qquad (10.38a)$$

$$\mathbf{C}_i^{approx} = \alpha \, \text{tr}\mathbf{C}_i\mathbf{I} + (1 - \alpha)\mathbf{C}_M \qquad (10.38b)$$

The value of the parameter $\alpha$ has to be determined to ensure that the approximation is as good as possible. One way to do that is to vary $\alpha$ and see how well the covariance estimate performs on a labelled set of data.

Another regularised covariance estimator is[25]

$$
\begin{aligned}
\mathbf{C}_i^{approx} &= \alpha\mathbf{C}_i + (1 - \alpha)\text{diag}\mathbf{C}_i & 0 \leq \alpha \leq 1 \\
&= (2 - \alpha)\mathbf{C}_i + (\alpha - 1)\mathbf{C}_M & 1 < \alpha \leq 2 \\
&= (3 - \alpha)\mathbf{C}_M + (\alpha - 2)\text{diag}\mathbf{C}_M & 2 < \alpha \leq 3
\end{aligned}
\qquad (10.39)
$$

Again, the optimum value for $\alpha$ would be found by checking performance with a labelled set of pixels.

It is interesting to examine the nature of the estimate in (10.39) for specific values of $\alpha$, noting the nature of the class conditional distributions that result and the likely forms of the discriminant functions. For example:

- For $\alpha = 0$, $\mathbf{C}_i^{approx} = \text{diag}\mathbf{C}_i$, meaning that each class is represented by the diagonal elements of its class covariance matrix, and that cross correlations are ignored. Consequently, the classes are assumed to be distributed hyperelliptically with axes parallel the spectral axes. A linear decision surface will result.
- For $\alpha = 1$, $\mathbf{C}_i^{approx} = \mathbf{C}_i$, meaning that each class is represented by its actual class conditional covariance matrix, generating quadratic decision surfaces

---

[25] See Landgrebe, 2003, loc. cit.

between the classes. This will give the full multivariate, normal maximum likelihood classification.

- For $\alpha = 2$, $\mathbf{C}_i^{approx} = \mathbf{C}_M$, meaning that all classes are assumed to have the same covariance matrix, equivalent to the global covariance, again generating linear decision surfaces.

- For $\alpha = 3$, $\mathbf{C}_i^{approx} = \mathrm{diag}\mathbf{C}_M$, meaning again that all classes have the same covariance matrix; in this case it consists just of the diagonal terms of the global covariance matrix. All class covariances will be identically hyperelliptical with axes parallel to the spectral axes, resulting in linear decision surfaces.

## 10.8   Bibliography on Feature Reduction

Possibly the most comprehensive treatment of feature reduction in remote sensing for data of high dimensionality, will be found in

D.A. Landgrebe, *Signal Theory Methods in Multispectral Remote Sensing*, Wiley, Hoboken, N.J., 2003.

An early but helpful treatment of Divergence and Bhattacharyya distance for feature selection is given by.

T. Kailath, The Divergence and Bhattacharyya distance measures in signal selection, *IEEE Transactions on Communication Theory,* vol. COM-15, 1967, pp. 52–60,

while

S.K. Jensen and F.A. Waltz, Principal components analysis and canonical analysis in remote sensing, *Proc. American Photogrammetric Society 45th Annual Meeting*, 1979, pp. 337–348

provides a good introductory treatment, with a minimum of mathematics. More mathematically detailed treatments are given by

H. Seal, *Multivariate Statistical Analysis for Biologists*, Methuen, London, 1964, and

N.A. Campbell and W. R. Atchley, The geometry of canonical variate analysis, *Systematic Zoology*, vol. 30, 1981, pp. 268–280.

Seal provides a set of hand calculations for two and three dimensional data. The development of Transformed Divergence will be found in[26]

---

[26] Many of the historical information notes and reports produced by the Laboratory for Applications of Remote Sensing (LARS) at Purdue University have been scanned and are available at http://www.lars.purdue.edu/home/References.html accessed 2021.

P.H. Swain, T.V. Robertson and A.G. Wacker, Comparison of divergence and B-distance in feature selection, *Information Note 020871*, Laboratory for Applications of Remote Sensing, Purdue University, West Lafayette, IN, 1971,

P.H. Swain and R.C. King, Two effective feature selection criteria for multispectral remote sensing, *Proc. 1st Int joint Conference on Pattern Recognition*, Washington, DC, Oct-Nov 1973, pp. 536–540, and

P.H. Swain and S.M. Davis, eds., *Remote Sensing: the Quantitative Approach*, McGraw-Hill, N.Y., 1978.

The last book also provides a detailed coverage of divergence and JM distance as feature selection criteria. Alternative feature selection procedures will be found in

S.B. Serpico, G. Moser and A.F. Cattoni, Feature reduction for classification purpose, Chapt 10 in C-I Chang (ed), *Hyperspectral Data Exploitation: Theory and Applications*, John Wiley & Sons, Hoboken, N.J., 2007.

More recent reviews of feature selection are

D. Singh, S. Appavu and E. Leavline, Literature review on feature selection methods for high-dimensional data, *International Journal of Computer Applications* (0975-8887) Vol. 136, No.1, February 2016, pp. 9–17

B. Rasti, D. Hong, R. Hang, P. Ghasimi, X. Kang, J. Chanussot and J. Benediktsson, Feature extraction for hyperspectral imagery, *IEEE Geoscience and Remote Sensing Magazine*, December 2020, pp. 60–88.

## 10.9   Problems

10.1   It can be shown[27] that the probability of making an error in labelling a pattern as belonging to one of two classes with equal prior probabilities is bound according to

$$\frac{1}{16}(2 - J_{ij})^2 \leq p_E \leq \frac{1}{4}(2 - J_{ij})$$

where $J_{ij}$ is the Jeffries-Matusita distance between the classes. Derive an expression for, and plot, the upper and lower bounds on classification accuracy for a two class problem, as a function of $J_{ij}$. An empirical relationship between classification accuracy and $J_{ij}$ is available in Swain and King.[28]

---

[27] See Kailath, loc. cit.

[28] See P. H. Swain and R. C. King, Two effective feature selection criteria for multispectral remote sensing, *Proc. First Int. Joint Conf. on Pattern Recognition*, November 1973, pp. 536–540.

10.2   Consider the training data given in problem 8.1. Suppose we want to use only one feature to characterise each spectral class. By computing pairwise transformed divergence ascertain the best feature to retain if:

      (a)  only classes 1 and 2 are of interest,
      (b)  only classes 2 and 3 are of interest,
      (c)  all three classes are of interest.

In each case estimate the maximum possible classification accuracy.

10.3   Using the same data as in problem 10.2 perform feature reduction, if possible, using the principal components transformation when the covariance matrix is generated using:

      (a)  only classes 1 and 2,
      (b)  only classes 2 and 3,
      (c)  all three classes.

10.4   Using the same data as in problem 10.2 compute a canonical analysis transformation involving all three classes and see whether the classes have better discrimination in the transformed axes compared with the original axes.

10.5   A particular image covers an agricultural region with several cover types. Suppose the mean vectors and covariance matrices have been found from training data for each class. Because of the nature of the land use, the region consists predominantly of fields that are large compared with the effective ground dimensions of a pixel. Within each field there is a degree of similarity among the pixels, owing to its use for a single crop type.

Suppose you delineate a field from the rest of the image and then compute the mean vector and covariance matrix for the pixels in that particular field. Describe how pairwise divergence, or Jeffries-Matusita distance, could be used to classify the complete *field* of pixels into one of the training classes.

10.6   The application of rotational transforms such as the principal components transformation and canonical analysis cannot improve intrinsic separability, which is the separability possible in the original data with all dimensions retained. Why?

10.7   The principal components transformation can be used for feature selection. What advantages and disadvantages does it have compared with canonical analysis?

10.8   Two classes of data have the statistics:

$$\mathbf{m}_1 = \begin{bmatrix} 10 \\ 20 \end{bmatrix} \quad \mathbf{C}_1 = \begin{bmatrix} 1 & 0 \\ 0 & 1 \end{bmatrix}$$

$$\mathbf{m}_2 = \begin{bmatrix} 10 \\ 20 \end{bmatrix} \quad \mathbf{C}_2 = \begin{bmatrix} 5 & 0 \\ 0 & 5 \end{bmatrix}$$

(a)  Will a minimum distance classifier work with this data?

(b)  Calculate the JM distance between the classes. Are the classes separable?

(c)  Assuming equal prior probabilities, classify the pixel vector $\mathbf{x} = [12 \ 30]^{\text{T}}$.

10.9  Both training and testing data are required for developing a Gaussian maximum likelihood classifier. What reason might there be for low classification accuracy on the training data? If the classification accuracy is high on the training data but low on the testing data, what could be the reason?

10.10  When displayed in image format, the correlation matrix for a hyperspectral image will show regions of high correlation existing in blocks, mostly down the diagonal. Is that because:

(a)  Adjacent bands of data are most likely to be highly correlated,

(b)  The maximum likelihood classifier requires the correlation matrix to look like that, or

(c)  The hyperspectral sensor is designed to make that happen for all images?

10.11  The block based maximum likelihood classifier characterised by (10.20) requires decisions to be taken about what blocks to use. From your knowledge of the spectral response of the three common ground cover types of vegetation, soil and water, recommend an acceptable set of block boundaries that might always be used with AVIRIS data.

10.12  Using the results of Problem 10.11, or otherwise, discuss how the canonical analysis transformation might take advantage of partitioning the covariance matrix into diagonal blocks.

10.13  Does partitioning the covariance matrix into blocks assist minimum distance classification?

10.14  A principal components transformation can be computed for the pixels in a given class, rather than over a whole image. By examining the correlation matrices in Fig. 10.13 which class would show the most compression of variance by a transformation?

10.15  Explain the structural differences in the correlation matrices for the stone and wheat classes in Fig. 10.13.

10.16  If you were to use the maximum likelihood classifier to produce a thematic map from an image data set with 6 bands, and wanted to reduce the number of features beforehand, would your chosen feature reduction method be

(a)  ReliefF,

(b)  The principal components transformation, or

(c)  Canonical analysis?

10.17  If you were to use a support vector classifier to produce a thematic map from an image data set with 200 bands, and wanted to reduce the number of features beforehand, would your chosen feature reduction method be

(a)  ReliefF,

(b)  The principal components transformation, or

(c)  Canonical analysis

10.18  Which of the following is a distribution free feature reduction technique?

(a)  Transformed divergence

(b)  Canonical analysis

(c)  Non-parametric discriminant analysis.

10.19  Explain the difference between distribution-based and distribution free measures of separability. When can distribution-based separability measures be used?

# Chapter 11
# Image Classification in Practice

**Abstract** Embedding supervised and unsupervised classification algorithms in methodologies is shown to be an effective way of producing reliable thematic maps from remotely sensed data. Examples are presented. Guidance is given on the selection of training and testing data, including how many pixels should be used in each case. The concept of the accuracy of a thematic map, in contrast to the performance of a classifier, is developed, noting that the two are the same only in special circumstances. The error matrix is described, as are the notions of producers and users accuracies and the Kappa coefficient. Decision tree classifiers are treated in some detail, including CART (Classification and Regression Trees) and forest classification. Hyperspectral image interpretation through library searching, and end-members and un-mixing, are also commented on. The coverage is concluded with a comparative assessment of maximum likelihood classification, the support vector machine and the convolutional neural network.

## 11.1 Introduction

In the previous chapters we have derived techniques that can be used to classify an image and create a thematic map. What we have to do now is to place those techniques into methodologies that make the classification task more efficient and accurate. Using an appropriate methodology is probably one of the most important considerations in applying classification to data recorded by imaging sensors and is often overlooked in practical image labelling. It is the role of this chapter to consider these operational aspects of thematic mapping, including how the accuracy of the final thematic map product can be assessed. We will consider accuracy determination in detail, including guidelines for choosing the number of samples needed to measure map accuracy effectively.

It is important to distinguish between classification techniques developed in the research laboratory and those applied in practice. While new algorithms evolve regularly, in pursuit of machine learning and mapping goals, many can be difficult to use in an operational sense, not because they don't perform well, but because

© The Author(s), under exclusive license to Springer Nature Switzerland AG 2022
J. A. Richards, *Remote Sensing Digital Image Analysis*,
https://doi.org/10.1007/978-3-030-82327-6_11

they have not reached operational maturity, or involve complex training processes. That can put them beyond the reach of a typical operational image analyst faced with a practical mapping task, who is more interested in outcomes than the novelty of an algorithm. There will be occasions in remote sensing, particularly when the number of bands is small, when classifiers as simple as the minimum distance rule will do the job perfectly. So, don't necessarily use a complicated classifier when a simple one will do the job just as well.

Further, we can sometimes generate better results by mixing algorithms and approaches; we demonstrate that in this chapter by illustrating a hybrid supervised-unsupervised classification methodology used since the earliest days of quantitative remote sensing.

Another classification methodology we will treat later in the chapter is the use of decision trees. Those structures have many benefits, including the ability to partition features into subsets for efficient scene labelling. We will also consider collections of decision trees (sometimes referred to as forest classifiers), to see how they perform in enhancing classification results in the manner of committee classifiers and the AdaBoost process. We also examine how scientific knowledge can be used in the classification of high spectral resolution data, and how sub-pixel class mixtures can be handled.

## 11.2   An Overview of Classification

First, consider a summary of the essential aspects of supervised and unsupervised classification as the foundations on which methodologies can be developed, and in order to highlight some practical aspects of operational thematic mapping.

### 11.2.1   Supervised Classification

#### 11.2.1.1   Selection of Training Data

The essence of supervised classification is that the analyst acquires beforehand a labelled set of training pixels for the classes of interest. Often that entails the analyst obtaining reference data such as aerial photographs, maps of the region of interest, or even hard copy products of the image data, from which skilled photointerpretation generates a set of acceptable training pixels. If the training data is to be taken from a map it is important that the scales of the map and image data be comparable. It is particularly useful if the image can be registered to the map beforehand so that selected polygons from the map can be laid over the image to identify pixels of given cover types for training.

Often the identification of training pixels, or more commonly training fields, will require field visits, which is an expensive component of the exercise, but one which is often used in practice to ensure good classification outcomes.

It is necessary to collect training data at least for all classes of interest and, preferably, for all apparent classes in the segment of image to be classified. In either case, and particularly if the selection of training data is not exhaustive or fully representative, it is prudent to use some form of threshold or limit on the classification so that pixels in the image that are not well represented in the training data are excluded in a trial classification. Such a limit can be imposed in maximum likelihood classification, for example, by the use of thresholds on the discriminant functions.[1] By limiting a classification in this way, pixels in the image that are not well characterised by the training data will not be classified. That will identify weaknesses in the selection of the training data which hopefully can be rectified; the image can then be reclassified. Repeated refinement of the training data, and reclassification, can be carried out using a representative portion of the image data. That is important if the image requiring analysis is very large.

The training data is used to estimate the parameters or other constants required to operate the chosen classification algorithm. If the algorithm involves explicit parameters, such as the mean vector and covariance matrix for the multivariate normal distribution, then the technique is called *parametric*. For algorithms which do not involve sets of parameters of those types the term *non-parametric* is used even though constants have to be estimated, such as the kernel parameters in the support vector machine approach. Once the appropriate parameters or constants have been estimated using the training data, the algorithm is then ready to be employed on unseen (testing) pixels—it is then said to be trained.

As a proportion of the full image to be analysed, the amount of training data will often be less than 1–5% of the image pixels. The learning phase, therefore, in which the analyst plays an important part in the labelling of pixels beforehand, is performed on a very small part of the image. Once trained on such a small image segment, the classifier is then asked to attach labels to *all* the image pixels. That is where a significant benefit occurs in thematic mapping.

### 11.2.1.2 Feature Selection

When undertaking a classification, it is sensible to use no more features than necessary to get results at the level of accuracy required. There are two reasons for this. First, the more features involved in a classification the more costly the exercise. In the case of the maximum likelihood classifier, for example, cost increases with the square of the number of features.

Secondly, and more importantly, the well-known Hughes effect or the overfitting problem, alerts us to the drop in classifier performance that can be experienced

---

[1] See Sect. 8.3.5.

when there are not enough training samples available relative to the number of features used in the classifier. Estimates of the maximum likelihood signatures in training are severely compromised if the dimensionality of the mean vector and covariance matrix is high compared with the number of pixels used. Clearly, this is most critical when using sensors with large numbers of bands, such as imaging spectrometers.

When considering the application of machine learning algorithms for thematic mapping it makes little sense to contemplate using all the bands available. Some exercises only require data in given regions of the spectrum. Moreover, adjacent bands tend to be highly correlated because of overlaps of the bandpass filters that define the individual band measurements, and also because of the spectral reflectance behaviours of the cover types. That is not to say that considerable diagnostic benefit is not available in the large numbers of bands from an imaging spectrometer, but in a practical exercise a consequence of injudiciously using a large number of bands, without some consideration beforehand of their relevance, is to prejudice the performance of the classification. That is why any classification methodology will include feature reduction as an early step. A range of procedures is presented in Chap. 10.

When using a maximum likelihood rule, feature selection can be guided by the application of separability measures such as those treated in Chap. 10. Those metrics have another benefit when dealing with classes that are assumed to be represented by multivariate normal distributions. By checking their statistical separability, we can assess whether any pair of classes are so similar in spectral space that a significant misclassification error would occur if they were both used. If they are too close, they should be merged.

### 11.2.1.3  Classifier Outputs and Accuracy Checking

The output from the supervised classification approach typically consists of a thematic map of class labels, often accompanied by a table of area estimates and, importantly, an *error matrix* which indicates by class the residual error, or accuracy, of the final product. We look at the error matrix in detail in Sect. 11.6.2.

## 11.2.2  Unsupervised Classification

Unsupervised classification is an analytical procedure based, generally, on clustering algorithms such as those treated in Chap. 9. Clustering partitions a sample of the image data in spectral space into a number of distinct clusters or spectral classes. It then labels all pixels of interest as belonging to one of those classes to produce a cluster map, albeit with purely symbolic labels at this stage and not labels that indicate ground cover types.

In contrast to the prior use of analyst-provided information in supervised classification, unsupervised classification is a segmentation of the spectral space in the absence of any information fed in by the analyst. Instead, analyst knowledge is used afterwards to attach class labels to the map segments established by clustering, often guided by the spatial distribution of the labels shown in the cluster map. Clearly this is an advantage of unsupervised classification. However, by comparison to most techniques for supervised classification, clustering is a time-consuming process. This can be demonstrated, for example, by comparing the multiplication requirements of the iterative clustering algorithm of Sect. 9.3 with the maximum likelihood decision rule of Sect. 8.3.3.

Suppose a particular exercise involves $N$ bands and $C$ classes or clusters. Maximum likelihood classification requires $CPN(N+1)$ multiplications[2] where $P$ is the number of pixels in the segment to be classified. Clustering the same dataset requires $PCI$ distance measures for $I$ iterations. Each distance calculation requires $N$ multiplications, so that the total number of multiplications for clustering is $PCIN$. Thus the speed comparison of the two approaches is approximately $(N+1)/I$.

Clustering would, therefore, need to be completed within 7 iterations for a 6-band data set to be faster than maximum likelihood classification. Frequently about 20 times this number of iterations is necessary to achieve an acceptable clustering. Training the classifier adds a loading to its time demand; however, a significant time loading should also be added to clustering to account for the labelling step. Often that is done by associating pixels with the nearest cluster using a Euclidean distance measure in the spectral space, including any pixels that were not used in clustering.

Because of the time demand of clustering algorithms, unsupervised classification is not often carried out with large image segments. Usually, a representative subset of data is employed for the actual clustering phase in order to segment the spectral space. That information is then used to assign all the image pixels to one of the clusters to create the unsupervised thematic map using a minimum distance assignment.

When comparing the time requirements of supervised and unsupervised classification it must be remembered that a large demand on the analyst's time is required for training the supervised procedure. That is necessary both for gathering reference data and for identifying training pixels using that data. The corresponding step in unsupervised classification is the labelling of clusters afterwards. While that still requires user effort to gather labelled prototype data, not as much may be needed as when training a supervised procedure. Data is only required for those classes of interest. Also, often only a handful of labelled pixels is necessary to identify a given class because we can use spatial cues to help in that process; for example, crop fields will show as polygons, while roads and rivers will look as elongated and sometimes straggly features.

---

[2] See Sect. 8.5.6.

By comparison, in supervised training sufficient training pixels for every class are required to ensure that reliable estimates of class signatures are generated; and the analyst has to be certain that those training pixels are as pure as possible. Gathering sufficient samples of roads and river systems can sometimes be challenging, and mixture classes on boundaries are regularly overlooked. That can be a particular problem with high dimensionality data such as that recorded by imaging spectrometers.

A final point that must be taken into account when contemplating unsupervised classification by clustering is that there is no facility for including prior probabilities of class membership. By comparison, many supervised algorithms allow prior knowledge to bias the result generated from spectral data alone.

Once classes have been generated for the purpose of unsupervised classification, it is of value to use separability measures to see whether some clusters are sufficiently similar spectrally that they should be combined. That is particularly the case when the classes are generated on a sample of data; separability checking and merging would be carried out before cluster maps are produced.

## 11.2.3   Semi-supervised Classification and Transfer Learning

When insufficient labelled training samples are available it is still possible to use supervised classification algorithms by employing one of a number of techniques that use unlabelled data to supplement known training pixels.[3] Most are known as semi-supervised training procedures or semi-supervised learning; we do not treat them any further in this coverage, but they feature strongly in the machine learning community.[4] Because of the potential for overfitting, they have become of significant interest for use in convolutional neural networks.[5] The hybrid methodology outlined in Sect. 11.4 is similar to semi-supervised learning in that hard to characterise classes are able to be labelled via clustering. Mixing supervised and unsupervised methods is one basis for semi-supervised approaches.

---

[3] See J.A. Richards and X. Jia, Using suitable neighbours to augment the training set in hyperspectral maximum likelihood classification, *IEEE Geoscience and Remote Sensing Letters*, vol 5. No. 4, October 2008, pp. 774–777 and R.G. Negri, S.J.S. Sant'Anna and L.V. Dutra, Semi-supervised remote sensing image classification methods assessment, *Proc. Int. Geoscience and Remote Sensing Symposium*, IGARSS2011, Vancouver, 24–29 July 2011, pp. 2939–2942.

[4] See J.E. van Engelen and H.H. Hoos, A survey on semi-supervised learning, *Machine Learning*, vol. 109, 2020, pp. 373–440, O. Chapelle, B. Schölkopf and A. Zien, eds., *Semi-Supervised Learning*, MIT Press, Cambridge, Mass., 2006, and X. Zhu and A. Goldberg, *Introduction to Semi-Supervised Learning*, Morgan and Claypool, CA, 2009.

[5] See X. Dai, X. Wu, B. Wang and L. Zhang, Semi-supervised scene classification for remote sensing images: A method based on convolutional neural networks and ensemble learning, *IEEE Geoscience and Remote Sensing Letters*, vol 16, No. 6 June 2019, pp. 869–873.

Another approach to avoid over-fitting when training deep learning algorithms, such as convolutional neural networks, has been to use transfer learning, in which the weights of a CNN are initialised with values from another CNN previously trained on a different data set.[6]

## 11.3  Effect of Resampling on Classification

It is much easier to work with image data that has been registered to a map grid using the techniques of Sect. 2.18. That requires an interpolation method with which to synthesise pixel values for placement on the map grid. The two most common interpolation procedures are nearest neighbour resampling and resampling by cubic convolution. In the former, the original image pixels are simply relocated onto a geometrically correct map grid whereas, in the latter, new pixel brightness values are created by interpolating over a group of 16 neighbouring pixels.

Clearly, it is also desirable to have thematic maps registered to a map base. That can be done by rectifying the image before classification or by rectifying the thematic map, in which case nearest neighbour resampling is the only interpolation option available.

An advantage in correcting the image beforehand is that it is often easier to relate reference data to the image if it is in correct geometric registration to a map. However, a drawback with doing this prior to classification is that some of the pixel brightness values may be changed by the interpolation process used. Since cubic convolution fits an interpolating function over 16 neighbouring pixels, the brightness values in the rectified product are not the original brightness values recorded by the imaging sensor. As a result, some may not classify well. By comparison, the brightness values of the pixels in a corrected image based on nearest neighbour resampling are the original pixel brightness values, simply relocated. In that case resampling cannot affect subsequent classification results.

The influence of resampling on classification accuracy has been considered since the earliest days of remote sensing image processing. Cubic convolution interpolation in particular, has been shown to have a major influence across boundaries such as that between vegetation and water, leading to uncertainties in classification.[7]

---

[6] D. Marmanis, M. Datcu, T. Esch and U. Stilla, Deep learning earth observation classification using ImageNet pretrained networks, *IEEE Geoscience and Remote Sensing Letters*, vol. 11, no. 1, January 2016, pp. 105–109 and P. Vinayaraj, R. Sugimoto, R. Nakamura and Y. Yamaguchi, Transfer learning with CNNs for segmentation of PALSAR-2 power decomposition components, *IEEE Journal of Selected Topics in Applied Earth Observation and Remote Sensing*, vol. 13, 2020, pp. 6352–6361.

[7] See F.C. Billingsley, Modelling misregistration and related effects on multispectral classification, *Photogrammetric Engineering and Remote Sensing*, vol. 48, 1982, pp. 421–430, B.C. Forster and J.C. Trinder, An examination of the effects of resampling on classification accuracy, *Proc. 3rd Australasian Conf. on Remote Sensing (Landsat84)*, Queensland, 1984, pp. 106–115, and

When images in a multitemporal sequence have to be classified to extract change information it is necessary to perform image to image registration. Since registration cannot be avoided, again nearest neighbour resampling should be used if classification is contemplated. It is most effective when the scales of the two products to be registered are comparable.

## 11.4  A Hybrid Supervised/Unsupervised Methodology

### 11.4.1  Outline of the Method

The methodology now to be covered can be used with any supervised classifier, but it was developed originally for the maximum likelihood rule. We will derive it from that perspective but note, when appropriate, its application using other supervised algorithms.

The strength of supervised classification based on the maximum likelihood procedure is that it minimises classification error for classes that are distributed in a multivariate normal fashion. It can also label data relatively quickly. Its major limitation is that the information classes of interest to the user will generally not be of that form; either the distribution of pixels in a given information class will be smeared out in spectral space or will appear as sets of moderately distinct clusters.

In order to apply the maximum likelihood rule we need to resolve an information class into acceptable sets of Gaussian modes. In the terminology of remote sensing that means finding the constituent set of spectral classes that are needed to model each information class so that good classifier performance will result. That is the step most often overlooked when using Gaussian maximum likelihood classification, as a consequence of which many users find that the maximum likelihood rule does not work as well as might otherwise be expected.

Resolving information classes into constituent sets of unimodal spectral classes is a task that can be handled by clustering on a representative subset of image data. Used for this purpose, unsupervised classification performs the valuable function of identifying the existence of all spectral classes, yet it is not expected to perform the entire classification. Consequently, the rather logical hybrid classification procedure outlined below has been shown to work well.[8] It consists of five fundamental steps:

J. Verdin, Corrected vs uncorrected Landsat 4 MSS data, *Landsat Data Users Notes*, issue 27, NOAA, Sioux Falls, June 4–8 1983.

[8] This procedure is developed in M.D. Fleming, J.S. Berkebile and R.M. Hofer, Computer aided analysis of Landsat 1 MSS data: a comparison of three approaches including a modified clustering approach, *Information Note 072475*, Laboratory for Applications of Remote Sensing, Purdue University, West Lafayette, Indiana, 1975. http://www.lars.purdue.edu/home/references/LTR_072475.pdf.

Step 1: Use clustering to determine the spectral classes into which the image resolves. For reasons of economy this is performed on a representative subset of data. Spectral class statistics are produced from this unsupervised step.

Step 2: Using available reference data associate the spectral classes, or clusters, from Step 1 with information classes. Frequently there will be more than one spectral class for each information class.

Step 3: Perform a feature selection to see whether all features (bands or channels) need to be retained for reliable classification.

Step 4: Using the maximum likelihood algorithm, classify the entire image into the set of spectral classes.

Step 5: Label each pixel in the classification with the information class corresponding to its spectral class and use independent testing data to determine the accuracy of the classified product.

We now consider some of the steps in detail and introduce some practical aspects. The method depends for its accuracy, as do all classifications, on the skills and experience of the analyst. Consequently, in practice it is not unusual to iterate over sets of steps as experience is gained with the particular problem at hand.

## 11.4.2 Choosing the Image Segments to Cluster

Clustering is applied to a subset of the total image to find suitable spectral classes. Although this will depend on experience, it is recommended that about 3–6 small regions, called candidate clustering areas, be chosen for the purpose. They should be well spaced over the image, located such that each one contains several of the information classes of interest, and such that all information classes are represented in the collection of clustering areas. An advantage of choosing heterogeneous regions to cluster, as against the apparently homogeneous training areas that are used for simple supervised classification, is that mixture pixels which lie on class boundaries will be identified as legitimate spectral classes that represent mixed information classes.

With most clustering procedures the analyst has to specify a set of parameters that controls the number of clusters generated. About 2–3 spectral classes per information class have been found to be useful in general; clustering parameters should be selected with that in mind. The number of clusters could be chosen conservatively high because unnecessary classes can be deleted or merged at a later stage.

It is of value to cluster each region separately as that has been found to produce cluster maps with more distinct boundaries than when all regions are pooled beforehand.

## 11.4.3   *Rationalising the Number of Spectral Classes*

When clustering is complete the spectral classes are then associated with information classes using the available reference data. It is then necessary to see whether any spectral classes can be discarded, or more importantly, whether sets of clusters can be merged, thereby reducing the number. Decisions about merging can be made on the basis of separability measures, such as those treated in Chap. 10.

During this rationalisation process it is useful to be able to visualise the locations of the spectral classes. For this a bispectral plot (or biplot) can be constructed. The bispectral plot is not unlike a two-dimensional scatter plot of the spectral space in which the data appears. However, rather than displaying the individual pixels, class means are shown by their spectral coordinates. The most significant pair of spectral bands would be chosen in order to see the relative locations of the cluster centres. Sometimes several biplots are produced using different band combinations.

## 11.4.4   *An Example*

We now present an example to illustrate some key features of the hybrid approach. Because of the simplicity of this illustration not all the steps outlined earlier are involved; but the example highlights the value of using unsupervised classification as a means for identifying spectral classes and for generating signatures of classes for which the acquisition of training data would be difficult.

Figure 11.1a shows a small image segment recorded by the HyVista HyMap sensor over the city of Perth in Western Australia. It is centred on a golf course. The obvious cover types are water, grass (fairways), trees, bare ground including bunkers (sand traps), a clubhouse, tracks and roads. Apart from a few cases, the distribution of cover types suggests that it might be hard to generate training fields for all classes of interest.

With the direction in which the image was recorded, north is to the right. Shown on the figure are three fields that were used as clustering regions. Inspection shows that those fields among them cover all the obvious information classes.

Rather than cluster each field independently, which is generally preferable, the three fields were aggregated and an Isodata clustering algorithm was applied to the group with the requirement to generate 10 clusters.[9] Because there were seven information classes, and some are clearly very homogeneous, it was felt that 10 clusters would be adequate. Five bands were chosen for the exercise: band 7 (visible green), band 15 (visible red), band 29 (near infrared), band 80 (mid infrared) and band 108 (mid infrared). The last two were chosen on the infrared maxima of the vegetation and soil curves, midway between the water absorption

---

[9] All processing for this example was carried out using the MultiSpec[©] package developed at Purdue University, West Lafayette, Indiana.

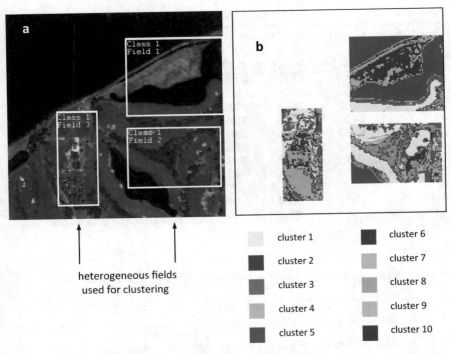

**Fig. 11.1 a** Image segment showing cluster regions, and **b** result of clustering those regions to generate 10 clusters

regions. It was felt that they would assist in discriminating among the bare ground and roadway classes.

Figure 11.1b shows the cluster maps produced for each of the three fields outlined in Fig. 11.1a. Table 11.1 shows the mean vectors for each of the 10 clusters, along with information class labels generated by observing the clusters against the image,[10] looking at their spatial orientation and distribution, and noting where they fall in the bispectral plot seen in Fig. 11.2. Of particular note is the ease with which signatures have been generated for the two elongated classes of tracks and roads. Although only the mean vectors are shown here, the clustering processor in MultiSpec generates the full covariance matrices that are used in the subsequent classification step.

The infrared versus red bispectral plot in Fig. 11.2 shows information class labels attached to the cluster means. As observed, there are two (spectral) classes of grass, and two of sparse vegetation. There is also a thin border class of mixed vegetation and water pixels; that often happens in practice and would not get picked

---

[10] Ordinarily these labels would be generated by using available reference data. In this case, photointerpretation of the original image easily reveals the information classes. This can be supplemented by other data for the region such as that available on Google Earth.

**Table 11.1** Cluster mean vectors and associated information class labels

| Cluster | Pixels | Cluster mean vector elements | | | | | Label |
|---|---|---|---|---|---|---|---|
| | | 7 (0.511 μm) | 15 (0.634 μm) | 29 (0.847 μm) | 80 (1.617 μm) | 108 (2.153 μm) | |
| 1 | 1957 | 1516.7 | 1074.9 | 6772.0 | 3690.6 | 1762.8 | Grass |
| 2 | 1216 | 1889.3 | 2275.3 | 4827.2 | 4766.3 | 3304.4 | Bare |
| 3 | 1643 | 1301.7 | 984.6 | 6005.4 | 3299.1 | 1574.9 | Grass |
| 4 | 1282 | 1592.5 | 1535.5 | 5503.9 | 4006.1 | 2275.6 | Sparse veg |
| 5 | 668 | 1012.9 | 830.8 | 4884.3 | 2519.4 | 1230.0 | Sparse veg |
| 6 | 649 | 1496.5 | 1708.4 | 3813.4 | 3369.0 | 2332.9 | Tracks |
| 7 | 950 | 707.7 | 540.2 | 3938.0 | 1600.1 | 722.6 | Trees |
| 8 | 669 | 1397.3 | 1509.3 | 2368.0 | 2135.2 | 1745.9 | Buildings/ roads |
| 9 | 462 | 642.1 | 625.2 | 2239.5 | 1194.4 | 656.9 | Water/veg |
| 10 | 2628 | 438.4 | 376.7 | 469.7 | 276.0 | 182.0 | Water |

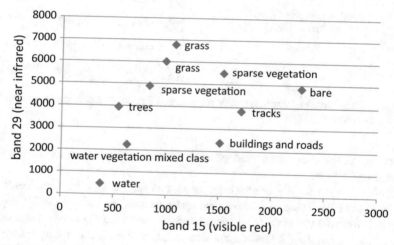

**Fig. 11.2** Near infrared versus visible red bispectral plot showing the cluster means and information class labels; note that two information classes each consist of two spectral classes

up in a traditional supervised classification exercise. Here it will be attributed to the water class.

Figure 11.3 shows the result of a maximum likelihood classification using the full set of signatures for all 10 clusters. Implicitly, that means we are using two spectral classes in the case of the grass information class and two spectral classes in the case of the sparse vegetation information class. In the classification map they have been coloured the same since the end user is not interested in the spectral class structure.

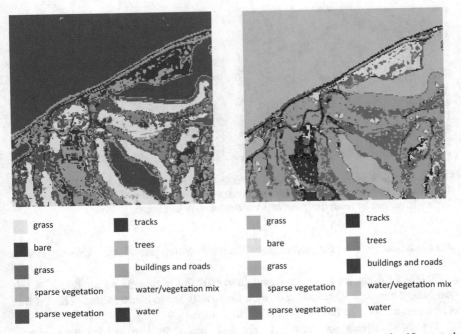

grass                     tracks            grass                     tracks

bare                      trees             bare                      trees

grass                     buildings and roads   grass                 buildings and roads

sparse vegetation         water/vegetation mix  sparse vegetation     water/vegetation mix

sparse vegetation         water             sparse vegetation         water

**Fig. 11.3** Thematic map produced by the maximum likelihood classifier using the 10 spectral classes generated by clustering, and re-coloured into the set of 7 information classes

## 11.4.5   Hybrid Classification with Other Supervised Algorithms

Although the method just demonstrated was devised for use with the maximum likelihood algorithm it can be used with other supervised approaches as well, taking advantage of the benefits derived from the unsupervised step—particularly the generation of training samples for mixture classes and spatially elongated classes, for which it might be difficult to extract meaningful training pixels. Here we show another simple example, in this case using the minimum distance rule.

The exercise involves classifying an arid region near the township of Bourke, Australia employed for growing cotton by the use of irrigation from a nearby river.[11] The task was to assess the area in hectares sown to cotton, as a surrogate for the amount of water used. Field agronomists had assessed the hectarage of cotton crops in the region but required corroborative evidence.

The image to be classified consists just of the visible red and the first of the two near infrared bands of a Landsat multispectral scanner image, recorded in February 1981. Although the region is very dry at that time of the year, apart from the crops

---

[11] See G.E. Moreton and J.A. Richards, Irrigated crop inventory by classification of satellite image data, *Photogrammetric Engineering and Remote Sensing*, vol. 50, 1984, pp. 729–737.

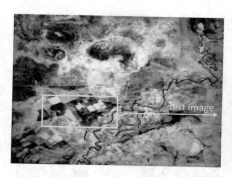

**Fig. 11.4** Near infrared Landsat MSS band of the region around the town of Bourke, NSW, Australia in February 1981, along with an extracted test region in which cotton crops and the Darling River can be seen; the green regions were chosen for clustering

there is a gallery (riparian) forest along the river, which provides another vegetation class.

Figure 11.4 shows a near infrared image of the region to be analysed. A test sub-image has been identified on which the results are to be evaluated. The Darling River can be seen to the south-east of the image.

The cotton fields are mostly in the test area (white in the left-hand image, indicating high IR response), along with an approximately triangular shaped crop in the bottom south-eastern corner, and some other scattered fields.

The four rectangular selections in the right-hand image sub-set, along with a sample of the lower triangular crop, were used to resolve the spectral space into spectral classes by clustering. Here, the simple single pass clustering algorithm (Sect. 9.9) was used and each of the five heterogenous regions was clustered separately. The results of the clustering are shown in Fig. 11.5 using a bispectral plot in terms of the means of the cluster centres found.

There were 34 clusters in total, which were then rationalised to the ten shown in the figure. That was done by associating the clusters with information classes using black and white and colour air photos, and photointerpretation of the image itself. Those ten grouped spectral classes were considered adequate to differentiate the image into its main cover types and thereby avoid any errors of commission which might lead to poor estimates of the area of the cotton crops.

When the minimum distance classifier was applied to the test image, using the ten rationalized spectral classes, it was found that the cotton crops accounted for 803 ha. The field agronomists had estimated 800 ha. It is not necessary to show a thematic map since the important result was the area of cotton in the test region. Nevertheless, one can be seen in Moreton and Richards.[12]

---

[12] ibid.

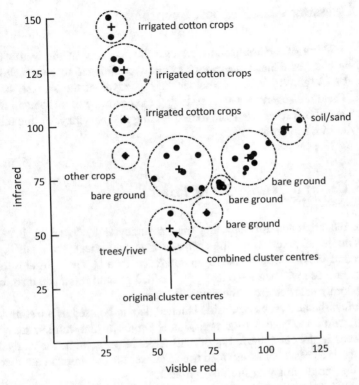

**Fig. 11.5** Near IR versus visible red bispectral plot, showing the original 34 cluster centres and the 10 rationalised classes, with their information class labels

## 11.5   Cluster Space Classification

Apart from the possibility of mixed pixel classes on information class boundaries, an assumption in the hybrid classification methodology just treated is that the spectral classes, by and large, map to single information classes. In other words, the overlap of a single spectral class into several information classes is assumed not to occur. When it is recalled that the spectral classes are a convenient segmentation of data in the spectral space, most likely generated by an unsupervised approach such as clustering, and that the information classes are simply user-defined labels that refer to regions on the ground, it is possible that the spectral classes and information classes might overlap substantially as illustrated in Fig. 9.15. Such a situation can be handled statistically, as a generalisation of the hybrid approach of Sect. 11.4, as shown in Sect. 9.14.

# 11.6   Assessing Classification Accuracy

At the completion of a classification exercise the accuracy of the results obtained needs to be checked. That is necessary to allow confidence to be attached to the results and will serve to indicate whether the objectives of the analysis have been achieved. There are several ways of assessing the accuracy of a thematic map, and there are several measures available for expressing that accuracy. In this section we consider the most common.

## 11.6.1   Use of a Testing Set of Pixels

The preferred approach for assessing map accuracy is to select an independent random sample of pixels from the thematic map and check their labels against actual ground classes determined from reference data.[13] That presents practical challenges because we then have to locate labelled ground reference data for those pixels randomly selected from the thematic map.

More often the analyst has set aside labelled data to be used as a testing set after the classification has been carried out. This is similar to the training set of pixels used to generate the classifier in the first instance. In most cases, the analyst labels as many pixels as practicable and then uses a subset for training and another subset for assessing the accuracy of the final product.

In principle, the testing data should be composed of pixels selected at random in order to avoid the interdependences of near neighbouring pixels. A difficulty that can arise with random sampling in this manner is that it is area-weighted. That is, large classes tend to be represented by a larger number of sample points than smaller classes; it is possible that some small classes may not be represented at all. To avoid the undesirable effect that that has on the assessment of accuracy of the smaller classes, it is necessary to ensure that those classes are represented adequately. An approach that is widely adopted is *stratified random sampling* in which the user first decides on a set of strata into which the image is divided. Random sampling is then carried out within each stratum. The strata could be any convenient area segmentation of the thematic map, such as grid cells. However, the most appropriate stratification to use is the actual thematic classes themselves. That effectively reduces any area bias in the random sampling that could lead to inappropriate accuracy estimation for the smaller classes.

---

[13] In the past, reference data was often called *ground truth*; that term is now less often used because sampling and interpretation inaccuracies can lead to errors in what might otherwise be regarded as a perfectly correct understanding of the ground labels. The term reference data is less dogmatic although we still tend to assume it is exact, unless we have reasons to believe differently.

**Table 11.2** Using an error matrix to summarise classifier performance and map accuracy, following the layout of Congalton and Green[14]

| | | reference data classes | | | |
|---|---|---|---|---|---|
| | | A | B | C | sum |
| thematic map classes | A | 35 | 2 | 2 | 39 |
| | B | 10 | 37 | 3 | 50 |
| | C | 5 | 1 | 41 | 47 |
| | sum | 50 | 40 | 46 | 136 |

overall accuracy = (35+37+41)/136 ≡ 83.1%

| producer's accuracies | | | user's accuracies | |
|---|---|---|---|---|
| A | 35/50 ≡ 70.0% | | A | 35/39 ≡ 89.7% |
| B | 37/40 ≡ 92.5% | | B | 37/50 ≡ 74.0% |
| C | 41/46 ≡ 89.1% | | C | 41/47 ≡ 87.2% |

## 11.6.2 The Error Matrix

Whichever of the approaches in Sect. 11.6.1 is used, it is common to express the results in the form of an *error matrix* (sometimes in the past called a *contingency matrix* or a *confusion matrix*) which lists the reference data classes by column and the classes indicated on the thematic map by row, as shown in Table 11.2. The cells in the table show the number of pixels that are common between a reference class and a map class. In an ideal result the table or matrix will be diagonal, indicating that for every reference class pixel the classifier has generated the correct label. For a poor classification the off-diagonal terms will be larger indicating that the classifier has had trouble correctly labelling the pixels from the reference data.

The column sums in the error matrix represent the total number of labelled reference pixels available per class. The row sums represent the total number of pixels labelled by the classifier as coming from a particular class in the set of pixels chosen to assess classification accuracy. Using those, we can define errors of *omission* and errors of *commission*. Errors of omission correspond to those pixels belonging to the reference class that the classifier has failed to recognise; they are therefore the off-diagonal terms down the column for a particular reference class. We can turn them into percentages by dividing the counts in those cells by the column sum. Errors of commission correspond to those pixels belonging to other reference classes that the classifier has placed in the class of interest; they are the off-diagonal terms across the row for a particular thematic map class. They can be turned into percentages by dividing the counts in those cells by the row sum.

---

[14] R.G. Congalton and K. Green, *Assessing the Accuracy of Remotely Sensed Data: Principles and Practices*, 2nd ed., CRC Press Taylor and Francis Group, Boca Raton Florida, 2009 and 3rd ed., 2019.

When interpreting an error matrix, it is important to understand that different indications of accuracy will result according to whether the number of correct pixels for a class (those on the diagonal) is divided by the total number of reference pixels for that class (the column sum) or the total number of pixels that the classifier attributes to the class (the row sum).

For example, consider class B in Table 11.2. As noted, 37 of the reference data pixels have been correctly labelled. This represents 37/40 ≡ 92.5% of the reference data pixels for the class. This is the probability that the classifier has labelled a pixel as class B given that the actual (reference data) class is B. It is often referred to as *producer's accuracy*[15] and is an indication of classifier performance.

A user of a thematic map produced by a classifier is often more interested in the likelihood that the actual class is B given at the pixel has been labelled B on the thematic map by the classifier; this is an indication of map accuracy. It is called *user's accuracy* and, for this example, is 37/50 ≡ 74.0%, indicating that only 74% of the pixels labelled B on the thematic map are correct, even though the classifier correctly handled almost 93% of the class B reference data. This distinction is important and leads one to believe that user's accuracy is the figure that should most often be adopted. Producer's and user's accuracies for all of the classes are indicated in Table 11.2.

Often the results of a classification exercise are expressed as a single figure of accuracy, independent of the class. In the case of the data in Table 11.2 we would say that the result is 113/136 ≡ 83.1% accurate. While that is an acceptable summary of the situation, it masks the fact that the classifier may handle some classes better than others. Knowing how the classifier performs on individual classes is important if the analyst is going to iterate through the results in order to refine the accuracy of the final product.

## 11.6.3   Quantifying the Error Matrix[16]

In Sect. 11.6.1 we noted two alternative methods that could be used to assess how a classifier performs and how accurate a map might be. In the first, samples are taken from the thematic map and checked against the 'true' labels on the ground. In the second, samples are taken of ground gathered reference data (a testing set) and used to check the classifier-generated map labels. The first checks the accuracy of the map and the second checks the performance of the classifier—they are not the same. The first is what the user wants but the second is easier in practice and generally is what is available when constructing the error matrix. The latter is equivalent to producer's accuracy, but the former is only equivalent to actual map accuracy when the distribution of testing pixels over the classes reflects the actual

---

[15] ibid.

[16] This section is based on J.A. Richards, Classifier performance and map accuracy, *Remote Sensing of Environment*, vol. 57, 1996, pp. 161–166.

class distributions on the ground, as the following analysis shows. This is not widely appreciated.

Suppose we use the variates $t$, $r$ and $m$ to represent respectively the true label for a pixel, its label in the reference data and its label on the thematic map; we will use the symbols $Y$ and $Z$ to represent any of the available class labels, i.e., $Y, Z \in \{A, B, C\}$.

If we sample the map and check the accuracy of those samples on the ground, then effectively we are estimating the map accuracy probabilities: $p(t = Z | m = Z)$. This is the likelihood that the actual label for a pixel is $Z$ if the map shows it as $Z$, which is what the user of the thematic map is interested in. More generally, $p(t = Y | m = Z)$ is the probability that $Y$ is the correct class if the map shows $Z$.

If, instead, we select reference pixels and check the map labels generated by the classifier then we are computing the classifier performance probabilities $p(m = Z | r = Z)$, or in general $p(m = Z | r = Y)$, which is the likelihood that the thematic map label is $Z$ for a pixel labelled as $Y$ in the reference data.

We have used the two different variates $t$ and $r$ to refer to the ground labels only because, in the case of checking the map by sampling from it, there is strictly no concept of reference data. We now assume they are the same, but that places an important constraint on the reference data set—i.e., the testing set. Its labelled pixels must be truly representative of the situation on the ground and, in particular, the number of pixels per class must be representative of their proportions on the region of the earth's surface being imaged. Random selections from the testing set will then yield a distribution by label which is the same as the prior probability of occurrence of the labels on the ground. If we make that assumption, then we can put

$$p(r = Y | m = Z) \equiv p(t = Y | m = Z)$$

and then use Bayes Theorem to relate the map accuracy and classifier performance probabilities

$$p(r = Y | m = Z) = \frac{p(m = Z | r = Y) p(r = Y)}{p(m = Z)}$$

In this last expression the prior probability $p(r = Y)$ represents the likelihood that class $Y$ exists in the region being imaged; $p(m = Z)$ is the probability that class $Z$ appears on the thematic map, which can also be generated from

$$p(m = Z) = \sum_{Y \in \{A, B, C\}} p(m = Z | r = Y) p(r = Y)$$

To reiterate, the $p(r = Y | m = Z)$ are what we are interested in because they tell us how accurately the map classes represent what's on the ground, but that requires the rather impractical step of identifying samples selected at random from the map for identification. In contrast, using previously labelled testing data we can find the $p(m = Z | r = Y)$, but that doesn't explicitly tell us about the accuracy of the map

**Table 11.3** Comparison of classifier performance, user's accuracies and map accuracies for different combinations of prior probabilities; the first set of priors is computed from Table 11.2

| Prior probabilities | Classifier performance | User's accuracies | Map accuracies |
|---|---|---|---|
| $p(A) = 0.368$ $p(B) = 0.294$ $p(C) = 0.338$ | $p(m = A\|r = A) = 0.700$ $p(m = B\|r = B) = 0.925$ $p(m = C\|r = C) = 0.891$ Average = 0.831 | Class A = 0.987 Class B = 0.740 Class C = 0.872 | $p(r = A\|m = A) = 0.897$ $p(r = B\|m = B) = 0.740$ $p(r = C\|m = C) = 0.872$ Average = 0.831 |
| $p(A) = 0.333$ $p(B) = 0.333$ $p(C) = 0.333$ | $p(m = A\|r = A) = 0.700$ $p(m = B\|r = B) = 0.925$ $p(m = C\|r = C) = 0.891$ Average = 0.831 | Class A = 0.987 Class B = 0.740 Class C = 0.872 | $p(r = A\|m = A) = 0.882$ $p(r = B\|m = B) = 0.777$ $p(r = C\|m = C) = 0.877$ Average = 0.838 |
| $p(A) = 0.900$ $p(B) = 0.050$ $p(C) = 0.050$ | $p(m = A\|r = A) = 0.700$ $p(m = B\|r = B) = 0.925$ $p(m = C\|r = C) = 0.891$ Average = 0.831 | Class A = 0.987 Class B = 0.740 Class C = 0.872 | $p(r = A\|m = A) = 0.993$ $p(r = B\|m = B) = 0.202$ $p(r = C\|m = C) = 0.328$ Average = 0.721 |

product. The concept of user's accuracy in the previous section is meant to be a surrogate, but that is only true if the testing set distribution reflects the relative proportions of classes on the ground as we will now show by example, using the data of Table 11.2.

Table 11.3 gives three sets of results for differing prior probability distributions for the ground class labels; here the results are left in the form of probabilities and not converted to percentages. In one, the distribution is the same as the proportion of classes in the testing set; in another equal prior probabilities are assumed; and in the third a distribution very different from the priors is used. As seen, the map accuracies are only the same as the user's accuracies when the testing set reflects the priors.

In Table 11.3 a number of average accuracies are reported. We need now to give some thought as to how the average is computed. In its ideal form the class-wise accuracy of a thematic map is expressed by the probabilities $p(r = Z|m = Z)$. As above, these are found by sampling the thematic classes on the map and seeing how many pixels correspond to the actual classes on the ground. Once those probabilities have been determined the average accuracy of the map should be expressed as

$$\text{map accuracy} = \sum_{Z \in \{A,B,C\}} p(r = Z|m = Z)p(m = Z)$$

in which the class-wise accuracies are weighted by the probability of occurrence of those classes in the thematic map. That is important to ensure that the user of a map is not misled by the disparity in the sizes of classes that might appear on a map. For example, even though a class which has a small area is highly accurate, it will not have a significant influence on the average accuracy, and certainly not as much influence as the larger classes. Using the reciprocity of joint probabilities, we can express the above formula for map accuracy in the form

$$\text{map accuracy} = \sum_{Z \in \{A,B,C\}} p(r = Z, m = Z)$$

$$= \sum_{Z \in \{A,B,C\}} p(m = Z, r = Z)$$

$$= \sum_{Z \in \{A,B,C\}} p(m = Z | r = Z) p(r = Z)$$

This last expression tells us that the true average map accuracy can be obtained from the classifier performance, provided the classifier probabilities are weighted by the prior probabilities.

There is one final lesson we need to learn from the entries in Table 11.3. The last example shows a case with extreme prior probabilities, which is indicative of a situation where we may have one large class and two smaller classes. The way this example has been set up is such that the classifier performs better on the smaller classes than it does on the large class. When the *map* accuracies are computed it is seen that the results for the two smaller classes are disappointingly low notwithstanding that the classifier worked well on those classes. The reason is that the errors the classifier made on the large class commit themselves to the smaller classes thereby causing the confusion. In practice, therefore, even though one might be interested in a set of smaller classes and be happy that a classifier performs well on them, it is nevertheless important that any larger classes are also well recognised by the classifier so that errors of commission do not distort the results for the smaller classes.

### 11.6.4   *The Kappa Coefficient*

The Kappa Coefficient is a measure of classifier performance derived from the error matrix but which, purportedly, is free of any bias resulting from chance agreement between the classifier output and the reference data.[17] It was proposed initially for checking the chance agreement between the results of two independent classifiers. In our case, we have only one classifier and a set of reference data but, nevertheless, the reference data is also regarded as a sample of the true situation on the ground for this purpose. Although, as discussed below, there is dispute over the efficacy of the method, it is nevertheless widely used in remote sensing to report classification results. We will demonstrate its use by reference to the error matrix in Table 11.2. If we look at the classifier output in Table 11.2, we see that

---

[17] The Kappa Coefficient is generally attributed to J. Cohen, A coefficient of agreement for nominal scales, *Educational and Psychological Measurement*, vol. 20, no. 1, 1960, pp. 37–46, although apparently its use has been traced back to the late 1800s. Its use in remote sensing is covered extensively in R.G. Congalton and K. Green, loc. cit.

The classifier places:
39/136 = 0.287 of the pixels in class A
50/136 = 0.368 of the pixels in class B
47/136 = 0.346 of the pixels in class C

The reference data places:
50/136 = 0.368 of the pixels in class A
40/136 = 0.294 of the pixels in class B
46/136 = 0.338 of the pixels in class C

The probability that they would *both* place a pixel at random in class A is $0.287 \times 0.368 = 0.106$; similarly, the probability that they would both place a pixel in class B is $0.368 \times 0.294 = 0.108$, and in C is $0.346 \times 0.338 = 0.117$. Overall, the probability that they place a pixel at random in the same class is the sum of the three probabilities, viz. $0.106 + 0.108 + 0.117 = 0.331$, which is the random chance of their agreeing on the label for a pixel. On the other hand, the probability of a correct classification determined from the agreement of the classifier output and the reference data is $(35 + 37 + 41)/136 = 0.831$.

Now, the Kappa Coefficient is defined, in words, as[18]

$$\kappa = \frac{\text{probability of correct classification} - \text{probability of chance agreement}}{1 - \text{probability of chance agreement}}$$

which, for this example, is $\kappa = (0.831 - 0.331)/(1 - 0.331) = 0.747$. We now need to do two things: first, express the Kappa Coefficient directly in terms of the elements of the error matrix and secondly gain some understanding of what certain levels of Kappa Coefficient mean.

The commonly used measure of the probability of correct classification is given by the sum of the diagonal elements of the error matrix divided by the global total. If we represent an entry in the error matrix by $n_{ij}$, the total number of pixels by $N$, and the number of classes by $M$ then the probability of correct classification is

$$p_o = \frac{1}{N} \sum_{i=1}^{M} n_{ii} \tag{11.1}$$

If we now define the sum over the rows of the error matrix and the sum over the columns, respectively, as

$$n_{+i} = \sum_{k=1}^{M} n_{ki} \qquad n_{i+} = \sum_{k=1}^{M} n_{ik} \tag{11.2a}$$

then the probabilities that the reference data and classifier respectively place a pixel at random into class $i$ are

$$p_{+i} = \frac{1}{N} \sum_{k=1}^{M} n_{ki} \qquad p_{i+} = \frac{1}{N} \sum_{k=1}^{M} n_{ik} \tag{11.2b}$$

---

[18] See Cohen, loc. cit.

so that the probability of their agreeing by chance on any of the available labels for a pixel is

$$p_c = \sum_{i=1}^{M} \left\{ \frac{1}{N} \sum_{k=1}^{M} n_{ki} \frac{1}{N} \sum_{k=1}^{M} n_{ik} \right\} \tag{11.2c}$$

By definition

$$\kappa = \frac{p_o - p_c}{1 - p_c} \tag{11.3a}$$

This is the ratio of (i) the agreement between the map and reference data, expressed as classification accuracy, minus the chance agreement (sometimes called the beyond-chance agreement) and (ii) the probability that there is no chance agreement between the map and the reference data. Said another way, it is the proportion of labels that are in agreement between the map and reference data after chance agreement is excluded. Substituting (11.1) and (11.2a–11.2c) into (11.3a) and multiplying throughout by $N^2$ gives

$$\kappa = \frac{N \sum_{i=1}^{M} n_{ii} - \sum_{i=1}^{M} \left\{ \sum_{k=1}^{M} n_{ki} \sum_{k=1}^{M} n_{ik} \right\}}{N^2 - \sum_{i=1}^{M} \left\{ \sum_{k=1}^{M} n_{ki} \sum_{k=1}^{M} n_{ik} \right\}} \tag{11.3b}$$

which is sometimes written as

$$\kappa = \frac{N \sum_{i=1}^{M} n_{ii} - \sum_{i=1}^{M} n_{+i} n_{i+}}{N^2 - \sum_{i=1}^{M} n_{+i} n_{i+}} \tag{11.3c}$$

We now turn our attention to the values of Kappa. How does the user of a thematic map, with accuracy assessed on the basis of the Kappa Coefficient, know if the result is good or not? That is not an easy question to answer in general and is one reason why some analysts still prefer to use classification accuracy, or even a presentation of the full error matrix, in favour of using Kappa. Based on empirical results over several authors, the guidelines on Kappa in Table 11.4 have been proposed, noting that its theoretical maximum is 1 (in the ideal case when there can be no chance agreement and there are no off diagonal elements in the error matrix)

| **Table 11.4** Suggested ranges for the Kappa coefficient | Kappa coefficient | Classification can be regarded as |
|---|---|---|
| | below 0.4 | Poor |
| | 0.41–0.60 | Moderate |
| | 0.61–0.75 | Good |
| | 0.76–0.80 | Excellent |
| | 0.81 and above | Almost perfect |

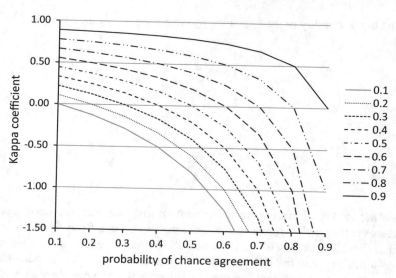

**Fig. 11.6** Kappa coefficient as a function of the probability of correct classification (in the legend) and probability of chance agreement

and its minimum can be large and negative. The ranges in Table 11.4 are by no means widely adopted. Three others are shown in Fig. 11.3 of Foody[19] indicating somewhat the arbitrariness of kappa as a map accuracy measure.

There is now significant concern that the Kappa Coefficient is a misleading measure[20] because (i) in an endeavour to be a single measure it masks importantly-different sources of error, (ii) its dependence on a comparison with chance agreement is not informative and (iii) being a ratio, similar values can be generated by differing combinations of the numerator and denominator. A simple illustration of this last point is shown in Fig. 11.6, which is a plot of (11.3a) for ranges of $p_0$ and $p_c$. Note the different combinations that correspond to the same value of Kappa, such as at 0.5.

As a result of these concerns a return to measures more directly related to the entries in the error matrix has been advocated. Two are *quantity disagreement* and *allocation disagreement*.[21] To derive these measures, we note that the expressions in (11.2b) can be written

---

[19] G.M. Foody, Explaining the unsuitability of the kappa coefficient in the assessment and comparison of the accuracy of thematic maps obtained by image classification, *Remote Sensing of Environment*, vol. 239, 2020, pp. 1–11.

[20] R. G. Pontius Jr and M. Millones, Death to Kappa: birth of quantity disagreement and allocation disagreement for accuracy assessment, *Int. J Remote Sensing*, vol. 32, no. 15, 2011, pp. 4407–4429 and G.M. Foody loc cit.

[21] R. G. Pontius Jr and M. Millones loc cit.

$$p_{+i} = \sum_{k=1}^{M} p_{ki} \qquad p_{i+} = \sum_{k=1}^{M} p_{ik}$$

in which the $p_{ij}$ are the error matrix entries expressed as proportions or probabilities; $p_{ij}$ is shorthand for the joint occurrence $p(m = i, r = j)$ when the relative class proportions in the reference data reflect the true proportions on the ground. Taking the difference between $p_{+i}$ and $p_{i+}$ is tantamount to the difference between the number of class $i$ pixels in the reference data and the number in the thematic map. Summing the absolute difference over all classes gives the discrepancy in the proportions of all classes between the reference data and the map[22]:

$$Q = \frac{1}{2} \sum_{i=1}^{M} |p_{+i} - p_{i+}| \qquad (11.4a)$$

That is called the *quantity disagreement*. As another interpretation it effectively measures the differences in the areas allocated to the classes in the reference data and map—it is one error measure. A second error measure is the *allocation disagreement* defined by[23]

$$A = \sum_{i=1}^{M} \min\{(p_{+i} - p_{ii}), (p_{i+} - p_{ii})\} \qquad (11.4b)$$

The first of the arguments in the minimum function is the proportion of class $i$ pixels in error in the map—the errors of commission, whereas the second is the proportion indicated from the reference data as errors of omission. This measure is intended to assess the aggregated misallocation of individual pixels for the same level of quantity agreement. If there is a specific error of commission there is a corresponding error of omission, which is recognised in the minimum operation of (11.4b).

From (11.1) $1 - p_o$ is the total error, or disagreement between the thematic map and the reference data. Interestingly it can be shown that[24]

$$1 - p_o = A + Q \qquad (11.4c)$$

indicating that the total disagreement can be disaggregated into quantity and allocation disagreements.

---

[22] The ½ is needed because operations involving both the row sums and columns sums will have a maximum of 2N.

[23] Pontius and Millones, loc. cit.

[24] ibid.

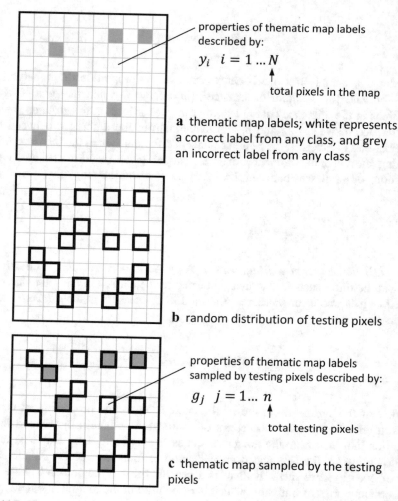

properties of thematic map labels
described by:

$$y_i \quad i = 1 \dots N$$

total pixels in the map

**a** thematic map labels; white represents
a correct label from any class, and grey
an incorrect label from any class

**b** random distribution of testing pixels

properties of thematic map labels
sampled by testing pixels described by:

$$g_j \quad j = 1 \dots n$$

total testing pixels

**c** thematic map sampled by the testing
pixels

**Fig. 11.7** Use of randomly distributed testing pixels for assessing thematic map accuracy

## 11.6.5   *Number of Testing Samples Required for Assessing Map Accuracy*

We now turn to the question of how many *testing* pixels should be chosen to ensure
that the assessed accuracy of the thematic map is a reliable estimate of its real
accuracy. Clearly, choosing too few samples per class will lead to a poor estimate of
map accuracy. To illustrate this point, a single testing sample from a particular class
can only indicate an accuracy of 100% or 0% depending on its match, or otherwise,
to the reference data. A larger sample will clearly give a more realistic estimate.

The problem we have to consider is shown in Fig. 11.7. In Fig. 11.7a we have
the pixels in a thematic map indicated as correctly (white) and incorrectly (grey)

labelled. Of course, we don't know which are correct and incorrect *a priori*; that's what we have to find using the labelled testing data. Figure 11.7b shows the locations of a random set of testing pixels, and Fig. 11.7c shows the testing pixels sampling the map. Note that some pixels in error have been detected while others have been missed. What we need to do is choose enough testing pixels so that the correct and incorrect labels sampled on the map are sufficient to reveal the map's accuracy to the precision required by the user.

If we select a pixel at random from the thematic map it can only be correct or incorrect. The chance of it being one or the other is described by a binomial probability distribution with properties established by the numbers of correct and incorrect labels in the map. The likelihood of our choosing a correctly labelled pixel will be high if there are few pixels in error in the map, and vice-versa.

For the thematic map we let $N$ be the total number of pixels, and $y_i, i = 1 \ldots N$ be a *property* of the $i$th pixel, which has the value 1 if the pixel is correctly labelled and 0 if it is incorrectly labelled. Thus, the sum $\sum_{i=1}^{N} y_i$ is the number of correctly labelled pixels, while

$$P = \frac{1}{N} \sum_{i=1}^{N} y_i \qquad (11.5)$$

is the *proportion* of pixels that are correctly labelled. This is the overall accuracy of the map,[25] which is what we want to determine using the testing pixels. It is also the mean of the binomial distribution describing the correct labels in the map.

For the testing set let $n$ be the total number of pixels, and $g_j, j = 1 \ldots n$ be the property of the $j$th testing set pixel which has the value 1 if a correctly labelled pixel is detected in the map and 0 if an incorrectly labelled pixel is found. Note $n \leq N$. The sum $\sum_{j=1}^{n} g_j$ is the number of correctly labelled pixels found by using the testing set, while

$$p = \frac{1}{n} \sum_{j=1}^{n} g_j \qquad (11.6)$$

is the *proportion* of pixels that are correctly labelled in the testing set and is thus an *estimate* of the accuracy of the map. Our task is to find the value of $n$ that makes $p$ in (11.6) an acceptable estimate of $P$ in (11.5).

The sum of the binomial random variables in (11.6) is itself a variate, as is $p$. Since the individual random variables come from the same underlying binomial distribution it can be shown that their expected value is the mean of that distribution, so that

---

[25] Here the accuracy is described by a proportion between 0 and 1, rather than the more usual form between 0 and 100%.

$$\mathcal{E}(p) = P$$

The variate $p$ also has a standard deviation that tells us the range about the mean $P$ within which the sample value $p$ is likely to occur. Since we are sampling a set of $n$ pixels from a finite number of pixels $N$ in the thematic map the variance of $p$ about its mean $P$ is[26]

$$var(p) = \frac{P(1-P)}{n}\frac{(N-n)}{(N-1)} \tag{11.7}$$

If we want confidence in the value of $P$ estimated from $p$ then we need to ensure that this variance is small. Before proceeding to do that, there are a couple of useful observations we can make from (11.7). First, if we just use one testing pixel then the variance is $P(1-P)$ which is in the range $(0, 0.25)$. We can conclude nothing about the map from the outcome of the single test, unless the map were near perfect ($P \approx 1$) or near imperfect ($P \approx 0$). That is the situation alluded to in the opening paragraph of this section. Secondly, if $n = N$, i.e., we test every pixel in the thematic map, then the variance in the estimate $p$ is zero, meaning that $p$ is exactly $P$, which is logical. Thirdly, if $N \gg n$, which is generally the case in remote sensing —i.e., the number of testing pixels is generally a small part of the overall scene— then (11.7) reduces to

$$var(p) = \sigma^2 = \frac{P(1-P)}{n} \tag{11.8}$$

The factor $\frac{(N-n)}{(N-1)}$ between (11.7) and (11.8) is sometimes called the *finite population correction*.

To a very good approximation[27] the variate $p$ can be assumed to be normally distributed about its mean $P$, as illustrated in Fig. 11.8, which shows graphically the range within which the estimate we generate from testing data is likely to occur. Note that 95% of the time our estimate is within $\pm 1.96$, or approximately $\pm 2$, standard deviations of the mean; so, with 95% confidence we can say the sampled map accuracy is in that range. If we were happy with a lower precision, then we can give a smaller range for the estimate of the map accuracy.

We are now in the position to use (11.8) to estimate the number of testing pixels $n$ needed to check the accuracy of a thematic map. We can be 95% confident it lies within two standard deviations of its true value. We now have to consider what range about the true value we are prepared to accept as an error, because that range specifies the value of the standard deviation. Suppose we are happy for the estimate to be within $\pm e$ of what we think is the true map accuracy: i.e., $p = P \pm e$. Then from (11.8) we have at the 95% confidence level and for $N \gg n$

---

[26] W.G. Cochran, *Sampling Techniques*, John Wiley & Sons, N.Y., 1961.
[27] ibid.

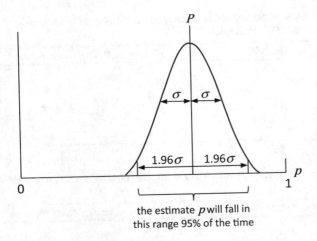

**Fig. 11.8** Showing the distribution of the map accuracy estimate from the testing data, $p$, about the true accuracy $P$ of the thematic map

$$2\sqrt{\frac{P(1-P)}{n}} = e$$

Rearranging gives as the minimum number of samples required

$$n = \frac{4P(1-P)}{e^2} \tag{11.9}$$

As an illustration, suppose we are happy for an estimate to be within ±0.04 of a true proportion which is thought to be about 0.85 (i.e., we are happy if the map accuracy estimated from the testing set is in the range 0.81–0.89) then from (11.9) at the 95% confidence level $n = 319$.

Thus, randomly selecting 319 testing pixels will allow a thematic map accuracy of about 85% to be checked with an uncertainty of ±4% with 95% confidence. Note that (11.7) and (11.8) can be used in percentage terms as well as proportions since we obtain percentages by multiplying proportions by 100; that applies also to the standard deviation, because it refers to a proportion or percentage as appropriate. Table 11.5 gives further examples for a range of likely thematic map accuracies, this time expressed in percentages.

We can make a couple of observations from (11.8) to help guide in the selection of the number of testing pixels. First, more samples will narrow the variance about the mean; however, since the variance is inversely proportional to $n$ there is a diminishing value in choosing many more samples than the minimum required. Secondly, the term $P(1-P)$ is largest in the vicinity of $P = 0.5$. Thus, the error (variance) is greatest when the accuracy of the map is poor, and correspondingly more testing samples are required, than when the overall thematic map accuracy is high.

**Table 11.5** Number of testing pixels required for evaluating thematic map accuracies, with an error ±4% and with 95% confidence

| Thematic map accuracy (%) | Number of testing pixels required |
| --- | --- |
| 70 | 525 |
| 75 | 469 |
| 80 | 400 |
| 85 | 319 |
| 90 | 225 |
| 95 | 119 |

**Table 11.6** Number of testing pixels required for class level accuracies, with an error ±10% with 95% confidence

| Thematic class accuracy (%) | Number of testing pixels required |
| --- | --- |
| 70 | 45 |
| 75 | 40 |
| 80 | 30 |
| 85 | 19 |

The results given above, and summarised in Table 11.5, have been developed by assuming that we are testing the overall accuracy of the thematic map, independent of class. We have made no specific reference to class or otherwise in this development, so the results apply equally at the category level if required. If there were a large number of classes in a thematic map, then clearly the requirements of Table 11.5 will lead to an excessively large number of testing pixels. Table 11.6 shows the number of testing pixels required for an individual class, in this case with 10% uncertainty in the accuracy of the class.[28]

### 11.6.6  Number of Testing Samples Required for Populating the Error Matrix

The calculations of the previous section were focused on the need to establish a sufficient number of testing pixels so that the overall accuracy of a thematic map could be assessed. Those calculations had their foundation in binomial statistics since, for each pixel in a thematic map, only two outcomes are possible—a correctly labelled pixel or an incorrectly labelled pixel. With reference to Table 11.2, those results are sufficient for evaluating the sum of the diagonal entries compared with the total number of testing pixels, but not sufficient for generating accurate estimates of the individual entries in the error matrix simultaneously.

---

[28] Taken from G.H. Rosenfield, K. Fitzpatrick-Lins and H.S. Ling, Sampling for thematic map accuracy testing, *Photogrammetric Engineering and Remote Sensing*, vol. 48, 1982, pp. 131–137.

When our interest is in the performance of a classifier when labelling a pixel into *each* of the available classes, testing data needs to be used to estimate the proportions of pixels by class. That means we need good estimates of *all* of the elements of the error matrix and, as would be expected, this will require more testing pixels overall.

Perhaps the simplest approach, particularly when stratified random sampling is used to avoid area bias in the results, is to choose samples within each class, using the guidelines in Table 11.6, although perhaps with a tighter error bound.

Another approach is to use the multinomial probability distribution to describe the multiple outcomes possible with simple, as against stratified, random sampling when using testing data to check the accuracy of the actual class to which a pixel is allocated in the thematic map.[29] The results parallel those of the binomial development of the previous section. When the total number of pixels in the thematic map is large, we obtain the following estimate of the necessary sample size, based on the tolerable error and expected mean for the $i$th class for the pixel:

$$n = \frac{BP_i(1 - P_i)}{e_i^2} \qquad (11.10)$$

$P_i$ is the population proportion for the class, $e_i$ is the error we can tolerate in the estimate of the accuracy of the proportion estimate for that class and $B$ is the upper $\beta$ percentile for the $\chi^2$ distribution with one degree of freedom, where $\beta$ is the overall precision needed, divided by the total number of classes.

A version of (11.10) can also be derived for cases when the number of pixels in the thematic map is not large enough to ignore a finite population correction.[30]

To find a satisfactory value for $n$, (11.10) would be evaluated for each class and the largest value of $n$ selected for the required size of the testing set, in the sense that that is the most demanding requirement. That number could then be divided by the total number of classes to find how many testing pixels per class are needed, noting that this is a simple and not stratified random sampling strategy, which assumes implicitly that the classes are comparable in size (numbers of pixels).

Following our observations with the binomial approach we can see that the worst case class in (11.10) would be one where the population proportion is 0.5. We can therefore derive a conservatively high estimate for $n$ by putting $P_i = 0.5$ in (11.10) to give the simple expression

$$n = \frac{B}{4e^2} \qquad (11.11)$$

in which we have also assumed the same tolerable error for each class.

---

[29] See R.D. Tortora, A note on sample size estimation for multinomial populations, *The American Statistician*, vol. 12, no. 3, August 1978, pp. 100–102.

[30] ibid.

As an example,[31] suppose we require our estimates of the proportions of each of 8 classes in the thematic map to be within the range $\pm 5\%$ and that we want our results to be at the 95% confidence level. Then $\beta$ is the upper $0.05/8 = 0.00625$ (0.625 percentile) of the distribution and has the value 7.568, giving

$$n = \frac{7.568}{4(0.05)^2} = 757$$

Therefore, we need about 757 testing pixels in total, with slightly fewer than 100 per class to get good estimates of the elements of the error matrix at the precision level specified. Although, based on simple random sampling we could assume about 100 per class in general, unless it was known beforehand that some classes are very different in size.

### 11.6.7   Placing Confidence Limits on Assessed Accuracy

Once accuracy has been estimated using testing data it is important to place some confidence on the actual figures derived. If the number of testing pixels has been determined using the guidance of the previous sections, then those limits have been set in the process. If not, we can use straightforward statistics to express the interval within which the true map accuracy lies, say, with 95% certainty. That interval can be determined using the expression derived from the normal distribution[32]

$$-z_{\alpha/2} < \frac{x - nP}{\sqrt{nP(1 - P)}} < z_{\alpha/2}$$

in which $n$ is the number of testing pixels, as before, $x(\equiv np)$ is the number that were correctly labelled, and $P$ is the thematic map accuracy, which we are estimating by $p = x/n$; $z_{\alpha/2}$ is the value of the normal distribution beyond which on both tails $\alpha$ of the population is excluded. As in the previous examples, if we want the normalised statistic $x - nP/\sqrt{nP(1 - P)}$ to lie in the 95% portion of the normal curve, then $z_{\alpha/2} = \pm 1.96 (\approx \pm 2)$. Using this value, it is a relatively straightforward matter to solve the two inequalities above to show that the extremes of $P$ estimated by $p$, at the 95% confidence level are

$$\frac{x + 1.921 \pm 1.960\sqrt{x(n - x)/n + 0.960}}{n + 3.842} \tag{11.12}$$

---

[31] Taken from Congalton and Green, loc. cit.

[32] See J.E. Freund, *Mathematical Statistics*, 5th ed., Prentice-Hall, Englewood Cliffs, N.J., 1992.

which for $n$ and $x$ large, and for reasonable accuracies, is approximated by

$$\frac{x \pm 1.960\sqrt{x(n-x)/n+0.960}}{n} = p \pm \frac{1.960}{n}\sqrt{x(n-x)/n+0.960}$$

Choosing an example of 400 testing pixels from Table 11.5 to assess an accuracy of 80% we would expect to find 320 of those pixels in agreement with the map. From (11.12) the bounds on the estimated map accuracy are $P = p \pm 0.039$ or, in percentage terms, the map accuracy is between 76 and 84%.

### 11.6.8  Cross Validation Accuracy Assessment and the Leave One Out Method

As an alternative to using a separate testing set of pixels, an effective method for assessing accuracy is based on cross validation. This involves taking a single labelled set of pixels and dividing it into $k$ separate, equally sized, subsets. One subset is put aside for accuracy testing and the classifier is trained on the pixels in the remaining $k - 1$ sets. The process is repeated $k$ times with each of the $k$ subsets excluded in rotation. At the end of those $k$ trials, $k$ different measures of classification accuracy have been generated. The final classification accuracy is the average of the $k$ trial outcomes.

A variation of the cross validation method is when each subset consists of a single pixel. In other words, one pixel from the training set is excluded and the classifier trained on the remainder. The pixel which has been excluded is then labelled. In this case there are as many trials as there are training pixels, in each case with a separate pixel left out during training. The average classification accuracy is then the average over the labelling of the pixels left out in each trial. Provided the original training pixels are representative, this method produces an unbiased estimate of classification accuracy.[33] This is called the Leave One Out (LOO) method.

## 11.7  Decision Tree Classifiers

Classifiers such as the maximum likelihood rule and the support vector machine are single stage processes. They make a single decision about a pixel when labelling it as belonging to one of the available classes, or it is left unclassified. Multistage classification techniques are also available, in which a series of decisions is taken to

---

[33] See D. A. Landgrebe, *Signal Theory Methods in Multispectral Remote Sensing*, John Wiley & Sons, Hoboken, N. J., 2003.

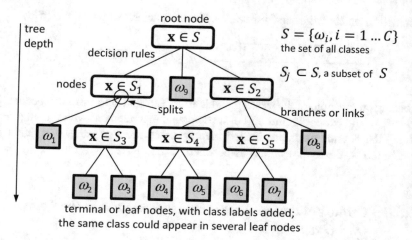

Fig. 11.9 The decision tree classifier in which a pixel is labelled into one of the available classes by a sequence of decisions, each of which narrows down the possibilities for membership

determine the most appropriate label for a pixel. The committee classifiers of Sect. 8.19 are examples.

The most commonly encountered multistage classifier is the decision tree, such as that shown in Fig. 11.9. Decision trees consist of a number of connected clas-sifiers (called decision nodes in the terminology of trees) none of which is expected to perform the complete segmentation of the image data set. Instead, each com-ponent classifier only carries out part of the task as indicated. The simplest is the binary decision tree in which each component classifier is expected to perform a segmentation of the data into one of two possible classes or groups of classes. It is the most commonly encountered tree in practice, and has the topologies shown in Fig. 11.10.

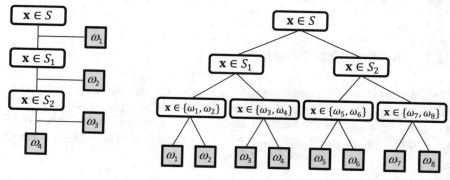

Fig. 11.10 Two versions of a binary decision tree

There is some ambiguity in the terminology of decision trees; in this treatment we adopt:

| | |
|---|---|
| Root node | This is where the tree commences |
| Decision node | Intermediate node (and the root node) |
| Terminal or leaf node | Final node, which usually represents a single class |
| Link or branch | Connection between nodes |
| Tree depth | Number of layers from the root node to the most distant leaf |
| Antecedent | Node immediately above a node of interest; sometimes called a parent node |
| Descendant | Node immediately following a node of interest; sometime called a child node |
| Split | The result of a decision to create new descendent nodes |

The advantages of the decision tree approach are that:

- different sets of features can be used at each decision node; this allows

    - feature subsets to be chosen that optimise segmentations
    - reduced feature subsets at individual decisions, so that the Hughes phenomenon might be avoided

- simpler segmentations than those needed when a decision has to be made among all available labels for a pixel in a single decision
- different algorithms can be used at each decision node
- different data types can be used at each decision node.

Decision tree design is usually not straightforward.[34] Sometimes the analyst can design a tree intuitively. For example, near infrared data might be used to segment between land and water bodies; subsequently, thermal infrared data might be used to map temperature contours within the water.[35] In principle, tree design involves finding the structure of the tree, choosing the subset of features to be used at each node, and selecting the decision rule to use at each node. If we restrict the range of possibilities for the last two requirements some automated procedures are possible, as developed in the following.

---

[34] See S.R. Safavian and D.A. Landgrebe, A survey of decision tree classifier methodology, *IEEE Transactions on Systems, Man and Cybernetics*, vol. 21, May 1991, pp. 660–674.

[35] See P. H. Swain and H. Hauska, The decision tree classifier: design and potential, *IEEE Transactions on Geoscience Electronics*, vol. GE-15, 1977, pp. 142–147.

## 11.7.1   CART (Classification and Regression Trees)

The CART[36] tree growing methodology is possibly the most commonly encountered because it restricts, and thus simplifies, the possible options for how the decision nodes function. Only one feature is involved in each decision step and a simple threshold rule is used in making that decision. As with other supervised classification procedures it uses labelled training data to construct the tree. Once the tree has been built it can then be used to label unseen data.

At each node in CART, including at the root node, a decision is made to split the training samples into two groups; the aim is to produce sub-groups that are purer class-wise than in the immediately preceding node.

The tree is developed in the following manner. All of the training data from all classes is fed to the root node. We then evaluate all possible binary partitions of the training pixels and choose that partition which minimises the class mixture in the two groups produced. For example, if there were five separate classes in the training set then we would expect the sub-groups to have pixels from fewer than five classes and, in some cases, hope that one sub-group might have pixels from one class only. We keep subdividing the groups as we go down the tree so that, ultimately, we end up with groups containing pixels from only one class—i.e., "pure" groups. That happens at the leaf nodes.

To be able to implement the process just described we have to have some way of measuring how mixed the training classes are in a particular group. We do that by using an impurity measure, several of which are used in practice. A common metric is the *Gini impurity*, or *Gini Index* defined at the $N$th node as

$$i(N) = \sum_j P_{\omega_j}\left(1 - P_{\omega_j}\right) \tag{11.13}$$

in which $P_{\omega_j}$ is the fraction of the training pixels at node $N$ that are in class $\omega_j$; $1 - P_{\omega_j}$ is the proportion not in class $\omega_j$. If all the pixels at the node were from a single class, then $P_{\omega_j} = 1$ so that $i(N) = 0$, indicating no impurity. If there were $M$ equally distributed classes in the training set then $i(N)$ is a maximum and equal to $1 - 1/M$, which is larger for larger $M$, as would be expected.

Another impurity measure is based on *entropy*, defined as

$$i(N) = -\sum_j P_{\omega_j} \log_2 P_{\omega_j} \tag{11.14}$$

Again, this is zero if all the training pixels are from the same class and is large when the group is mixed.

---

[36] See R.O. Duda, P.E. Hart and D.G. Stork, *Pattern Classification*, 2nd ed., John Wiley & Sons, N.Y., 2001, and L. Breiman, J.H. Friedman, R.A. Olshen and C.J. Stone, *Classification and Regression Trees*, Chapman and Hall/CRC, Roca Baton Florida, 1998.

**Fig. 11.11** Two-dimensional data with three classes used for generating a binary decision tree by the CART procedure

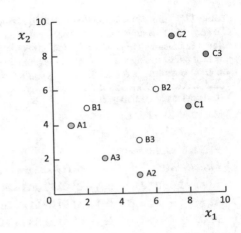

In splitting the training pixels as we go down the tree, we are interested in that split which gives the greatest drop in impurity from the antecedent to the descendent nodes—in other words, the split that generates the purest descendent groups. We can measure the reduction in impurity by subtracting the impurities of the descendent nodes from the impurity of their antecedent node, weighted by the relative proportions of the training pixels in each of the descendent nodes. Let $N$ refer to a node and $N_L$ and $N_R$ refer to its left and right descendents; let $P_L$ be the proportion of the training pixels from node $N$ that end up in $N_L$. Then the reduction in impurity in splitting $N$ into $N_L$ and $N_R$ is

$$\Delta i(N) = i(N) - P_L i(N_L) - (1 - P_L) i(N_R) \tag{11.15}$$

To see how this is used in building a decision tree consider the training data shown in Fig. 11.11. This consists of three classes, each of which is described by two features (bands). The Gini impurity is used. Table 11.7 shows the original impurity for the complete set of data and the subsequent drops in impurity with various candidate splits. Not all possible splits are given because the number of combinations is excessive; only those that are clearly the most favoured are shown. The table is segmented by successive layers in the decision tree as it is built, showing splits by layer until the leaf nodes are reached. There are several split options later in the tree; only two are given to demonstrate that trees are often not unique but will still segment the data as required. The resulting segmentations of the training set and the corresponding decision trees are shown in Fig. 11.12.

One of the problems with splitting based on the simple thresholding of individual features is that quite complicated trees can be generated compared with what should be possible if more flexibility is introduced into the decision functions and thus the decision boundaries in the spectral space. For example, inspection of Fig. 11.11 suggests that the data could easily be split into the three classes by two inclined linear surfaces, one between class A and B pixels, and the other between class B and C pixels. While it is feasible to develop a tree design methodology that

**Table 11.7** Impurity calculations and splits leading to the decision trees of Fig. 11.12 based on the single feature shown in the left descendent column in each case; only the most likely splits are shown to illustrate the process; shaded boxes highlight the greatest reduction in impurity and thus the best splits, noting that two equally favourable splits are possible at the second stage leading to different outcomes and thus illustrating that the resulting tree is not unique

| Original unsplit training set | | | | | |
|---|---|---|---|---|---|
| A1 A2 A3 B1 B2 B3 C1 C2 C3 | | | $i(N) = 0.667$ | | |
| | | | | | |
| **First split candidates** | | | | | |
| *left descendent* | | *right descendent* | $i(N_L)$ | $i(N_R)$ | $\Delta i(N)$ |
| A1 A2 A3 B1 B2 B3 | $(x_1)$ | C1 C2 C3 (leaf node) | 0.500 | 0 | 0.334 |
| A2 A3 | $(x_2)$ | A1 B1 B2 B3 C1 C2 C3 | 0 | 0.612 | 0.191 |
| C2 C3 | $(x_2)$ | C1 A1 A2 A3 B1 B2 B3 | 0 | 0.612 | 0.191 |
| A1 | $(x_1)$ | A2 A3 B1 B2 B3 C1 C2 C3 | 0 | 0.656 | 0.084 |
| **Second split candidates from A1 A2 A3 B1 B2 B3 \| C1 C2 C3 first split** | | | | | |
| B1 B2 | $(x_2)$ | A1 A2 A3 B3 | 0 | 0.375 | 0.250 |
| A2 A3 | $(x_2)$ | A1 B1 B2 B3 | 0 | 0.375 | 0.250 |
| A1 | $(x_1)$ | A2 A3 B1 B2 B3 | 0 | 0.480 | 0.100 |
| **Third split from B1 B2 \| A1 A2 A3 B3 second split** | | | | | |
| A1 A3 | $(x_1)$ | A2 B3 | 0 | 0.500 | 0.125 |
| **Fourth split from A1 A3 \| A2 B3 third split** | | | | | |
| A2 (leaf node) | $(x_2)$ | B3 | 0 | 0 | 0.500 |
| **Third split from A2 A3 \| A1 B1 B2 B3 second split** | | | | | |
| A1 (leaf node) | $(x_1)$ | B1 B2 B3 (leaf node) | 0 | 0 | 0.375 |

implements linear decisions of that nature at each node, it is sometimes simpler to transform the data prior to tree growth.[37] For example, if the data of Fig. 11.11 is used to generate its principal components, the principal axes will provide a simpler tree. Figure 11.13 shows the principal axes of the data from Fig. 11.11 along with a decision tree generated with the CART methodology.

Once a tree has been grown it can be examined to see if it can be simplified by pruning; that involves removing nodes or sets of nodes such that the tree is simpler but still gives acceptable accuracy on a testing set of pixels, i.e., so that it still generalises well. Several strategies for pruning exist,[38] including just working upwards through the tree by layers and noting the drop in generalisation accuracy.

---

[37] See Duda, Hart and Stork, loc. cit.

[38] See L. Breiman, J.H. Friedman, R.A. Olshen and C.J. Stone, *Classification and Regression Trees*, Chapman and Hall/CRC, Roca Baton Florida, 1998.

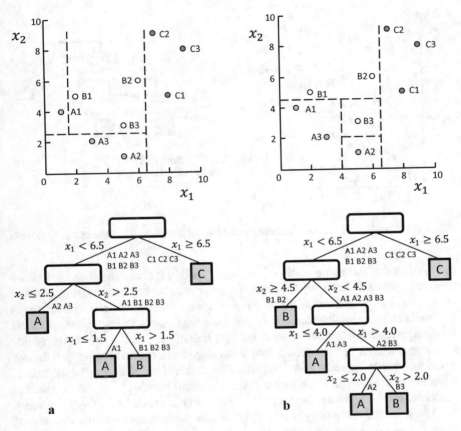

**Fig. 11.12** Two alternative tree segmentations of the training data in Fig. 11.11

## 11.7.2  Random Forests

In Sect. 8.19 we considered committee classifiers as typical of an ensemble approach that led to strong classification decisions using component classifiers that in themselves may not perform well. We can also form ensembles of decision trees with the same goal in mind. One particularly successful decision tree committee is the Random Forest.[39] As its name implies it is a collection of trees (a "forest") that are somehow random in their construction.

In common with other supervised classifiers, we assume that we have available a labelled set of training pixels. Those pixels are not used as a complete set as would be the case with single stage supervised classifiers. Instead, bootstrapped samples

---

[39] See L. Breiman, Random forests, *Machine Learning*, vol 45, 2001, pp. 5–32, and T. Hastie, R. Tibshirani and J. Friedman, *The Elements of Statistical Learning: Data Mining, Inference and Prediction*, Springer Science + Business Media, N.Y., 2009.

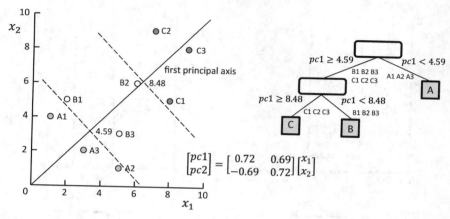

**Fig. 11.13** Simpler decision tree generation after principal components transformation of the data

are used in the following manner.[40] If there are $K$ pixels in the training set, we randomly select $K$ *with replacement*. In other words, the first training pixel is selected and then returned to the set; then a second pixel is selected, and so on. In this manner it is possible for one or more training pixels to be replicated in the bootstrapped sample chosen to train the first tree in the random forest. Using this training set a CART style decision tree is developed. Before that can happen though a decision has to be made as to the feature set that will be used to grow the tree. If there are $N$ features in the spectral domain a small number[41] $n \ll N$ is selected randomly for the first tree. That small set is used at each node in the CART process. Typically, the Gini impurity is employed to find the best split and the best one of the $n$ features to use in that split.

We then need to assess how well the (first) tree generalises. The classic approach is to have a separate testing set of pixels which would be run through the tree to see what errors are made. In the case of a random forest, however, those pixels in the original training set that are not picked up in the bootstrapped sample used to develop a particular tree can be used as testing pixels. It turns out that taking a sample $K$ with replacement from the available pixels, leaves out about one third of the original set. They are the pixels that are used to check tree performance.

Clearly, the first tree, trained in this manner, would not be expected to perform well. As a consequence, a second tree is grown using another bootstrapped sample from the available training pixels along with a second random selection of features, but with the same dimension $n$ as used in the growth of the first tree. A third tree is developed in the same manner. The newly grown trees are tested using the pixels from the original training set left over after the bootstrapped samples were chosen for training. We now have three trees capable of performing a classification on the

---

[40] See also Sect. 8.19.1.

[41] A suggested value for $n$ is $\sqrt{N}$. For hyperspectral data sets this will most likely be too big.

same data set. In order to combine their results, we use a majority vote logic, in that the actual label allocated to an unknown pixel is given by the most favoured label among the trees. Sometimes this is called *modal logic*.

The process just described is continued through the addition of as many randomly generated trees as necessary in order to reduce classification error and thus produce results with the desired accuracy. It would not be uncommon for several hundreds to thousands of trees to be generated randomly in pursuit of such a goal.[42]

There are two requirements for the random forest methodology to work well. First, the trees generated randomly have to be uncorrelated; the choice of the bootstrapped training sets (with replacement) provides that. Secondly, the individual trees should be strong classifiers. Generally, classifier strength will increase with the number of features used at each decision node. However, that increases substantially the complexity of tree growth so that the number of features is nevertheless kept small and weak classifiers are generally used. A review including a commentary on the application of random forests to high dimensional remote sensing image data will be found in Belgiu and Dragut.[43]

### 11.7.3 Progressive Two-Class Decision Classifier

Another tree classifier is shown in Fig. 11.14. It makes sequential binary decisions, operating similar to the one-against-one multiclass strategy of Sect. 8.17. As indicated in the figure, at the first decision node a separation is made between classes 1 and 2; pixels from all other classes will not be separated and may appear in both subsets. At the left-hand node in the second layer, class 2 pixels are not considered any further since they were split off in the first decision; instead, a split is now made between class 1 and another class, in this case class 3. In the right-hand node in the second layer class 3 is also split off. Ultimately, all pixels will be split into the constituent set of classes as a result of the progressive set of decisions, each based on two of the training classes. In Fig. 11.14 the overbars indicate which classes are not considered any further at any decision node, having been handled earlier.

Since pairs of classes are considered at each node, and not class subsets, the algorithm to be used, and the set of features for separating that pair, can be chosen optimally for those classes and could thus be different at each decision node.

[42] See P.O. Gislason, J.A. Benediktsson and J.R. Sveinsson, Random forest classification of multisource remote sensing and geographic data, *Proc. Int. Geoscience and Remote Sensing Symposium IGARSS2004*, Alaska, 20–24 Sep 2004, pp. II: 1049–1052, and J.S. Ham, Y. Chen, M. M. Crawford and J. Ghosh, Investigation of the random forest framework for classification of hyperspectral data, *IEEE Transactions on Geoscience and Remote Sensing*, vol. 43, no. 3, March 2005, pp. 492–501.

[43] M. Belgiu and L. Dragut, Random forest in remote sensing: a review of applications and future directions, *ISPRS Journal of Photogrammetry and Remote Sensing*, vol. 114, 2016, pp. 24–31.

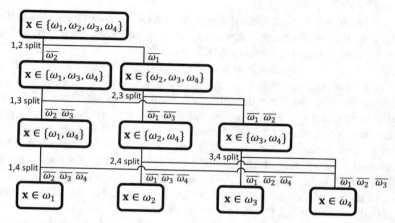

**Fig. 11.14** Example of a four layer progressive two class decision classifier

The method has been tested on four difficult-to-separate classes, leading to an overall accuracy of 72.5% when using the maximum likelihood classifier as the decision rule, compared with a performance of 64.4% when the four classes are treated in a single step.[44]

## 11.8   Image Interpretation Through Spectroscopy and Spectral Library Searching

One of the great benefits of recording many, finely spaced spectral samples for a pixel using an imaging spectrometer is that a scientific approach to interpretation can be carried out, instead of the machine learning route to image understanding; this was noted in Sect. 10.1. Effectively, the latter looks for separable patterns in the data in spectral or feature space, whereas a scientific analysis seeks to associate observed spectral features with known chemical characteristics. That is the basis of spectroscopy used in many fields and is why hyperspectral sensors are known as imaging spectrometers.

Absorption features in recorded spectra, seen as localised dips, usually provide the information needed for identification, and are referred to as *diagnostically significant features*. Characterisation and recognition of those features is of paramount importance when taking an expert spectroscopic approach to the analysis of hyperspectral imagery. They are described by their locations along the spectrum, and their relative depths and widths. Feature identification by the analysis of

---

[44] See X. Jia and J.A. Richards, Progressive two-class decision classifier for optimization of class discrimination, *Remote Sensing of Environment*, vol. 63, 1998, pp. 289–297.

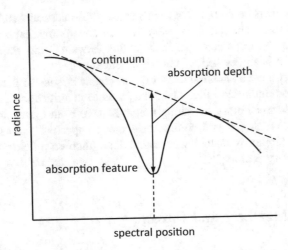

**Fig. 11.15** Continuum removal to improve the characterisation of diagnostically significant features

spectral properties has been used in many remote sensing applications, particularly for soils and rocks.[45]

The absorption features result from photon interaction with the atomic structure of the chemicals that make up the material being imaged. To be able to quantify them it is necessary, first, to separate them from the background continuum of the spectrum that results from light transmission and scattering.

The importance of continuum removal is illustrated in Fig. 11.15. Often the continuum will not be horizontal, which makes the measurement of the properties of an absorption feature, especially its depth, difficult. If the continuum in the vicinity of the feature is defined by a line of best fit between those parts of the spectrum either side of the feature, then a reasonably consistent measure of band depth results.

In such an approach to interpretation the spectrum is generally divided into several spectral regions, usually under the guidance of a domain expert. Absorption features are then detected in each of those regions. An unknown pixel is labelled into a class for which its diagnostically significant features match those of prototype spectra for that same class stored in a spectral feature library.

---

[45] See Imaging Spectroscopy Special Issue, *Remote Sensing of Environment*, vol. 113, 2009 for several examples which demonstrate the value of the spectroscopic approach; two more recent studies which demonstrate the scientific approach with imaging spectroscopy are Y-Q Wan, Y-H Fan, and M-S Jin, Application of hyperspectral remote sensing for supplementary investigation of polymetallic deposits in Huaniushan ore region, northwestern China. *Scientific Reports*, vol. 11, 2021, pp. 40, http://doi.org/10.1038/s41598-020-79864-0, and G. E. Graham, R.F. Kokaly, K.D. Kelley, T.M. Hoefen, M.R. Johnson and B.E. Hubbard, Application of imaging spectroscopy for mineral exploration in Alaska: a study over porphyry Cu deposits in the Eastern Alaska Range, *Economic Geology*, vol. 113, no. 2, 2018, pp. 489–510.

The process can be complicated when pure substances are embedded in mixtures. Also, some materials have very similar absorption features. Those complexities make the approach less than straightforward[46] but quite sophisticated techniques have been devised to make it practical.

Not all spectra have identifiable absorption characteristics so that the continuum removal method outlined in Fig. 11.15 may lead to identification failures for some cover types. Recent library searching methods have been based on features from both continuum-included and continuum-removed spectra. Thematic mapping is carried out by similarity matching recorded data, often on a field rather than pixel basis, with library prototypes.[47]

## 11.9  End Members and Unmixing

A challenge that has faced image analysts throughout history of remote sensing has been the need to handle mixed pixels.[48] They represent a mixture of cover types or information classes within the same pixel; they occur whenever there are indistinct boundaries between cover types and whenever the classes of interest to the user exist implicitly in mixtures, such as in the analysis of geological regions.

Early on, several efforts were directed to resolving the proportions of pure cover types within mixed pixels by assuming that the measured radiance is a linear combination of the radiances of the "pure" constituents in each of the imaging wavebands used. With low spectral resolution data that approach generally did not meet with a great deal of success because most cover types are not well differentiated in the small number of wavebands of the instruments available in the 1980s and 1990s. However, with hyperspectral data, the prospect for uniquely characterising a vast number of cover types, and thus differentiating them from each other spectroscopically, suggests that the mixing approach be revisited as a means for establishing mixture proportions of pure cover types. This has particular relevance in mineral studies where abundance maps for the minerals of interest could be produced, based on the proportions determined for all the pixels in an image.

If we make the assumption that the total radiance recorded for a given pixel is a linear combination of the radiances of its constituents, the process can be developed mathematically in the following manner. This assumption is supportable if we accept that the incident energy is scattered only once to the sensor from the landscape and does not undergo multiple scatterings from among, say, foliage

---

[46] See R.N. Clark, G.A. Swayze, K.E. Livio, R.F. Kokaly S.J. Sutly, J.B. Dalton, R.R. McDougal and C.A. Gent, Imaging spectroscopy: earth and planetary remote sensing with the USGS Tetracorder and expert systems, *J. Geophysical Research*, vol. 108, E12, pp. 5.1–5.44, 2003.

[47] See B.D. Bue, E. Merenyi and B. Csatho, Automated labeling of materials in hyperspectral imagery, *IEEE Transactions on Geoscience and Remote Sensing*, vol. 48, no. 11, November 2010, pp. 4059–4070.

[48] They were even referred to as "mixels" in some image processing systems.

components. Clearly that assumption will be violated in some situations, in which case the following simple approach may not work well.

Assume there are $M$ pure cover types in the image of interest. In the nomenclature of mixing models these are referred to as *end members*. We assume they are known to us, perhaps because of regions of pure cover in the image.

Let the proportions of the various end members in a pixel be represented by $f_m, m = 1 \ldots M$. They are the unknowns in the process which we need to find, based on observing the hyperspectral reflectance of the pixel.

Let $r_n, n = 1 \ldots N$, be the observed reflectance of the pixel in the $n$th band recorded by the sensor, and $a_{mn}$ be the spectral reflectance in the $n$th band for the $m$th end member. Then, as above, we assume

$$r_n = \sum_{m=1}^{M} f_m a_{mn} + \xi_n$$

in which $\xi_n$ accounts for any errors in band $n$. This equation says that the observed pixel reflectance in each hyperspectral band is the weighted sum of the reflectances of the end members in that band. The extent to which it does not work exactly is provided for in the error term. In matrix form the mixing equation can be expressed

$$\mathbf{r} = \mathbf{A}\mathbf{f} + \boldsymbol{\xi}$$

in which $\mathbf{f}$ is a column vector of mixing proportions, of size $M$. $\mathbf{r}$ is the spectral reflectance vector and $\boldsymbol{\xi}$ is the error vector; they are column vectors of size $N$. $\mathbf{A}$ is an $N \times M$ matrix of end member spectral responses, by column.

Spectral unmixing, as the process is called, involves finding a set of end member proportions that minimise the error vector $\boldsymbol{\xi}$. On the assumption that the correct set of end members has been chosen, the problem is then one of solving the error free equation

$$\mathbf{r} = \mathbf{A}\mathbf{f}$$

Normally there are more equations than unknowns so that simple inversion to find the vector of mixing proportions is not possible. Instead, a least squares solution is found by using the Moore–Penrose pseudo inverse

$$\mathbf{f} = \left(\mathbf{A}^{\mathsf{T}}\mathbf{A}\right)^{-1}\mathbf{A}^{\mathsf{T}}\mathbf{r}$$

There are two constraints that the mixing proportions must satisfy, and that need to be taken into account during the inversion process. The first is that the proportions are all positive and less than one, and the second is that they must sum to unity:

$$0 \leq f_m \leq 1 \text{ and } \sum_{m=1}^{M} f_m = 1$$

In practice, these constraints are sometimes found to be violated, particularly if the end members are derived from average cover type spectra or the end member selection is poor.[49]

In this development we have assumed that the end member matrix **A** is known. Sometimes that will not be the case and the unmixing process needs not only to find the mixing proportions but an acceptable set of end members as well. There are several methods by which that can be approached, including the adoption of independent component analysis or independent factor analysis[50] and use of the Gaussian mixture model approach of Sect. 8.4.

## 11.10   Is There a Best Classifier?

This question has been in the minds of image analysts ever since alternative classifiers became available. It drives research in classifier design to find so-called better algorithms, in which comparative results are often quoted to demonstrate how new algorithms perform better than their predecessors. Is it conceivable that newer algorithms and methodologies will always be better[51] particularly if algorithms are compared in fair, unbiased trials? In this section we consider that question and draw the conclusion that, provided a particular algorithm does not suffer any theoretical deficiencies, if properly applied it should deliver results almost as good as any other classifier. Outcomes are driven more by how the algorithm is used—i.e., by methodologies—rather than by any inherent superiority of the algorithm itself.

### *11.10.1   Segmenting the Spectral Space*

As noted in Sect. 3.3 classification is a mapping from measurement space to a set of labels in the form of a thematic map. Most algorithms perform this mapping using

---

[49] See H.N. Gross and J.R. Schott, Application of spectral mixture analysis and image fusion techniques for image sharpening, *Remote Sensing of Environment*, vol. 63, 1998, pp. 85–94.

[50] See J.M.P. Nascimento and J.M.B Dias, Unmixing hyperspectral data: independent and dependent component analysis, in C-I Chang, ed., *Hyperspectral Data Exploitation*, John Wiley & Sons, Hoboken, N.J., 2007, which also contains an excellent overview of unmixing in general.

[51] See G. Wilkinson, Results and implications of a study of fifteen years of satellite image classification experiments, *IEEE Transactions on Geoscience and Remote Sensing*, vol. 43, 2005, pp. 433–440 for an interesting demonstration that, in general, performance had not materially improved to 2004 despite new algorithms and methodologies becoming available.

the measurement attributes of the pixels, although some processes exploit context information as well, as we saw in Sect. 8.23. Here we will not add that complication and focus just on the pixel-specific classification task.

How large is the measurement (i.e., spectral) space? For a sensor with $C$ channels and a radiometric resolution of $b$ bits per measurement there are $2^{bC}$ cells or individual sites in the discrete spectral space. That can be an enormous number, but for the moment suppose we had an instrument with just two bands, and one bit of radiometric resolution (allowing two levels of brightness). The measurement space then consists of just four cells. Conceivably, based on training data or some other form of reference information, we could attach a label to each of those four cells so that any subsequent measurement could be classified by reference into which cell in the spectral space the measurement fell. That is the basis of the table look up classifier of Sect. 8.9.

It could be argued that the ideal classifier is one that is able to attach a label uniquely to each of the cells in the spectral space. Clearly, the huge size of that space for most modern instruments, along with the fact that many cells are unoccupied, means that labelling at the individual cell level is totally impracticable. As a consequence, we use algorithms that segment the space rather than label each of its cells. The suitability of a classifier is then very much related to how well the spectral space is segmented.

In the very early days of pattern recognition the linear discriminant function was a common algorithm. The available training data was used to locate a hyperplane between the training classes thereby segmenting the measurement space into regions that were associated with the training classes. Whether that classifier then *generalised* well, by performing accurately on testing data, depended on how good the location of the hyperplane was. By comparison to labelling each of the individual cells in the measurement space, we can see that there is a trade-off between ease of training (finding a hyperplane rather than trying to label every cell) and performance. More generally, we can describe the trade-off as between how easily a classifier can be trained and used, and the performance it delivers. Such trade-offs are important in operational thematic mapping.

Clearly, the basic hyperplane of the previous paragraph would not be used now, except in the simplest of circumstances, because it only implements linear decisions. Modern classifiers segment the data space into much more flexible regions so that good generalisation is possible. The neural network uses many linear decisions in such a way that very complex piecewise-linear hyper-surfaces are implemented. The Gaussian maximum likelihood approach sets up quadratic hyper-surfaces which, on first sight, would suggest their power of generalisation might be quite limited compared with neural networks, for example. But when information classes are represented by sums of Gaussian spectral classes quite flexible separating boundaries emerge as illustrated in Fig. 11.16; that is why the hybrid method of Sect. 11.4 works well. The support vector machine, while still implementing simple linear separating hyper-surfaces, performs well because it projects the data into a higher dimensional space in which linear separation is possible.

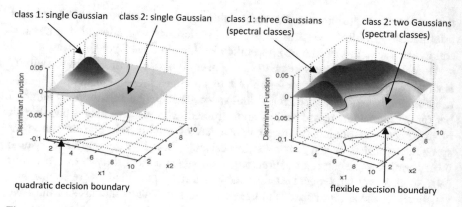

**Fig. 11.16** Showing the decision boundary between **a** two information classes each with one spectral class and **b** two information classes, one of which has three and the other two spectral classes; for clarity, the distributions for the second class are shown plotted negatively, but that does not affect the separating boundary. See J. A. Richards and N. G. Kingsbury, Is there a preferred classifier for operational thematic mapping? *IEEE Transactions on Geoscience and Remote Sensing*, 52, no 2, May 2014, pp. 2715–2725

## 11.10.2    Comparing the Classifiers

From the above perspective we could infer that each of those well-recognised approaches, when used properly, performs acceptably and that, on the basis of generalisation accuracy, one would probably not be preferred over any other. However, there are other comparators that we should take into account, particularly from an operational perspective. To consider those we will continue commenting on the maximum likelihood rule, the support vector machine and the convolutional neural network, examining the methodologies that would be employed in practice with each algorithm. We commence that process by summarising in Table 11.8 the key consideration in using each method.

Against this outline, Table 11.9 summarises the practical considerations in each case. As is to be expected there are always trade-offs. Where one approach excels, the others may have limitations. Nonetheless, it is important in comparative studies that the methodologies summarised Table 11.8 are used for each procedure. Too often, one sees comparisons that conclude that certain approaches are superior to others and yet care has not been taken to ensure that the algorithmic methodologies have been properly applied. The very reason for placing an algorithm within its operational methodology is to ensure that the measurement space is segmented in such a manner as to ensure that the algorithm performs as well as possible. Even moderately primitive algorithms such as the parallelepiped classifier can be made to perform well provided the data space is appropriately segmented.

**Table 11.8**  Methodological comparison of three supervised classifiers

| Approach | Methodology |
| --- | --- |
| Gaussian maximum likelihood | 1. Obtain labelled training data<br>2. Apply feature reduction. This will be necessary to avoid poor generalisation if the Hughes phenomenon is encountered. With hyperspectral data that can be a non-trivial task, but mechanisms such as the block diagonalisation approach of Sect. 10.4 can render the maximum likelihood rule viable for high dimensional data sets<br>3. Resolve multimodality (using unsupervised clustering). Information classes have now been resolved into several (spectral) sub-classes<br>4. Compute the statistics (mean vector and covariance matrix) for each spectral class<br>5. Apply the discriminant function of (8.7) to perform multiclass labelling into the spectral classes<br>6. Label unseen pixels (generalisation) on the basis of posterior probabilities. Prior probabilities can be incorporated at this step<br>7. Generate a thematic map by mapping information classes to spectral classes |
| Support vector machine | 1. Obtained labelled training data<br>2. Determine the multiclass strategy to be used<br>3. Determine the kernel function to be used<br>4. Possibly use feature selection, although the SVM can handle data of hyperspectral dimensions<br>5. Train the machine through optimisation, while also optimising the kernel parameter(s) and the regularisation parameter. This may require many trials using grid searching to find the best parameters. This step can be time consuming<br>6. Generate the thematic map, via a tree of binary decisions |
| Convolutional neural network | 1. Obtain labelled training data<br>2. Choose a topology; this may not be evident at the start and many trials may be required to find the one best suited to the problem at hand; transfer learning can help expedite this step. The choices here relate to the kernel sizes to use, the number of filters to employ, whether pooling layers will be incorporated and how deep the network should be<br>3. Train the network iteratively<br>4. Refine the network through pruning unwanted hidden layer nodes<br>5. Generate the (multiclass) thematic map |

**Table 11.9** Qualitative comparison of the three common classifier types

|  | Maximum likelihood classifier | Support vector machine | Convolutional neural network |
|---|---|---|---|
| Training | Simple, although sets of normal distributions may need to be found per information class for good results | Grid search procedures are needed to determine kernel and regularisation parameters | Extensive trial and error may be needed to find best network topology, although transfer learning can shorten the task |
| Training time | Fast | Can be long | Can be very long; depends on the number of weights and offsets to be found |
| Classification time | Good but depends quadratically on dimensionality | Fast; depends linearly on dimensionality | Fast; depends linearly on dimensionality |
| Multiclass | Is a multiclass classifier | Sets of binary SVMs need to be embedded in a decision tree for multiclass operation | Is a multiclass classifier |
| Hyperspectral | Challenge with high dimensional data because of the need to estimate second order parameters; requires a feature reduction approach | Handles hyperspectral data because it is based on linear decisions | Handles hyperspectral data because it is based on linear decisions |
| Context classification | Is a point classifier and requires post processors like label relaxation to embed context information | Is a point classifier, and doesn't easily interface to other post-processing statistical techniques | Spatial context sensitivity is a key feature because of the convolutional basis of the algorithm |
| Posterior probabilities | The algorithm generates class posterior probabilities prior to maximum selection | Hard classifier, so posterior probabilities unavailable | Surrogate posterior probabilities can be generated with the *softmax* operation |

There is some very sobering guidance provided by the No Free Lunch Theorem from the field of machine learning,[52] which we state as

> If the goal is to obtain good generalisation performance, there are no context-independent or usage-independent reason to favour one learning or classification method over another. If one classification algorithm seems to outperform another in a particular situation, it is a consequence of its fit to the particular pattern recognition problem, not the general superiority of the algorithm.

This does not imply that there are no poor classifiers, but it reminds us that the classifiers we have come to consider as our benchmarks in remote sensing do not inherently perform well—they perform well when the analyst has used them properly in a given task, which is the central message of this section. Overall, there is probably no best classifier.

There is a final point that is material in some circumstances. If classifier results are to be incorporated into some other statistical process, such as a multisource statistical analysis procedure, a Markov random field model or post classification processing like relaxation labelling, then the posterior probabilities generated from the maximum likelihood rule are already in the form required for use in those other techniques. Also, the set of posterior probabilities can be used before the maximum selection step in the maximum likelihood rule to give measures of relative likelihoods of the less favoured classes. By contrast, the outputs from the support vector and neural network approaches do not interface well with those other procedures, nor do they naturally provide relative measures of likelihood, unless mechanisms such as *Softmax* are used (see Sect. 8.21.6).

## 11.11   Bibliography on Image Classification in Practice

An important early paper that seemed to be the first to address a real image classification methodology, and which is now available in readily accessible form, is

> M.D. Fleming, J.S. Berkebile and R.M. Hofer, Computer aided analysis of Landsat 1 MSS data: a comparison of three approaches including a modified clustering approach, information note 072475, Laboratory for Applications of Remote Sensing, Purdue University, West Lafayette, Indiana, 1975.
>
> http://www.lars.purdue.edu/home/references/LTR_072475.pdf.

A comprehensive recent review of semi-supervised learning techniques, from a machine learning rather than remote sensing perspective, will be found in

> J.E. van Engelen and H.H. Hoos, A survey on semi-supervised learning, *Machine Learning*, vol. 109, 2020, pp. 373–440.

---

[52] See R.O. Duda, P.E. Hart and D.G. Stork, *Pattern Classification*, 2nd ed., John Wiley & Sons, N.Y., 2001. The name for the theorem is attributed to David Haussler.

An easily read account of the sampling and quantitative procedures used to assess the performance of a classifier and the accuracy of a thematic map is given in

R.G. Congalton and K. Green, *Assessing the Accuracy of Remotely Sensed Data: Principles and Practices*, 2nd ed., CRC Press Taylor and Francis Group, Boca Raton FL, 2009 and its 3rd edition published in 2019.

The following supplements that treatment with the necessary statistical material

W.G. Cochran, *Sampling Techniques*, John Wiley & Sons, N.Y., 1961.

Other recognised historical accounts on assessing accuracy are

R.M. Hord and W. Brooner, Land-use map accuracy criteria, *Photogrammetric Engineering and Remote Sensing*, vol. 42, 1976, pp. 671–677,

J.L. van Genderen, B. F. Lock and P. A. Vass, Remote sensing: statistical testing of thematic map accuracy, *Remote Sensing of Environment*, vol. 7, 1978, pp. 3–14,

G. H. Rosenfield, K. Fitzpatrick-Lins and H.S. Ling, Sampling for thematic map accuracy testing, *Photogrammetric Engineering and Remote Sensing*, vol. 48, 1982, pp. 131–137, and

S.V. Stehman, R.L. Czaplewski, Design and analysis for thematic map accuracy assessment, fundamental principles, *Remote Sensing of Environment*, vol. 64, 1998, pp. 331–344.

Although the kappa coefficient has been a popular measure of map accuracy, its efficacy is now seriously questioned, and alternative metrics are proposed, for which see

G. M. Foody, Explaining the unsuitability of the kappa coefficient in the assessment and comparison of the accuracy of thematic maps obtained by image classification, *Remote Sensing of Environment*, vol. 239, 2020, pp. 1–11.

R. G. Pontius Jr and M. Millones, Death to Kappa: birth of quantity disagreement and allocation disagreement for accuracy assessment, *Int. J Remote Sensing*, vol. 32, no. 15, pp. 4407–4429, 2011.

As well as in Congalton and Green above, a significant discussion of the kappa coefficient, including its variants, will be found in

W.D. Hudson and C.W. Ramm, Correct formulation of the Kappa coefficient of agreement, *Photogrammetric Engineering and Remote Sensing*, vol. 53, 1987, pp. 421–422.

Decision tree material will be found in many of the standard image processing textbooks. A specialized treatment is

L. Breiman, J.H. Friedman, R.A. Olshen and C.J. Stone, *Classification and Regression Trees*, Chapman and Hall/CRC, Roca Baton Florida, 1998.

Some significant research results of historical importance are

P.H. Swain and H. Hauska, The decision tree classifier: design and potential, *IEEE Transactions on Geoscience Electronics*, vol. GE-15, 1977, pp. 142–147.

E.M. Rounds, A combined nonparametric approach to feature selection and binary decision tree design, *Pattern Recognition*, vol. 12, 1980, pp. 313–317.

B. Kim and D. A. Landgrebe, Hierarchical classifier design in high-dimensional, numerous class cases, *IEEE Transactions on Geoscience and Remote Sensing*, vol. 29, no. 4, July 1991, pp. 518–528.

S.R. Safavian and D.A. Landgrebe, A survey of decision tree classifier methodology, *IEEE Transactions on Systems, Man and Cybernetics*, vol. 21, May 1991, pp. 660–674.

Y. Ikura and Y. Yasuoka, Utilisation of a best linear discriminant function for designing the binary decision tree, *Int. J. Remote Sensing*, vol. 12, 1991, pp. 55–67.

The interesting field of random forest classifiers is covered in

L. Breiman, Random Forests, *Machine Learning*, vol 45, 2001, pp. 5–32, and

T. Hastie, R. Tibshirani and J. Friedman, *The Elements of Statistical Learning: Data Mining, Inference and Prediction*, Springer Science+Business Media, N.Y., 2009,

while their application to remote sensing problems is demonstrated in

P.O. Gislason, J.A. Benediktsson and J.R. Sveinsson, Random forest classification of multisource remote sensing and geographic data, *Proc. Int. Geoscience and Remote Sensing Symposium*, Alaska, 20–24 Sep. 2004, pp. II: 1049–1052

J.S. Ham, Y. Chen, M.M. Crawford and J. Ghosh, Investigation of the random forest framework for classification of hyperspectral data, *IEEE Transactions on Geoscience and Remote Sensing*, vol. 43, no. 3, March 2005, pp. 492–501.

Spectroscopic and library matching techniques for the analysis of hyperspectral data are covered in

R.N. Clark, G.A. Swayze, K.E. Livio, R.F. Kokaly S.J. Sutly, J.B. Dalton, R.R. McDougal and C.A. Gent, Imaging spectroscopy: earth and planetary remote sensing with the USGS Tretracorder and expert systems, *J. Geophysical Research*, vol. 108, E12, pp. 5.1–5.44, 2003, and

Imaging Spectroscopy Special Issue, *Remote Sensing of Environment*, vol. 113, 2009

An excellent review of spectral unmixing procedures, with references to many of the important works in this area, and the need for care in selecting end members, will be found in

B. Somers, G. P. Asner, L. Tits and P. Coppin, End member variability in spectral mixture analysis: a review, *Remote Sensing of Environment*, vol. 115, 2011, pp. 1603–1616, while

J.M.P. Nascimento and J.M.B Dias, Unmixing hyperspectral data: independent and dependent component analysis, in C-I Chang, ed., *Hyperspectral Data Exploitation*, John Wiley & Sons, Hoboken, NJ, 2007,

## 11.12   Problems

11.1   What is the difference between an *information class* and a *spectral class*?
The notion of spectral class has more relevance to optical and thermal
remote sensing imagery than to other data types such as radar. When
analysing co-registered data sets it is often necessary to determine the class
structures for each data set independently, before they are combined into the
information classes of interest to the user. In such cases the term *data class*
may be more relevant than spectral class; see Sect. 12.3.1.

11.2   Four analysts use different qualitative methodologies for interpreting
spectral imagery. They are summarised below. Comment on the merits and
shortcomings of each approach and indicate which one you think is most
effective.

Analyst 1

1. Chooses training data from homogeneous regions for each cover type.
2. Develops statistics for a maximum likelihood classifier.
3. Classifies the image.

Analyst 2

1. Performs a clustering of the whole image and attaches labels to each
   cluster type afterwards.

Analyst 3

1. Chooses several regions within the image, each of which includes more
   than one cover type.
2. Clusters each region and identifies the cluster types.
3. Uses statistics from the clustering step to perform a maximum likelihood
   classification of the whole image.

Analyst 4

1. Chooses training fields within apparently homogeneous regions for each
   cover type.
2. Clusters those regions to identify spectral classes.
3. Uses statistics from the clustering step to perform a maximum likelihood
   classification of the whole image.

11.3   For the method you have identified as preferable in Prob. 11.2 comment on
how separability measures would be used to advantage. What if the data
were of hyperspectral dimensions?

11.4   The spectral classes used with the maximum likelihood decision rule in
supervised classification are assumed to be representable by single multivariate normal probability distributions. Geometrically, this implies that
they will have hyperellipsoidal distributions in the spectral domain. Do you

think that clustering by the iterative moving means algorithm will generate spectral classes of that nature? Given this observation, how might you best generate spectral classes for maximum likelihood classification, the minimum distance classifier and parallelepiped classification?

11.5  How can thresholds be used to help in the identification of spectral classes when using the maximum likelihood decision rule?

11.6  Is there a preferred kernel for use with the support vector machine when classifying high dimensional remote sensing image data?

11.7  Just before the labelling step when generating a thematic map using a classifier, each pixel has associated with it a set of measures that guide the allocation of the preferred label. For example, in the case of maximum likelihood classification it is the posterior probability. What measures of typicality would apply in the case of the support vector machine and the neural network?

11.8  Suppose a particular image contains just two cover types—vegetation and soil. A pixel identification exercise is carried out to label each pixel as either soil or vegetation and thus generate an estimate of the proportion of vegetation in the region being imaged. For homogeneous regions the labelling exercise is straightforward. However, the image will also contain a number of mixed pixels so that end member analysis would be considered as a means for resolving their soil/vegetation proportions. Is the additional work warranted if the approximate proportion of vegetation to soil is 1:100, 50:50, or 100:1?

11.9  This question relates to the effect on classification accuracy of resampling that might be used to correct imagery before analysis. For simplicity, consider a single line of infrared image data over a region that is vegetated to the left and water to the right. Imagine the vegetation/water boundary is sharp. Resample your line of data onto a grid with the same centres as the original, using both nearest neighbour and cubic convolution interpolation. Clearly, in the latter case the interpolation is just along the line, rather than over a neighbourhood of 16 surrounding pixels. Comment on the results of classifying each of the resampled pixels given that a classifier would have been trained on classes that include those with responses between vegetation and water.

11.10  Sometimes the spectral domain for a particular sensor and scene consists of a set of distinct clusters. An example would be a near infrared versus red spectral domain of an image in which there are regions of just water, sand and vegetation. The spectral domain would then have three distinct groups of pixels. More often than not, particularly for images of natural vegetation and soil, the spectral domain will be a continuum because of the differing degrees of mixing of the various cover types that occur in nature. One is then led to question the distinctness and uniqueness, not only of spectral classes, but information classes as well. Comment on the issues involved in the classification of natural regions both in terms of the definition of the set of information classes and how spectral classes might be identified to assist in training.

11.11  Manually design a simple decision tree that could be used efficiently with ETM+ data for classification into deep water, shallow water, green vegetation and soil.

11.12  A particular study requires the mapping of water temperature in the effluent from an industrial complex located on the side of a river. There is no interest in the land itself. Using ETM+ data design a strategy, most likely using a decision tree that, first, separates the soil and water and then maps the temperature variation within the water.

11.13  The error matrix is a very comprehensive summary of the accuracy of the classes on a map produced from a classification exercise. Perhaps its only drawback is the number of elements it contains, particularly for a large number of classes. That is why in many cases we turn to single measures such as total classification accuracy and Kappa coefficient. Discuss the benefits and disadvantages of single metrics as against retaining the complete error matrix.

11.14  Generate a CART decision tree for the two class data shown in Fig. 10.5. Then perform a principal components transformation and repeat the exercise.

11.15  To determine the accuracy of a thematic map, is it better to check the map labels against available reference (ground truth) data or check the performance of the classifier used to produce the map against the available reference data. Why?

11.16  Is imaging spectroscopy

(a) a method where images are examined for spectral content in a laboratory,
(b) a technique for recording sampled pixel spectra in image format, or
(c) imaging with two completely different image types—such as radar and thermal data?

11.17  Does the principal components transformation improve the intrinsic separability of an image data set?

11.18  How might you obtain training data for use with a support vector classifier for a mixed vegetation/water class boundary, and for a river with dimensions of about 2–3 pixels wide?

# Chapter 12
# Multisource Image Analysis

**Abstract** Noting that more information about the landscape can generally be found by using a range of remote sensing data types together, techniques are covered for so-called multisource image analysis. They include those methods which have a statistical basis, those based on the Theory of Evidence and those which adopt an expert systems approach. Combined optical and radar imagery is used as an example of the value of processing disparate data types together. A discussion then follows on how to operationalise multisource analysis to maximum benefit, noting that the different data types may have different qualities, they may not be available at the same time and they each may have their own optimal methods for analysis. Rather than data fusion, fusion at the label (information class) level is suggested as the most effective methodology.

## 12.1 Introduction

Many applications in remote sensing can be handled using a single source of image data and, for years, simple multispectral imagery was seen to be sufficient for straightforward applications. However, the ability to interpret what is on the ground is often enhanced considerably when more than one type of spatial data is employed. For example, a combination of multispectral or hyperspectral data with radar imagery has been shown to be a particularly valuable data set.

The reason that multiple sources are of value is that each typically measures different things about the earth's surface. Multispectral measurements tend to be dominated by properties such as plant pigmentation and soil mineralisation, by moisture content and by microscopic physical properties such as the cellular structure of plants and the scattering properties of suspensions in water. On the other hand, hyperspectral data is much more sensitive to fine spectral detail such as the resonant absorptions and other spectroscopic properties of matter that can be detected in the wavelength range of the sensor; typically, that allows a more detailed biochemical understanding of the plant, soil or other surface cover being imaged.

© The Author(s), under exclusive license to Springer Nature Switzerland AG 2022
J. A. Richards, *Remote Sensing Digital Image Analysis*,
https://doi.org/10.1007/978-3-030-82327-6_12

The surface properties that affect radar imagery are quite different. Radar reflectivity is typically dominated by gross geometric detail such as the smoothness or roughness of a surface, or the adjacency of reflecting faces, and by the complex dielectric constant of the surface material—which, in turn, is established by bulk water content.

It is because different properties are sensed by different imaging modalities that multisource data in combination often gives superior results when seeking to describe the surface cover types being imaged.

When undertaking multisource analysis, we have to be mindful of the fact that each data type has its own preferred means for analysis, a characteristic often overlooked in multisource, or multisensor, applications. For example, hyperspectral imagery can be interpreted very effectively using library searching methods in which recorded single pixel spectra are compared against prototype laboratory measurements. Multispectral imagery has been interpreted successfully for many years using standard supervised classification techniques. Radar imagery is not handled well by those procedures. Instead, because of the different surface properties being imaged, and the complexity of the radar signal, analytical procedures unique to radar have been devised.[1] They include modelling the interaction between the incident microwave radiation and the landscape, a form of maximum likelihood classification that is matched to the complexity of the radar signal, and methods based on a mathematical decomposition of the received radar signal into components that can be associated with fundamental scattering behaviours. We return to this in Sect. 12.6.

It is the purpose of this chapter to present some of the more common techniques used for addressing the multisource analysis task quantitatively. Sometimes they are numerically based but at other times they involve the manipulation of data in the form of labels. Such a case occurs when an existing map forms one of the sources.

Analysis of multisource data by photointerpretation is also possible but depends on the human analyst being skilled in the visual interpretation of all the different data types available. We do not treat that approach here.

Clearly the data sets to be analysed must first be geometrically registered. If that has not been done, then the procedures of Chap. 2 need to be employed. A word of caution is in order here: the accuracy that can be achieved from the analysis of multisource data can be influenced by the accuracy of the registration process almost as much as by the effectiveness of the analytical procedures employed.

## 12.2   Stacked Vector Analysis

A straightforward, but not very satisfactory, way to classify mixed data is to form extended pixel vectors by stacking together the individual vectors that describe the separate data sources. This stacked vector will be of the form

---

[1] See J.A. Richards, *Remote Sensing with Imaging Radar*, Springer, Berlin, 2009.

$$\mathbf{X} = \left[\mathbf{x}_1^T, \mathbf{x}_2^T, \ldots \mathbf{x}_s^T\right]^T \tag{12.1}$$

where $S$ is the total number of individual data sources with corresponding data (column) vectors $\mathbf{x}_1, \mathbf{x}_2, \ldots \mathbf{x}_S$. The stacked vector $\mathbf{X}$ can, in principle, be analysed using any of the standard classification techniques. That presents a number of difficulties, particularly if statistical methods are to be used, including the incompatible statistics of the different data types. Secondly, as alluded to in Sect. 12.1, this approach does not take advantage of the fact that different data types have their own preferred methods for analysis. At best, the stacked vector technique could be regarded as a simple yet not necessarily very effective approach. At worst, it will not take full advantage of the information offered in the different sets of measurements.

## 12.3 Statistical Multisource Methods

If the different datasets are technically not too different from each other, such as sets of spectral measurements from complementary sensors, then a statistical approach can be considered. Several of the more common statistical methods are presented in the following.

### 12.3.1 Joint Statistical Decision Rules

The single data source decision rule in (8.1) can be restated for the multisource data described in (12.1) as

$$\mathbf{X} \in \omega_i \text{ if } p(\omega_i|\mathbf{X}) > p(\omega_j|\mathbf{X}) \quad \text{for all } j \neq i \tag{12.2}$$

As with a single source analysis we can apply Bayes' theorem to give

$$\mathbf{X} \in \omega_i \text{ if } p(\mathbf{X}|\omega_i)p(\omega_i) > p(\mathbf{X}|\omega_j)p(\omega_j) \quad \text{for all } j \neq i$$

To proceed further we need to find or estimate the class conditional joint source probabilities $p(\mathbf{X}|\omega_i) = p(\mathbf{x}_1, \mathbf{x}_2, \ldots \mathbf{x}_S|\omega_i)$. To make that possible we assume statistical independence among the sources so that

$$p(\mathbf{X}|\omega_i) = p(\mathbf{x}_1|\omega_i)p(\mathbf{x}_2|\omega_i)\ldots p(\mathbf{x}_S|\omega_i) = \prod_{s=1}^{S} p(\mathbf{x}_s|\omega_i)$$

in which the $p(\mathbf{x}_s|\omega_i)$ are the class conditional distribution functions for each of the data sources individually. They could be called source specific class conditional distribution functions.

It is unlikely that the assumption of independence is valid, but it is usually necessary in order to perform multisource statistical classification, allowing the multisource decision rule to be written

$$\mathbf{X} \in \omega_i \text{ if } p(\omega_i) \prod_{s=1}^{S} p(\mathbf{x}_s|\omega_i) > p(\omega_j) \prod_{s=1}^{S} p(\mathbf{x}_s|\omega_j) \quad \text{for all } j \neq i \qquad (12.3)$$

An important consideration when classifying multiple data sources is whether each of the available sources has the same quality; in other words, the analyst has to trust each data source equally if the above decision rule is to be used. In practice it is likely that some sources will be less reliable than others and should be weighted down in a joint decision-making process. That can be achieved by adding powers to the source specific class conditional probabilities to give the decision rule

$$\mathbf{X} \in \omega_i \text{ if } p(\omega_i) \prod_{s=1}^{S} p(\mathbf{x}_s|\omega_i)^{\alpha_s} > p(\omega_j) \prod_{s=1}^{S} p(\mathbf{x}_s|\omega_j)^{\alpha_s} \quad \text{for all } j \neq i$$

where the $\alpha_s$ are a set of weights chosen to enhance the influence of some sources (those most trusted) and to diminish the influence of other, perhaps noisy sources.

There are several problems with the joint statistical approach, in common with the stacked vector method of the previous section. First, the class conditional distribution function for each source must be obtainable. Secondly, all the sources have to be available at the same time in order to apply the decision rule.

Finally, in (12.3) the information classes must be consistent over the sources—in other words, the set of information classes appropriate to one source must be the same as those for the other sources. For example, we would not ordinarily use the joint statistical approach if one of the sources were radar and another multispectral because it is then highly unlikely that the information classes would be the same. Even if the sources are compatible—all optical, for example—the spectral classes (the individual Gaussian modes for maximum likelihood) must be the same for each source in the way (12.3) is expressed, which would be highly unlikely.

It is possible to modify (12.3) to allow for differences in spectral classes. It is useful in this case to call the spectral classes *source-specific data classes*. Using Bayes' theorem in (12.3) we get, after removing non-discriminating terms,

$$\mathbf{X} \in \omega_i \text{ if } p(\omega_i)^{1-S} \prod_{s=1}^{S} p(\omega_i|\mathbf{x}_s) > p(\omega_j)^{1-S} \prod_{s=1}^{S} p(\omega_j|\mathbf{x}_s) \quad \text{for all } j \neq i \qquad (12.4)$$

Suppose by clustering or some other means we have identified $M_s$ data classes $d_{sm}, m = 1 \ldots M_s$ for data source $s$. The source specific posterior probabilities in

(12.4) can be derived in the following manner. Note that the joint probability of an information class and a data vector in source $s$ can be expressed as the sum over the joint occurrence of the data classes in the source with the information class and data vector:

$$p(\omega_i, \mathbf{x}_s) = \sum_{m=1}^{M_s} p(\omega_i, d_{sm}, \mathbf{x}_s)$$

so that

$$p(\omega_i|\mathbf{x}_s)p(\mathbf{x}_s) = \sum_{m=1}^{M_s} p(\omega_i, d_{sm}, \mathbf{x}_s)$$

$$= \sum_{m=1}^{M_s} p(\mathbf{x}_s|d_{sm}, \omega_i)p(d_{sm}, \omega_i)$$

giving

$$p(\omega_i|\mathbf{x}_s) = \sum_{m=1}^{M_s} p(\mathbf{x}_s|d_{sm}, \omega_i)p(d_{sm}|\omega_i)p(\omega_i)/p(\mathbf{x}_s) \qquad (12.5)$$

After $p(\mathbf{x}_s)$ has been removed as non-discriminating, (12.5) can be substituted into (12.4) to provide a decision rule that allows for the possibility of different data (spectral) class sets for different data sources. Note that (12.5) depends on the bridge between the data classes and information classes via the second term inside the summation, and the distribution function for the measurement vectors within the data classes of the information class of interest. Both of those terms can be generated once the data classes have been identified in the source-specific measurement data. This form of decomposition into data classes, which conceivably could overlap information class boundaries, is not unlike the technique of cluster space classification.[2]

## 12.3.2  Committee Classifiers

Although they do not need necessarily to have a statistical basis, it is helpful at this stage to consider the concept of committee classifiers because some further multisource statistical methods depend on that concept.

We met committees of classifiers in Sect. 8.19. For convenience we reproduce in Fig. 12.1 two versions of the generic committee structure. In one case the same data

---

[2] See Sect. 9.14.

is provided to all individual classifiers in the committee whereas, in the other, individual data sources are handled by their own classifiers. In the first case the committee is used to obtain a better estimate for the label using a single data source and is not unlike the boosting technique of Sect. 8.19.2. Of course, that configuration could also be used for multisource classification if the single input vector were the stacked vector of (12.1). More often, though, the form of committee used for multisource analysis is that shown in Fig. 12.1b. In principle, each classifier could be optimised for handling one particular data type.

It is a feature of committee classifiers that there is a chairman or decision-maker, whose role it is to consider the outputs of the individual classifiers and make a decision about the most appropriate class label for a pixel.

As noted in Sect. 8.19 the decision-maker can use one of several logics in coming to a decision:

- *majority vote logic* in which the decision-maker chooses the class most recommended by the committee members.
- *veto logic* in which all classifiers have to agree about the class membership of a pixel before the decision-maker will label it.
- *seniority logic* in which the decision-maker consults one particular classifier first (the most senior). If that classifier is able to allocate a class label to a pixel, then the decision-maker chooses that label. Otherwise, the decision-maker consults the next most senior classifier, and so on, until the pixel can be labelled.

### 12.3.3 Opinion Pools and Consensus Theory

Closely related to the committee classifier concept is the use of opinion pools. They depend on finding the single source posterior probabilities and then combining them arithmetically or geometrically (logarithmically). The *linear opinion pool* computes a group membership function, similar to a joint posterior probability, of the form

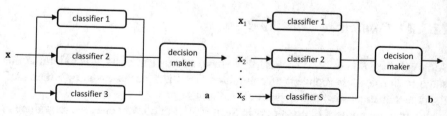

**Fig. 12.1 a** A committee of classifiers in which all data is fed to each classifier; **b** a committee of classifiers in which each classifier handles a separate data source

$$f(\omega_i|\mathbf{X}) = \sum_{s=1}^{S} \alpha_s p(\omega_i|\mathbf{x}_s)$$

in which the $\alpha_s$ are a set of weighting constants that sum to unity. They control the relative influences of each source in the final value of the group membership function, and thus in the labelling of a pixel. A limitation of this rule—known generically as a consensus rule—is that one data source tends to dominate decision-making.[3] Another consensus rule which doesn't suffer that limitation is the multiplicative version

$$f(\omega_i|\mathbf{X}) = \prod_{s=1}^{S} p(\omega_i|\mathbf{x}_s)^{\alpha_s}$$

Note that if one source posterior probability is zero then the membership function is zero and, irrespective of the recommendation from any of the other sources, the group recommendation is zero for that class-pixel combination. In other words, one very weak source can veto a decision.

Taking the logarithm of the multiplicative rule gives the *logarithmic opinion pool* consensus rule

$$\log\{f(\omega_i|\mathbf{X})\} = \sum_{s=1}^{S} \alpha_s \log\{p(\omega_i|\mathbf{x}_s)\}$$

### 12.3.4   *Use of Prior Probabilities*

In the decision rule of (8.3) and the discriminant function of (8.4) the prior probability term tells us the probability with which the class membership could be assessed based on any information we have about a pixel prior to considering the available remotely sensed information. In its simplest form we assume that the prior probabilities represent the relative abundances of ground cover types in the scene being analysed. Prior class membership can be obtained from other sources of information as well. In the case of the Markov Random Field approach to incorporating spatial context, developed in Sect. 8.23.5, the prior term is a neighbourhood conditional prior probability.

---

[3] See J.A. Benediktsson, J.R. Sveinsson and P.H. Swain, Hybrid consensus theoretic classification, *IEEE Transactions on Geoscience and Remote Sensing*, vol. 25, no. 4, July 1997, pp. 833–843.

We could also use the prior probability to incorporate the effect of another data source, such as topography, when carrying out classification based on spectral data.[4]

### 12.3.5  Supervised Label Relaxation

The probabilistic label relaxation algorithm of Sect. 8.23.4 can also be used to refine the results of a spectrally-based classification by bringing in the effect of another data source while developing spatial neighbourhood consistency. The updating rule in (8.72) can be modified for this purpose. Although heuristic, it has been seen to perform well when embedding topographic data into a classification.[5] Known as supervised relaxation, the updating rule at the $k$th iteration for checking that the correct class is $\omega_i$ on pixel $m$ is

$$p_m^{k+1}(\omega_i)^* = p_m^k(\omega_i)Q_m^k(\omega_i) \quad \text{for embedding spatial context}$$
$$p_m^{k+1}(\omega_i) = p_m^{k+1}(\omega_i)^* \phi_m(\omega_i) \quad \text{for incorporating another data source}$$

where $\phi_m(\omega_i)$ is the probability that $\omega_i$ is the correct class for pixel $m$ from the additional data source. After the application of these two equations the rule in (8.72) is used.

## 12.4   The Theory of Evidence

The mathematical Theory of Evidence[6] is an interesting heuristic technique devised to allow sets of inferences about a particular labelling proposition to be merged. It is attractive in multisource image analysis because it does not require the original data variables necessarily to be in numerical form; as a result, map-like data can be one of the sources of evidence considered. More importantly, it allows the user to ascribe degrees of trust or uncertainty to each data type when the combination is carried out. That is particularly attractive when the analyst has differing degrees of confidence in the data sets available for analysis. Although the original data does not necessarily need to be numerical, the combination step involves numerical

---

[4] See A.H. Strahler, The use of prior probabilities in maximum likelihood classification of remotely sensed data, *Remote Sensing of Environment*, vol. 10, 1980, pp. 135–163, and L. Bruzzone, C. Conese, F. Maselli and F. Aoli, Multisource classification of complex rural areas by statistical and neural network approaches, *Photogrammetric Engineering and Remote Sensing*, vol. 63, 1997, pp. 523–533.

[5] See J.A. Richards, P.H. Swain and D.A. Landgrebe, A means of utilising ancillary information in multispectral classification, *Remote Sensing of Environment*, vol. 12, 1982, pp. 463–477.

[6] The fundamental reference is G. Shafer, *A Mathematical Theory of Evidence*, Princeton UP, N.J., 1976.

manipulation of quantitative measures of evidence. The bridge between those measures and the original data is left largely to the user. Accordingly, there is substantial room for the user to choose how to represent the data; that flexibility is one of the criticisms of the technique, since it appears to lack a firm analytical basis.

### 12.4.1 The Concept of Evidential Mass

The foundation of the technique involves assigning belief, represented by a so-called *mass* of evidence, to various labelling propositions for a pixel following the analysis of a given source of remotely sensed data. The total mass of evidence available for allocation over the set of possible labels for the pixel is unity. To this extent, it seems to be similar to the distribution of probabilities. There is, however, a significant difference in that mass can also be allocated to uncertainty about the labelling, as we will see shortly.

To see how mass might be distributed consider a classification from a single source of image data that has led to a labelling into one of three classes $\{\omega_1, \omega_2, \omega_3\}$. It is a requirement that the set of classes be exhaustive, in that they cover all possibilities. Suppose that the classification algorithm employed tells us that the three labels for a given pixel have likelihoods in the ratios 2:1:1. Suppose, also, that we are a little uncertain about the classification, either as a result of the quality of the data used, or perhaps our concern about the effectiveness of the classification algorithm. Owing to this uncertainty, suppose we are only willing to commit ourselves to classify the pixel with about 80% confidence. In other words, we are 20% uncertain about the labelling, even though we are reasonably happy with the relative likelihoods. Using the symbolism of the Theory of Evidence the distribution of the unit mass of evidence over the three possible labels, and our uncertainty about the labelling, is expressed:

$$m(\langle \omega_1, \omega_2, \omega_3, \theta \rangle) = \langle 0.40, 0.20, 0.20, 0.20 \rangle \tag{12.6}$$

where the symbol $\theta$ is used to signify the uncertainty. Effectively, uncertainty represents the set of all possible labels; the mass associated with uncertainty has to be allocated somewhere so it is allocated to the set as a whole. Equation (12.6) tells us that the mass of evidence assigned to label $\omega_1$ as being the correct one for the pixel is 0.4, etc.[7]

We now define two further evidential measures. The *support* for a labelling proposition is the sum of the mass assigned to the proposition and any of its subsets. Subsets are considered later. The *plausibility* of the proposition is one minus the total support for any contradictory propositions. Support is considered to be the

---

[7] This is in contrast to the probability 0.5 that $\omega_1$ is the correct class for the pixel as result of the classification step, in the absence of uncertainty.

minimum amount of evidence in favour of a particular labelling for a pixel whereas plausibility is the maximum possible evidence in favour of the labelling. The difference between the measures of plausibility and support is called the *evidential interval*. The true likelihood that the label under consideration is correct for the pixel is assumed to lie somewhere between the support and plausibility. For the above example, the supports, plausibilities and evidential intervals are:

$$s(\omega_1) = 0.40 \qquad p(\omega_1) = 0.60 \qquad p(\omega_1) - s(\omega_1) = 0.20$$
$$s(\omega_2) = 0.20 \qquad p(\omega_2) = 0.40 \qquad p(\omega_2) - s(\omega_2) = 0.20$$
$$s(\omega_3) = 0.20 \qquad p(\omega_3) = 0.40 \qquad p(\omega_3) - s(\omega_3) = 0.20$$

In this simple case the evidential intervals for all labelling propositions are the same and equal to the mass allocated to uncertainty in the process or data as discussed above, i.e., $m(\theta) = 0.20$. We have used the symbol $p$ to represent plausibility in order to avoid confusion with probability.

Consider another example involving four possible spectral classes, one of which represents our belief that the pixel is in either of two classes. This will demonstrate that, in general, the evidential interval is different from the mass allocated to uncertainty. In this case suppose the mass distribution is:

$$m(\langle \omega_1, \omega_2, \omega_1 \vee \omega_2, \omega_3, \theta \rangle) = \langle 0.35, 0.15, 0.05, 0.30, 0.15 \rangle$$

where $\omega_1 \vee \omega_2$ represents ambiguity in that, for the pixel under consideration, while we are prepared to allocate 0.35 mass to the proposition that it belongs class $\omega_1$ and 0.15 mass that it belongs to class $\omega_2$, we are prepared also to allocate some additional mass to the fact that it could belong to either of those two classes and not to any others.

For this example, the support for $\omega_1$ is 0.35 (being the mass attributed to the label) whereas the plausibility of $\omega_1$ being the correct class for the pixel is one minus the support for the set of contradictory propositions. There are two—$\omega_2$ and $\omega_3$. Therefore, the plausibility for the correct label being $\omega_1$ is 0.55, and the evidential interval is 0.2 (different now from the mass attributed to uncertainty). Support given to the mixture class $\omega_1 \vee \omega_2$ is 0.55, being the sum of the masses attributed to that class and its subsets.

To see how the Theory of Evidence works with multisource data return now to the simple example given by the mass distribution in (12.6). Suppose there is available a second source which can also be labelled into the same set of classes. Again, there will be some uncertainty in the labelling process. For a particular pixel suppose the mass distribution after analysing the second data source is

$$\mu(\langle \omega_1, \omega_2, \omega_3, \theta \rangle) = \langle 0.20, 0.45, 0.30, 0.05 \rangle \tag{12.7}$$

This shows that the second source of remotely sensed data, after analysis, favours $\omega_2$ as the correct label for the pixel, whereas analysis of the first data source favours

$\omega_1$. To resolve this problem the Theory of Evidence allows the two mass distributions to be merged in order to combine the evidences and thus come up with a label which is jointly preferred. At the same time overall uncertainty should be reduced because more evidence is available. This is done through the mechanism of the orthogonal sum in the next section.

## 12.4.2 Combining Evidence with the Orthogonal Sum

The *orthogonal sum* is illustrated graphically in Fig. 12.2. Also called Dempster's orthogonal sum, it is performed by constructing a unit square and partitioning it vertically in proportion to the mass distribution from one source and horizontally in proportion to the mass distribution from the other source. The areas of the rectangles so formed are calculated. One rectangle is formed from the masses attributed to uncertainty ($\theta$) in both sources; that is considered to be the remaining uncertainty in the labelling after the evidences from both sources have been combined. Rectangles formed from the masses attributed to the same class have their resultant area (mass) assigned to the class. Rectangles formed from the product of mass assigned to a particular class in one source and mass assigned to uncertainty in the other source have their resultant mass attributed to the specific class. Similarly, rectangles formed from the product of a specific class, say $\omega_2$, and an ambiguity, say $\omega_1 \vee \omega_2$, are allocated to the specific class. Rectangles formed from different classes in the two sources are contradictory and are not used in computing merged evidence. In order that the resulting mass distribution sums to unity a normalising denominator is computed as the sum of the areas of all the rectangles that are not contradictory. For the current example this factor is 0.47.

After the orthogonal sum has been computed, the resultant merged mass distributions are:

$$m(\omega_1) = (0.08 + 0.02 + 0.04)/0.47 = 0.298$$
$$m(\omega_2) = (0.09 + 0.01 + 0.09)/0.47 = 0.404$$
$$m(\omega_3) = (0.06 + 0.01 + 0.06)/0.47 = 0.277$$
$$m(\theta) = 0.01/0.47 \qquad\qquad = 0.021$$

We conclude that class two is recommended jointly. The reason is that source two favours class two and had less uncertainty; although source one favoured class one, its higher level of uncertainty meant that it was not as significant in influencing the outcome.

**Fig. 12.2** Graphical illustration of the Dempster orthogonal sum for merging evidences from two data sources; the shaded squares represent contradictory and thus unused evidence, while the evidences from the white squares (their areas) are allocated to the classes indicated

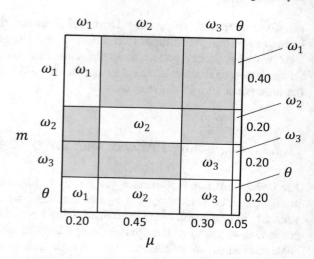

The orthogonal sum can also be expressed in algebraic form.[8] If two mass distributions are denoted $m_1$ and $m_2$ then their orthogonal sum is:

$$m_{12}(z) = \mathcal{H} \sum_{x \cap y = z} m_1(x)m_2(y)$$

where

$$\mathcal{H}^{-1} = \sum_{x \cap y \neq \phi} m_1(x)m_2(y)$$

in which $\phi$ is the null set. When applying these formulas, it is important to recognise that

$$(x \vee y) \cap y = y \text{ and } \theta \cap y = y$$

For more than two sources of data, the orthogonal sum can be applied repetitively since the expression is both commutative (the order in which the sources are considered is not important) and associative (it can be applied to any pair sources and then a third source, or equivalently can be applied to a different pair and then the third source).

---

[8] See T. D. Garvey, J.D. Lowrance and M.A. Fisher, An inference technique for integrating knowledge from disparate sources, *Proc. 7th Int. Conf. Artificial Intelligence*, Vancouver, 1981, pp. 319–325, and T. Lee, J.A. Richards and P.H. Swain, Probabilistic and evidential approaches for multisource data analysis, *IEEE Transactions on Geoscience and Remote Sensing*, vol. GE-25, no. 3, May 1987, pp. 283–293.

## *12.4.3   Decision Rules*

After the orthogonal sum has been applied the user can compute the joint support and plausibility for each possible class for a pixel. Maps could then be produced showing the distribution of supports or plausibilities for each category. This might be particularly appropriate in a situation where classes are not well-resolved and the user is interested in the relative likelihoods of each class.[9]

More often, the analyst will be interested in producing a thematic map in the same manner as when traditional statistical or numerical classification techniques are applied. That requires a decision rule. There are several candidates including a comparison for each pixel of supports, a comparison of the plausibilities, or both.[10] Generally, a maximum support decision rule would be used, although if the plausibility of the second most favoured class is higher than the support for the preferred label, the decision might be regarded as having a degree of risk.

## 12.5   Knowledge-Based Image Analysis

The statistical and evidential multisource techniques treated above have their limitations. Most are restricted to data that is inherently in numerical form, such as that from optical and radar sensors, along with quantifiable terrain data like digital elevation maps. Yet often we need to incorporate into a multisource analysis spatial data types that are non-numerical. These include geological, soil and planning maps, and even prior thematic maps of the same region. The Theory of Evidence can incorporate non-numerical image data, but it still requires the data to be expressed in a quantifiable form so that the numerical manipulation of evidence is possible.

Clearly a different approach is needed when one wishes to contemplate analysis using combinations of numerical and non-numerical imagery, or maps. Knowledge-based methods offer promise in this regard. In this section we outline some of the fundamental aspects of expert systems as knowledge-based processing tools. There are many knowledge processing systems available; we focus here on the expert system approach because it is easily related to general concepts in remote sensing.[11]

It will become clear during this treatment that the technique allows a more general treatment of the concept of information class than the statistical and evidential methods covered above. In particular, we can allow for different information

---

[9] For a geological example see W.L. Moon, Integration of geophysical and geological data using evidential belief function, *IEEE Transactions on Geoscience and Remote Sensing*, vol. 28, no. 4, July 1990, pp. 711–720.

[10] See Lee et al., 1987, loc. cit.

[11] Much of the research on expert systems as a method for knowledge processing was carried out during the 1980s and early 1990s. Readable references that go further than the material presented here include N. Bryant, *Managing Expert Systems*, John Wiley & Sons, Chichester, 1998, and E.C. Payne and R.C. McArthur, *Developing Expert Systems*, John Wiley & Sons, N.Y., 1990.

class types for each data set and can even allow for a final set of information classes that are not definable from any individual data type on its own. That concept will be elaborated towards the end of this treatment.

## 12.5.1 Emulating Photointerpretation to Understand Knowledge Processing

To develop an appreciation of a knowledge-based approach it is of value to return to the comparison of the attributes of photointerpretation and quantitative analysis given in Table 3.1. However, rather than making the comparison solely on the basis of a single source of data, consider now that the data to be analysed consists of three subsets: an optical image, a radar image of the same region and a soil map. Standard methods for quantitative analysis are not effective in drawing inferences about cover types using two quite different numerical sources such as optical and radar data because those sources, as noted earlier, sense different properties of the earth's surface.

Consider now how a skilled photointerpreter might approach the problem, making the assumption that the same person is an expert at interpreting optical data and radar data, a case not always found in practice. The photointerpreter would not work at the individual pixel level but, more likely, would concentrate on regions. Suppose a particular region was seen to be predominantly pink on a standard false colour composite print of the optical data, leading the photointerpreter to infer initially that the region is vegetated. Whether it is grassland, a crop or a forested region is not yet clear. The photointerpreter might now refer to the radar imagery. If the tone of that region is dark, then the analyst would assume that the area is almost smooth at the radar wavelength being used. Combining this understanding with the inference from the optical data source, the photointerpreter is then led to think that the region is either grassland or a low-level crop. The analyst might then note that the soil type is not that normally associated with agriculture; he or she would then conclude that the region is some form of natural grassland.

In practice the process may not be so straightforward and the photointerpreter may need to refer backwards and forwards over the data sets in order to finalise an interpretation, particularly if the optical and radar tones were not uniform over the region. For example, some spots on the radar imagery might be very bright; the photointerpreter would probably regard those as indicating shrubs or trees, consistent with the overall region being labelled as natural grassland. The photointerpreter can also easily account for differences in data quality when making a decision about the most likely label for a pixel by placing more reliance on the data sources that are seen to be most accurate or relevant to the particular exercise. Unreliable or marginally relevant data would weigh less in the mind of the analyst during the decision-making process.

The question we should ask at this stage is how the photointerpreter is able to form those inferences so readily. Even apart from the ease with which the analyst can undertake spatial processing, involving the use of cues such a shape and texture, the key to the analyst's success lies in his or her knowledge of how to interpret the various data types. For example, the analyst's knowledge of spectral reflectance characteristics allows the interpretation of optical data while knowledge of radar response is crucial to how the analyst views radar imagery. The analyst also makes composite qualitative judgements, as in the above example, by noting that pink on optical data and dark on radar imagery implies a low-level vegetation type.

We are led therefore to consider whether the knowledge possessed by an expert, such as a skilled photointerpreter, can be given to and used by a machine that emulates how the photointerpreter thinks and deduces facts about the landscape. If we can emulate the photointerpreter's approach we would have available an automated analytical procedure capable of handling mixed data types and, unlike the human analyst, able to work repeatedly at the pixel level if necessary. Fundamental to this approach, of course, is capturing the knowledge possessed by the expert. In the classical expert system approach that is done through the tool of the production rule considered in Sect. 12.5.3.

Although we will pursue this approach to demonstrate the value of qualitative (human) reasoning in image analysis, and the effectiveness of properly captured photointerpretative knowledge, the ultimate value of such an approach is limited by the difficulty in recording effectively the thought processes used by an analyst, since analysts may not always be able to express verbally how they come to certain analytical decisions. This has led recently to a focus on cognitive task analysis as a means for capturing an expert's reasoning skills.[12]

## 12.5.2   The Structure of a Knowledge-Based Image Analysis System

We could represent the structure of a traditional supervised classification approach to the analysis of image data in the manner shown in Fig. 12.3a. The data to be analysed is fed to a processor which is also supplied with the algorithms (maximum likelihood rule, support vector machine, CNN, etc.) In addition, the machine is given training data from which it can generate the parameters required by the chosen algorithm. The algorithm is then applied pixel-by-pixel to produce labels dependent only on the class signatures and the characteristics of the data. While some expert knowledge has been supplied by the user concerning the choice of the algorithm and the selection of reference data with which to train the classifier, that has been limited by comparison to the sort of expert knowledge that is a major feature of knowledge-based processing system.

The comparable diagram for a knowledge-based approach is shown in Fig. 12.3b. Again, the data to be analysed is fed to the processor, but so is a detailed set of rules which captures the knowledge from experts in the field of the analysis. That knowledge is central to the analysis of the data and is processed in what is

---

[12] See A.R. White, Human expertise in the interpretation of remote sensing data: a cognitive task analysis of forest disturbance, *Int. J. of Earth Observation Geoinformation*, vol. 74, 2019, pp. 37–44.

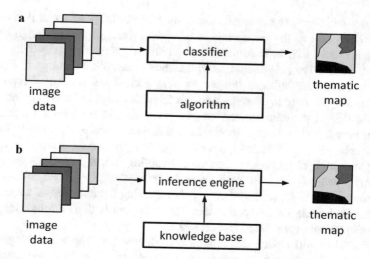

**Fig. 12.3 a** Traditional image analysis system, and **b** knowledge-based image analysis system

called an *inference engine* which not only applies the knowledge but also keeps track of decisions being made about the class memberships of pixels. That entails resolving conflicts when knowledge elements are in dispute and incorporates mechanisms for strengthening decisions made about particular pixels if new knowledge comes to hand during analysis.

### 12.5.3 Representing Knowledge in a Knowledge-Based Image Analysis System

We need a mechanism for capturing the knowledge held in the minds of experts. There are several ways that can be done, the simplest and perhaps most common of which is to use what are called *production rules*, or simply rules. They are developed by interviewing an expert, and typically take the form

**if** condition **then** inference

What we have shown here as *condition* is also called the *antecedent* because it is the expression which precedes the decision step. Sometimes the *inference* is called the *consequent*.

Some simple examples of this type of production rule, generally but not necessarily applied at the pixel level, are

1. **if** visible red response < near infrared response **then** green vegetation
2. **if** radar tone is black **then** shadow
3. **if** near infrared response < visible green response **then** water

As seen in these illustrations, "condition" is a logical expression (sometimes referred to as a clause) which can be either true or false. If it is true, then the inference is assumed to be correct or *justified*. Otherwise, the rule provides no information for or against the correct label for a pixel or region being assessed. Apart from a simple logical expression the condition can be a *compound* logical statement in which several components are linked through the logical **or** and **and** operations. In their simplest form these compound statements are

**if** condition 1 **and** condition 2
**then** inference is justified only if both conditions are true

**if** condition 1 **or** condition 2
**then** inference is justified if either condition is true

In both cases more than two conditions can appear in the antecedent. Expert systems can also use the logical **not** operation, defined by

**not** condition is true **if** condition is false, and vice versa.

Each single rule can be thought of as providing one item of knowledge about the region or pixel being considered. Some rules are not totally conclusive, such as the first in the set of examples above, even though it is acceptable from a knowledge of the spectral reflectance characteristics of vegetation. The same is true for the third example. The second example, concerning a radar shadow, is more conclusive and an analyst might be quite justified in assuming black regions are shadows in radar imagery without appealing to any other source of information or knowledge. In contrast, the first and third rules would probably not be used on their own; while they are indicative of the cover type, the final decision would probably need some form of corroborative information.

A knowledge base in such an analysis system might contain many hundreds of rules of these types, obtained from experts in particular fields. When image data is presented to the inference engine for analysis, the engine goes through the rule base and applies those rules which are relevant in a particular exercise. It checks support for or against various labelling propositions. Some rules will offer strong support while others will be weak. Several classes for a particular pixel might be supported among the rules; procedures are then needed for resolving among them. That is taken up below.

An example of a simple rule base for segmenting optical imagery into just vegetation, water and "other" cover types is

**if** near infrared response/visible red response > threshold **then** vegetation
**if** near infrared response/visible green response < 1 **then** water
**if not** water **and not** vegetation **then** other

In the first rule above a parameter is used—the threshold. That requires a numerical value which almost certainly will be scene dependent. It could be provided before the analysis starts by the user entering it manually or, alternatively, a small training region of vegetation could be used from which the value could be estimated. Many of the rules encountered in remote sensing image analysis will require parameters such as these.

The rules illustrated here rely on spectral or radar pixel-specific knowledge. In many expert systems spatial constraints are also used[13] as a source of knowledge, and appropriate rules are developed. Even spectrally-derived rules might not just rely on simple expressions and comparisons of bands. Spectral contrasts, such as the brightness in a given band compared to the total image brightness have also been used.[14]

One of the real strengths of the rule-based approach is that it facilitates multi-source decisions by merging interpretations from individual data sources. As an example, the following set of rules allows a simple integration of radar and optical image data, and information from a soil type map. It is based on optical data having been classified beforehand into the three cover types of vegetation, water, and soil. It is assumed that three classes of soil are shown on the soil map: loam, sand, clay. We also assume that the radar data has been previously differentiated into specular (mirror-like) surfaces, diffuse (moderately rough) surfaces, volume scatterers (vegetation canopies in this case) and corner reflectors. The last are strong reflectors given by the double reflection mechanism of a vertical surface, such as wall, standing on a horizontal surface, such as the ground. The rule set is:

**if** soil **and** specular surface **then** bare ground
**if** soil **and** corner reflector effect **then** urban
**if** vegetation **and** specular surface **then** low-level vegetation
**if** vegetation **and** diffuse surface or volume **then** trees or shrubs
**if** low-level vegetation **and** loam **then** crops
**if** low-level vegetation **and** sand **then** grassland
**if** low-level vegetation **and** clay **then** grassland
**if** water **and** specular surface **then** lake
**if** water **and** diffuse surface **then** open water

## 12.5.4 Processing Knowledge—The Inference Engine

The inference engine can be simple if the expert system is specific to a particular application. More often it is complex if a powerful and general knowledge processing system is required. In the simple example of the previous section all the mechanism has to do is check which of the rules gives a positive response for a pixel and then label the pixel accordingly. More generally, when large rule sets are used,

---

[13] See J. Ton, J. Sticken and A.K. Jain, Knowledge-based segmentation of Landsat images, *IEEE Transactions on Geoscience and Remote Sensing*, vol. 29, no. 2, March 1991, pp. 222–232.
[14] See S.W. Wharton, A spectral knowledge based approach for urban land cover discrimination, *IEEE Transactions on Geoscience and Remote Sensing*, vol. 25, no. 3, May 1987, pp. 272–282.

the inference engine has to keep track of all those that infer a particular cover type and those which contradict that inference. Similarly, rules that recommend other cover types for the pixel have to be processed. Finally, the inference engine has to make a decision about the correct class by weighting all the evidence from the rules. It also has to account for redundant reasoning and circular arguments, and has to assess whether long reasoning chains carry as much weight in decision-making as inferences involving only a single decision step. In addition, uncertainties and data quality, missing data and missing rules need to be accommodated in the inference process. We provide in the next section some insight into how such an inference engine is constructed, although the general case is quite complex.[15]

## 12.5.5 Rules as Justifiers of a Labelling Proposition

Production rules are sometimes referred to as *justifiers* since they provide a degree of justification or evidence in favour of a particular labelling proposition. Categorising decision rules in this manner is the basis of several qualitative reasoning systems.[16] Rules can be placed into one of four types of varying strength of justification, illustrated by the examples given in the following.

| | |
|---|---|
| *Conclusive rule* | In this rule if the condition (antecedent) is true then the justification for the inference is absolute (conclusive). For example **if** radar tone is black **then** shadow |
| *Prima facie rule* | In this rule if the condition (antecedent) is true then there is reason to believe that the inference is true, although if the condition is false it cannot be concluded that the inference is false. For example **if** near infrared response/visible red response > 2 **then** vegetation |
| *Contingent rule* | In this rule if the condition (antecedent) is true then support is provided to other prima facie reasons to believe that the inference is true. These types of rule are not sufficient in themselves to justify the inference. For example **if** near infrared response > visible red response **then** vegetation |

(continued)

---

[15] See A. Srinivasan, An artificial intelligence approach to the analysis of multiple information sources in remote sensing. *Ph.D. Thesis*, The University of New South Wales, Kensington, Australia, 1990, and A. Srinivasan and J.A. Richards, Analysis of GIS spatial data using knowledge-based methods, *Int. J. Geographic Information Systems*, vol. 7, no. 6, 1993, pp. 479–500.

[16] Srinivasan, loc. cit.

(continued)

*Criterion*            This is a special prima facie justifier for which a false condition
                       provides prima facie justification to disbelieve the inference. For
                       example

                       **if** near infrared response < visible green response **then** water
                       noting that

                       **if** near infrared response > visible green response **then** definitely
                       not water

   Such a structuring of justifications is not unlike the strengths of reasoning used
by photointerpreters. In some cases, evidence would be so overwhelming that the
cover type must be from a particular class. In other cases, the evidence might be so
slight as simply to suggest what the cover type might be; indeed, the photointerpreter might withhold making a decision in such a case until some further evidence
is available.

## 12.5.6   Endorsing a Labelling Proposition

Once the rules, or justifiers, have been applied the inference engine has to make a
decision about whether a particular labelling proposition is supported or not. That is
the role of the *endorsement*, which is the final level of justification for an inference.
   Given a set of justifiers for and against a particular inference, the following are a
set of endorsements that could be used in coming to a final decision.

| | |
|---|---|
| *The inference **is definitely true*** | if there is at least one conclusive justifier in support |
| *The inference **is likely to be true*** | if there is some net prima facie evidence in support |
| *The inference **is indicated*** | if, in the absence of prima facie justification, there are some net contingent justifiers in support |
| *A labelling proposition **is null*** | if all justifiers for the proposition are balanced by those against the proposition |
| *A proposition **is contradicted*** | if the proposition has conclusive justifiers for and against it |
| *A labelling proposition **is unknown*** | if nothing is known about the proposition |

If an endorsement falls into any of the last three categories, the pixel would be left
unclassified. Complements of these endorsements also exist.
   After all the relevant rules in the knowledge base have been applied to a pixel
under consideration, each of the possible labels for the pixel will have some level of
endorsement. That with the strongest endorsement is chosen as the label most
appropriate for the pixel. Endorsements for other labels, although weak, may still

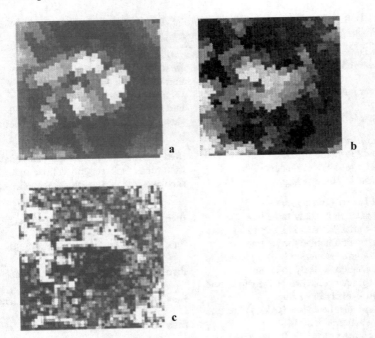

**Fig. 12.4** **a** Landsat MSS visible red band, **b** Landsat MSS near infrared band, and **c** SIR-B radar image of a small area in Sydney

have value. For example, two endorsements that "grassland is likely to be true" and "soil is indicated" are fully consistent; the cover type may in fact be sparse grassland, which the analyst would infer from the pair of endorsements.

## 12.5.7 An Example

Figure 12.4 shows Landsat Multispectral Scanner visible red and near infrared image data along with an L band SIR-B synthetic aperture radar image for a small urban area in Sydney's north-western suburbs. The Landsat data is unable to distinguish between urban areas and areas cleared for development. On the other hand, the radar data provides structural information, but no information on the actual cover type. Using a knowledge-based approach[17] we are able to analyse the images jointly and develop a cover type map that resolves classes which are confused in either the Landsat or radar data alone.

The expert system uses the rules in Table 12.1. The conditions tested in the combination rules are endorsements from the single source knowledge-base analyses. When applied, the rules yield the thematic map of Fig. 12.5; Table 12.2

---

[17] Srinivasan and Richards, loc. cit., and Srinivasan, loc. cit.

**Table 12.1** Rule base for analysing the data of Fig. 12.4; in the single source analyses the rules need to be trained to establish what is meant by low, moderate and high

---

*For the Landsat multispectral scanner data source*

| | |
|---|---|
| **if** near infrared/visible red ≈1 | **then** contingent support for urban |
| | **and** contingent support for soil |
| **if** near infrared/visible red is moderate | **then** contingent support for urban |
| | **and** contingent support for vegetation |
| **if** near infrared/visible red is high | **then** prima facie support for vegetation |

*For the SIR-B radar data source*

| | |
|---|---|
| **if** radar response is low | **then** prima facie support specular behaviour |
| **if** radar response is moderate | **then** prima facie support volume scattering |
| **if** radar response is high | **then** prima facie support corner reflector |

*To combine the data sources*

| | |
|---|---|
| **if** vegetation is likely to be true **and** corner reflector is likely to be true | **then** prima facie support for woody vegetation |
| **if** vegetation is likely to be true **and** volume scattering is likely to be true | **then** prima facie support for vegetation |
| **if** vegetation is likely to be true **and** specular behaviour is likely to be true | **then** prima facie support for grassland |
| **if** soil is likely to be true **and** specular behaviour is likely to be true | **then** prima facie support for cleared land |
| **if** vegetation is indicated **and** corner reflector is likely to be true | **then** prima facie support for residential |
| **if** vegetation is indicated **and** volume scattering is likely to be true | **then** prima facie support for residential |
| **if** urban is likely to be true **and** corner reflector is likely to be true | **then** prima facie support for buildings |
| **if** vegetation is indicated **and** vegetation is not likely to be true **and** specular behaviour | **then** contingent support for grassland |

---

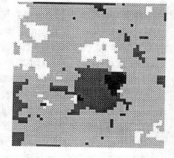

**Fig. 12.5** Thematic map produced by knowledge-based analysis of the data in Fig. 12.4. Classes are: black = soil, dark grey = grassland, light grey = woody vegetation, white = urban (cleared land, buildings, residential)

**Table 12.2** Results of combined optical and radar analysis

| Classes from rule-based classification (%) | Ground classes | | | |
|---|---|---|---|---|
| | Cleared land | Grassland | Woody vegetation | Urban |
| Cleared land | 82.5 | 2.5 | 0.0 | 0.0 |
| Grassland (likely) | 2.5 | 57.2 | 6.6 | 0.8 |
| Grassland (indicated) | 5.0 | 20.8 | 0.0 | 3.5 |
| Woody vegetation | 2.5 | 16.1 | 88.1 | 13.5 |
| Residential | 0.0 | 2.1 | 5.2 | 70.6 |
| Buildings | 0.0 | 0.7 | 0.0 | 10.7 |
| Soil | 7.5 | 0.0 | 0.0 | 0.9 |

Overall accuracy = 77.35%, area weighted 81.5%

summarises the results quantitatively for which a careful photointerpretation of the data, and local knowledge, provided the necessary ground truth. Figure 12.5 shows that the classifier is able to distinguish between grassland and woody vegetation owing to the structural information present in the radar image data. Note also that not all bright regions in the radar image are classified as urban. Some are actually rows of trees; the confusion has been resolved using the land-cover information present in the Landsat image.

Although simple, this example demonstrates how readily multisource data can be handled using a knowledge-based approach to analysis and interpretation.

# 12.6 Operational Multisource Analysis[18]

Given the ready availability of data types now available, and the fact that the user generally acquires spatial data over a network, there are a number of fundamental requirements that an operational thematic mapping schema should satisfy. Apart from being able to cope with interpreting disparate data types, which could include existing maps and prior thematic maps, we should also

- account for relative data quality
- allow for each data source to be analysed separately, in time and location
- accept that the thematic classes from a combined data set might be different from the classes achievable with any data set on its own, and
- recognise that many image data types have their own preferred methods for analysis.

---

[18] This section is based in part on J.A. Richards, Analysis of remotely sensed data: the formative decades and the future, *IEEE Transactions on Geoscience and Remote Sensing*, vol. 43, no. 3, March 2005, pp. 422–432.

Meeting these conditions can be difficult when using any (fusion) technique that depends on combining data prior to analysis since the data sources would all have to be available simultaneously, the class definitions would need to be consistent over the sources and the analysis techniques would need to be common over the data types.

Consider the last two requirements. Depending on the preferences of the analyst, optical data can be classified with a high degree of accuracy with well-known supervised methods. When the data is of hyperspectral dimension those same techniques could still be used, with appropriate feature reduction techniques applied beforehand, but that ignores the scientific spectroscopic information available with high spectral resolution data. Hyperspectral imagery is often better analysed using spectroscopic knowledge and library searching techniques, including those which seek to identify and name specific absorption features.

Radar imagery, on the other hand, does not respond well to the sorts of classifiers regularly applied to optical remote sensing data partly, but not fully, because of the overlaid speckle present in radar imagery. Instead, radar data is best analysed by techniques designed specifically to identify the types of radar response found in practice. Those techniques include Wishart-based classifiers, eigenvalue decomposition methods, and their derivatives.[19]

When undertaking multisource analysis by combining the data sets beforehand we deny the application of analysis techniques that are optimal to given data types and thus we potentially prejudice the quality of the outcome by forcing some information class labels relevant to one particular data type on to a data set for which those labels have no meaning. For example, turbid water is a label that could be attached to an optical pixel but not to a radar pixel.

That leads to a consideration of the differences in the thematic classes from one data type to the next, and the consequent fact that the information classes able to be identified from multisource data, properly analysed, could even be different from the information classes found with any single data type on its own. These points are illustrated in Fig. 12.6 for optical and radar data. While simple, it nevertheless serves to illustrate the point that some class labels are effectively only achievable after the data sources have been processed separately, and the results combined; that is, some outcomes are more readily achieved if we process labels, rather than data. Effectively that is label or knowledge fusion rather than data fusion.

As result of these types of consideration, the decompositional analysis methodology presented in Fig. 12.7 is important. It can meet all the requirements in the dot points above and allows the analysis technique to be applied to a given data source to be one that is optimal for that data type. Importantly, after analysis, the results are all available in a common vocabulary—labels rather than numerical measurements. The combination module then only has to process label-like data rather than mixtures of data types that are often incommensurate. It can jointly process labels using the knowledge-based methods of the previous section. An

---

[19] See J.A. Richards, *Remote Sensing with Imaging Radar*, Springer, Berlin, 2009.

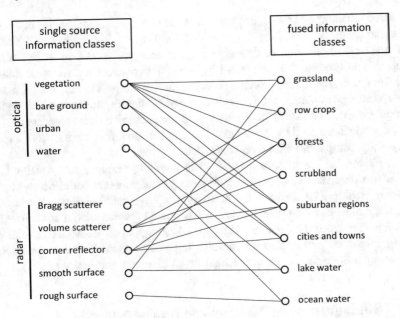

**Fig. 12.6** Simple illustration of how more complex information classes can result from the fusion of simpler information classes generated from single sources of remote sensing image data

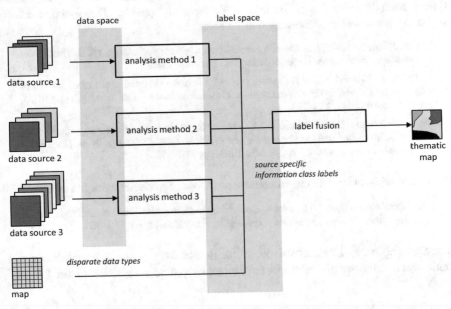

**Fig. 12.7** A multisource analysis methodology based on label fusion

associated advantage of this approach is that map data, including previous thematic maps, can easily be joined into the combination process as depicted in figure.

A practical implication of decomposing the analyses as shown in Fig. 12.7 is that each data type can be interpreted by separate experts, if necessary at different places and at different times. Such a model opens up the possibility of a knowledge broker who would source the individual analyses, having understood beforehand the requirements of a client; the broker would then combine those separate analyses into the product required by the client. The computational framework for this model is also fully compatible with web-based data serving.

One matter that has to be addressed when handling multiple data sources in the manner just described is where in the process spatial context should be embedded. There are two options. Perhaps the better is to apply spatial context algorithms at the individual data source level because the degrees to which spatial context will be important vary from data type to data type. Although such individual spatial processing might be preferable, the simpler approach would be to embed spatial consistency among the labels in the final fused thematic map output.

## 12.7   Bibliography on Multisource Image Analysis

Multisource analysis has been of interest ever since the early days of thematic mapping from remotely sensed image data even though many of the earlier analytical procedures tended to be ad hoc or problem specific. The foundations of statistical multisource processing will be found in

T. Lee, Multisource Context Classification Methods in Remote Sensing, *Ph.D. Thesis*, The University of New South Wales, Kensington, Australia, 1986,

T. Lee, J.A. Richards and P.H. Swain, Probabilistic and evidential approaches for multisource data analysis, *IEEE Transactions on Geoscience and Remote Sensing*, vol. GE-25, no. 3, May 1987, pp. 283–293,

J.A. Benediktsson, P.H. Swain and O.K. Ersoy, Neural network approaches versus statistical classification of multisource remote sensing data, *IEEE Transactions on Geoscience and Remote Sensing*, vol. 28, no. 4, July 1990, pp. 540–552

the last of which also introduces consensus theoretic methods, which are refined in

J.A. Benediktsson and P.H. Swain, Consensus theoretic classification methods, *IEEE Transactions on Systems, Man and Cybernetics*, vol. 22, 1992, pp. 688–704.

The thesis above by Lee was also one of the first treatments of evidential theory as a multisource analysis procedure in remote sensing. The standard evidential theory book is

G. Shafer, *A Mathematical Theory of Evidence*, Princeton UP, N.J., 1976.

Application-specific treatments using evidential methods for multisource analysis are

W. Moon, Integration of geophysical and geological data using evidential belief function, *IEEE Transactions on Geoscience and Remote Sensing*, vol. 28, no. 4, July 1990, pp. 711–720,

P. Gong, Integrated analysis of spatial data from multiple sources using evidential reasoning and artificial neural network techniques for geologic mapping, *Photogrammetric Engineering and Remote Sensing*, vol. 62, 1996, pp. 513–523.

Evidential reasoning has also been used as an approach to spatial context classification, by treating the spatial information as a separate data source:

J.A. Richards and X. Jia, A Dempster-Shafer approach to context classification. *IEEE Transactions on Geoscience and Remote Sensing*, vol. 45, no. 5 pt. 2, May 2007, pp. 1422–1431.

The foundations of expert systems and reasoning are covered in a number of key texts, including

N. Bryant, *Managing Expert Systems*, John Wiley & Sons, Chichester, 1998,

E.C. Payne and R.C. McArthur, *Developing Expert Systems*, John Wiley & Sons, N.Y., 1990,

R. Frost, *Introduction to Knowledge Base Systems*, McGraw-Hill, N.Y., 1986,

J.L. Pollack, *Knowledge and Justification*, Princeton UP., N.J., 1974,

D. Nute, Defeasible Reasoning: A Philosophical Analysis in Prolog, in J.H. Fetzwer ed., *Aspects of Artificial Intelligence*, Kluwer, Dordrecht, 1988.

The first application of defeasible reasoning as the basis for an expert system in remote sensing image analysis is in

A. Srinivasan, An Artificial Intelligence Approach to the Analysis of Multiple Information Sources in Remote Sensing. *Ph.D. Thesis*, The University of New South Wales, Kensington, Australia, 1990, and

A. Srinivasan and J.A. Richards, Analysis of GIS spatial data using knowledge-based methods, *Int. J. Geographic Information Systems*, vol. 7, no. 6, 1993, pp. 479–500.

Application of various expert system approaches are given in

J. Ton, J. Sticken and A.K. Jain, Knowledge-based segmentation of Landsat images, *IEEE Transactions on Geoscience and Remote Sensing*, vol. 29, no. 2, March 1991, pp. 222–232,

S.W. Wharton, A spectral knowledge based approach for urban land cover discrimination, *IEEE Transactions on Geoscience and Remote Sensing*, vol. 25, no. 3, May 1987, pp. 272–282,

R.A. Schowengerdt, A general purpose expert system for image processing, *Photogrammetric Engineering and Remote Sensing*, vol. 55, 1989, pp. 1277–1284.

E. Binaghi, I. Gallo, P. Madella and A. Rampini, Approximate reasoning and multistrategy learning for multisource remote sensing data interpretation, in C.H. Chen, ed., *Information Processing for Remote Sensing*, World Scientific, Singapore, 1999.

Because of the importance of the multisource analysis problem several special issues have appeared in the past. These include

Special Issue on Data Fusion, *IEEE Transactions on Geoscience and Remote Sensing*, vol. 46, no. 5, May 2008, which includes a good definition of important terms in its editorial,

Special Issue on Data Fusion, *IEEE Transactions on Geoscience and Remote Sensing*, vol. 37, no. 3, May 1999,

Special Issue on the Workshop on Analytical Methods in Remote Sensing for Geographic Information Systems, *IEEE Transactions on Geoscience and Remote Sensing*, vol. GE-25, no. 3 May 1987.

Not unexpectedly, multisource image processing features in fields apart from remote sensing. For an overview see

R.S. Blum and Z. Liu, *Multi-Sensor Image Fusion and Its Applications*, CRC Taylor and Francis., Boca Raton, Florida, 1996.

## 12.8  Problems

12.1  Suppose you have been asked to produce a thematic map for a given region of natural vegetation in which is embedded a number of crop fields. You have been told that you have both optical and radar image data available for the task. The client has asked you to recommend whether you would choose to use a stacked vector method with a traditional classifier algorithm, or whether you would use a knowledge-based approach. Outline the considerations that you would include in your report to the client.

12.2  Develop a set of production rules that might be used to smooth a thematic map. The rules are to be applied to the central labelled pixel in a 3 × 3 window. Assume the map has five classes and that map segments with fewer than 4 pixels are unacceptable to the user.

12.3  Assume you have two optical image data sources available, but with different spatial resolutions. After geometric registration the pixels in one image are nine times the size of the pixels in the other. Suppose both images have been classified into the same set of spectral classes. Outline how you might write a set of production rules to fuse the two data sets, noting that the high resolution data is less reliable than the low resolution data.

12.4  Develop a set of production rules that might be applied to Landsat ETM+ imagery to create a thematic map with five classes: vegetation, deep clear water, shallow or muddy water, dry soils and wet soils. To do this you may need to refer to a source of information on the spectral reflectance characteristics of those cover types in the ranges of the thematic mapper bands.

12.5  Compare the relative attributes and disadvantages of the following methods for multisource image interpretation:

multisource statistical analysis
consensus theoretical analysis
evidential reasoning
expert systems analysis

12.6  Suppose a future remote sensing program consists of 100 microsatellites orbiting the earth with substantial overlap of coverage either simultaneously, or with time, as the satellites pass over a given region of the earth's surface. If the satellites were designed for thematic mapping, there are two possible means by which the data recorded can be used. One is to down link the data from each satellite, which is then processed in ground-based facilities; the other is to undertake thematic mapping on the satellites and then downlink the results. Compare the two methods. Consider also the desirability, or otherwise, of the satellites communicating thematic information among themselves in order to map the landscape better.

12.7  When using the Theory of Evidence as a multisource analysis tool the analyst needs to generate mass distributions from each available source. If one source is optical, a method for getting an effective mass distribution, prior to the attribution of uncertainty, is by a statistical analysis of the source. Suppose that analysis is carried out by the maximum likelihood rule. If you were to generate the mass distribution would you use the prior probabilities, the class conditional posterior probabilities, or the class conditional distribution functions to generate the mass distributions?

12.8  What is the difference between theoretical, heuristic and ad hoc? The statistical multisource technique of Sect. 12.3.1 is regarded as having a theoretical basis. The Theory of Evidence is generally thought of as a heuristic method. Does that make it any less valuable as a technique?

12.9  This is a variation on Problem 12.3. A rule-based analysis system is a very effective way of handling multi-resolution image data. For example, rules could be applied first to the pixels of the low resolution data to see whether there is a strong endorsement for any of the available labels. If so, the high spatial resolution data source need not be consulted, and data processing effort is saved. If, however, the rule-based system can only give weak support to any of the available labels on the basis of the low resolution data, then it could consult the high resolution source to see whether the smaller pixels can be labelled with a higher degree of confidence. That could be the case in urban regions where some low resolution pixels may be difficult to classify because they are a mixture of vegetation and concrete. High-resolution imagery may be able to resolve those classes. In some other urban regions, such as golf courses which are large areas of vegetation, the low resolution data might be quite adequate. Using the techniques of Sect. 12.5 develop a set of rules for such a multi-resolution problem. Your solution should not go beyond the low resolution data source if a pixel has likely or definite endorsement.

# Appendices

**Abstract** Material on satellite altitudes and orbits, the binary number system and vector and matrix algebra are presented in three appendices. The last summarises the range of matrix and vector operations that are routinely encountered in quantitative remote sensing image analysis. Two final appendices summarise some fundamental statistical concepts of importance in remote sensing, including the normal or Gaussian distribution, and a more theoretically sound, but less often used, derivation of maximum likelihood classification than that developed in Chap. 8.

© The Editor(s) (if applicable) and The Author(s), under exclusive license to
Springer Nature Switzerland AG 2022
J. A. Richards, *Remote Sensing Digital Image Analysis*,
https://doi.org/10.1007/978-3-030-82327-6

# Appendix A
# Satellite Altitudes and Periods

Remote sensing satellites are generally launched into circular, or near circular, orbits. By equating centripetal acceleration in a circular orbit with the acceleration resulting from earth's gravity it can be shown that the orbital period at an orbital radius $r$ is given by[1]

$$T = 2\pi\sqrt{r^3/\mu} \text{ s} \tag{A.1}$$

where $\mu$ is the earth gravitational constant[2] of $3.986 \times 10^{14}$ m$^3$s$^{-2}$. The corresponding orbital angular velocity is

$$\omega_o = \sqrt{\mu/r^3} \text{ rad s}^{-1} \tag{A.2}$$

The orbital radius $r$ is the sum of the earth radius $r_e$ and the altitude of the satellite above the earth $h$

$$r = r_e + h$$

in which $r_e = 6.378$ Mm. This gives the effective forward velocity of a satellite in orbit as

$$v = r\omega_o = (r_e + h)\sqrt{\mu/(r_e + h)^3} \text{ m s}^{-1} \tag{A.3a}$$

while its velocity over the ground (sub-nadir), ignoring earth rotation, is given by

---

[1]See K.I. Duck and J.C. King, Orbital Mechanics for Remote Sensing, Chap. 16 in R.N. Colwell ed., *Manual of Remote Sensing*, 2nd ed., American Society of Photogrammetry, Falls Church, Virginia, 1983. An alternative treatment, which expresses the equations in terms of the acceleration due to gravity at the earth's surface, is given in C. Elachi, *Introduction to the Physics and Techniques of Remote Sensing*, Wiley-Interscience, N.Y., 1987.

[2]This is the product of the universal gravitational constant G ($6.674 \times 10^{-11}$ m$^3$ kg$^{-1}$ s$^{-2}$) and the mass of the earth ($5.972 \times 10^{24}$ kg).

© The Editor(s) (if applicable) and The Author(s), under exclusive license to
Springer Nature Switzerland AG 2022
J. A. Richards, *Remote Sensing Digital Image Analysis*,
https://doi.org/10.1007/978-3-030-82327-6

**Table A.1** Altitudes and periods of some typical satellites

| Satellite | h (km) | T (min) | T (h) | T (day) |
|---|---|---|---|---|
| Landsat 1,2,3 | 920 | 103.4 | 1.72 | 0.072 |
| Landsat 4,5,7,8 | 705 | 98.9 | 1.65 | 0.069 |
| SPOT 1,2,3 | 832 | 101.5 | 1.69 | 0.071 |
| SPOT 7 | 695 | 98.8 | 1.65 | 0.067 |
| Sentinel 2 | 786 | 100.0 | 1.67 | 0.069 |
| WorldView 4 | 617 | 97.0 | 1.61 | 0.067 |
| Geostationary | 35,786 | 1436.1 | 23.93 | 0.997 |
| Moon | 377,612[a] | 39,467.5 | 657.79 | 27.408 |

[a] The lunar altitude in these calculations was determined by subtracting the earth radius from the average of the apogee and perigee of the lunar orbit

$$v = r_e \omega_o = r_e \sqrt{\mu/(r_e + h)^3} \text{ m s}^{-1} \tag{A.3b}$$

The actual velocity over the earth's surface, taking into account the earth rotation, depends on the orbit's inclination and the latitude at which the velocity is of interest. The orbital inclination is measured as an angle $i$ anticlockwise from the equator on an ascending pass—called the ascending node—or clockwise on a descending pass. If the earth rotational velocity at the equator is $v_e$ in the easterly direction, and the latitude is $\varphi$, then the surface velocity of a point on the earth directly under the satellite will be $v_e \cos \varphi \cos i$. That will add or subtract to the surface velocity of the satellite to give the equivalent ground track velocity

$$v_s = r_e \sqrt{\mu/(r_e + h)^3} \pm v_e \cos \varphi \cos i \text{ m s}^{-1} \tag{A.4}$$

The positive sign applies when the component of earth rotation opposes the satellite motion (on descending nodes for inclinations less than 90° or for ascending nodes with inclinations greater than 90°); otherwise, the negative sign is used.

Provided the assumptions of circular orbits and a spherical earth are acceptable the equations above can be used to derive some numbers of significance in remote sensing.

Table A.1 shows orbital period as a function of altitude for a number of satellite programs; included are comparable calculations for the moon and for a geostationary satellite. A geostationary orbit is one that has a period of 24 h such that if the satellite were placed above the equator travelling in the same direction as earth rotation it would appear stationary above the nadir point. Telecommunications satellites and some weather satellites occupy geostationary orbits.

Consider now the calculation of the time taken for the Landsat 3 satellite to acquire a 185 km frame of image data. That can be found by determining the local velocity. The orbital inclination of the Landsat satellite is approximately 99° and its altitude is 920 km. At Sydney, Australia the latitude is 33.8° S. From (A.4) this gives $v_s = 6.393 \text{ km s}^{-1}$, so that the 185 km frame requires 28.9 s to record.

# Appendix B
# Binary Representation of Decimal Numbers

In digital data handling we refer to numbers in binary format, because computers represent data in that form. In the binary system numbers are arranged in columns that represent powers of 2, while in the decimal system numbers are arranged in columns that are powers of 10. Whereas we can count up to 9 in each column in the decimal system, we can only count up to 1 in each binary column. From the right, the columns are called $2^0$, $2^1$, $2^2$, etc., indicating the highest decimal number that can be represented by each column, so that the decimal numbers between 0 and 7 have the binary versions:

| Decimal | Binary | | | |
|---------|--------|--------|--------|--------|
| | $2^2$ | $2^1$ | $2^0$ | |
| 0 | 0 | 0 | 0 | |
| 1 | 0 | 0 | 1 | |
| 2 | 0 | 1 | 0 | |
| 3 | 0 | 1 | 1 | i.e., 2 + 1 |
| 4 | 1 | 0 | 0 | |
| 5 | 1 | 0 | 1 | i.e., 4 + 1 |
| 6 | 1 | 1 | 0 | i.e., 4 + 2 |
| 7 | 1 | 1 | 1 | i.e., 4 + 2 + 1 |

Digits in the binary system are referred to as *bits*. In the example above it can be seen that eight different decimal values (0...7) can be represented with three bits, but it is not possible to represent decimal numbers beyond 7. To represent the 16 decimal numbers 0–15 it is necessary to have a binary word with 4 bits; in that case 1111 is equivalent to decimal 15. The numbers of decimal values that can be represented by different numbers of bits, up to levels that can be encountered in remote sensing imagery, are:

© The Editor(s) (if applicable) and The Author(s), under exclusive license to Springer Nature Switzerland AG 2022
J. A. Richards, *Remote Sensing Digital Image Analysis*, https://doi.org/10.1007/978-3-030-82327-6

| Bits | Decimal levels | Ranges | Bits | Decimal levels | Ranges |
|------|----------------|--------|------|----------------|--------|
| 1 | 2 | 0,1 | 9 | 512 | 0…511 |
| 2 | 4 | 0…3 | 10 | 1024 | 0…1023 |
| 3 | 8 | 0…7 | 11 | 2048 | 0…2047 |
| 4 | 16 | 0…15 | 12 | 4096 | 0…4095 |
| 5 | 32 | 0…31 | 13 | 8192 | 0…8191 |
| 6 | 64 | 0…63 | 14 | 16,384 | 0…16,383 |
| 7 | 128 | 0…127 | 15 | 32,768 | 0…32,767 |
| 8 | 256 | 0…255 | 16 | 65,536 | 0…65,635 |

Eight bits are referred to as a *byte*, which is a fundamental data unit used in computers.

# Appendix C
# Essential Results from Vector and Matrix Algebra

## C.1 Matrices and Vectors, and Matrix Arithmetic

As noted in Sect. 3.5 the pixels in an image can be plotted in a rectangular coordinate system according to the brightness values in each of the recorded bands of data. A vegetation pixel would appear as shown in Fig. C.1 if visible red and near infrared measurements were available. The pixel can then be described by its coordinates (10, 40). The coordinate space is referred to as the *spectral domain* or as a *spectral space*, and it will have as many dimensions as there are bands of data. For multispectral measurements the dimensionality would be less than 10 whereas for hyperspectral measurements the dimensionality would be around 200 or more.

Mathematically we can represent the set of measurements for a pixel in the form of a *column vector*, in which the entries are the individual pixel brightness values arranged down the column in ascending order of band

$$\mathbf{x} = \begin{bmatrix} x_1 \\ x_2 \\ \vdots \\ x_N \end{bmatrix} \tag{C.1}$$

$\mathbf{x}$ is the symbol for a column vector and $N$ is the dimensionality of the spectral domain—the number of bands. The vector is said to be $N \times 1$ dimensional. For the simple example of Fig. C.1, we have the $2 \times 1$ vector

$$\mathbf{x} = \begin{bmatrix} 10 \\ 40 \end{bmatrix}$$

Sometimes we might want to create a new vector $\mathbf{y}$ from the existing vector $\mathbf{x}$. That can be done by way of linear combinations of the original vector elements which, in the two-dimensional case, is written

© The Editor(s) (if applicable) and The Author(s), under exclusive license to
Springer Nature Switzerland AG 2022
J. A. Richards, *Remote Sensing Digital Image Analysis*,
https://doi.org/10.1007/978-3-030-82327-6

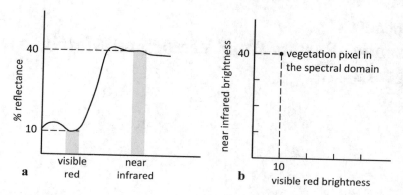

**Fig. C.1** **a** Spectral reflectance characteristic of vegetation, and **b** typical response plotted in the near IR, visible red spectral domain

$$y_1 = m_{11}x_1 + m_{12}x_2$$
$$y_2 = m_{21}x_1 + m_{22}x_2$$

These are summarised as

$$\begin{bmatrix} y_1 \\ y_2 \end{bmatrix} = \begin{bmatrix} m_{11} & m_{12} \\ m_{21} & m_{22} \end{bmatrix} \begin{bmatrix} x_1 \\ x_2 \end{bmatrix}$$

or symbolically

$$\mathbf{y} = \mathbf{Mx}$$

in which $\mathbf{M}$ is called a matrix of the numbers $m_{11}$, etc. For this example, the matrix has dimensionality of $2 \times 2$. Movement between the matrix–vector form and the original set of equations requires an understanding of how vectors and matrices are multiplied. The result of the multiplication is obtained by multiplying the column entries of the vector, one by one, with the row entries of the matrix and adding the products, one row at a time. The result of each of those operations is a new vector element. That is illustrated in Fig. C.2, along with the multiplication of two matrices, which follows the same pattern.

Figure C.2 introduces the *row vector*. Column vectors have their elements arranged down a column, whereas row vectors have their elements arranged across a row. The difference is important because row vectors enter into multiplication in a different way, as illustrated.

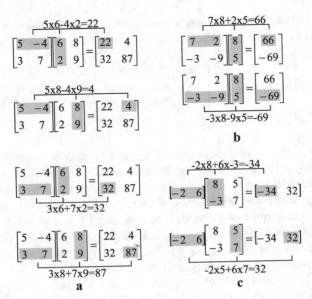

**Fig. C.2** Illustrating the steps involved in matrix multiplication: **a** two matrices, **b** a column vector post-multiplying a matrix and **c** a row vector pre-multiplying a matrix

The product of a row vector and a column vector will also be different depending on the order in which they appear. If the row vector is on the left-hand side the result is a simple scalar; if it is on the right-hand side the result is a matrix, as seen in the following

$$[4 \quad -3]\begin{bmatrix} 9 \\ 7 \end{bmatrix} = 15 \tag{C.2a}$$

$$\begin{bmatrix} 9 \\ 7 \end{bmatrix}[4 \quad -3] = \begin{bmatrix} 36 & -27 \\ 28 & -21 \end{bmatrix} \tag{C.2b}$$

The order in which matrices are multiplied is also important. **AB** will give a different result from **BA**, except in special circumstances. We say that **A** "pre-multiplies" **B** in **AB** whereas **B** "post-multiplies" **A**. Although the above examples were computed with 2 dimensional vectors and matrices, the patterns are the same for any order so long as the dimensionality of the vector matches the relevant dimension of the matrix. For example, a $3 \times 12$ matrix (3 rows and 12 columns) can only be *post-multiplied* by a 12 element column vector and can be *pre-multiplied* by a 3 element row vector.

We use a double subscript notation to describe the elements of a matrix, in which the first refers to the row to which the element belongs and the second to its column:

$$\mathbf{M} = \begin{bmatrix} m_{11} & m_{12} & m_{13} & \cdots \\ m_{21} & m_{22} & m_{23} & \cdots \\ m_{31} & m_{32} & m_{33} & \cdots \\ \vdots & \vdots & \vdots & \ddots \end{bmatrix}$$

The elements, referred to generically as $m_{ij}$, can be real or complex. In optical remote sensing, only real elements are usually encountered, while complex entries occur often in radar remote sensing. The dots in this expression simply mean the matrix can be of any size, as determined by the problem being considered. If the matrix has as many rows as columns, then it is called a *square matrix*.

Elements that lie on the same row and column, $m_{ii}$, are called *diagonal* elements and together define the *diagonal*, or principal diagonal, of the matrix. All the other elements are referred to as *off-diagonal* elements.

If $\mathbf{M}$ and $\mathbf{N}$ are two matrices then their addition and subtraction is given by adding or subtracting their elements, as shown in the following $2 \times 2$ examples:

$$\text{For } \mathbf{M} = \begin{bmatrix} m_{11} & m_{12} \\ m_{21} & m_{22} \end{bmatrix} \text{ and } \mathbf{N} = \begin{bmatrix} n_{11} & n_{12} \\ n_{21} & n_{22} \end{bmatrix}, \text{ then}$$

$$\mathbf{M} \pm \mathbf{N} = \begin{bmatrix} m_{11} \pm n_{11} & m_{12} \pm n_{12} \\ m_{21} \pm n_{21} & m_{22} \pm n_{22} \end{bmatrix}$$

## C.2 The Trace of a Matrix

The trace of a matrix is the sum of its diagonal terms, which for an $N \times N$ square matrix is expressed:

$$\text{trace}\mathbf{M} = \text{tr}\mathbf{M} = \sum_{i=1}^{N} m_{ii} \tag{C.3}$$

## C.3 The Transpose of a Matrix or a Vector

If the elements of a matrix are rotated about the diagonal, the *transpose* of the matrix results. The transpose is represented by a superscript T (or sometimes t), so that if

$$\mathbf{M} = \begin{bmatrix} m_{11} & m_{12} & m_{13} \\ m_{21} & m_{22} & m_{23} \\ m_{31} & m_{32} & m_{33} \end{bmatrix}$$

$$\mathbf{M}^{\mathrm{T}} = \begin{bmatrix} m_{11} & m_{21} & m_{31} \\ m_{12} & m_{22} & m_{32} \\ m_{13} & m_{23} & m_{33} \end{bmatrix}$$

Vectors can also be transposed by rotating around their first element thus transforming a row vector into a column vector and vice versa. If

$$\mathbf{g} = \begin{bmatrix} g_1 \\ g_2 \\ g_3 \end{bmatrix}$$

Then

$$\mathbf{g}^{\mathrm{T}} = \begin{bmatrix} g_1 & g_2 & g_3 \end{bmatrix}$$

Note that

$$\mathbf{g}^{\mathrm{T}}\mathbf{g} = g_1^2 + g_2^2 + g_3^2 = |\mathbf{g}|^2$$

In other words that operation gives the square of the magnitude of the vector. How can vectors have magnitude? To illustrate: the vector $\mathbf{g}$ could be the set of spectral measurements for a pixel in three dimensional spectral space. Its magnitude is its overall brightness, or the length of the vector drawn from the origin to the point in the three dimensional space using the vector elements as coordinates. That can be seen by applying Pythagoras' Theorem to the simple example in Fig. C.1b.

Note that for two column vectors $\quad \mathbf{g} = \begin{bmatrix} 4 \\ -3 \end{bmatrix}$ and $\mathbf{h} = \begin{bmatrix} 9 \\ 7 \end{bmatrix}$

then

$$\mathbf{g}^{\mathrm{T}}\mathbf{h} = 15 = \text{scalar}$$

while

$$\mathbf{h}\mathbf{g}^{\mathrm{T}} = \begin{bmatrix} 36 & -27 \\ 28 & -21 \end{bmatrix}$$

The vector transpose can be used to evaluate the dot product of two vectors:

$$\mathbf{g}.\mathbf{h} = \mathbf{g}^{\mathrm{T}}\mathbf{h} = \mathbf{h}^{\mathrm{T}}\mathbf{g}$$

## C.4 The Identity Matrix

The identity matrix is a square matrix which is zero everywhere except down its diagonal, on which each element is 1. Multiplication of a vector by the identity matrix, which has the symbol $\mathbf{I}$, leaves the vector unchanged. Thus

$$
\begin{bmatrix} 1 & 0 & 0 \\ 0 & 1 & 0 \\ 0 & 0 & 1 \end{bmatrix} \begin{bmatrix} g_1 \\ g_2 \\ g_3 \end{bmatrix} = \begin{bmatrix} g_1 \\ g_2 \\ g_3 \end{bmatrix} \text{ and } \begin{bmatrix} g_1 & g_2 & g_3 \end{bmatrix} \begin{bmatrix} 1 & 0 & 0 \\ 0 & 1 & 0 \\ 0 & 0 & 1 \end{bmatrix} = \begin{bmatrix} g_1 & g_2 & g_3 \end{bmatrix}
$$

or symbolically $\mathbf{Ig} = \mathbf{g}$ and $\mathbf{g}^T\mathbf{I} = \mathbf{g}^T$ as appropriate. Similarly, multiplication of any matrix by the identity matrix leaves the matrix unchanged. Thus

$$\mathbf{MI} = \mathbf{M}$$

The identity matrix is the matrix equivalent of the real number "1".

## C.5 The Determinant

The determinant of the square matrix $\mathbf{M}$ is expressed

$$
|M| = \det \mathbf{M} = \begin{vmatrix} m_{11} & m_{12} & m_{13} & \cdots \\ m_{21} & m_{22} & m_{23} & \cdots \\ m_{31} & m_{32} & m_{33} & \cdots \\ \vdots & \vdots & \vdots & \ddots \end{vmatrix}
$$

It is a scalar quantity that, in principle, can be computed in the following manner. In all but the simplest cases, this approach is not efficient.

First, we define the *cofactor* of a matrix element. The cofactor of the element $m_{ij}$ is the determinant of the matrix formed by removing the $i$th row and $j$th column from $\mathbf{M}$ and multiplying the result by $(-)^{i+j}$. Thus, the cofactor of $m_{21}$ is

$$
\mathcal{M}_{21} = - \begin{vmatrix} m_{12} & m_{13} & m_{14} & \cdots \\ m_{32} & m_{33} & m_{34} & \cdots \\ m_{42} & m_{43} & m_{44} & \cdots \\ \vdots & \vdots & \vdots & \ddots \end{vmatrix}
$$

The classical method for evaluating the determinant is to express it in terms of the cofactors of its first row (or of any row or column). For a square matrix of size $N \times N$ this expansion is

$$|\mathbf{M}| = \sum_{j=1}^{N} m_{1j}\mathcal{M}_{1j}$$

The cofactors in this expression can be expanded in terms of *their* cofactors, and so on until the solution is found. The case of a $2 \times 2$ matrix is simple:

$$\det \mathbf{M} = \begin{vmatrix} m_{11} & m_{12} \\ m_{21} & m_{22} \end{vmatrix} = m_{11}m_{22} - m_{12}m_{21}$$

For matrices of larger dimensions this method for evaluating the determinant is grossly inefficient computationally and numerical methods are adopted.

If the determinant of a matrix is zero, the matrix is called *singular*.

## C.6 The Matrix Inverse

A matrix multiplied by its inverse gives the identity matrix. The inverse is represented by adding the superscript $-1$ to the matrix symbol. Thus

$$\mathbf{M}^{-1}\mathbf{M} = \mathbf{I} \tag{C.4}$$

The solution to the simultaneous equations summarised in matrix form as $\mathbf{Mg} = \mathbf{c}$ can be found by pre-multiplying both sides by $\mathbf{M}^{-1}$ to give

$$\mathbf{g} = \mathbf{M}^{-1}\mathbf{c}$$

provided the inverse matrix can be found. As with determinants, finding inverses is not a trivial task and approximations and numerical methods are generally used. However, also like determinants, there are theoretical expressions for the matrix inverse. It can be defined in terms of the *adjoint* (more recently called the *adjugate*) of the matrix, which is the transposed matrix of cofactors:

$$adj\mathbf{M} = \begin{bmatrix} \mathcal{M}_{11} & \mathcal{M}_{12} & \mathcal{M}_{13} & \cdots \\ \mathcal{M}_{21} & \mathcal{M}_{22} & \mathcal{M}_{23} & \cdots \\ \mathcal{M}_{31} & \mathcal{M}_{32} & \mathcal{M}_{33} & \cdots \\ \vdots & \vdots & \vdots & \ddots \end{bmatrix}^{\mathrm{T}} = \begin{bmatrix} \mathcal{M}_{11} & \mathcal{M}_{21} & \mathcal{M}_{31} & \cdots \\ \mathcal{M}_{12} & \mathcal{M}_{22} & \mathcal{M}_{32} & \cdots \\ \mathcal{M}_{13} & \mathcal{M}_{23} & \mathcal{M}_{33} & \cdots \\ \vdots & \vdots & \vdots & \ddots \end{bmatrix}$$

with which the inverse of $\mathbf{M}$ is

$$\mathbf{M}^{-1} = \frac{adj\mathbf{M}}{|\mathbf{M}|} \tag{C.5}$$

From this we see that the matrix must be non-singular to have an inverse—i.e., its determinant must not be zero.

## C.7 The Eigenvalues and Eigenvectors of a Matrix

The equation $\mathbf{Mg} = \mathbf{c}$ can be interpreted as the transformation of the column vector $\mathbf{g}$ by the matrix $\mathbf{M}$ to form a new column vector $\mathbf{c}$. We now ask ourselves whether there is any particular vector, say $\mathbf{g}_1$, for which multiplication by a scalar will produce the same transformation as multiplication by the matrix. In other words, can we find a $\mathbf{g}_1$ such that

$$\lambda \mathbf{g}_1 = \mathbf{Mg}_1 \tag{C.6}$$

where $\lambda$ is a constant, which is sometimes complex.[3] We can introduce the identity matrix into this equation without changing its meaning:

$$\lambda \mathbf{I} \mathbf{g}_1 = \mathbf{Mg}_1$$

so that we can then re-arrange the equation to read

$$(\mathbf{M} - \lambda \mathbf{I})\mathbf{g}_1 = 0 \tag{C.7}$$

Equation (C.7) is actually a short hand version of the set of homogeneous simultaneous equations in the unknown components[4] of $\mathbf{g}_1$

$$(m_{11} - \lambda)g_{11} + m_{12}g_{21} + m_{13}g_{31}\ldots = 0$$

$$m_{21}g_{11} + (m_{22} - \lambda)g_{21} + m_{23}g_{31}\ldots = 0$$

For a set of homogeneous simultaneous equations to have a non-trivial solution the determinant of the coefficients of the unknowns must be zero, viz.

$$|\mathbf{M} - \lambda \mathbf{I}| = 0 \tag{C.8}$$

This is called the *characteristic equation* of the matrix $\mathbf{M}$. It consists of a set of equations in the unknown $\lambda$. By solving (C.8) the values of $\lambda$ can be found. They can then be substituted into (C.7) to find the corresponding vectors $\mathbf{g}_1$. The $\lambda$ are referred to as the *eigenvalues* (or sometimes proper values or latent roots) of the

---

[3]This has nothing to do with wavelength, which conventionally has the same symbol.
[4]Note that we have indexed the components of the vector using a double subscript notation in which the first subscript refers to the component and the second refers to the vector itself, in this case $\mathbf{g}_1$. Later we will have a $\mathbf{g}_2$, etc.

matrix $\mathbf{M}$ and the corresponding vectors $\mathbf{g}_1$ are called the *eigenvectors* (proper vectors or latent vectors) of $\mathbf{M}$. As a simple example consider the matrix

$$\mathbf{M} = \begin{bmatrix} 6 & 3 \\ 4 & 9 \end{bmatrix}$$

Substituting into (C.8) gives

$$\begin{vmatrix} 6 - \lambda & 3 \\ 4 & 9 - \lambda \end{vmatrix} = 0$$

i.e. $(6 - \lambda)(9 - \lambda) - 12 = 0$

or $\lambda^2 - 15\lambda + 42 = 0$ \hspace{2cm} (C.9)

which has the roots 11.275 and 3.725. In (C.9) it is interesting to note that the coefficient of $\lambda$ is the trace of $\mathbf{M}$ and the constant term is its determinant. Substituting the first eigenvalue into (C.7) gives

$$-5.275g_{11} + 3g_{21} = 0$$

so that

$$g_{11} = 0.569g_{21} \hspace{2cm} (C.10a)$$

That result was obtained using the first row equation in (C.7). If we used the second equation we would get

$$4g_{11} - 2.275g_{21} = 0$$

which again gives

$$g_{11} = 0.569g_{21}$$

In other words, the two equations represented by (C.7) for this two dimensional example are not independent.

Substituting the second eigenvalue into either equation in (C.7) shows

$$4g_{12} + 5.275g_{22} = 0$$

so that

$$g_{12} = -1.319g_{22} \qquad \text{(C.10b)}$$

Note that the eigenvectors are not completely specified; only the ratio of the elements is known. This is consistent with the fact that a non-trivial solution to a set of homogeneous equations will not be unique.

The eigenvalues for this example are both (all) real. A matrix for which all the eigenvalues are real is called a *positive definite* matrix. If they could also be zero, the matrix is called *positive semi-definite*. In optical remote sensing the eigenvalues are generally all real.

Even though we commenced this analysis based on matrices that transform vectors, the concept of the eigenvalues and eigenvectors of a matrix is more general and finds widespread use in science and engineering.

## C.8 Diagonalisation of a Matrix

If we compute all the eigenvalues of a matrix $\mathbf{M}$ and construct the diagonal matrix

$$\Lambda = \begin{bmatrix} \lambda_1 & 0 & 0 & \cdots \\ 0 & \lambda_2 & 0 & \cdots \\ 0 & 0 & \lambda_3 & \cdots \\ \vdots & \vdots & \vdots & \ddots \end{bmatrix}$$

then (C.6) can be generalised to

$$\mathbf{G}\Lambda = \mathbf{M}\mathbf{G} \qquad \text{(C.11)}$$

in which $\mathbf{G}$ is a matrix formed from the set of eigenvectors of $\mathbf{M}$:

$$\mathbf{G} = [\mathbf{g}_1 \ \mathbf{g}_2 \ \mathbf{g}_3 \ \cdots]$$

Provided $\mathbf{G}$ is non-singular, which it will be if the eigenvalues of $\mathbf{M}$ are all distinct, then (C.11) can be written

$$\Lambda = \mathbf{G}^{-1}\mathbf{M}\mathbf{G} \qquad \text{(C.12)}$$

which is called the *diagonal form* of $\mathbf{M}$. Alternatively

$$\mathbf{M} = \mathbf{G}\Lambda\mathbf{G}^{-1} \qquad \text{(C.13)}$$

This last expression is very useful for computing certain functions of matrices. For example, consider $\mathbf{M}$ raised to the power $p$:

$$\mathbf{M}^p = \mathbf{G \Lambda G}^{-1}.\mathbf{G \Lambda G}^{-1}.\mathbf{G \Lambda G}^{-1} \ldots \mathbf{G \Lambda G}^{-1} = \mathbf{G \Lambda}^p \mathbf{G}^{-1}$$

The advantage of this approach is that the diagonal matrix $\mathbf{\Lambda}$ raised to the power $p$ simply requires each of its elements to be raised to that power.

# Appendix D
# Some Fundamental Material
# from Probability and Statistics

## D.1 Conditional Probability and Bayes' Theorem

In this appendix we outline some of the fundamental statistical concepts commonly used in remote sensing. Remote sensing terminology is used, and it is assumed that the variables involved are discrete rather than continuous.

Along with vector and matrix analysis, and calculus, a sound understanding of probability and statistics is important in developing a high degree of skill in quantitative remote sensing. This is necessary, not only to appreciate algorithm development and use, but also because of the role of statistical techniques in dealing with sampled data. The depth of treatment here is sufficient for a first level appreciation of quantitative methods. A more detailed treatment can be obtained from standard statistical texts.[5]

The expression $p(\mathbf{x})$ is interpreted as the probability that the event $\mathbf{x}$ occurs. In the case of remote sensing, if $\mathbf{x}$ is a pixel vector, $p(\mathbf{x})$ is the probability that a pixel can be found at position $\mathbf{x}$ in the spectral domain.

Often, we will want to know the probability that an event occurs conditional on some other event or circumstance. That is written as $p(\mathbf{x}|y)$ which is interpreted as a probability that $\mathbf{x}$ occurs given that $y$, regarded as a *condition*, is specified previously or is already known. For example, $p(\mathbf{x}|\omega_i)$ is the probability of finding a pixel at position $\mathbf{x}$ in the spectral domain, given that we are only interested in those from class $\omega_i$; in other words, it is the probability that a pixel from class $\omega_i$ exists at position $\mathbf{x}$. The $p(\mathbf{x}|y)$ are referred to as *conditional probabilities*. The available conditions $y$ form a complete set. In the case of remote sensing $\omega_i, i = 1 \ldots M$ is the complete set of classes used to describe the image data in a given classification exercise.

---

[5]See J.E. Freund, *Mathematical Statistics*, 5th ed., Prentice Hall, Englewood Cliffs, N.J., 1992, and C.M. Bishop, *Pattern Recognition and Machine Learning*, Springer Science + Business Media, N. Y., 2006.

© The Editor(s) (if applicable) and The Author(s), under exclusive license to
Springer Nature Switzerland AG 2022
J. A. Richards, *Remote Sensing Digital Image Analysis*,
https://doi.org/10.1007/978-3-030-82327-6

If we know the set of $p(\mathbf{x}|\omega_i)$, which are called *class conditional probabilities*, we can determine $p(\mathbf{x})$ in the following manner. Consider the product $p(\mathbf{x}|\omega_i)p(\omega_i)$ in which $p(\omega_i)$ is the probability that class $\omega_i$ pixels occur in the image[6]; it is the probability that a pixel selected at random will come from class $\omega_i$. The product is the probability that a pixel at position $\mathbf{x}$ in the spectral domain *is* an $\omega_i$ pixel, because it describes the probability of a pixel at that position as coming from class $\omega_i$, multiplied by the probability that class $\omega_i$ exists. The probability that a pixel from *any* class can be found at position $\mathbf{x}$ is the sum of the probabilities that pixels would be found there from all of the available classes. In other words

$$p(\mathbf{x}) = \sum_{i=1}^{M} p(\mathbf{x}|\omega_i)p(\omega_i) \tag{D.1}$$

The product $p(\mathbf{x}|\omega_i)p(\omega_i)$ is called the *joint probability* of the "events" $\mathbf{x}$ and $\omega_i$. It is interpreted as the probability that a pixel occurs at position $\mathbf{x}$ *and* that the class *is* $\omega_i$. This is different from the probability that a pixel occurs at position $\mathbf{x}$ from class $\omega_i$ because it also takes into account the likelihood that class $\omega_i$ pixels actually occur in the image.

The joint probability is written

$$p(\mathbf{x}, \omega_i) = p(\mathbf{x}|\omega_i)p(\omega_i) \tag{D.2a}$$

We can also write

$$p(\omega_i, \mathbf{x}) = p(\omega_i|\mathbf{x})p(\mathbf{x}) \tag{D.2b}$$

where $p(\omega_i|\mathbf{x})$, is the conditional probability that expresses the likelihood that the class is $\omega_i$ given that we are examining a pixel at position $\mathbf{x}$ in the spectral domain. That is referred to as the *posterior probability* because it describes the likelihood of finding a pixel from class $\omega_i$ given that we have used all information available to us, in this case the remote sensing measurements.

Because the order of two events occurring simultaneously is irrelevant (D.2a) and (D.2b) are equivalent so that, after rearrangement, we have

$$p(\omega_i|\mathbf{x}) = \frac{p(\mathbf{x}|\omega_i)p(\omega_i)}{p(\mathbf{x})} \tag{D.3}$$

which is known as Bayes' theorem.

---

[6]$p(\omega_i)$ is called the *prior probability* because, in principle, it is the probability with which we could guess class membership in the absence of any information, other than a knowledge of the priors.

## D.2 The Normal Probability Distribution

### D.2.1 The One Dimensional Case

The class conditional probabilities $p(\mathbf{x}|\omega_i)$ are frequently modelled in remote sensing by a normal probability distribution. In the case of a one dimensional spectral space this is described by

$$p(x|\omega_i) = (2\pi)^{-1/2}\sigma_i^{-1}\exp\left\{-\tfrac{1}{2}(x - m_i)^2/\sigma_i^2\right\} \tag{D.4}$$

in which $x$ is the single spectral variable, $m_i$ is the mean value of the measurements $x$ from class $\omega_i$ and $\sigma_i$ is their standard deviation, which describes the scatter of the values of $x$ about the mean. The square of the standard deviation is called the *variance* of the distribution. The mean is also referred to as the expected value of $x$ since, on the average, it is the value of $x$ that would be observed in many trials.

The variance of the normal distribution is found as the expected value of the squared difference of $x$ from its mean. A simple average of the squared difference gives a biased estimate; an unbiased estimate for class $\omega_i$ pixels is given by

$$\sigma_i^2 = \frac{1}{q_i - 1}\sum_{j=1}^{q_i}(x_j - m_i)^2 \tag{D.5}$$

where $q_i$ is the number of pixels from class $\omega_i$ used to compute the variance and $x_j$ is the $j$th of those pixels.

### D.2.2 The Multidimensional Case

The one-dimensional case just outlined is seldom encountered in remote sensing, but it serves as a basis for inducing the nature of the multidimensional, or multivariate, normal probability distribution without the need for detailed theoretical development. Sometimes the bivariate case—that in which $\mathbf{x}$ is two-dimensional— is used as an illustration of how the multivariate case appears.[7]

Let us now examine (D.4) and see how it can be modified to accommodate a multidimensional $x$. First $x$ must be replaced by its vector counterpart $\mathbf{x}$. Similarly, the one dimensional mean $m_i$ must be replaced by its multivariate vector counterpart $\mathbf{m}_i$. The variance $\sigma_i^2$ in (D.4) must be modified, not only to take account of

---

[7]See P.H. Swain and S.M. Davis, eds., *Remote Sensing: The Quantitative Approach*, McGraw-Hill, N.Y., 1978.

multidimensionality, but also to include the effect of correlations among the spectral bands. That role is filled by the covariance matrix $\mathbf{C}_i$ is defined by[8]

$$\mathbf{C}_i = \mathcal{E}\left\{(\mathbf{x} - \mathbf{m}_i)(\mathbf{x} - \mathbf{m}_i)^{\mathrm{T}}\right\} \tag{D.6a}$$

where $\mathcal{E}$ is the expectation operator and the superscript T is the vector transpose operation. As in the one dimensional case an unbiased estimator for the covariance matrix for class $\omega_i$ is

$$\mathbf{C}_i = \frac{1}{q_i - 1} \sum_{j=1}^{q_i} (\mathbf{x}_j - \mathbf{m}_i)(\mathbf{x}_j - \mathbf{m}_i)^{\mathrm{T}} \tag{D.6b}$$

Inside the exponent in (D.4) the variance $\sigma_i^2$ appears in the denominator. In its multivariate extension the covariance matrix is inverted and inserted into the numerator of the exponent. Also, the squared difference between $x$ and $m_i$ is written using the vector transpose expression $(\mathbf{x} - \mathbf{m}_i)^{\mathrm{T}}(\mathbf{x} - \mathbf{m}_i)$. Together, these allow the exponent to be recast as $-\frac{1}{2}(\mathbf{x} - \mathbf{m}_i)^{\mathrm{T}}\mathbf{C}_i^{-1}(\mathbf{x} - \mathbf{m}_i)$.

We now turn our attention to the pre-exponential term. First, we need to obtain a multivariate form for the reciprocal of the standard deviation. That is achieved first by using the determinant of the covariance matrix as a measure of its size, giving a single number measure of variance, and then taking its square root. Finally, the term $(2\pi)^{-\frac{1}{2}}$ is replaced by $(2\pi)^{-\frac{N}{2}}$, leading to the complete form of the multivariate normal distribution for $N$ spectral dimensions

$$p(\mathbf{x}|\omega_i) = (2\pi)^{-N/2}|\mathbf{C}_i|^{-0.5}\exp\left\{-\frac{1}{2}(\mathbf{x} - \mathbf{m}_i)^{\mathrm{T}}\mathbf{C}_i^{-1}(\mathbf{x} - \mathbf{m}_i)\right\} \tag{D.7}$$

---

[8]Sometimes the covariance matrix is represented by $\Sigma$; however, that often causes confusion with the sum operation, and so is avoided in this treatment.

# Appendix E
# Penalty Function Derivation
# of the Maximum Likelihood Decision Rule

## E.1 Loss Function and Conditional Average Loss

The derivation of maximum likelihood classification followed in Sect. 8.3 is generally regarded as acceptable for remote sensing applications and is used widely. However, it is based on the understanding that misclassifying a pixel into any class is no better or worse than misclassifying it into any other class. The more general approach presented here allows the user to specify the importance of some labelling errors compared with others.[9] For example, in a crop classification involving two sub-classes of wheat it would probably be less of a problem if a wheat pixel were wrongly classified into the other sub-class than it would if it were classified as water.

To develop the general method we introduce the penalty, or loss, function

$$\lambda(i|k) \quad k = 1 \ldots M \tag{E.1}$$

where $M$ is the number of classes. This is a measure of the penalty or loss incurred when a classifier erroneously labels a pixel as belonging to class $\omega_i$ when in reality the pixel is from class $\omega_k$. It is reasonable to expect that $\lambda(i|i) = 0$ for all $i$: in other words, there is no penalty in a correct classification. There can be $M^2$ distinct values of $\lambda(i|k)$.

The penalty incurred by erroneously labelling a pixel at position $\mathbf{x}$ in the spectral domain into class $\omega_i$, when in fact the true class is $\omega_k$, is

$$\lambda(i|k)p(\omega_k|\mathbf{x}).$$

---

[9]See N.J. Nilsson, *Learning Machines*, McGraw-Hill., N.Y., 1965, R.O. Duda, P.E. Hart and R.G. Stork, *Pattern Classification,* 2nd ed., John Wiley & Sons, N.Y., 2001, P.H. Swain and S.M. Davis, eds., *Remote Sensing: The Quantitative Approach*, McGraw-Hill, N.Y., 1978.

© The Editor(s) (if applicable) and The Author(s), under exclusive license to
Springer Nature Switzerland AG 2022
J. A. Richards, *Remote Sensing Digital Image Analysis*,
https://doi.org/10.1007/978-3-030-82327-6

$p(\omega_k|\mathbf{x})$ is the posterior probability that $\omega_k$ is the correct class for pixels at $\mathbf{x}$. Averaging this over all possible $\omega_k$ we have the average loss, called the *conditional average loss*, incurred when labelling a pixel incorrectly as coming from class $\omega_i$:

$$L_{\mathbf{x}}(\omega_i) = \sum_{k=1}^{M} \lambda(i|k)p(\omega_k|\mathbf{x}) \tag{E.2}$$

This is a measure of the accumulated penalty incurred given that the pixel could have belonged to any of the available classes, and that we have penalty functions relating all the classes to the wrong choice, $\omega_i$.

A suitable decision rule for deciding the correct class for a pixel is that corresponding to the smallest conditional average loss:

$$\mathbf{x} \in \omega_i \text{ if } L_{\mathbf{x}}(\omega_i) < L_{\mathbf{x}}(\omega_j) \quad \text{for all } j \neq i \tag{E.3}$$

Implementation of (E.3) is referred to as the Bayes' optimal algorithm.

Because the posterior probabilities $p(\omega_k|\mathbf{x})$ are generally not known we use Bayes' theorem in (E.2) to give

$$L_{\mathbf{x}}(\omega_i) = l_{\mathbf{x}}(\omega_i)/p(\mathbf{x})$$

in which

$$l_{\mathbf{x}}(\omega_i) = \sum_{k=1}^{M} \lambda(i|k)p(\mathbf{x}|\omega_k)p(\omega_k) \tag{E.4}$$

Since $p(\mathbf{x})$ is common to all classes, class membership depends on $l_{\mathbf{x}}(\omega_i)$ alone.

## E.2 A Particular Loss Function

Suppose $\lambda(i|k) = 1 - \Phi_{ik}$ with $\Phi_{ii} = 1$ and $\Phi_{ik}(k \neq i)$ to be defined. Then (E.4) can be expressed

$$l_{\mathbf{x}}(\omega_i) = \sum_{k=1}^{M} p(\mathbf{x}|\omega_k)p(\omega_k) - \sum_{k=1}^{M} \Phi_{ik}p(\mathbf{x}|\omega_k)p(\omega_k)$$
$$= p(\mathbf{x}) - g_i(\mathbf{x})$$

with

$$g_i(\mathbf{x}) = \sum_{k=1}^{M} \Phi_{ik}p(\mathbf{x}|\omega_k)p(\omega_k) \tag{E.5}$$

Again $p(\mathbf{x})$ does not aid discrimination and can be removed from the conditional average loss expression, leaving $l_\mathbf{x}(\omega_i) = -g_i(\mathbf{x})$. Because of the minus sign in this expression, we can decide the least cost labelling of a pixel at position $\mathbf{x}$ on the basis of maximising the *discriminant function* $g_i(\mathbf{x})$:

$$\mathbf{x} \in \omega_i \text{ if } g_i(\mathbf{x}) > g_j(\mathbf{x}) \quad \text{for all } j \neq i \tag{E.6}$$

As a special case we adopt the Kronecker delta function for $\Phi_{ik}$, i.e. $\Phi_{ik} = \delta_{ik}$ where

$$\delta_{ik} = 1 \text{ for } k = i$$
$$= 0 \text{ for } k \neq i$$

Thus, the penalty for misclassification is not class dependent, and (E.5) becomes

$$g_i(\mathbf{x}) = p(\mathbf{x}|\omega_i)p(\omega_i)$$

The decision rule in (E.6) then reduces to

$$\mathbf{x} \in \omega_i \text{ if } p(\mathbf{x}|\omega_i)p(\omega_i) > p(\mathbf{x}|\omega_j)p(\omega_j) \quad \text{for all } j \neq i \tag{E.7}$$

which is the classification rule of (8.3), called the *unconditional maximum likelihood decision rule*.

# Index

## A
Absorption features, 36, 45, 48, 50, 404, 488–490, 526
Abundance map, 490
Accuracy
  confidence limits, 478
  cross validation, 479
  error matrix, 447, 450, 463, 464, 467–471, 476–478, 502
  Kappa coefficient, 447, 467–470, 498, 502
  leave one out method, 479
  map, 217, 462–467, 470, 472–476, 478, 479, 498
  producer's, 464
  user's, 464, 466
Activation function, 316–318, 324–326, 332–334, 336, 345, 366
Active remote sensing, 4
AdaBoost, 313, 315, 448
Aerosol Optical Depth, 40
Aerosol Optical Thickness, 40, 45
Aliasing, 235, 250, 261
Allocation disagreement, 470, 471, 498
Aperiodic function, 229
Ascending node, 536
Atmosphere
  atmospheric constituents, 31, 41, 43, 45–47
  effect on imagery, 29, 35, 39, 42
  optical thickness, 40, 43, 45
Atmospheric correction
  5S, 47
  6S, 47
  ACORN, 48, 49, 79
  ATREM, 47–49, 79
  broad waveband systems, 42, 43, 50

dark subtraction, 50
empirical line method, The, 51, 52
FLAASH, 48, 49, 79
flat field method, The, 50
haze removal, 50
HITRAN, 46
log residuals, 52, 53, 84
MODTRAN4, 48
narrow waveband systems, 45
water vapour, 43, 47, 49
Atmospheric scattering
  aerosol, 39, 40, 45
  diffuse, 18, 41
  Mie, 39, 42–44, 50
  non-selective, 40
  Rayleigh, 39, 40, 42–44
Atmospheric transmittance, 40, 41, 52
Atmospheric windows, 2, 3

## B
Backpropagation through time. *See* Neural network
Bagging, 313, 315
Band ratios, 173
Basis functions, 244–247, 251, 255, 304–309, 311
Bathymetry, 126
Binary numbers, 533
Binomial distribution, 473
Bispectral plot, 456–458, 460, 461
Bit, 5, 28, 42, 44, 105, 154, 158, 159, 187, 208, 258, 311, 365, 379, 381, 382, 493, 537, 538
Boosting, 313, 315, 508
Byte, 538

© The Editor(s) (if applicable) and The Author(s), under exclusive license to
Springer Nature Switzerland AG 2022
J. A. Richards, *Remote Sensing Digital Image Analysis*,
https://doi.org/10.1007/978-3-030-82327-6